D1274537

AIR-BREATHING FISHES

EVOLUTION, DIVERSITY, AND ADAPTATION

AIR-BREATHING FISHES

EVOLUTION, DIVERSITY, AND ADAPTATION

Jeffrey B. Graham

Center for Marine Biotechnology and Biomedicine
and Marine Biology Research Division
Scripps Institution of Oceanography
University of California, San Diego
La Jolla, California

Academic Press

San Diego London Boston New York Sydney Tokyo Toronto

Cover Photograph: Periophthalmus barbarus from west Africa.
(Photograph by Susan R. Green of the Scripps Institution of Oceanography, La Jolla, California.)

This book is printed on acid-free paper. ∞

Academic Press
a division of Harcourt Brace & Company
525 B Street, Suite 1900, San Diego, California 92101-4495, USA
http://www.apnet.com

Academic Press Limited
24-28 Oval Road, London NW1 7DX, UK
http://www.hbuk.co.uk/ap/

Library of Congress Cataloging-in-Publication Data

Graham, Jeffrey B.
Air-breathing fishes : evolution, diversity, and adaptation / by Jeffrey B. Graham.
p. cm.
Includes bibliographical references and index.
ISBN 0-12-294860-2 (alk. paper)
1. Air-breathing fishes. 2. Air-breathing fishes--Respiration.
I. Title.
QL615.073 1997
573.2'17--dc21 96-47586
CIP

PRINTED IN THE UNITED STATES OF AMERICA
97 98 99 00 01 02 EB 9 8 7 6 5 4 3 2 1

For my family,
especially Grace and Bill, and young Tate

Contents

CHAPTER

7

Blood Respiratory Properties

CHAPTER

8

Metabolic Adaptations

CHAPTER

9

Synthesis

Preface

Exquisite structural design and functional detail, particularly the specializations developed among living organisms as a result of natural selection, have always fascinated me. In light of this, my attraction to questions about the origin of vertebrate air breathing and the evolution of tetrapods from the fishes is not surprising. The transition from aquatic to aerial respiration and from living in water to living on land required dramatic alterations in morphology, physiology, and ecology. These changes have long interested most biologists, particularly paleontologists and comparative physiologists (of which I am one), and the vertebrate transition to land, moreover, has implications for a range of human endeavors extending from medicine to literature and philosophy.

Whereas the connection between fish air breathing and tetrapod evolution is generally appreciated, there is far less familiarity with the physical and biological details underlying this connectivity and even less awareness of the role air breathing has had—quite apart from the origin of tetrapods—in the evolution and natural history of many fishes.

My interest in exploring these aspects of air breathing, however, was frustrated by the lack of a synthetic comparative treatment of this subject. In 1986 I decided to write a book on the air-breathing fishes and began this project in 1987 with the aid of a Guggenheim Fellowship.

Because I had been doing air-breathing fish research for a number of years, I was confident that I had some important things to say. These fishes are, after all, most interesting experimental subjects. One can, for example, manipulate their surroundings and induce air gulping or emergence from water. It is also possible to trick an air-breathing fish into hiding under a dark cover only to therein conceal an inverted funnel and use this to trap and measure the gas exhaled by the fish or to use this same ruse to induce the fish to rest on a platform suspended from a balance and thereby measure its buoyancy.

More than these laboratory observations, however, it was perceptions about air-breathing fishes gained through countless hours in the field that I wanted to communicate through my book. I have, for example, watched an amphibious fish emerge from water and skip up the face of a slippery rock to seize a piece of food thrown in its direction. I have dug synbranchid eels out of their dry season mud burrows, and I have experimentally manipulated the gulping frequency of synchronously air-breathing fishes in an hypoxic forest pool. I have come across loricariid catfish trapped in dry season puddles just barely large enough to hold them.

Air-breathing fishes and field observations are a special milieu; the field data one gathers not only are relevant to the species involved, but also provide a tangible basis for imagining the circumstances that led to air breathing in the Paleozoic. Perhaps a special insight is born in the solitude of a tropical forest stream at dawn or comes with the exhilaration of pursuing fast-moving blennies across a wave-pounded reef—and being swept out to sea in the process. Whatever it is, I have become convinced that the study of air-breathing fishes, in addition to being extraordinarily interesting, permits one to transcend time and to experience, as nearly as can be done, what it must have been like when the first vertebrate air breathers began to make their way to the water's surface.

Jeffrey B. Graham

Acknowledgments

A Guggenheim Fellowship enabled me to start this book and to spend well over a month studying the air-breathing fishes in the British Museum (Natural History). I thank Gordon Howes and Alwynn Wheeler of the museum staff for an especially productive stay; I now view that period in London as a unique, enriching experience. The Guggenheim Fellowship also enabled me to study fish in the Musée Royal de l'Afrique Centrale, Tervuren, Belgium, and I thank Dirk Thys van den Audenaerde for that opportunity.

Much of the literature cited in this book actually resides in the Scripps Institution of Oceanography (SIO) library, and this is a tribute to the foresight of a long line of dedicated librarians. The SIO library is also the home of the Carl Hubbs library, and access to this collection proved essential to my research. As for the references that needed to be gathered from the outside, I thank SIO reference librarians Phyllis C. Lett and Valerie A. Quate for invaluable assistance.

I am indebted to a number of individuals for help with various parts of this project. My graduate students, Nancy Aguilar, Heidi Dewar, Peter Fields, Keith Korsmeyer, N.C. Lai, and Bill Lowell, were continually on the lookout for new references and were always available for discussions. Many individuals shared their insights with me, and I especially thank Troy Baird, George Bartholomew, Beth Brainerd, Jim Cameron, Kathy Dickson, Jack Gee, Malcolm Gordon, Peter Hochachka, Don Kramer, Mike Hedrick, Steve Katz, Karel Liem, Ken Olson, Mike Robinson, Stan Rand, Dick Rosenblatt, Ira Rubinoff, Roberta Rubinoff, Dirk Thys van den Audenaerde, Ralph Shabetai, Neal Smatresk, Neal Smith, and the late Ethelwynn Trewavas, the late Humphrey Greenwood, and the late Martin Moynihan. Special thanks are also expressed to the Smithsonian Tropical Research Institute, which supported my early research on air-breathing fishes. A grant from the University of California Academic Senate aided in final manuscript preparation.

The following individuals reviewed and commented on drafts of chapters and sections of the manuscript: Georgina de Alba, Nancy Aguilar, Beth Brainerd, Kathy Dickson, Malcolm Gordon, Mike Hedrick, Don Kramer, N.C. Lai, Ken Olson, and Stan Weitzman. Patricia Fisher, Don Ward, and Rose Yamasaki aided in the preparation of some tables and figures. Janet K. Duermeyer deserves special mention and thanks for her significant assistance in copyediting the entire manuscript and for tirelessly entering, checking, and rechecking the references. All of the errors that persist in this book are attributable to me.

Finally, I want to remember the late Freddy Robison, the young man who first told me about a fish that came out of water and then took me out on the rocks to meet *Mnierpes*.

CHAPTER

1

The Biology of Air-Breathing Fishes

It began as such things always begin—in the ooze of unnoticed swamps, in the darkness of eclipsed moons. It began with a strangled gasping for air.

L. Eiseley, *The Immense Journey*, 1957

INTRODUCTION

This is a book about air-breathing fishes. Its objectives are to describe the diversity of these fishes, compare their structural, physiological, biochemical, and behavioral specializations for both aerial and aquatic respiration, and examine selective pressures leading to the appearance of air breathing among various groups.

A goal of this first chapter is to establish the rationale for this book through a two-part essay linking fish air breathing, vertebrate evolution, and comparative physiology. Another goal is to set the stage for later discussions by reviewing physical principles governing the suitability of air and water as respiratory media, examining the habitats of extant air-breathing fishes, and defining terms and concepts related to aerial and aquatic gas exchange and bimodal (i.e., simultaneously in two respiratory media, air and water) respiration. An operational definition of fish air breathing is provided along with a discussion of the types of observations, experimental evidence, and background knowledge required to evaluate the presence of air breathing in a species. Lastly, this chapter presents a general classification of the types of air-breathing fishes.

Vertebrate Evolution and Air Breathing

Air breathing is an ancient vertebrate specialization. As indicated by Figure 1.1, this adaptation is thought to have appeared in fishes during the Late Silurian Period (438–408 million years before present, mybp), long before the evolution of amphibians and the invasion of land by the tetrapods (Gardiner, 1980; Panchen, 1980; Little, 1983, 1990; Gordon and Olson, 1994; Long, 1995). Amphibians were derived from the lobe-finned fishes (Sarcopterygii) probably during or before the Middle Devonian Period (400–370 mybp). There has been some recent controversy as to which group of the lobe-fins was actually ancestral to the tetrapods (Marshall and Schultze, 1992; Ahlberg and Milner, 1994). The prevailing view is that fishes very similar to those in the extinct family Panderichthyidae were the closest relatives of the early tetrapods which, following a 10–20 million year period of aquatic evolution, gained a solid foothold on land by the Early Carboniferous Period (360–330 mybp) (Ahlberg and Milner, 1994; Long, 1995). Figure 1.1, however, indicates that, in addition to the lobe-finned fishes, air breathing was present in the other major Devonian fish group, the ray fins (Actinopterygii, Barrell, 1916; Goodrich, 1930; Romer, 1958a; Halstead, 1968; Thomson, 1980; Bray, 1985). It has also been speculated that air breathing was present in certain placoderms (Wells and Dorr, 1985) and possibly even the acanthodians (Schaeffer, 1965), the Silurian group ancestral to the Osteichthyes.

Most accounts of vertebrate evolution describe the

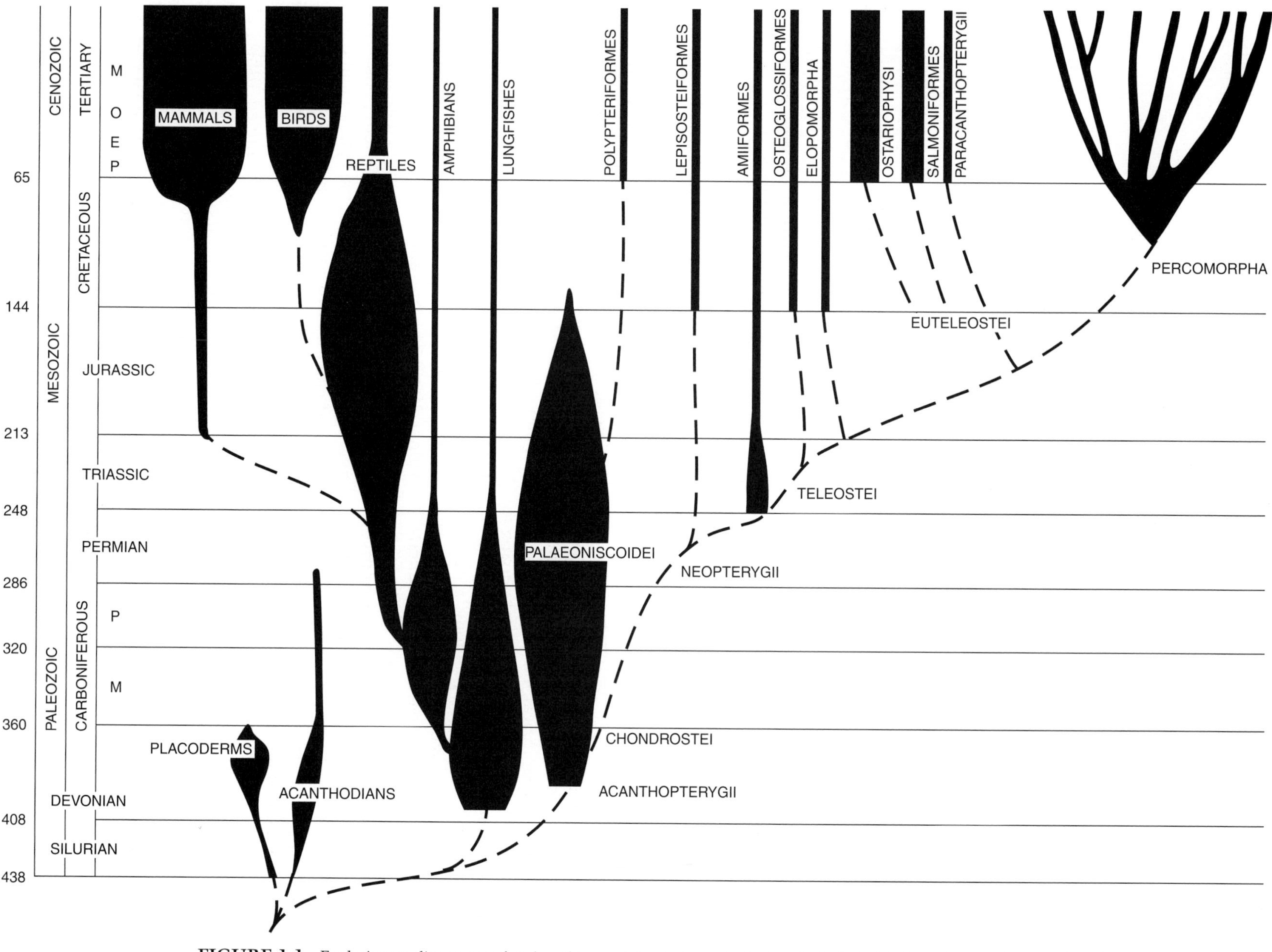

FIGURE 1.1 Evolutionary divergence of air-breathing vertebrates in geologic time. Included with the major vertebrate air breathers (mammals, birds, reptiles, amphibians) are the principal groups of fishes in which air breathing is documented or was likely present. Relative abundance and diversity of each group is reflected in the band width.

early air-breathing fishes and stress the importance of aerial respiration in the origin of the tetrapods (Smith, 1959; Schmalhausen, 1968; Randall *et al.,* 1981a). Following this, however, the focus of these treatments usually shifts to the tetrapods themselves and to the changes taking place in the phyletic progression from amphibians to mammals. Such accounts thus rarely consider the evolution of fishes beyond the Paleozoic and as a result succeed, more often than not, in conveying the impression that both fish evolution and the importance of air breathing to fishes ended with the appearance of the amphibians. Similarly, comparative surveys of air-breathing fish respiratory adaptation (Singh, 1976; Randall *et al.,* 1981a; Munshi, 1985; Munshi and Hughes, 1992; Randall, 1994) have not taken the phyletic histories of fishes into account and thus most often treat both the primitive and modern air-breathing fishes in the same light, as inferior grades of the mammalian specialization, evolutionary curiosities, or both.

The objectives of this book are the documentation of the persistence of air breathing in fishes and demonstration of the influence air breathing has exerted on this group. As is evident in Figure 1.1, air breathing was not an exclusive discovery of the "prototetrapod line" of the lobe-fins but was also present in Paleozoic Era (570–245 mybp) actinopterygians and now occurs in many extant species. Included among extant air breathers are direct descendants of Paleozoic and Mesozoic (245–65 mybp) forms (e.g., bichirs, bowfins, and gars), as well as many teleosts (Osteoglossiformes, Elopiformes, and Euteleostei), the largest group of extant bony fishes which originated in the Mesozoic Era but came into prominence (particularly the Percomorpha) during or since the Late Cretaceous Period (108–65 mybp).

Air breathing has therefore persisted throughout the evolutionary history of the fishes and has played a fundamental role in this group's evolution. It can be argued, for example, that the Paleozoic appearance of air breathing dramatically shaped the evolutionary course taken by actinopterygians, both by influencing the survival of certain lineages and by setting into motion the sequence of events leading to the evolutionary transition of the primitive piscine lung into a gas bladder, a structure of paramount importance to the adaptive radiation of modern bony fishes (Greenwood and Norman, 1963; Lauder and Liem, 1983a,b; Liem, 1989a). The frequent occurrence of air breathing among the Euteleostei (Figure 1.1) additionally reflects the continued presence of factors selecting for this specialization.

The study of the modern air-breathing fishes therefore assumes added importance because of its potential for drawing parallels between the ecological and environmental factors affecting extant forms and the factors that likely led to the early origin of air breathing. The modern air breathers also enable us to examine the numerous ways in which air breathing and related adaptations are integrated into each species' life history. This is also relevant to considerations of the early origins of air breathing because in the final analysis an air-breathing fish, if it is to survive and contribute to the next generation, must have the capacity to live a wholly functional life as a fish (Liem, 1989b, 1990). This was as true for the air-breathing fishes of the Silurian and Devonian as it is for all extant species.

A Contemporary View of Air Breathing: From Fossils to Swamps and Shower Curtains

Man's place in nature is partly defined by the circumstances of his origin. A remote but fundamental aspect of this was the evolution of air breathing among fishes during the Silurian and the tetrapod invasion of land during the Late Devonian and Carboniferous. The important place of early air breathers in vertebrate and human evolutionary history is seen in contemporary cartoons depicting the likely events during a critical period of the transition to land (Figure 1.2). Although these often portray fishes (i.e., there are dorsal fins and a distinct broad tail) as leading the terrestrial charge, it was a diverse group of fishlike amphibians that actually invaded the land (Chapter 9).

FIGURE 1.2 Drawing by Robert Day; © 1966. *The New Yorker Magazine, Inc.*

There is a general public awareness of air-breathing fishes. Many popular aquarium pets (bettas, gouramys, and *Plecostomus*) are air breathers. (Some might suggest that air breathing evolved in these forms to enable them to survive in poorly kept aquaria.) Many of us have taken a shower behind a curtain decorated with fishes that, while otherwise swimming along happily, are engaged in the curious and highly atypical piscine behavior of releasing long streams of gas bubbles. Is this artistic license or do some shower-curtain designers purposely illustrate the release of air breaths?

Most biologists acquire an appreciation of the contribution made by fish air breathing to tetrapod origins during their undergraduate training. In my case this took place in a comparative anatomy class and I remember, as a result of newly acquired insight, looking somewhat reverently at our preserved lungfish specimen. I vividly recall the account that Professor Richard Etheridge gave our class about an experience he had while collecting fishes in South America. After depositing poison in a pool, he and his colleagues marveled as some fishes crawled out of the water and moved to another pond. "It was as if," Etheridge said, "we were seeing things as they must have been in the Devonian." Later that same semester, on one of my first dates with my future wife, Rosemarie, we took a 300 mile Sunday drive just to see a living lungfish.

Human evolutionary affinities with fishes have been the subject of numerous works by philosophers and zoologists, and volumes notable for treatment of both the scientific principles and philosophical implications of this evolutionary transition have been written by Eiseley (1957), Smith (1959), Gregory (1963), and others. Eiseley's essay "The Snout," part of which is quoted in the epigraph of this chapter, is a lucid, poignant, and all-encompassing view of air breathing and vertebrate evolution. Rather than to review these implications or search for more of them, my objectives in this book are to describe the likely combinations of environmental conditions and adaptations that brought about the first vertebrate transition to air breathing as well as the subsequent appearance of this capability in other groups of fishes. This book will also emphasize that the study of air-breathing fishes is an easily exploited area of inquiry that, while giving information about the biology of modern species, provides insight into a range of subjects, from the origin of aerial respiration to the control of bimodal gas exchange, and the evolution of integrated systemic function in the vertebrates.

The scientific importance and validity of this approach was clearly established for me when, as a graduate student, I heard a lecture by the late Per Scholander on the bradycardic diving responses of mudskippers *(Periophthalmus)*. During that lecture I realized that although the first vertebrate transition to air breathing and the tetrapod invasion of land lay in the distant past, the tools needed to make both discoveries and inferences about the range of environmental conditions and the organismic adaptations or preadaptations that favored this are at hand, in the form of the modern air-breathing fishes and amphibians (Gordon, 1970; Gordon *et al.*, 1961, 1965, 1969, 1970). In addition, events somewhat like those that influenced tetrapod evolution in the Devonian and Carboniferous (Figure 1.1) occur regularly in many habitats. Metamorphosing frogs, for example, make the transition from a swimming, water-ventilating tadpole to an amphibious, bimodally breathing juvenile in a matter of days (Burggren and Doyle, 1986a,b; Burggren and Pinder, 1991; Burggren and Infantino, 1994). Many air-breathing fishes live in habitats that are permanently deoxygenated or that become so at night or seasonally, and to survive they must periodically surface to breathe air. On the other hand, species like Scholander's mudskippers have evolved highly amphibious behaviors and some are averse to immersion (Bandurski *et al.*, 1968; Graham, 1976a).

A few years after Scholander's lecture had solidified my interest in air-breathing fishes, I had an opportunity to study them first hand when, while conducting doctoral research in Panama, I became interested in the tropical rockskipper *Mnierpes macrocephalus* (Graham, 1970, 1973). Later, while working in eastern Panama with colleagues Donald Kramer and Elpidio Pineda, I had an experience somewhat like that recounted by Etheridge. We came across a small, isolated pool buried deeply in a dense growth of brush. It seemed lifeless except for the brief periods when, in close unison, a group of air-breathing fish would quickly break the surface for air and then disappear back into the depths (Kramer and Graham, 1976). Clearly the terrestrial biota surrounding this pool differed vastly from the Devonian (i.e., the terrestrial biota did not begin its expansive development until the Carboniferous). We were, after all, surrounded by lush and, as I distinctly recall, spiny vegetation and had the company of both biting insects and aquatic leeches. Nevertheless, and in spite of these differences, it was exhilarating to feel, at least momentarily, like privileged time travelers, reliving events as they must have been played out in the early dawn of vertebrate air breathing.

ENVIRONMENTAL FACTORS AFFECTING AIR-BREATHING FISHES

The physical differences between water and air and the effect of these on respiration, the ecological and physical factors affecting the O_2 concentrations in different aquatic habitats, and a perspective spanning 400 million years of fish evolution all contribute to an understanding of factors important to the origin of air-breathing fishes.

Differences between Air and Water as Respiratory Media

Physical differences between air and water as respiratory media have been discussed by many writers (Dejours, 1981, 1988; Piiper, 1982). Relative to air, the higher viscosity of water imposes a greater frictional resistance and its greater density requires more work to overcome inertia. Water also contains much less O_2 than air and because its diffusion constants are much less, gases diffuse more slowly through water than air. Correspondingly, the energetic cost for fish gill ventilation is usually much greater than for lung ventilation (Graham *et al.*, 1987; Dejours, 1994).

Air contains nearly 20.95% O_2 (0.2095 atmospheres, atm); hence a liter of air holds almost 210 ml of O_2. At sea level, atmospheric pressure is 760 mm Hg ([= torr] or, in SI units, 101.3 kilopascals [Pa = 1 newton per square meter]) and, assuming that atmospheric air is water-vapor saturated and at 25 C, its oxygen partial pressure (PO_2) is calculated to be 154 torr (= 20.6 kPa) as follows:

$$PO_2 = (760 - vH_2O)0.2095 \qquad (1.1)$$
$$154 = (760 - 24.9)0.2095$$

where vH_2O is the water vapor pressure at 25 C.

The volume of O_2 (or any gas) dissolved in water depends on both its partial pressure and its solubility as described by Henry's law:

$$Vg = \alpha PO_2 \qquad (1.2)$$

where Vg is the O_2 concentration (ml/liter) in water, α is its solubility coefficient (ml/liter atm), and PO_2 is expressed in atmospheres. If water is in equilibrium with atmospheric air (i.e., air saturated) and at sea level, its partial pressure (after correction for water vapor pressure, and assuming 25 C), would be the same as that of water-vapor saturated air, or in this case 154 torr (0.209 atm). PO_2 thus indicates the driving force for the diffusion of O_2, and because air and water are in equilibrium in this example, there is no net diffusion between them (Steen, 1971; Dejours, 1981, 1988).

It is α, the gas solubility coefficient that determines the concentration of a gas in water. The αO_2 of water is small and declines with both increasing water temperature and salinity (Dejours, 1981; Piiper, 1982, 1989). Whereas air contains nearly 210 ml O_2/liter, the dissolved O_2 concentrations in air-saturated (normoxic) freshwater and seawater (35‰) at 25 C are 6.04 and 4.95 ml/liter (2.6 and $2.2 \cdot 10^{-4}$ mM), respectively. This ranges from 35 to 42 times less than in air!

The partial pressure and solubility coefficient of another respiratory gas, CO_2, also determine its concentration in water. CO_2 is about 30 times more soluble in water than is O_2; however, because CO_2 has a much lower atmospheric partial pressure than O_2 ($3.6 \cdot 10^{-4}$ atm versus 0.21 atm), then, in accordance with Henry's law, the CO_2 content of air-saturated water is well below that of O_2. Whereas air-equilibrated freshwater (25 C) contains 6.04 ml/liter O_2, the total CO_2 content of this water is about 0.3 ml/liter (Dejours, 1988, 1994).

Aquatic CO_2 concentration can be increased by bubbling this gas into water, and this is a common technique for studying the effects of hypercapnia (an elevated CO_2) on air-breathing fish behavior and respiration. Equilibrating freshwater (20 C) with a mixture of 5% CO_2 (37 torr) and 95% air, for example, will raise total CO_2 to 43 ml/liter, which is over 6 times the air-saturated O_2 level (Figure 1.3).

Air Breathing in Salt and Freshwaters

The different O_2 contents of air and water have profoundly affected animal gas exchange (Piiper, 1982; Dejours, 1988, 1994; Graham, 1988; Piiper and Scheid, 1992). Could the above-determined 1.1 ml/liter difference in the O_2 content of salt and freshwater be a

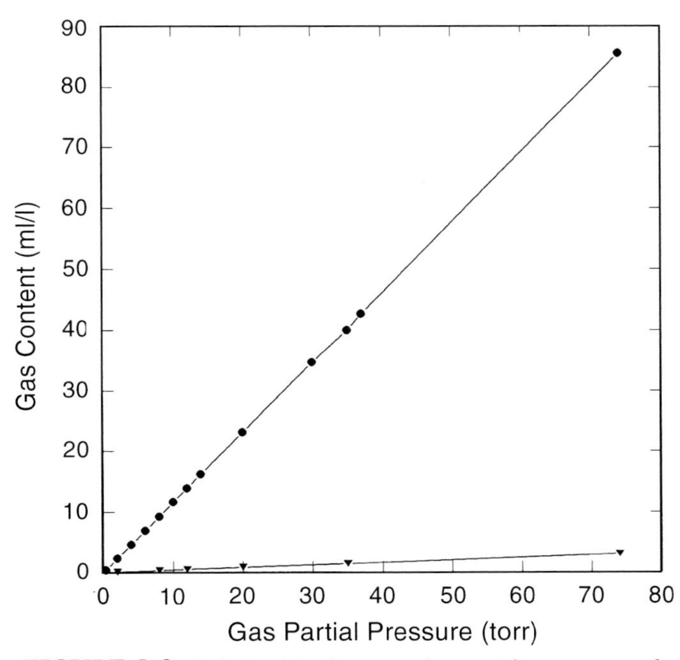

FIGURE 1.3 Relationships between the partial pressures and contents of CO_2 (dots) and O_2 (triangles) in freshwater at 20 C.

causal factor in fish air breathing? Probably not. Even though seawater contains about 20% less dissolved O_2, both air-saturated saltwater and freshwater have the same PO_2. Thus, saltwater is not hypoxic relative to freshwater, and this means, assuming osmotic stresses are compensated, that fish hemoglobin would saturate with O_2 just as readily in a gill ventilated with air-saturated saltwater as in one ventilated by air-saturated freshwater. This, however, does not mean that ventilating in water with a lower O_2 content does not have its price. Assuming all other things are equal, the lower O_2 content of saltwater would require a fish to compensate by ventilating more water, extracting more O_2 from each pulse of ventilated water (which could be done with more gill area), or both (Graham *et al.*, 1978a). Packard (1974) has hypothesized that air breathing evolved in saltwater (see Chapter 9). The now prevailing view, that many of the early lobe-finned fishes occurred in saltwater (Bray, 1985; Campbell and Barwick, 1987, 1988; Carroll, 1988), again raises questions about how salt and freshwater habitats may have affected the origin of air breathing (Thomson, 1980, 1993; and Chapter 9). Nevertheless, in virtually all experimental and natural circumstances where the level of ambient O_2 has been empirically determined to be a factor in fish air breathing, it is invariably hypoxia (i.e., a low PO_2) and not αO_2 that is the driving force (Krogh, 1941; Steen, 1971; Graham *et al.*, 1978a; Holeton, 1980).

Conditions Affecting Aquatic O_2 Content

The O_2 content of water is subject to much greater variability than air. The sources for the O_2 contained in natural water bodies are photosynthesis by phytoplankton and macroalgae and O_2 diffusion from the atmosphere. Sunlight is needed for photosynthesis and physical conditions such as wind and vertical density currents in the water are vital for the equilibration of atmospheric O_2 with water as well as the transport of dissolved O_2 to depth (Graham *et al.*, 1978a; Dunn, 1983). However, in some habitats these mixing processes are not effective and nearly all of the O_2 contained in water can be depleted by organismic respiration (biological oxygen demand). Conditions of aquatic hypoxia (i.e., a PO_2 less than saturation) can occur chronically in water bodies where there is little flow, vertical mixing, and photosynthesis (a usual result of shading rather than nutrient limitation) and an excess of organic decomposition and organismic respiration. Some swamps are chronically hypoxic whereas tropical streams and flood plains may be so seasonally (Carter and Beadle, 1931; Carter, 1933;

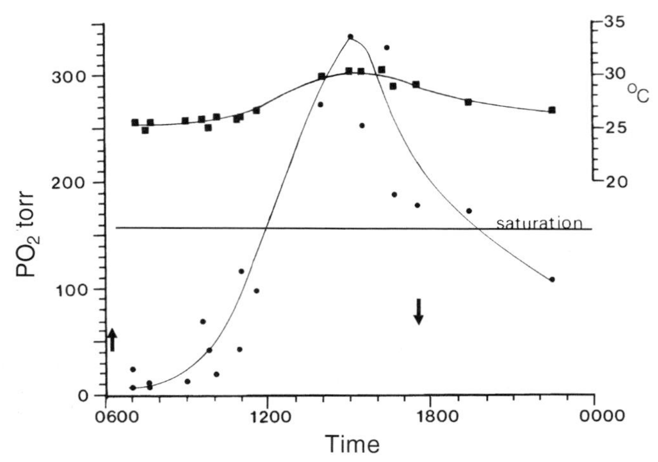

FIGURE 1.4 Surface (0–5 cm) water temperatures (squares) and oxygen tensions (dots) at different times in Ocelot Pond, near Gamboa, Panama. Data collected from 18 March to 9 April, 1981. Surface water pH was always between 4.5 and 5.5. Maximum depth of water was about 40 cm and temperature and oxygen data indicated no stratification. Horizontal line is PO_2 at air saturation. Curved lines fitted by eye. Arrows indicate approximate times of sunrise and sunset. (From *Env. Biol. of Fishes,* volume 12[4], pp. 291–301, figure 2; Graham, 1985; with kind permission of Kluwer Academic Publishers.)

Willmer, 1934; Beebe, 1945; Kramer *et al.*, 1978; Cala, 1987a). On the other hand, some isolated pools may become hyperoxic (i.e., a PO_2 greater than air-saturation) during daylight hours, when photosynthetic O_2 production vastly exceeds respiration, but then hypoxic or even nearly anoxic and hypercarbic at night (Figure 1.4 and Truchot and Duhamel-Jouve, 1980; Graham, 1985, 1990).

The Habitats of Extant Air-Breathing Fishes

Modern air-breathing fishes occur in a variety of freshwater and marine habitats. Marine air breathers occur on tropical coral reefs, in rocky intertidal zones, and in marshes and bays. A variety of factors, from the adoption of an amphibious behavior to exploit resources at the air–water interface, to the periodic aerial exposure imposed by low tides, and even aquatic hypoxia in certain habitats have all played a role in selection for air breathing among marine species (Graham, 1976a; Gibson, 1969, 1982; Sayer and Davenport, 1991; Bridges, 1993a,b).

Among freshwater fishes the impetus for air breathing seems to have come primarily from the effects of aquatic hypoxia (Carter, 1957; Johansen, 1970; Dunn, 1983). Air breathing has evolved in fish that occur under ice and other hypoxic environments (Klinger *et al.*, 1982; Magnuson *et al.*, 1983, 1985) as well as in many tropical lowland habitats (Johansen, 1970). In fact, most of the modern air-breathing fishes occur in tropical freshwaters that are subject to seasonal drought. The lack of rain stagnates and warms these waters and often crowds fish in isolated pools. Photosynthesis in these waters is not sufficient to maintain normoxia because they are usually shaded by surface-growing macrophytes or dense stands of trees. Thus, limited O_2 production, heating, and the unusually large biological O_2 demand imposed by crowding and decomposing organic material can lead to hypoxia and even anoxia.

It is in these types of habitats, where extreme shifts in O_2 (and water availability) often occur regularly (i.e., annually during the tropical dry season or even daily in some swamps, Figure 1.4), that natural selection has worked repeatedly, among the diverse tropical freshwater fish fauna, to bring about a number of successful and strikingly different specializations for air breathing. This same scenario, set in the habitats and environmental conditions of the Late Silurian, likely led to the development of air breathing among the early fishes as well as to the Devonian appearance of tetrapods (Barrell, 1916; Goodrich, 1930; Inger, 1957; Romer, 1958a,b, 1966; Thomson, 1980, 1993; Bray, 1985; Long, 1995).

WHAT IS AN AIR-BREATHING FISH?

Air-Breathing versus Non-Air-Breathing Fishes

Two terms: "water-breathing" and "air-breathing," are often used to describe fish respiratory mechanisms. As used throughout this book, fish air breathing means the process of gas exchange, for purposes of respiration, directly with the aerial environment. An air-breathing fish may swim to the water surface, gulp air and then dive, or it may crawl onto land and either gulp air or passively exchange gases with the atmosphere across several exposed respiratory surfaces. Air breathing can involve the aerial exchanges of both respiratory O_2 and CO_2, although not necessarily at the same rate or across the same surface.

"Air breathing" is a useful term in that it signifies a marked departure from the "typically piscine" respiratory mode. However, except to make a gross distinction, "water-breathing" is not a useful term to describe normal fish respiration. The reason for this is that a dichotomy between "air breathers" and "water breathers" does not exist among fishes because all of the air breathers respire bimodally to some extent. Thus, all air breathers are also "water breathers;" that is, they all ventilate their gills or other aquatic respiratory surfaces to sustain some amount of gas exchange (as well as nitrogen excretion and ionic regulation). Also, with their capacity to respire aquatically, some air-breathing fishes can survive indefinitely without aerial O_2. Thus, in order to contrast the two groups, the terms "air breather" and "non-air breather" are more apropos.

The Terminology of Aerial Respiration

Physical differences in air and water together with environmental conditions and biotic factors all influence the relative partitioning of aerial and aquatic respiration in bimodally respiring fishes (Johansen, 1972; Perry and Wood, 1989). Parameters of bimodal respiration that have been most frequently examined include:

i. The relative volumes of air and water that are ventilated (the symbol $\dot{V}$ is used to indicate a ventilated-rate volume, while f denotes the frequency of either gill ventilation or air breathing)
ii. The partitioning of O_2 consumption and CO_2 release between air and water (where gas consumption or release is denoted by $\dot{V}$ in the case of volumes or by $\dot{M}$ in the case of molar amounts, e.g., $\dot{V}O_2$ and $\dot{V}CO_2$) and
iii. The respiratory exchange ratio of these gases in each phase (e.g., $R_{air} = \dot{V}CO_2/\dot{V}O_2$).

What Constitutes Fish Air Breathing?

Air breathing is but one of several adaptive responses utilized by fishes dwelling in habitats where O_2 supplies may be severely depleted. Others are behavioral avoidance of hypoxic areas, which in some cases may include a seasonal migration (Cala, 1987a) or aquatic surface respiration (ASR), in which a fish swims up to ventilate its gills with the thin layer of air-saturated surface water (Kramer, 1983a). Experiments by Breder (1941) showed that fish in hypoxic water instinctively swam to shallower water and eventually to the surface itself to obtain more O_2 by ASR. This behavior, in the company of uncritical observation, has led to claims for air breathing among some unlikely species (Chapter 2).

The principal challenge to a general understanding of the biology of fish air breathing is the establishment of criteria to decide what does and does not constitute this activity. Chapter 2 will discuss species that have in the past been described as air breathers but for which there is no evidence that this trait is present. Much of this confusion is due to incomplete information about the species in question, and Table 1.1

TABLE 1.1 Different Natural Fish Activities Likely Associated with Aerial Respiration, with Selected Examples of Genera or Families Fitting Each Category

I. Leaving water
  A. Crawling on land to migrate, feed, find a mate, burrow, defend a territory, avoid turbulence
    1. Many amphibious species: *Periophthalmus, Mnierpes, Gobionellus, Andamia, Alticus*
    2. Some freshwater species are amphibious or migrate: *Hoplerythrinus, Monopterus, Synbranchus, Anabas, Clarias, Anableps, Rivulus*, and some *Fundulus*
    3. Some intertidal marine species emerge from hypoxic water: *Blennius, Clinocottus, Helcogramma*
  B. Laying eggs above the water line
    1. Freshwater: *Brycon, Copella*
    2. Marine: *Leuresthes, Hubbesia* (only females emerge), *Mallotus* (some populations)
  C. Tolerating air exposure
    1. During low tide: *Tomicodon, Anoplarchus, Xiphister, Apodicthys* and others
    2. During the dry season: *Monopterus, Protopterus, Mastacembelus, Clarias* and others
    3. Surviving in the stalls of a tropical fish market: various species
  D. Flying through air
    1. Freshwater: *Gastropelecus*
    2. Marine: *Cypselurus, Exocetus*

II. Gulping air at the surface
  A. Nearly all species with physostomous gas bladders (Chapter 2): Clupeidae, Cyprinidae
  B. Bottom or demersal fish in hypoxic water: *Achirus* (possibly aquatic surface respiration, ASR)
  C. During surface feeding: *Poecilia, Mugil*
  D. During ASR: *Cyprinus, Carassius*, some Gobiidae and Eleotrididae

lists a variety of fish activities that, in the absence of experimental data, would suggest air breathing. Even experimental data may be difficult to apply to the question of what constitutes air breathing because, found among the various non-air-breathing fishes are species that can survive long periods in and exchange respiratory gases with air, emerge naturally, and gulp air (Table 1.1). As we shall see, the designation of a fish as an air breather depends as much or more upon documentation of its natural behavior as it does on experimental data demonstrating aerial gas exchange.

Anyone who has visited a tropical fish market knows that some fishes can survive air exposure for long periods, longer in fact than the exposure tolerances of many species considered specialized for an amphibious existence; they are simply hardy fish that, while not prone to terrestrial excursions, are very tolerant of emersion (Nakamura, 1994). But, most of these market fish lack a specialized air-breathing organ (ABO) and are not known to naturally gulp air or crawl onto land. However, if we were to buy one of these fish and put it in a respirometer, some level of aerial gas exchange could probably be measured. Indeed, this was precisely the type of experiment that was carried out by Nakamura (1994)! Thus, the abilities to survive out of water for a long period and even to exchange respiratory gases with air do not in themselves make a species an air breather. Carter (1957) stressed that aerial gas exchange can occur across practically any fish body surface provided it has sufficient respiratory gas permeability and that a diffusion gradient exists. Another point made by Carter simply underscores part of the problem in defining air-breathing species; not all air-breathing fishes need or use an ABO in this process. Finally, many fishes naturally leave water for a variety of reasons (Table 1.1) but are not considered air breathers. Also, others may gulp air at the water surface but not absorb the contained O_2 for respiration.

How then is it possible to differentiate between fish air breathing and behaviors suggesting this action? Critical observation under natural conditions and controlled experimentation are obviously important in demonstrating air breathing. However, both these approaches may be neither feasible nor required in many cases. To define air breathing and to aid investigators trying to detect its presence in fishes, I have drafted the following set of questions. These emphasize the adaptive function that air breathing must play in the life history of a species. I have been influenced in this approach both by Carter's (1957) examination of air breathing diversity and by Kramer's (1983b, 1988) analyses of the adaptive significance of various fish respiratory adaptations:

i. Does the action or behavior in question likely serve some other function in addition to putative aerial gas exchange?
ii. Does it occur regularly in all individuals of a population and throughout the life of each fish?
iii. Or, if not the latter, is initiation of the activity linked to ontogenetic changes, such as attainment of a certain size or migration to a new habitat?
iv. Is the action essential for the individual's survival?
v. Does the frequency of the action increase when the fish is exposed to conditions not conducive to normal aquatic respiration or when fish activity increases?

Answers to these questions require simple but complete observations or rudimentary experimental techniques. They also require fundamental facts about the natural history of the species and its behavior. They do, however, enable the use of reductionist methods to test for the occurrence and likely importance of aerial gas exchange to a species. A recurrent theme of this book, and one that will be stressed in Chapter 2, is that our suspicions about the air-breathing capacity of many species and some entire groups of fish await both field and laboratory observation. In many cases these observations will be done by investigators living and working in regions where the unstudied species occur naturally. Armed with the above criteria and the minimum equipment required to document air breathing in a fish (i.e., aquaria, some glassware, methods to experimentally modify aquatic gas tension such as a sheet of plastic, and perhaps a portable O_2 meter), biologists around the world can, in time, detail our knowledge of air breathing in species for which there is little present information.

However, even with such criteria in place, certain problem areas remain and it is not always possible to establish that a particular action is or is not related to aerial gas exchange. In such cases only certain types of experimental data may be useful. How do we treat, for example, the frequent sorties made by the freshwater and marine fliers and the emergence of grunion *(Leuresthes)*, capelin *(Mallotus)* (Leggett and Frank, 1990; Sleggs, 1990) and other species during breeding (reviewed by Kramer, 1978c)? It might be possible to measure a small rate of gas exchange for them in a respirometer, but the time constants required for stable measurements of gas flux vastly exceed normal fish exposure periods. A "bradycardic dive response" has been demonstrated in both the grunion and flying fish *(Cypselurus)* during brief air exposure (Garey, 1962). Bradycardia is the usual reflex response to an interrupted O_2 supply and suggests that little of this gas is captured during the brief aerial sojourn.

It is also difficult to ascertain whether or not air gulping at the water surface constitutes air breathing. Certain species gulp air in hypoxia to increase buoyancy and become more proficient at ASR (Gee, 1986; Kramer, 1987; Gee and Gee, 1991, 1995). Air breathing in several groups may have evolved via the "air-gulping for positive-buoyancy route" (Graham *et al.*, 1977; Gee and Gee, 1991, 1995; and Chapter 9) and some species capable of ASR may actually be evolving toward air breathing, as suggested by their capacity to make respiratory use of the O_2 contained in air bubbles that are inadvertently inhaled with the ventilated stream of surface water, or in air that is gulped to increase buoyancy during ASR (Liem, 1980; Burggren, 1982; Gee and Gee, 1991, 1995). Some fish inhale air to regulate the volume of a physostomous gas bladder (i.e., connected to the esophagus, see Chapter 2) and thus serve its functions in both buoyancy and sound reception. But, as has been demonstrated empirically, most of the physostomous fishes are not air breathers (Johansen, 1970; Fange, 1976).

THE TYPES OF AIR-BREATHING FISHES

Table 1.2 classifies the variety of fish air breathing on the basis of the behavioral and habitat conditions that control it. There are two main categories of air-breathing fishes, amphibious and aquatic.

Amphibious Air Breathers

These fish breathe air mainly during the periods they are out of water. They emerge from water for various reasons: to feed, rest, orientate, escape predators, and even court and hold territories (Graham, 1976a; Gibson, 1969, 1982; Sayer and Davenport, 1991; Bridges, 1993a). Amphibious air breathers are mainly found in the littoral zones of marine habitats. Many of these (e.g., mudskippers, *Periophthalmus*) are highly active on land whereas others appear to endure air exposure while stranded during low tide and awaiting the return of water.

Several freshwater fishes are amphibious and most of these are also proficient aquatic air breathers (Table 1.1). Some migrate across the land apparently in search of new bodies of water. Others inadvertently become amphibious air breathers when they are stranded by the disappearance of habitat water, as commonly occurs in certain areas during the tropical dry season. These fishes often become confined in mud burrows until rains come, and to endure low water conditions as well as the absence of food, they enter hypometabolic or even estivating states.

Aquatic Air Breathers

These fish, which remain in water and surface periodically to gulp air, are divided into two air-breathing types: facultative and continuous (Table 1.2). Facul-

TABLE 1.2 Classification of Air-Breathing Fish Types Based on Behavior and Factors Affecting Respiration, with Selected Examples for Each Group

I. Amphibious air breathers
   A. Active on land (volitional exposure): *Periophthalmus, Mnierpes, Entomacrodus, Andamia*
   B. Inactive on land (enforced exposure)
      1. Endure brief exposure: *Tomicodon, Blennius, Xiphister, Pholis*
      2. Estivators: *Protopterus, Synbranchus, Mastacembelus*

II. Aquatic air breathers
   A. Facultative: *Ancistrus, Hypostomus*
   B. Continuous
      1. Obligatory: *Arapaima, Anabas*
      2. Non-obligatory: *Hoplosternum, Piabucina, Erythrinus*

tative air breathers do not normally breathe air in normoxic water but need to adopt this mode when exposed to conditions unfavorable for aquatic respiration (hypoxia, hypercapnia) or in response to increased O_2 requirements (as a result of changes in water temperature or activity).

Continuous air breathers take air breaths at more or less regular (but not highly predictable) intervals, at all times, and under all aquatic conditions from hyperoxia to hypoxia. Their frequency of air breathing is regulated by factors affecting their need for aerial O_2 such as aquatic hypoxia, water temperature, and activity level, although the use of inspired gas for buoyancy and digestive functions may also be important. Continuous air breathers can be further separated into two types: obligatory and non-obligatory (Table 1.2). Obligatory air breathers are not able to survive on the quantity of O_2 obtained by aquatic respiration, even in normoxic water, and thus always need supplemental aerial oxygen. By contrast, non-obligatory, continuous air breathers do not require air breathing to survive while in normoxic water.

In the early research on air-breathing fishes, much emphasis was given to determining whether or not a species was an obligatory air breather (Day, 1868, 1877; Dobson, 1874). These experiments usually compared the survival times of fish in open aquaria to those in aquaria covered by a screen barrier that prevented air access (Day, 1868). The early results for the same species very often differed widely because there was little understanding of the importance of variables such as water O_2 content and temperature or because of the methods used to restrain fish below the surface. In many experiments purported to show obligatory air breathing, the fish were denied air access while held in stagnant, unaerated water and succumbed for these reasons, as did those held in small, semi-enclosed containers, where local O_2 may have been depleted (Ghosh, 1934; Hora, 1935).

Table 1.2 distinguishes different types of fish air breathing; however, because various combinations of biological and environmental factors all influence respiratory partitioning, this classification is far from an exact science. It cannot be used to invariably designate a particular species as one type of air breather as opposed to another. *Synbranchus* and *Monopterus*, for example, are in the same family (Synbranchidae) but differ in their air-breathing requirements (*Synbranchus* is facultative, *Monopterus* is obligatory). However, depending upon environmental conditions, these fishes may also be either amphibious or aquatic air breathers and they can also enter a semi-estivation state in mud (Rosen and Greenwood, 1976; Bicudo and Johansen, 1979). There are also different degrees of estivation, and some swamp-dwelling fishes, the Mastacembelidae and Galaxiidae, for example, can endure long confinement in moist mud even though lacking specialized air-breathing surfaces. Another example is the threshold PO_2 that triggers facultative air breathing. This is not always a constant for a species because of the effects of variables like temperature and body size on O_2 demand and because of likely allometric changes in respiratory organs during ontogeny (Graham and Baird, 1982, 1984). *Amia*, for example, is a facultative air breather at 20 C but an obligatory air breather at 30 C (Johansen *et al.*, 1970a). Small *Protopterus* are facultative air breathers whereas larger ones are obligatory air breathers (Jesse *et al.*, 1967). Finally, regardless of whether or not a fish breathes air continuously in normoxic water, air breathing in hypoxic water is obligatory for most species.

SUMMARY AND OVERVIEW

The appearance of air breathing among Paleozoic fishes was the critical first step in the evolution of vertebrate terrestriality and the proliferation of the tetrapods. The capacity for air breathing was present in at least two groups of Paleozoic fishes, the Actinopterygii and Sarcopterygii, and may have also occurred in certain placoderms and possibly even in acanthodians. In addition to its occurrence in Paleozoic bony fishes, air breathing has been independently and frequently acquired in a variety of actinopterygians during the 400 million year evolutionary history of this group. Included among the ranks of extant air-breathing fishes are forms like those that first appeared in the Paleozoic (lungfishes and the bichirs), groups that evolved into prominence in the Mesozoic (gars) or just prior to teleosts *(Amia)*, as well as a broad diversity of modern teleostean air breathers of Cretaceous to Recent origin.

Modern air-breathing fishes occur in a variety of habitats, both marine and freshwater, and in both tropical and temperate latitudes. Among these forms, the most specialized air-breathers are found in tropical freshwaters where the regular occurrence of seasonal droughts and hypoxia has selected for a large diversity of aerial respiratory adaptations. The habitat conditions that led to the evolution of air breathing in the Late Silurian fishes are thought to have been like those of modern tropical freshwaters, with the regular occurrence of aquatic hypoxia being a primary factor.

Several criteria have importance in making the sometimes fine distinction between air-breathing and non-air-breathing fishes. Air-breathing fishes are those that naturally rely upon aerial gas exchange during some time or phase of their existence to obtain O_2 and thus sustain metabolism in O_2-deficient habitats or while out of water, or to increase their metabolic scope for activity. Whereas actions such as emergence and air gulping at the water surface, and the capacities to tolerate air exposure and to even exchange respiratory gases with air would all seem indicative of air breathing, these criteria are not absolute, just as the absence of an ABO does not necessarily mean a fish cannot breathe air. Thus, designation of a fish as an air breather requires knowledge of its behavior and ecology as well as physiological data.

On the basis of their behavior, physiology, and ecology, air-breathing fishes can be separated into amphibious and aquatic forms. The former breathe air while emersed and their ranks range from highly active species such as mudskippers to less active forms that are inadvertently stranded by low tide or drought. Aquatic air breathers remain in water and surface for air gulps. Some aquatic air breathers do so more or less regularly and thus are termed continuous air breathers. Among continuous air breathers are some forms (obligate air breathers) that cannot survive in normoxic water without aerial respiration. In contrast to the continuous air breathers, some species need to breathe air only when aquatic conditions do not favor aquatic respiration and these are termed facultative air breathers.

CHAPTER

2

Diversity and Natural History

As a rule fishes are so adapted to the aquatic habitat that under normal conditions it is impossible for them to emerge into air. The direct cause of emergence into dry land cannot be either the search for food material, intense competition or flight from pursuing predators. None of these or other factors could have led to positive changes if there had not been some form or organizational preconditioning in the form of adaptations for aerial respiration and locomotion on the ground. A fundamental cause is found in the contrast of ecological conditions by which aerial respiration as well as temporary emergence from water is accomplished in modern fishes.

I.I. Schmalhausen,
The Origin of Terrestrial Vertebrates, 1968

INTRODUCTION

Awareness that a large number of fishes were capable of breathing air prompted efforts to categorize them in an interpretable manner. The first of these attempts was by Rauther (1910) who in "Die akzessorischen Atmungsorgane der Knochenfische," reviewed both the aerial and aquatic respiratory adaptations of fishes. Rauther compared the habitat diversity of the air breathers and detailed their ABO (air-breathing organ) structure. Subsequent workers (Das, 1927; Carter and Beadle, 1931; Hora, 1935; Leiner, 1938; Carter, 1957; Saxena, 1963; Johansen, 1970; Gans, 1970; Dehadrai and Tripathi, 1976; Munshi, 1985; Munshi and Hughes, 1992) drew up lists of the air-breathing fishes, usually grouping them on the basis of their type of ABO. The compilation of Carter and Beadle (1931) was significant because it focused attention on the diversity of the air-breathing fishes and their environments, on the attendant circulatory and behavioral adaptations related to this specialization, and on the potential application of studies of fish air breathing to broader problems in vertebrate physiology and evolution.

Categorizations of fishes based on ABO structure, however, cannot differentiate homologous structures and those arising through convergence. Also, many air-breathing fishes lack ABOs. A classification of the air-breathing fishes must therefore be sufficiently broad to allow considerations of ABO structure and permit estimation of air-breathing fish diversity and the frequency of the independent origin of this specialization. Information of this type can be obtained by phylogenetic analysis, the strategy adopted for this chapter.

Figure 2.1 presents a phylogenetic arrangement of the 49 known families having air-breathing species. This figure, based on the phylogeny described by Lauder and Liem (1983a,b), shows the monophyletic origin of the early fish stock, the derivation of the tetrapods from the primitive lobe-fins (Sarcopterygii), and the progressive evolution of the ray-finned fishes (Actinopterygii) from the primitive Infraclass Cladistia (Order Polypteriformes) to the modern Series Percomorpha. A generalized morphology of each family is illustrated in Figure 2.2 and Table 2.1

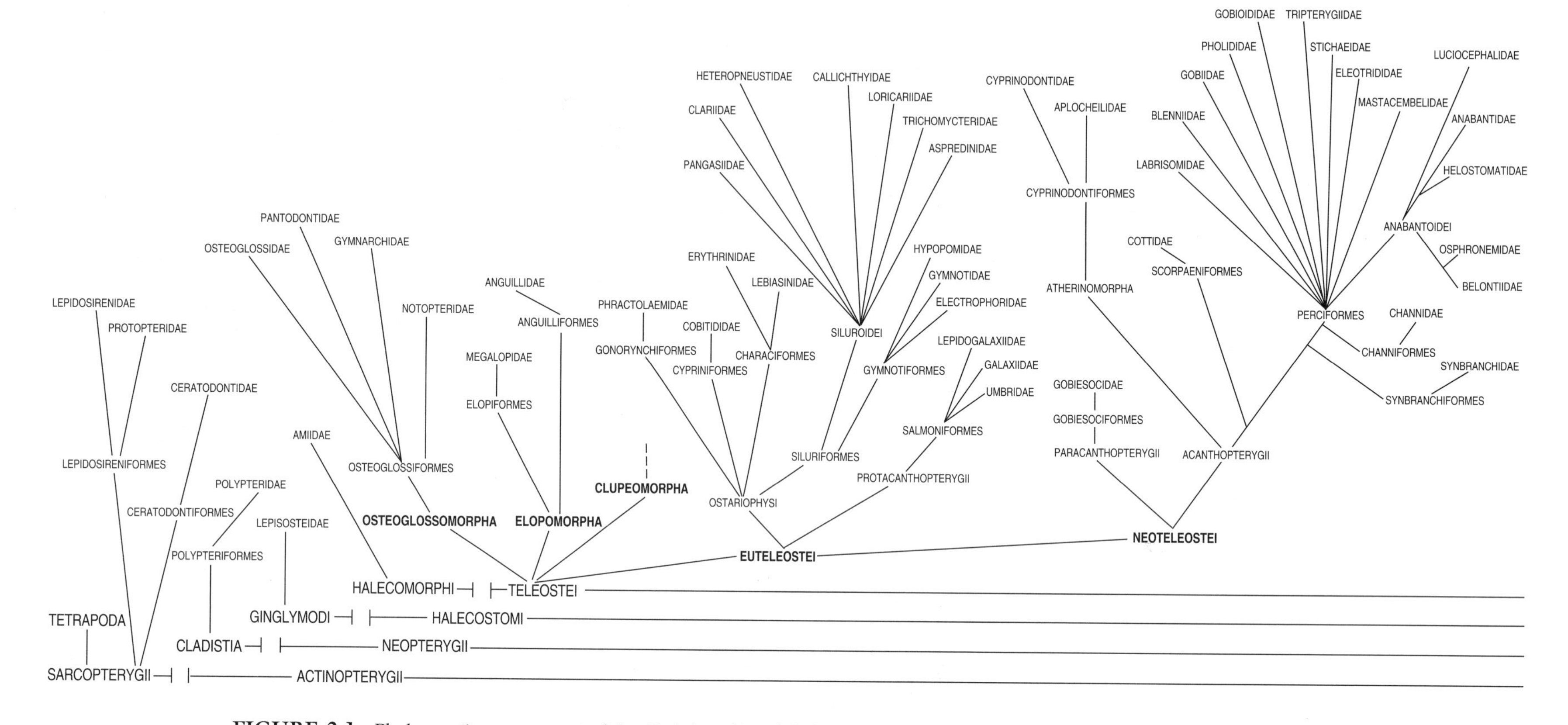

FIGURE 2.1 Phylogenetic arrangement of the 49 air-breathing fish families within the bony fishes (Class Osteichthyes). The progressive taxonomic groupings defining the phylogeny are: Subclasses Sarcopterygii and Actinopterygii; Infraclasses Cladistia and Neopterygii; Divisions Ginglymodi and Halecostomi; Subdivisions Halecomorphi and Teleostei; and the four teleost Infradivisions, Osteoglossomorpha, Elopomorpha, Clupeomorpha, Euteleostei, and the Neoteleostei.

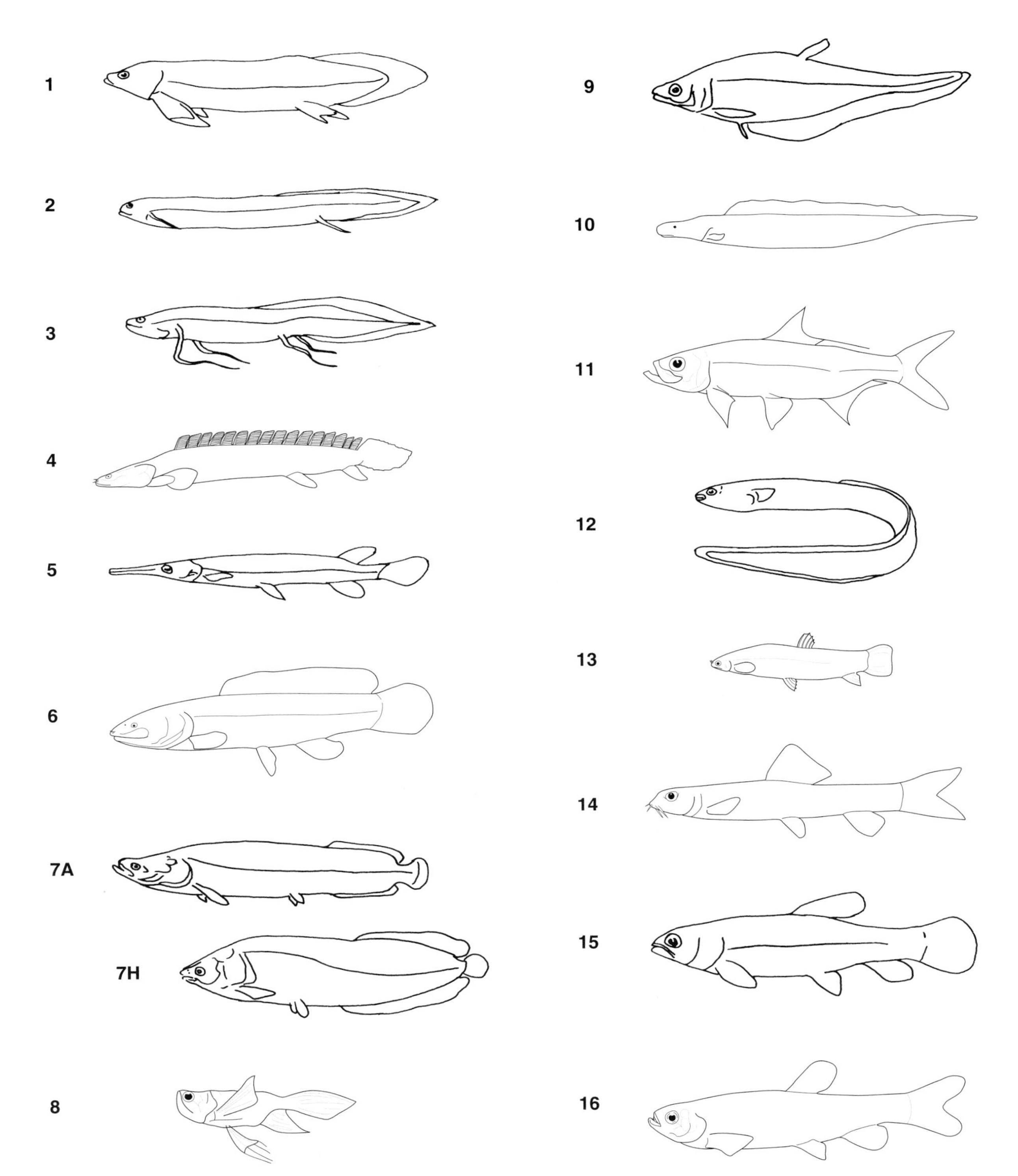

FIGURE 2.2 Representative illustrations of the fish families containing air-breathing species. **1,** Ceratodontidae; **2,** Lepidosirenidae; **3,** Protopteridae; **4,** Polypteridae; **5,** Lepisosteidae; **6,** Amiidae; **7,** Osteoglossidae; *Arapaima,* (A), *Heterotis* (H); **8,** Pantodontidae; **9,** Notopteridae; **10,** Gymnarchidae; **11,** Megalopidae; **12,** Anguillidae; **13,** Phractolaemidae; **14,** Cobitididae; **15,** Erythrinidae; **16,** Lebiasinidae; **17,** Pangasiidae; **18,** Clariidae; *Clarias* (C), *Heterobranchus* (H); **19,** Heteropneustidae; **20,** Aspredinidae; **21,** Trichomycteridae; **22,** Callichthyidae; **23,** Loricariidae; **24,** Hypopomidae; **25,** Gymnotidae; **26,** Electrophoridae; **27,** Umbridae; *Umbra* (U), *Dallia* (D); **28,** Lepidogalaxiidae; **29,** Galaxiidae; **30,** Gobiesocidae; **31,** Aplocheilidae; **32,** Cyprinodontidae; **33,** Cottidae; **34,** Stichaeidae; **35,** Pholididae; **36,** Tripterygiidae; **37,** Labrisomidae; **38,** Blenniidae; *Alticus* (Al), *Andamia* (A); **39,** Eleotrididae; **40,** Gobiidae; **41,** Gobioididae; **42,** Mastacembelidae; **43,** Anabantidae; *Anabas* (A), *Ctenopoma* (C), *Sandelia* (S); **44,** Belontiidae; *Belontia* (B), *Trichogaster* (T); **45,** Helostomatidae; **46,** Osphronemidae; **47,** Luciocephalidae; **48,** Channidae; **49,** Synbranchidae. (Redrawn from various sources including *Fishes of the World*, Nelson, 1994; and reprinted by permission of Wiley-Liss, Inc., a subsidiary of John Wiley & Sons, Inc.)

(continues)

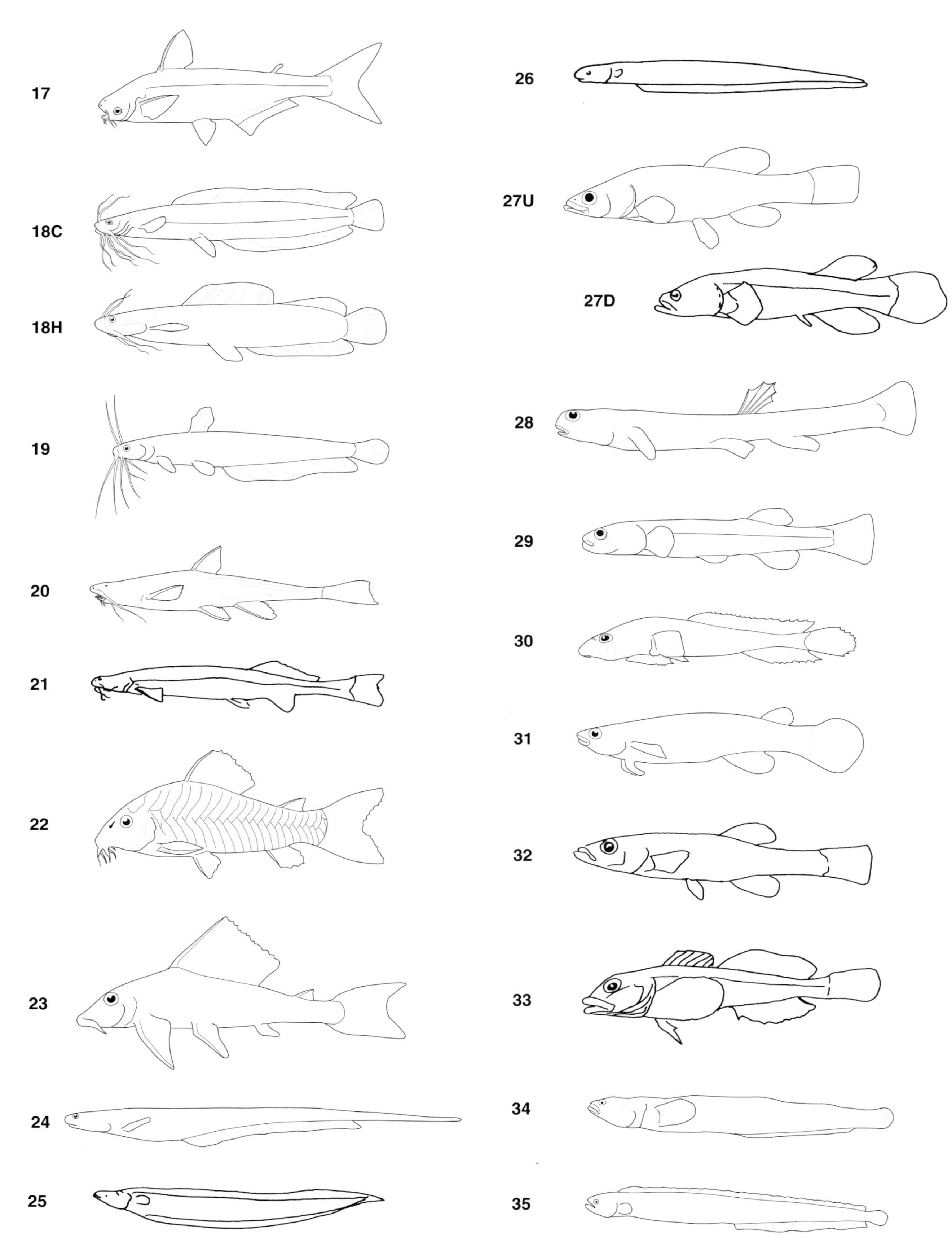

FIGURE 2.2 *(continued)*

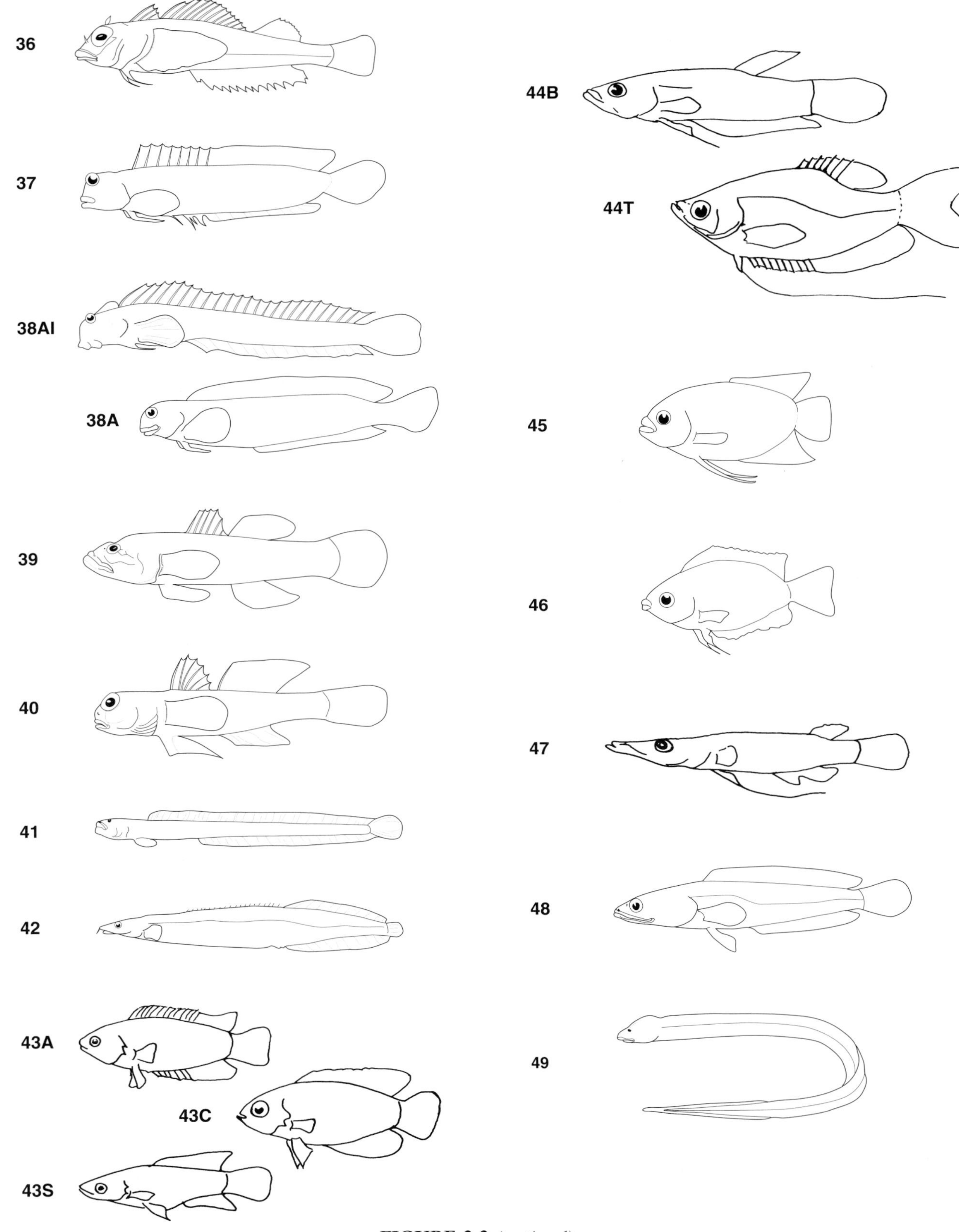

FIGURE 2.2 *(continued)*

TABLE 2.1 Fishes, Listed in Phylogenetic Order, for Which Air Breathing Has Been Documented or Implicated by Behavioral or Other Observation[a]

- **SUPERORDER**
- **Order**
- **Suborder**
  - **Family**
    - ***Genus species***

- CLASS OSTEICHTHYES
- SUBCLASS SARCOPTERYGII
- SUPERORDER CERATODONTIMORPHA
- Ceratodontiformes
  - Ceratodontidae
    - *Neoceratodus forsteri*
- Lepidosireniformes
  - Lepidosirenidae
    - *Lepidosiren paradoxa*
  - Protopteridae
    - *Protopterus* (4)
- SUBCLASS ACTINOPTERYGII
- INFRACLASS CLADISTIA
- Polypteriformes
  - Polypteridae
    - *Polypterus* (10–12)[b]
    - *Erpetoichthys (Calamoichthys) calabaricus*
- INFRACLASS NEOPTERYGII
- DIVISION GINGLYMODI
- Lepisosteiformes
  - Lepisosteidae
    - *Lepisosteus (4)*[b]
    - *Atractosteus* (3)[b]
- DIVISION HALECOSTOMI
- SUBDIVISION HALECOMORPHI
  - Amiiformes
    - Amiidae
      - *Amia calva*
- SUBDIVISION TELEOSTEI
- INFRADIVISION OSTEOGLOSSOMORPHA
- Osteoglossiformes
  - Suborder Osteoglossoidei
    - Osteoglossidae
      - *Arapaima gigas*
      - *Heterotis niloticus*
    - Pantodontidae
      - *Pantodon buchholzi*
  - Suborder Notopteroidei
    - Notopteridae
    - (Sf) Notopterinae
      - *Notopterus* (3)
    - (Sf) Xenomystinae
      - *Papyrocranus afer*
      - *Xenomystus nigri*
  - Suborder Mormyridae
    - Gymnarchidae
      - *Gymnarchus niloticus*
- INFRADIVISION ELOPOMORPHA
- Elopiformes
  - Suborder Elopoidei
    - Megalopidae
      - *Megalops* (2)
- Anguilliformes
  - Suborder Anguilloidei
    - Anguillidae
      - *A. anguilla*
- INFRADIVISION EUTELEOSTEI
- SUPERORDER OSTARIOPHYSI
- Gonorynchiformes
  - Suborder Knerioidei
    - Phractolaemidae
      - *Phractolaemus ansorgii*
- Cypriniformes
  - Cobitididae
    - *Misgurnus* (2)[b]
    - *Lepidocephalichthys* (2)[b]
    - *Acanthophthalmus* (2+)[b,e]
    - *Cobitis barbatula*
- Characiformes
  - Erythrinidae
    - *E. erythrinus*
    - *Hoplerythrinus unitaeniatus*
  - Lebiasinidae
    - *Piabucina festae*
    - *Lebiasina bimaculata*
- Siluriformes
  - Pangasiidae
    - *Pangasius* (4+)[e]
  - Clariidae[b]
    - *Clarias* (36)
    - *Heterobranchus* (4)
    - *Dinotopterus* (4)
  - Heteropneustidae
    - *Heteropneustes* (2)
  - Aspredinidae
    - *Bunocephalus* (2+)[e]
  - Trichomycteridae
    - *Pygidium striatum*
    - *Eremophilus mutisii*
  - Callichthyidae
    - *C. callichthys*
    - *Hoplosternum* (2)[b]
    - *Brochis* (2)[b]
    - *Corydoras* (23)[b]
    - *Dianema longibrabus*
    - *D. urostriata*
  - Loricariidae
    - *Hypostomus (Plecostomus)* (3)[b]
    - *Ancistrus* (2)
    - *Sturisoma citurense*
    - *Loricaria (Rineloricaria) uracantha*
    - *Loricariichthys*
    - *Pterygoplichthys* (2)[b]
    - *Chaetostoma fischeri*
    - *Lasiancistrus* sp.
    - *Pseudoancistrus* sp.
    - *Acanthicus* sp.
    - *Rhinelepis strigosa*
- Gymnotiformes
  - Suborder Sternopygoidei
    - Hypopomidae
      - *Hypopomus* (3)
  - Suborder Gymnotoidei
    - Gymnotidae
      - *Gymnotus carapo*
    - Electrophoridae
      - *Electrophorus electricus*
- SUPERORDER PROTACANTHOPTERYGII
- Salmoniformes
  - Suborder Esocidae
    - Umbridae
      - *Umbra* (4)
      - *Dallia pectoralis*
- Suborder Lepidogalaxioidei
    - Lepidogalaxiidae
      - *Lepidogalaxias salamandroides*
  - Suborder Salmonoidei
    - Galaxiidae
      - *Neochanna* (3)
      - *Galaxias* (5)
      - *Galaxiella* (2)[e]
- INFRADIVISION NEOTELEOSTEI
- SUPERORDER PARACANTHOPTERYGII
- Gobiesociformes
    - Gobiesocidae
      - *Sicyases sanguinieus*
      - *Phallerodiscus funebris*
      - *Gobiesox pinniger*
      - *Tomicodon* (3)
      - *Lepadogaster* sp.
- SUPERORDER ACANTHOPTERYGII
- Cyprinodontiformes
    - Aplocheilidae
      - *Rivulus* (5)
    - Cyprinodontidae
      - *Fundulus* (4)
- Scorpaeniformes
    - Cottidae
      - *Clinocottus* (3)
      - *Taurulus bubalis*
- Perciformes
  - Suborder Zoarcoidei
    - Stichaeidae
      - *Xiphister* (2)[b]
      - *Anoplarchus purpurescens*
      - *Cebidichthys violaceus*
      - *Chirolophis polyactocephalus*[e]
    - Pholididae
      - *Xererpes fucorum*
      - *Pholis* (3)[b]
      - *Apodichthys flavidus*
  - Suborder Blennioidei
    - Tripterygiidae
      - *Helcogramma medium*
    - Labrisomidae
      - *Mnierpes macrocephalus*
      - *Dialommus fuscus*
    - Blenniidae
      - *Andamia* (4+)
      - *Alticus* (7+)
      - *Salarias* (8+)
      - *Blennius* (7+)
      - *Coryphoblennius galerita*
      - *Istiblennius* 1 (25)[e]
      - *Entomacrodus* 4 (22)[e]
  - Suborder Gobioidei

(continues)

TABLE 2.1 *(continued)*

- Eleotrididae
  - *Dormitator latifrons*
  - *Oxyeleotris marmorata*[c]
- Gobiidae
  - *Periophthalmus* (12)[d]
  - *Boleophthalmus* (5)[d]
  - *Periophthalmodon* (3)[d]
  - *Scartelaos* (4)[d]
  - *Gillichthys* (2)
  - *Quietula guaymasiae*
  - *Gobionellus sagittula*
  - *Pseudapocryptes lanceolatus*
  - *Apocryptes bato*
  - *Favonigobius exquisitus*
  - *Pseudogobius dorum*
  - *Arenigobius* (2)
  - *Cryptocentroides cristatus*
  - *Mugilogobius* (2)
  - *Chlamydogobius* (2+)
- Gobioididae
  - *Taeniodes rubicundus*
- Suborder Mastacembeloiodei
  - Mastacembelidae
    - *Mastacembelus* (2)
    - *Macrognathus aculeatus*
- Suborder Anabantoidei
  - Anabantidae
    - *Anabas testudineus*
    - *Ctenopoma* (21)[b]
    - *Sandelia* (2)
  - Belontiidae
  - (Sf) Belontiinae
    - *Belontia* (2)[b]
  - (Sf) Macropodinae
    - *Macropodus* (3)[b]
  - (Sf) Trichogasterinae
    - *Colisa* (4)[b]
    - *Trichogaster* (4)[b]
  - (Sf) Sphaerichthyinae
    - *Sphaerichthys* (3)[b]
    - *Parasphaericthys ocellatus*
  - (Sf) Ctenopinae
    - *Betta* (16)[b]
    - *Ctenops nobilis*
    - *Malpulutta kretseri*
    - *Parosphromenus* (4)[b]
    - *Pseudophromenus* (2)[b]
    - *Trichopsis* (3)[b]
  - Helostomatidae
    - *Helostoma temmincki*
  - Osphronemidae
    - *Osphronemus goramy*
- Suborder Luciocephaloidei
  - Luciocephalidae
    - *Luciocephalus pulcher*
- Channiformes
  - Channidae
    - *Channa* (12–14+)[b]
- Synbranchiformes
  - Synbranchidae[b]
    - *Ophisternon* (6)
    - *Synbranchus* (2)
    - *Monopterus (Amphipnous)* (6)

[a]Species names are listed if one species of a genus is known to air breathe. If more than one species of a genus is an air breather, the number is specified in parenthesis and the species are named in the text or in either Tables 2.2 or 2.3. Other designations: Species identity is unknown, sp.; subfamily (Sf).
[b]Listed in Table 2.2.
[c]See text.
[d]Listed in Table 2.3.
[e]Observations very incomplete.

lists the air-breathing genera and species in families having more than just a few air breathers.

Figure 2.1 shows that all of the extant air-breathing fishes are in the class Osteichthyes. There are no air-breathing cyclostomes. The only report of air breathing in elasmobranchs is an insufficiently documented account by George (1953) who observed aquarium specimens of two species of dogfish *(Chiloscyllium)* to occasionally gulp air or aspirate it through their spiracles and then to release same from the gill slits. Periodic air gulping by the sand tiger shark *(Odontaspis taurus)* for purposes of buoyancy control (Herald, 1961, as *Carcharinus taurus;* S.H. Gruber, *in litt.*) needs evaluation for a possible respiratory function.

Figure 2.1 also indicates that air breathing is relatively more common among the primitive families but has often been acquired independently (i.e., not present in a common ancestor) by the more modern families. In different families, air breathing ranges from rare-to-common-to-universal (Table 2.2); however, there is presently not enough information about the air-breathing species in certain families to make more than an educated guess about the relative occurrence of air breathing.

THE FAMILIES OF AIR-BREATHING FISHES

The following account of the air-breathing fishes provides general information on geographical distribution, ecology, and habitats of the species, as well as details about ABO structure, air-breathing behavior and physiology, and whether or not air breathing is obligatory (Chapter 1). The diverse approaches brought to the study of air breathing in different species are also detailed in order to bring to light the full spectrum of ideas and information embodied in this area of research. The final section of this chapter summarizes air-breathing fish diversity and discusses the species for which this specialization has been reported but not fully documented.

Lungfishes (Subclass Sarcopterygii)

The three genera (six species) of lungfish and the coelacanth *(Latimeria)* are the only living sarcopterygian (lobe-finned) fishes. They are relics of a group that dominated in the Late Paleozoic Era and gave rise to the tetrapods (Long, 1995). Among modern sarcop-

TABLE 2.2 Complete Species Lists for Air-Breathing Families Containing Too Many Species to Be Conveniently Listed in Either Table 2.1 or the Text

Polypteridae
- *Polypterus bicher*
- *P. ansorgei*
- *P. endlichari*
- *P. katangae*
- *P. lapradei*
- *P. palmas*
- *P. congicus*
- *P. ornatipinnis*
- *P. weeksii*
- *P. retropinnis*
- *P. delhezi*
- *P. senegalus*
- *Erpetoichthys (Calamoichthys) calabaricus*

Lepisosteidae
- *Lepisosteus osseus*
- *L. oculatus*
- *L. platyrhincus*
- *L. platostomus*
- *Atractosteus tristoechus*
- *A. tropicus*
- *A. spatula*

Cobitididae
- *Misgurnus fossilis*
- *M. anguillocaudatus*
- *Lepidocephalichtys guntea*
- *L. thermalis*
- *Acanthophthalmus kuhlii*
- *A. semicinctus*

Clariidae
- *C. Heterobranchoides mellandi*
- *C. H. ngamensis*
- *C. H. anguillaris*
- *C. H. loangwensis*
- *C. H. lazera*
- *C. H. mossambicus*
- *C. H. gariepinus*
- *C. H. capensis*
- *C. H. senegalensis*
- *C. H. nyansis*
- *C. H. longibarbis*
- *Clarias Clarias hilgendorfi*
- *C. C. alluaudi*
- *C. C. eupogon*
- *C. C. werneri*
- *C. C. pachynema*
- *C. C. salae*
- *C. C. bythipogon*
- *C. C. angolensis*
- *C. C. theodorae*
- *C. C. amplexicauda*
- *C. C. liberiensis*
- *C. C. esamesae*
- *C. C. macromystax*
- *C. C. jaensis*
- *C. C. cameronensis*
- *C. C. walkeri*
- *C. maclareni*[c]
- *C. dumerili*[c]
- *C. batrachus*[c]
- *C. macrocephalus*[c]
- *Clarias Allabenchelys submarginatus*
- *C. A. dumerili*
- *C. A. carsonii*
- *C. A. phillipsi*
- *C. A. poensis*
- *C. A. longior*
- *C. A. laticeps*
- *C. A. laeviceps*
- *Heterobranchus longifilus*
- *H. bidorsalis*
- *H. boulengeri*
- *H. isopterus*
- *Dinotopterus ilesi*
- *D. loweae*
- *D. longibarbis*
- *D. gigas*

Callichthyidae
- *Hoplosternum thoracatum*
- *H. littorale*
- *Brochis splendens*
- *B. coeruleus*
- *Corydoras aeneus*
- *C. agassizi*
- *C. arcuatus*
- *C. barbatus*
- *C. cochui*
- *C. elegans*
- *C. graphi*
- *C. griseus*
- *C. hastatus*
- *C. julii*
- *C. leopardus*
- *C. macropterus*
- *C. melanistius*
- *C. myersi*
- *C. nattereri*
- *C. paleatus*
- *C. punctatus*
- *C. rabauti*
- *C. reticulatus*
- *C. schultzei*
- *C. treitli*
- *C. trilineatus*
- *C. undulatus*
- *Dianema* 2 sp.

Loricariidae
- *Hypostomus (Plecostomus) plecostomus*
- *H. emarginatus*
- *H. punctatus*
- *Ancistrus chagresi*
- *A. anisitsi*
- *Pterygoplichthys multiradiatus*
- *P. pardalis*

Stichaeidae
- *Xiphister atropurpureus*
- *X. mucosus*

Pholidae
- *Pholis laeta*
- *P. ornata*
- *P. schultzi*

Anabantidae
- *Ctenopoma acutirostris*
- *C. ansorgii*
- *C. argentoventer*
- *C. breviventralis*
- *C. brunneus*
- *C. congicum*
- *C. ctenotis*
- *C. damasi*
- *C. davidae*
- *C. fasciolatum*
- *C. kingsleyae*
- *C. lineatus*
- *C. maculatum*
- *C. multispinis*
- *C. muriei*
- *C. nanum*
- *C. nigropannosum*
- *C. ocellatus*
- *C. oxyrhynchus*
- *C. pellegrini*
- *C. pethericì*

Belontiidae
- *Belontia signata*
- *B. hasselti*
- *Macropodus opercularis*
- *M. cupanus*
- *M. chinensis*
- *Colisa chuna*
- *C. fasciata*
- *C. lalia*
- *C. labiosa*
- *Trichogaster trichopterus*
- *T. pectoralis*
- *T. leeri*
- *T. microlepis*
- *Sphaerichthys osphromenoides*
- *S. acrostoma*
- *S. vaillanti*
- *Betta bellica*[a]
- *B. cocina*[a]
- *B. fasciata*[a]
- *B. foerschi*[a]
- *B. imbellis*[a]
- *B. smaragdina*[a]
- *B. splendens*[a]
- *B. anabatoides*[b]
- *B. balunga*[b]
- *B. brederi*[b]
- *B. macrostoma*[b]
- *B. picta*[b]
- *B. pugnax*[b]
- *B. rubra*[b]
- *B. taeniata*[b]
- *B. unimaculata*[b]
- *Parosphromenus filamentosus*
- *P. parvulus*
- *P. dreissneri*
- *P. paludicola*
- *Pseudophromenus cupanus*
- *P. dayi*
- *Trichopsis pumilis*
- *T. vittatus*
- *T. harrisi*

Channidae
- *Channa punctatus*
- *C. argus*
- *C. striatus*
- *C. gaucha*
- *C. maculata*
- *C. marulius*
- *C. asiatica*
- *C. africanus*
- *C. pleurophthalmus*
- *C. obscurus*
- *C. lucius*
- *C. micropeltes*
- *C. orientalis*[b]
- *C. melasoma*

Synbranchidae
- *Ophisternon (Furmastix) bengalense*
- *O. gutturale*
- *O. candidum*
- *O. afrum*
- *O. infernale*
- *O. aenigmaticum*
- *Synbranchus marmoratus*
- *S. madeirae*
- *Monopterus albus*
- *M. boueti*
- *M. cuchia*
- *M. fossorius*
- *M. indicus* type 1
- *M. indicus* type 2

[a]Bubble-nest builder.
[b]Mouth brooder.
[c]Subgenus uncertain.

terygians only the lungfishes are air breathers and, as their name indicates, a lung is used for this purpose (Chapter 3). The bimodal breathing capability of these fishes led early workers to classify them as "dual breathers" (e.g., Dipnoans, Dipnoi, or Dipneusti), and place them in a separate subdivision of the Sarcopterygii.

The lungfish fossil record extends back to the Devonian (408 million years before present [mybp]) with fossils of estivating fish in cocoons *(Gnathorhiza)* dating from the Permian (286 mybp). The modern lungfishes are distributed on three continents and are classified into three separate families and genera. The African (Protopteridae, *Protopterus*) and South American (Lepidosirenidae, *Lepidosiren*) lungfishes comprise one lineage which, often termed the mainline of dipnoan evolution, appears largely unchanged from the ancestral *Dipterus* of the Carboniferous Period (360–286 mybp). The Australian lungfish (Ceratodontidae) is a descendant of the fossil form *Ceratodus* which occurred on all continents from the Triassic to Cretaceous Periods (245–144 mybp).

A volume (Bemis *et al.*, 1987) on the biology, evolution, and relationships of the lungfishes includes chapters on the circulatory and respiratory adaptations of extant species (Burggren and Johansen, 1987), on the natural history of the Australian and African species (Kemp, 1987; Greenwood, 1987) and an annotated bibliography for the group (Conant, 1987).

Ceratodontidae. The Australian lungfish, *Neoceratodus forsteri,* is the most primitive extant lungfish, having large, fleshy fins and thick scales. This fish is restricted to a small geographic area—the Burns and Mary Rivers in eastern Australia—where it frequents the more open waters and deep pools (Kemp, 1987) and is nocturnally active. It reaches a maximum length of about 1.5 m (40 kg body weight) and is mainly carnivorous (earthworms, snails, shrimp, tadpoles, frogs) but will also eat plant material (Merrick and Schmida, 1984).

Unlike the African and South American lungfishes, *Neoceratodus* is not known to burrow or estivate and takes air breaths more or less continuously at low frequencies. It ventilates its lung using the same buccal force-pump mechanism found in the other lungfishes and amphibians (Chapter 3; Grigg, 1965; McMahon, 1969). The lung itself, however, is less specialized for respiration than that of either *Lepidosiren* or *Protopterus.* This organ is also positioned in the dorsal part of the coelomic cavity and may, according to Grigg (1965, also see Lenfant and Johansen, 1968), have greater importance in buoyancy control than in respiration. Grigg (1965) further suggests that aerial respiration by *Neoceratodus* may occur more to sustain elevated activity than to ensure survival in either warm or hypoxic water. This fish can survive forced air exposure for a brief time; however, its inability to release CO_2 aerially makes its blood acidotic which reduces the O_2 transport-capacity of blood hemoglobin (Hb) (i.e., the rise in blood PCO_2 interferes with Hb–O_2 binding; this is the Bohr effect, Chapter 7) and reduces total O_2 consumption ($\dot{V}O_2$) (Lenfant *et al.*, 1970).

According to Merrick and Schmida (1984) larval *Neoceratodus* hatch 21 days after spawning, have thin bodies, and spend most of their time on the benthic substrate, lying on their side. Also, young fish reportedly survive better in habitats with sloping banks where they can periodically emerge from water. Young *Neoceratodus* can also respire cutaneously and use well-developed ciliated cells to drive a current over their body surface to aid respiration and other functions (Whiting and Bone, 1980; Kemp, 1987, 1996). Air gulping develops at about 25 mm length (several weeks after hatching) and this is similar to both *Protopterus* and *Lepidosiren.* The body of an adult *Neoceratodus* is covered by thick scales. Bemis and Northcutt (1992) report rich concentrations of hairpin vascular loops in the snout epidermis. These loops may be homologous with the cosmine pore canal system of fossil lungfish, long thought to house sensory organs. They also resemble the integumentary vascular loops found in certain cutaneously respiring fishes (Chapter 3) suggesting a possible respiratory role.

Lepidosirenidae. The Lepidosirenidae *(Lepidosiren)* and the Protopteridae *(Protopterus)* have common features insofar as their life history, early development, and air breathing are concerned. These include external gills in young fish, the presence of highly partitioned, paired lungs that originate from the ventral surface of the esophagus, and similar circulatory specializations for air breathing. The two genera have similar lung anatomies and buccal force-pump mechanisms (Bishop and Foxon, 1968; Jesse *et al.*, 1967; McMahon, 1969). Johansen *et al.* (1968a) considered *Lepidosiren* to be "further advanced" towards pure air breathing than *Protopterus,* because it has a greater degree of heart septation (Chapter 4) and because its aerial O_2 uptake makes up a slightly greater percentage of its total $\dot{V}O_2$ (Burggren and Johansen, 1987 and Chapter 5).

Lepidosiren paradoxa occurs throughout the Amazon and Parana river basins of South America (Lowe-McConnell, 1987). Less detail is known about its biology and, reflecting this, the Bemis *et al.* (1987) volume does not contain a natural history chapter on *Lepidosiren.* Pettit and Beitinger (1980) reviewed the major types of experiments carried out on this fish.

Unlike *Neoceratodus*, *Lepidosiren* frequents swamps and shallow water habitats where hypoxia is common; it also takes refuge in burrows. This fish reaches a maximum size of about 1 m and feeds on a variety of aquatic vertebrates and invertebrates. Its paired fins (and those of *Protopterus* as well) are reduced to thin, flexible wisps which, while aiding minimally in locomotion, have tactile and chemosensory capabilities useful in orientation and prey capture in muddy water.

Lepidosiren does not actively emerge from water and is thus not naturally amphibious (Chapter 1). In some areas it becomes confined in a moist mud burrow during the dry season. Although it can survive for several months in such conditions and has been reported to form a cocoon similar to that of *Protopterus* (Kerr, 1900), there are no details and the estivation of *Lepidosiren* has not been examined in equivalent detail (Carter and Beadle, 1930). During experimental air exposure the $\dot{V}O_2$ and blood pH of *Lepidosiren* drop because of respiratory CO_2 build-up (Johansen and Lenfant, 1967; see Chapter 7). However, *Lepidosiren* endures this better than *Neoceratodus*, partly because of a smaller Bohr effect (Lenfant *et al.*, 1970; and Chapter 7).

Male *Lepidosiren* guard eggs and developing larvae within nests and during the course of this duty develop vascular filaments on their paired fins. These structures have been illustrated by several authors (Kerr, 1900, plate 12, figure 2C; Krogh, 1941, figure 10; Sterba, 1963, figure 1191), and while it is generally known they form on the pelvic fins, a report of their occurrence on the pectoral fins (Agar, 1908) has gone almost unnoticed.

The role of the fin filaments in respiration is strongly suggested by their gill-like structure and by their exclusive appearance on nest-guarding males. However, considerable debate still surrounds the question of their role in respiration (Foxon, 1933a,b; Cunningham and Reid, 1932, 1933; Kramer *et al.*, 1978). The older, accepted view is that the organs serve the auxiliary respiration of the adult enabling it to continuously guard the nest and not have to leave periodically for air. A different view, that the fin filaments permit the "emission" of aerially obtained O_2 into the nest water to oxygenate the developing brood, was advanced by Cunningham (1932) and Cunningham and Reid (1932, 1933). These workers measured changes in the O_2 content of water in glass tubes containing air-breathing *Lepidosiren* and showed that males with well-developed filaments actually exuded O_2 to water; whereas, both females and males that either lacked or had rudimentary filaments consumed O_2 from water. In spite of this elegant and highly convincing demonstration of the filamental fins' role in nest oxygenation, Foxon (1933b) vigorously defended the idea that these structures were auxiliary gills. Foxon's scientific critique, however, failed to undermine the evidence of Cunningham and Reid and his letter to *Nature* was little more than an appeal for retaining the conventional view and contained tones of what seems to be either envy or antipathy. Nevertheless, and in spite of the well conceived and difficult field experiments done by Cunningham and Reid, the exchange with Foxon left the function of the fin filaments in doubt.

Additional studies with more sophisticated instrumentation are now feasible and should be done to confirm the O_2-emission findings of Cunningham and Reid (perhaps by placing submersible O_2 electrodes at several places in the nest). New experiments could provide data for O_2 flux in relation to the air-breathing behavior of the nest-guarding male (which according to the two different hypotheses should either increase or decrease), and allow examination of the adult's response to manipulations of nest O_2 conditions.

Lepidosiren larvae have an external gill, a primitive characteristic shared with the African lungfish, the Polypteridae (also a primitive group, Figure 2.1), and larval amphibians, but not present in *Neoceratodus*. With growth, the external gill is lost and a greater dependence upon air breathing gradually occurs. Very young and juvenile *Lepidosiren* are not obligatory air breathers. Adults by contrast are obligatory air breathers and obtain over 90% of their O_2 requirement aerially (Sawaya, 1946; Johansen and Lenfant, 1967).

Although the lung of *Lepidosiren* functions for the bulk of its O_2 uptake, most respiratory CO_2 is shed aquatically via the gills (Sawaya, 1946; Johansen and Lenfant, 1967). The skin may also play a role in CO_2 release because *Lepidosiren* has small gills that are, in order to prevent the transbranchial loss of aerially obtained O_2, chronically hypoperfused during air breathing (Johansen *et al.*, 1968a; and Chapter 4). Cunningham (1932) demonstrated both transcutaneous CO_2 elimination and O_2 uptake for *Lepidosiren* in water and estimated the R (the respiratory gas-exchange ratio, $R = \dot{V}CO_2/\dot{V}O_2$) of this surface to be over 10.

Protopteridae. The four African lungfishes *(Protopterus dolli, P. annectens, P. amphibius,* and *P. aethiopicus)* are distributed across the tropical parts of the continent, inhabiting swamps, shallow pools, deep lakes, and rivers. These fish are voracious predators and are also an important food to native Africans, being fished in the major lakes and rivers and hunted in their dry season burrows (Greenwood, 1987). Hypoxic and hypercarbic conditions occur naturally in the many lungfish habitats and all species become obligatory air breathers at a young age (Johansen *et al.*, 1976; Babiker, 1979; Greenwood, 1987). In addition, some of the habitats occupied by these fishes may be subject to complete seasonal drying and to endure this, African lungfishes have the capacity to burrow, envelop themselves in a mucus cocoon, and estivate in dry mud for considerable periods (up to four years, Smith, 1931, 1959; DeLaney *et al.*, 1974). (Fish held in cocoons brought into the laboratory have remained viable for up to six years [Lomholt, 1993].)

Physiological and ecological aspects of lungfish estivation have been reviewed by Smith (1931), DeLaney *et al.* (1974), Fishman *et al.* (1987), Greenwood (1987), and by Lomholt (1993). Estivation is stimulated by a variety of factors including environmental drying and reductions in food supply. As water disappears from its habitat, the lungfish burrows vertically into the mud and hollows a space for its body. Silt eventually plugs the opening of the burrow, and the enclosed fish covers itself with a mucus cocoon that will prevent dehydration as the surrounding mud drys. Except for a small breathing hole contiguous with the mouth and burrow opening, the cocoon is sealed. Estimates of estivating lungfish metabolism range from 1 to 20% of the resting rate in water, and its nitrogenous waste products, excreted as ammonia by fish in water, are converted to urea and stored in the body for the duration (Smith, 1930; Fishman *et al.*, 1987). Estivation is terminated soon after the return of water to the habitat. Smith (1931, 1959) speculated that water, by flooding the burrow opening, imposes asphyxic stress on the fish which in turn responds by surfacing for respiration. On the other hand, Lomholt (1993) has observed estivating fish to move in their cocoons and suggests that they may be awake rather than semi-conscious. Soon after arousal all urea stores are voided and the fish returns to its normal pattern of ammonotelism. The transition from ammonotelism to ureotelism during estivation has broad evolutionary significance in that *Protopterus* (and all other sarcopterygians) possess the enzymes of the ornithine-urea cycle and thus synthesize urea in a manner similar to terrestrial vertebrates (Brown *et al.*, 1966; Mommsen and Walsh, 1989; and Chapter 8).

All four *Protopterus* species can be induced to estivate in the laboratory; however, this is not the uniform response of all species to habitat drying (Greenwood, 1987). In certain areas, lungfish may endure dry periods in mud burrows and not estivate (i.e., form a cocoon). Factors ranging from soil type to the historical severity of the dry periods in an area may influence whether or not a species estivates (Fishman *et al.*, 1987; Greenwood, 1987).

Male African lungfish guard their nests and are able to oxygenate them by agitating the surface water (Greenwood, 1987; Lowe-McConnell, 1987); however, they lack the filamentous fins of *Lepidosiren.* Young *Protopterus* behaviorally orientate to aquatic O_2 gradients and have external gills, although there are specific differences in the length of time these persist and remain functional. Smith (1931) reported that all *P. aethiopicus* lose external gills by the time they reach 10 cm length whereas gills were still present on *P. annectens* up to 80 cm long. Smith also found that even with gills, small *P. aethiopicus* have functional lungs and are obligatory air breathers. This contrasts with the report that post-larval *P. aethiopicus* and *P. annectens* are not obligatory air breathers (Greenwood, 1987).

McMahon (1969) investigated the air-ventilation of *P. aethiopicus* in water and showed that inspiration is driven by the buccal force-pump and that expiration occurs passively through both lung decompression and elastic recoil. In contrast, an investigation of lung ventilation in cocoon-dwelling *P. amphibius* (Lomholt *et al.*, 1975) suggested inspiration was by a suctional force. Data for this, however, are not convincing and the idea was rejected by DeLaney and Fishman (1977) (also see Lomholt, 1993).

Burggren and Johansen (1987) reviewed the extensive literature on lungfish cardiorespiratory physiology (Chapters 4 and 6). Studies with three species (*P. aethiopicus,* Lenfant and Johansen, [1968]; *P. amphibius,* Johansen *et al.* [1976]; and *P. annectens,* Babiker, [1979]) indicate that post-larval and juvenile fishes obtain about 70% of their total $\dot{V}O_2$ aquatically; whereas, adults acquire only 10 to 15% of total O_2 via this mode with both the skin and gills involved (Lenfant and Johansen, 1968; McMahon, 1970). Aquatic gas exchange, however, does function for the bulk of respiratory CO_2 release and opinions differ (but there are few data) as to the relative importance of the skin versus gills in this process (Lenfant and Johansen, 1968; McMahon, 1970).

One facet of air-breathing fish biology that has been pursued with *Protopterus* is the effect of air breathing on intermediary metabolism (Chapter 8). Dunn *et al.* (1981, 1983) found that the relatively ready access to O_2 provided by air breathing does not manifest itself at the muscle tissue level; both the red and white fibers of *P. aethiopicus* were poised mainly for anaerobic activity as indicated by their fine structure, enzyme profile, low vascularity, low number of mitochondria, and low lipid content.

Subclass Actinopterygii

Infraclass Cladistia

Polypteriformes

Polypteridae. The polypterids are the most primitive family of the living ray-finned fishes and form a separate infraclass, Cladistia. Polypterids are considered to be the only surviving remnant of the diverse paleonisciform fishes (Chapter 1), a dominant group of the Paleozoic that extended into the Triassic (245–208 mybp). The polypterid fossil record, however, can only be traced back to the Middle Cretaceous (110 mybp). All the fossil and extant polypterids occur in Africa. This family contains one species of rope, or reed fish, *Erpetoichthys (Calamoichthys) calabaricus,* and from 10 to 12 species of bichir, *Polypterus* (Daget, 1950 and Table 2.2). These fishes have a thick armor covering of ganoid scales. *Erpetoichthys* lacks pelvic fins and, as its common name suggests, has an ophidian shape. Polypterids range in maximum length from 40 to 90 cm. They frequent shallow areas of tropical rivers, streams, and lakes and feed on frogs, fishes, and invertebrates (Lowe-McConnell, 1975, 1987).

In common with the larva of *Protopterus* and *Lepidosiren* and of most amphibians, *Polypterus* and *Erpetoichthys* larva have a large external gill that aids in aquatic respiration. Both genera have paired lungs that, as in lungfish, originate from the ventral pharynx. Early observations suggested that *Polypterus* rarely, if ever, breathed air, but this activity is now well documented by observation and both anatomical and physiological investigations (Smith, 1931; Magid, 1966, 1967; Magid *et al.*, 1970; Babiker, 1984a). It seems likely that air breathing occurs in all species of *Polypterus* (Poll and Deswattines, 1967; Lowe-McConnell, 1987).

A surprising result of Babiker's (1984a) experiments with *P. senegalus* was that this fish did not begin air breathing until it reached the relatively large body size of about 22 g. This is relatively late in ontogeny compared to most other air-breathing fishes and more data are needed for this and other *Polypterus* species. Once air breathing begins, dependency on this mode increases with body size but never becomes obligatory. Nevertheless, *P. senegalus* is an efficient air breather and has the capacity to breathe air exclusively in hypoxic water (i.e., reaching the point of reducing its aquatic-ventilation and $\dot{V}O_2$ to virtually zero, Babiker, 1984a). Brainerd *et al.* (1989) showed that muscles intrinsic to the lung wall of *Polypterus* actively force expiration which is then followed by passive lung aspiration during muscle relaxation. This confirmed the earlier observations on *Erpetoichthys* by Purser (1926) and, by suggesting the mechanism for rapid inspiration, adds new impetus to Magid's (1966) observation that *P. senegalus* can inhale rapidly via its spiracles (Chapter 3).

Studies by Pettit and Beitinger (1981, 1985), Sacca and Burggren (1982), and Beitinger and Pettit (1984) determined that *E. calabaricus* respires both aerially and aquatically but is not an obligatory air breather. This species is capable of regular terrestrial excursions, which it does volitionally, and while on land is capable of feeding on insects as well as negotiating minor obstacles. Also, and like *Polypterus,* the ability of *Erpetoichthys* to air breathe makes it largely insensitive to the microhabitat limits that aquatic hypoxia places on non-air-breathing fishes (Beitinger and Pettit, 1984). Pettit and Beitinger (1985) further determined that this fish could tolerate 6–8 hours air exposure without adverse effects. A surprising result of the Sacca and Burggren (1982) study was that even with its thick, armored plate covering of ganoid scales, *E. calabaricus* was able to take up 32% of its total $\dot{V}O_2$ via its skin.

Finally, a different facet of air-breathing behavior in *Polypterus* was revealed by Svensson (1933) who reported (for either *P. lapradei* or *P. senegalus*) that

> when one fish is seen or heard splashing at the surface [i.e., air breathing], several others are almost always observed on the same spot immediately afterwards.

As subsequent pages will show, the phenomenon of synchronous air breathing has been documented for a number of fishes (Kramer and Graham, 1976).

Infraclass Neopterygii

The Neopterygii is an assemblage of extant and fossil fishes, formerly described as Holosteans, that succeeded the Paleonisciformes in dominance and gave rise to the teleosts (Lauder and Liem, 1983a,b, and Chapter 1). The living neopterygians include two groups, the gars (Division Ginglymodi) and *Amia* (Division Halecostomi), all of which are air breathers. These fishes use their physostomous gas bladder as an ABO. (Differences between a lung and a respiratory gas bladder are discussed in Chapter 3. Physostomous denotes the presence of a tube, the pneumatic duct, connecting the gas bladder with the gut and through which the organ can be either inflated or deflated. Physoclistous refers to the absence of a pneumatic duct.) This class of ABO, considered to be derived from the lung (Liem, 1988, 1989a), is found in neopterygians and some lower teleosts.

Division Ginglymodi

Lepisosteiformes

Lepisosteidae. The gars are heavily armored, predaceous fishes that occur in shallow, fresh, and brackish waters in eastern North America, Central America, and Cuba (Wiley, 1976). There are two genera of gars, *Lepisosteus* (4 species) and *Atractosteus* (3) (Table 2.2). Fossil gars occur in Africa, Europe, and India and date from the Early Cretaceous (140 mybp) but the group is thought to have evolved in the Permian (286–245 mybp). Gars are voracious predators (in some regions they are fished using dead grackles and chickens as bait) and reach a maximum size of between 1 and 3 m *(A. spatula).*

Air-breathing studies to date have been done mostly, if not exclusively, on species of *Lepisosteus* (Potter, 1927; Wiley, 1976) with only the anatomical and behavioral observations of Poey (1856, 1858) available for *A. tristoechus* in Cuba. (Hematological data have also been obtained for this genus, Chapter 7.) Gars are continuous, aquatic air breathers that use a single, bilobed gas bladder as an ABO. The gas bladder originates on the dorsal side of the esophagus and contains numerous septations that provide a large surface area and dense vascularization for gas exchange (Chapter 3). Sheets of striated muscle invested in the organ's septa enable volitional control of volume (i.e., buoyancy regulation) and also aid in air expiration (Crawford, 1971; Wiley, 1976). The internal "cellular" (= alveolar or trabeculated) structure of the gas bladder was first detailed by Van der Hoeven (1841) and Hyrtl (1852a,b) described the organ's blood circulation.

Observations by Poey (1856, 1858) confirmed that the gas bladder is used for air breathing, and Wilder (1875) coined the term "break" (also used by Day in 1877) to describe the surfacing for air by gars and by *Amia* (see following). Mark (1890) analyzed gas exhaled by *L. osseus* and Potter (1927) first documented the gas bladder's role in gar respiration by showing that blockage of the pneumatic duct with melted paraffin did not affect the survival of *L. osseus* in normoxic water (i.e., air breathing is not obligatory) but permitted only brief survival in foul water. Potter (1927) further confirmed the function of the ABO in both O_2 uptake and CO_2 release by showing differences in the respiratory gas contents at different phases of the air-gulping cycle. Rahn *et al.* (1971) demonstrated that temperature affected both the partitioning of aerial and aquatic respiration and the aerial gas exchange ratio of *L. osseus.* These workers also estimated total ABO and tidal volumes and ABO surface area. Landolt and Hill (1975) found that total gill area in three species of *Lepisosteus* was reduced compared to more active fish species, although not reduced to the extent seen in some obligatory air breathing species.

Smatresk and colleagues provided numerous details regarding the air-breathing physiology, acid–base balance, and the control of respiration in *Lepisosteus.* Separate experiments determined that increased temperature, a drop in aquatic O_2 tension, and transfer to 50% seawater all increased the dependence of *L. oculatus* on aerial respiration (Smatresk and Cameron, 1982a,b,c). In low O_2 tensions this fish can obtain all of its respiratory O_2 from its ABO but, because its gills lack anatomical shunts to prevent transbranchial O_2 loss, ABO ventilation and perfusion must also be significantly elevated to meet O_2 requirements (Smatresk and Cameron, 1982a). Experiments with *L. osseus* (Smatresk *et al.*, 1986) show that gill ventilation and air-breathing frequencies are subject to two O_2-sensitive controllers, one internal and the other external (Chapter 6). The internal controller sets the organism's need for O_2 as determined by its O_2 store and metabolic rate. The external controller determines the amount of branchial ventilation that can be used to obtain the needed O_2, a variable determined primarily by aquatic O_2 tension. Reflex air-ventilation responses in gar are triggered by changes in ABO volume and are modulated by both slow and rapid-adapting mechanoreceptors of ABO wall deformation (Smatresk and Azizi, 1987).

The effects of PO_2 on the air-breathing frequency and the movements of *L. oculatus* were examined by Hill *et al.* (1972, 1973). The air-breathing frequency of this fish averaged 190 breaths/day in O_2 tensions ranging from normoxia down to 80 torr (18 C) and increased to 480/day at lower tensions. Fish denied access to air behaviorally avoided hypoxic water and selected water with near normoxic O_2 tensions, particularly at elevated temperatures. Movements by fish with access to air, however, were independent of water O_2 content. Another behavioral aspect of gar respiration was investigated by Smith and Kramer (1986) who found that exposure to a potential predator (Great Blue Heron, *Ardea herodius*) increased the aquatic-ventilatory frequency and decreased the air-breathing frequency of *L. platyrhincus.* Synchronous (i.e., nearly simultaneous) air breathing in groups of *L. osseus* and *L. oculatus* was observed by Hill (1972).

Division Halecostomi

Subdivision Halecomorphi

Amiiformes

Amiidae. The bowfin, *Amia calva,* is the only extant species in the Subdivision Halecomorphi, the group of primitive bony fishes most closely related to the Subdivision Teleostei (Figure 2.1). There are at least seven extinct genera in the family Amiidae, the fossil record of which extends back to the Late Jurassic (150 mybp). Fossil *Amia* are first seen in the Lower Cretaceous (100 mybp) of North America and Europe.

Amia calva is a nocturnally active predator that reaches a maximum length of nearly 1 m and frequents waters with sluggish flow in the eastern part of North America. Early studies of its air breathing were done by Wilder (1875, 1877) and Potter (1927). *Amia* uses a vascularized physostomous gas bladder as an ABO. This organ has many features in common with that of *Lepisosteus* (Crawford, 1971); it is single but bilobed, originates embryologically from double dorsal pharyngeal outpocketings, and occurs in the dorsal part of the coelomic cavity. The ABO contains mechanoreceptors capable of transducing the rate and degree of deflation and inflation (Milsom and Jones, 1985), and it is wrapped by a double layer of striated muscle (Potter, 1927; Crawford, 1971) and can thus be squeezed to reduce volume, both for exhalation and for buoyancy regulation.

Bevalander (1934; also see Crawford, 1971) found that the secondary lamella of *Amia* gills are fused to interfilamental supports that form a sieve-like structure. He suggested this was adaptive in preventing gill collapse during air exposure and in promoting respiration in hypoxia. Although some branchial gas exchange does occur in air-exposed *Amia* (Daxboeck *et al.*, 1981), this fish does not survive long in air, is not known to be naturally amphibious, and cannot be induced to estivate (McKenzie and Randall, 1990). *Amia* is a hardy fish and can survive overwinter under ice, in isolated mudholes, and has even been found in the mud of plowed fields where it was apparently transported during seasonal flooding (Dence, 1933; Neill, 1950).

Amia is considered a continuous air breather, however, aerial exchange is also facultative in that the rate depends upon temperature, O_2, and activity level, and becomes obligatory under certain conditions (Johansen *et al.*, 1970a; Crawford, 1971; Horn and Riggs, 1973; Randall *et al.*, 1981b). The ABO is the primary site of O_2 uptake; whereas, the gills function mostly for CO_2 exchange, and respiratory partitioning varies with temperature, aquatic O_2 tension, and activity. Experiments by Johansen *et al.* (1970a) suggested that *Amia* could channel blood through nonexchange shunt circuits in either its ABO or branchial surfaces depending upon the prevailing respiratory mode at the time. This was not confirmed in experiments by Randall *et al.* (1981b) who actually documented the transbranchial loss of aerially obtained O_2 to hypoxic water. Examination of the microvascular gill anatomy of *Amia* by Olson (1981) did not reveal shunt pathways. Olson did, however, estimate that the effective exchange surface of *Amia* gills was reduced by 30 to 50% from the potential maximum because of the encasement of some lamellar channels within the gill filaments and the interfilamental supports described by Bevalander (1934) (Chapter 5). Olson suggested that the redistribution of cardiac output to gills with more occluded surfaces was the likely mechanism used by *Amia* to minimize O_2 loss to hypoxic water. Studies of respiratory control in *Amia* have differentiated between the neurophysiological mechanisms for glottal regulation during apnea and ABO ventilation (Davies *et al.*, 1993). Other investigations have documented the effects of aerial and aquatic hypoxia and other factors on respiratory partitioning (Hedrick and Jones, 1993) and air-breathing pattern (Hedrick *et al.*, 1994), and demonstrated the absence of a central chemosensory mediated respiratory response (Hedrick *et al.*, 1991). Observations by Hedrick and his colleagues (Hedrick and Katz, 1991; Hedrick and Jones, 1993; Hedrick *et al.*, 1994) documented two types of air breathing in *Amia*. The type I breath, which features both inhalation and exhalation, is the normal pattern. Type II breaths, characterized by inhalation without exhalation, serve to refill the gas bladder and preserve buoyancy (Chapters 3 and 6).

Subdivision Teleostei

Teleosts are the largest and most specialized assemblage of the bony fishes. According to Nelson (1994) this subdivision contains 21,000 species (i.e., about 96% of all the bony fishes, sharks, and rays) placed in 35 orders, 409 families and 3876 genera. The teleost fossil record goes back to the Late Cretaceous (100–65 mybp). The Teleostei consists of four major infradivisions, Osteoglossomorpha, Elopomorpha, Clupeomorpha, and Euteleostei (Figure 2.1). Air breathing has thus far been documented in all but one of these groups, the Clupeomorpha. Both the osteoglossomorph and the elopomorph air breathers use a physostomous gas bladder as an ABO. The trend among most higher teleosts is for the gas bladder's ABO function to be replaced by other organs or body surfaces (Liem, 1989a).

Infradivision Osteoglossomorpha

Osteoglossiformes

Although not a large group (six families, 26 genera, and 206 species, Nelson, 1994), the osteoglossiform fishes are diverse and highly specialized. Species in two families have an electrogenic capability, and the gas bladder of some species is adapted for as many as four different, however, simultaneous functions: Sound production, sound reception, buoyancy regulation, and aerial respiration.

Osteoglossidae. This family occurs in the tropical freshwaters of South America, Africa, Asia, and

Australia. There are four genera and six species all of which are primarily carnivorous, although some are also capable of filter feeding. Two species, *Arapaima gigas* of the Amazon Basin and *Heterotis niloticus* of tropical Africa, are air breathers using a highly vascularized and subdivided physostomous gas bladder. The ABO and circulatory anatomy of *Heterotis niloticus* were described in a classic paper by Hyrtl (1854a). The curious helical organ found on the fourth gill arch of this species was long thought to be an accessory respiratory organ (Rauther, 1910). Subsequent studies, however, showed it was part of a plankton-feeding mechanism (Lowe-McConnell, 1987). *Heterotis niloticus* grows to about 1 m length and is probably an obligatory air breather; however, there have been no studies of its air-breathing behavior or physiology. It has been observed swimming in small groups whose members frequently take their air breaths at more or less the same time (synchronous air breathing).

Considerably more is known about the South American counterpart of *Heterotis,* the pirarucu, *Arapaima gigas.* One of the largest freshwater fishes in the world (to 4.5 m, 200 kg), *Arapaima* is an obligatory air breather that obtains as much as 95% of its total $\dot{V}O_2$ from air (Sawaya, 1946). It has a well-developed ABO that is extremely large and actually envelops the kidneys (Chapter 3). Much of our knowledge of the behavior and natural history of this fish was provided by Lüling (1964a) who made observations on feeding habits, parental care, development, the ontogeny of air breathing, and air-breathing frequency and behavior. Young *Arapaima,* which are shepherded by the adult male, swim in schools, begin air breathing eight to nine days after hatching (18 mm length), and, because they are in schools, often take their air breaths simultaneously. Anatomical studies by Lüling also showed that the ABO of *Arapaima* is quite similar to that of *Heterotis.*

A large body of research on the air breathing of *Arapaima* was obtained during the 1976 *R. V. Alpha Helix* Expedition to the Amazon (Can. J. Zool. 56 [1978], Comp. Biochem. Physiol. 62A [1979]). One tack taken in these studies was to compare *A. gigas* and the related but non-air-breathing osteoglossid *O. bicirrhosum* to determine ways that air breathing alters the biochemistry and physiology. A notable finding was that the Hb–O_2 dissociation curve of *O. bicirrhosum* is left-shifted (i.e., has a higher O_2 affinity) relative to that of *A. gigas* (Johansen *et al.,* 1978a; Powers *et al.,* 1979, Chapter 7).

Hochachka *et al.* (1978a,b) found that the O_2 access, afforded by air breathing, poised many pirarucu tissues for aerobic metabolism to a greater extent than those of *Osteoglossum.* Stevens and Holeton (1978a) found respiratory partitioning in *A. gigas* to vary with aquatic O_2 tension; air breathing accounted for as little as 50% of total $\dot{V}O_2$ in air-saturated water but 100% in anoxic water. Randall *et al.* (1978a) estimated aerial CO_2 expiration to be 37% of total and postulated that non-equilibrium between CO_2 and bicarbonate ion in plasma circulating from the gills and ABO to be important in driving the high aerial R of *A. gigas.* Whereas a buccal force-pump is used to ventilate most fish ABOs, Farrell and Randall (1978) reported that *A. gigas* employs a suction force (negative pressure) to fill its ABO and that a special diaphragm-like structure is important for this. A subsequent analysis by Greenwood and Liem (1984) disproved these assertions.

Pantodontidae. This family contains one species, the butterfly-fish, *Pantodon buchholzi* which occurs in the backwaters of rivers, small ditches, and ponds of tropical west Africa (Lowe-McConnell, 1987). Popular in the aquarium trade, *Pantodon* is positively buoyant and commonly floats at the water surface where it feeds on small fish as well as insects that either fall to the water surface or fly not far above it. The butterfly fish reaches about 12 cm in length. As implied by its common name, this fish can leap from water and is able to glide as far as 2 m using its enlarged pectoral fins (Lowe-McConnell, 1975).

Pantodon is an obligatory air breather and uses its physostomous gas bladder as an ABO as well as for sound detection (Schwartz, 1969). In addition to containing a respiratory epithelium, this organ has numerous alveolar lobules that penetrate into the ribs, the vertebral centra, and parapophyses to lessen specific gravity (Chapter 3). Schwartz (1969) found that air breathing is the dominant respiratory mode of this fish and that air-breathing frequency is dictated by fish activity and by ambient O_2 tensions, primarily those in its air supply. Whereas exposure to an N_2 atmosphere caused *Pantodon* to increase air-breathing frequency by fivefold and to nearly double its aquatic ventilation frequency, aquatic hypoxia elevated air-breathing by only 20% and did not affect aquatic ventilation. The aquatic ventilation rate of a butterfly fish in normoxic water undergoes a cyclic change, from a 15–17% drop after an air breath is taken, to maximal frequencies just prior to the next air breath (Schwartz, 1969).

Notopteridae. The featherback knifefishes occur in fresh and brackish water lagoons and swamps of tropical west Africa, India, and southeast Asia. Greenwood (1963) reports that there are two African species, *Papyrocranus afer* and *Xenomystus nigri* and three species of Asian *Notopterus (notopterus, chitala,* *boreensis).* Notopterids are nocturnally active and feed on insect larvae, worms, and small fishes. Some species reach a length of 1 m *(N. notopterus)* while others do not grow to more than 30 or 60 cm. These fishes have elongated compressed bodies with a tapered tail and a long, narrow anal fin. Waves progressing along the fin enable forward and backward motion. The name featherback refers to the small erect dorsal fin (absent in *Xenomystus*). According to Greenwood (1963) air breathing, with the gas bladder serving as the ABO, is likely present in all of these fishes.

Air-breathing behavior and physiology have been documented for the Indian species *N. notopterus* and *N. chitala.* Dehadrai (1960, 1962a) described the gas-bladder morphology and confirmed its respiratory function by measuring gas contents under different conditions. In addition to respiration, this organ carries out three other functions: Sound production, sound reception, and buoyancy control. No other fish ABO is known to be so diversified. Dehadrai (1962a) also described ventilation behavior and compared air-breathing frequencies in different experimental conditions. He showed that both *N. notopterus* and *N. chitala* are continuous air breathers that survive indefinitely in nearly anoxic water by breathing air. His observations also suggested these fishes to be obligatory air breathers; however, his methods lacked the detail required to conclude this unequivocally. Dehadrai (1962a) additionally noted synchronous air breathing (page 347),

> When a group of four *Notopterus* was kept together in an aquarium, the fishes rose simultaneously to the surface to take in air.

Ghosh *et al.* (1986) examined respiratory partitioning in *N. chitala* and determined the effects of preventing air access. This fish normally takes air breaths every six to eight minutes and its use of aerial O_2 accounts for an average of 70% of its total $\dot{V}O_2$. Fish denied air access had twice the gill-ventilation frequency as did bimodally breathing fish. Nevertheless, fish without air access had a total $\dot{V}O_2$ that was less than half that of the bimodal group. Ghosh *et al.* (1986) determined that although *N. chitala* is a continuous air breather, this is not obligatory because this fish can reduce metabolic demand in the face of limited ambient O_2 supplies.

Gymnarchidae. The one species in this family, *Gymnarchus niloticus,* occurs in tropical Africa and the Nile River. The body shape and swimming motions of this fish are similar to the notopterids. It grows to a maximum length of about 1 m and is a nocturnally active predator on small fishes, using electrolocation for this purpose. *Gymnarchus* breathes air with a highly developed, lung-like gas bladder. Early studies by Förg (1853), Hyrtl (1856), and Assheton (1907) describe its ABO and vascular anatomy. Assheton (1907) observed that newly hatched *Gymnarchus* larva have long, filamentous external gills and are attached to a large vascular yolk sac which effectively anchors them to the substrate. Air breathing begins about 10 days after hatching. Other than ventilatory observations (Liem, 1988, 1989a), there are no studies of the air-breathing behavior or physiology of this species.

Infradivision Elopomorpha

The elopomorphs are a diverse group consisting of 25 families, 157 genera, and 633 species (Nelson, 1994). Air breathing is relatively rare, occurring in species of only two families, the Megalopidae and the Anguillidae.

Elopiformes

Megalopidae. The tarpon, *Megalops atlanticus* and the ox eye herring, *M. cyprinoides* are aquatic air-breathing fishes that frequent tropical and subtropical marine and brackish waters. These fishes are carnivorous, and the fossil record for them extends from the Early Cretaceous (140 mybp). *M. atlanticus* occurs in coastal waters of the tropical Atlantic and Caribbean and extends its range along eastern North America in summer. This fish has also transited the Panama Canal and established a Pacific population. It is a prized sport fish, reaches a maximum size of over 2 m, and feeds on small fishes. *M. cyprinoides,* which does not grow as large, occurs throughout the Indo-West Pacific from east Africa to Tahiti and from Japan to Australia.

Most observations related to air breathing have been made for *M. atlanticus*. Its air breathing, described by Babcock (1951), involves a characteristic "surface-rolling behavior," during which large bubbles exit the operculum. Babcock (1951, pages 50–59) also illustrated and generally described the respiratory gas bladder of *M. atlanticus,* several aspects of its air-breathing behavior were studied by Shlaifer and Breder (1940) and Shlaifer (1941). There have not been any physiological studies of the respiration of either *M. atlanticus* or *M. cyprinoides.* The respiratory gas bladder of *M. cyprinoides* was described and illustrated by de Beaufort (1909), and Merrick and Schmida (1984) report it is an obligatory air breather.

Tarpon eggs are fertilized in coastal waters where, after development, leptocephali larva remain for nearly two months before making their way into brackish and freshwater habitats. In these environments the leptocephali "shrink" and metamorphose into juvenile fish which feed mainly on insects and crustaceans and return to the sea after about nine months.

Air breathing enables young fish to survive the

hypoxia encountered in their fresh and brackish backwater refugia. Fish as small as 5–8 cm length have been seen "rolling" at the surface of canals and ditches in Florida (Wade, 1962). Atlantic tarpon between 10 and 12 cm are obligatory air breathers (Shlaifer, 1941) that must take air breaths about every five minutes depending upon both aquatic O_2 level and activity state (Shlaifer and Breder, 1940).

Tarpon air breathing was also discovered to be subject to social facilitation; undisturbed groups observed both in laboratory tanks and in ponds demonstrated a "marked tendency to rise (for air) in groups" (Shlaifer and Breder, 1940). Experiments by Shlaifer (1941) further demonstrated that the stimulus for surfacing did not necessarily have to be another fish as young tarpon could be induced to "follow" various objects (a wooden spatula, rubber hose, glass rod, and a "model" tarpon were all effective) to the surface for an air breath. Also, receptivity to behavioral cues leading to this behavior, while low just after a breath, increased with the time the breath was held (Shlaifer and Breder, 1940; Shlaifer, 1941).

Anguilliformes

Anguillidae. During the phases of its life history when it is in freshwater, the European eel *Anguilla anguilla* is a capable amphibious air-breather during overland excursions. This fish grows to over 1 m in length and is a carnivore. It has been observed to gulp air, and gas exchange via the gas bladder, the pneumatic duct, or both has been suggested (Mott, 1951; Fange, 1976; see Chapter 3).

Krogh's (1904a) experiments demonstrated that considerable quantities of O_2 and CO_2 were exchanged across the skin of *A. anguilla* in both air and water. Berg and Steen (1965) reported that cutaneous exchange in *A. vulgaris* was sufficient to sustain its metabolism at cool temperatures, a finding confirmed by the recent work of Le Moigne *et al.* (1986 for *A. anguilla*). *Anguilla* is known to estivate in mud burrows during the winter (Day, 1868; Ultsch, 1989). *A. anguilla* adapts to hypoxic water by increasing its blood Hb–O_2 affinity (Wood and Johansen, 1972), but it is unknown if this property changes during emersion. Considering there are about 16 species of *Anguilla*, it is likely that these have behavior patterns and air-breathing capabilities similar to those of *A. anguilla.*

Infradivision Euteleostei

Superorder Ostariophysi

The Ostariophysans are a diverse group containing four orders (Gonorynchiformes, Characiformes, Cypriniformes, and Siluriformes) and numbering over 6000 species. Ostariophysans are united by skeletal similarities, the presence of the alarm reaction (Schreckstoffe), and, in most, the presence of a Weberian apparatus connecting the gas bladder to the inner ear. The latter is rudimentary in the Gonorynchiformes but well developed in the other orders (Lauder and Liem, 1983a; Liem, 1988, 1989a).

Air-breathing species occur in all Ostariophysan orders (Figure 2.1) and, while air breathers do not constitute a high proportion of the total species, their respiratory specializations are diverse. The diversity of ABO structure found in this group suggests a period of evolutionary radiation when diverse selective forces operated on gas-bladder function.

Gonorynchiformes

Phractolaemidae. This family contains only one species, *Phractolaemus ansorgei.* Discovered in 1900, *P. ansorgei* occurs in tropical west Africa in the habitats where air-breathing fishes flourish: Small streams, swamps, flooded forests, and even lungfish nests (Poll, 1932). This fish, which grows to a maximum length of 15 cm, is heavily scaled. It prefers a muddy substrate into which it can burrow. It has a small mouth, but its jaws are protrusible and function for scraping microalgae from substrate (Thys van den Audenaerde, 1959, 1961; and pers. comm.), feeding on detritus, and for digging small invertebrates out of the mud (Lowe-McConnell, 1975).

Thys van den Audenaerde (1961) provided a detailed description of the anatomy of *P. ansorgei* and reviewed aspects of its biology and taxonomic position. Its air-breathing behavior was initially described by amateur aquarists who noted periodic surfacings with bubble ejection from the mouth or opercles during ascent. There was early disagreement over whether the gut or the large pharyngeal diverticulum of this fish could be the site of aerial gas exchange. Although there still have been no quantitative studies of its air-breathing behavior or physiology, investigations by Thys van den Audenaerde (1961) leave little doubt that the long, aleveoli-rich gas bladder of *P. ansorgei* functions for air breathing. It is unknown if this fish is an obligate air breather; however, this is suggested by its particularly small gills. Thys van den Audenaerde (1961) also described modifications in the circulatory system that favored air breathing and was of the opinion that these, together with the structural complexity found in the ABO of *Phractolaemus,* provided a degree of air-breathing proficiency rivaling the most specialized fishes.

Cypriniformes

Cobitididae. The cobitids, or loaches, are bottom-dwelling fishes that have elongated small bodies (rarely longer than 30 cm long). They occur in a variety of habitats, from mountain streams to standing waters, and some species find protection from predators by burrowing in the soft substrate. Their diet consists mainly of worms and insect larvae. Loaches have a broad temperate, subtropical, and tropical distribution extending across Eurasia, into parts of Africa, and Malaysia. There are 21 genera and at least 175 species in this family. The greatest diversity of loaches is seen in tropical Asia and several of the brightly colored species are popular aquarium pets. Loaches in the genus *Misgurnus* are called weatherfish because they exhibit restless behavior whenever atmospheric pressure changes (Herald, 1961), presumably because this affects gas volume in their ABO and gas bladder.

Day's (1868) early experiments with Indian loaches verified air breathing in *Platacanthus* (= *Lepidocephalicthys*). He also discovered loaches "suspended in summer sleep" in the mud of dry ponds. Air breathing has thus far been described in at least four genera; *Misgurnus, Cobitis, Acanthophthalmus*, and *Lepidocephalichthys* (= *Lepidocephalus*). Air breathing can also be suspected for some species of *Botia;* however, there are no detailed accounts of this. Jeuken's (1957) dissertation reviewed early research on air breathing in this family (also see Hora, 1935; Kuttil, 1963; Yadav and Singh 1980; Kramer and McClure, 1982; and McMahon and Burggren, 1987). Although Day (1868) stated that loaches use a special air receptacle near the vertebrae as an ABO, it was established very early that the gut serves this function (Erman, 1808; Babak and Dedek, 1907; Calugareanu, 1907; Lupu, 1911, 1914; Wu and Chang, 1945). These fishes appear to be continuous but not obligatory air breathers; however, this has not been examined in all species known to breathe air. McMahon and Burggren (1987) established that *M. anguillicaudatus* has an ABO tidal volume of nearly 85% of ABO capacity and that about 50% of inspired O_2 is extracted from each breath. They also found that elevated water temperature and aquatic hypoxia both increased aerial respiration rate.

In all of the fish examined, the expiration of old air from the vent occurs simultaneously or immediately after buccal inspiration of a new air breath. There have been several histological examinations of the gut primarily to determine how this organ is simultaneously involved in the disparate functions of digestion and respiration (Jeuken, 1957; Yadav and Singh, 1980; McMahon and Burggren, 1987; Moitra *et al.*, 1989). While air-breathing fishes most often have relatively reduced gill areas, Singh *et al.* (1981) found, rather surprisingly, that gill surfaces in *Lepidocephalus guntea* were larger than most other air-breathing species (Chapter 3) and as great as some non-air breathers.

Characiformes

The vast majority of characoids are not air breathers; this specialization is known thus far in two families—the Erythrinidae and Lebiasinidae.

Erythrinidae. Erythrinids are freshwater inhabitants of South and Central America. They often live in shallow streams and swamps and feed on insects, decapods, and other fishes (Lüling, 1964b, 1969; Knoppel, 1970). There are three genera, five described species, and several undescribed species. Most species are less than 30 cm maximum length. Two species, *E. erythrinus* and *Hoplerythrinus unitaeniatus*, are continuous—but not obligatory—aquatic air breathers using a modified region of the posterior chamber of the gas bladder as an ABO (Kramer, 1978a). The remaining genus, *Hoplias*, is a non-air breather highly adapted for life in hypoxic water (Dickson and Graham, 1986; Mattias *et al.*, 1996). Erythrinids are known to migrate overland (Saul, 1975). Beebe (1945) found *Erythrinus* in pit traps dug in the forest floor at considerable distances from water.

The early account of this family by Cuvier and Valenciennes (1846, as part of the Characidae) included a discussion (page 485) of the probable respiratory function of the specially structured posterior chamber of the gas bladder in some species. Cuvier and Valenciennes termed this a "cellular gas bladder" because of the trabeculated, compartmentalized matrix of respiratory epithelium resembling alveoli:

> Le tiers antérieur de cette seconde vessie offre des parois celluleuses; les cellules sont déterminées par des brides transversales nombreuses, serrées, parallèles, entre elles, et perpendiculaires aux grandes brides tangentes à la surface du cône.

This functional interpretation exposed a conceptual limitation of these times as Müller (1842) argued that, based on its circulatory anatomy (i.e., with blood arriving in a systemic artery and leaving in a systemic vein), the erythrinid gas bladder could hardly serve a respiratory function because veins could not transport O_2-laden blood.

Jobert (1878) was the first to observe air-breathing behavior in an erythrinid (probably *H. unitaeniatus*). He also demonstrated the essential role of air gulping for survival in hypoxic water and did manipulative experiments such as ligating the pneumatic duct and measuring ABO–CO_2 content. Nevertheless, uncertainties lingered about erythrinid air breathing and the role played by the gas bladder, and it fell to Carter and Beadle (1931) to determine this unequivocally. Working probably with *H. unitaeniatus*, these workers verified that inspired air did in fact pass into the gas bladder and determined that the ABO–O_2 level of fish held without access to air decreased significantly while that of CO_2 increased somewhat. They also determined that air breathing was not obligatory.

A study notable for the attention it has received over the years was that by Willmer (1934) who graphically depicted the effects of aquatic O_2 and CO_2 on the air-breathing responses of but one specimen of the "yarrow" *(Hoplerythrinus).* His figure 7 (see Figure 6.1), depicting O_2/CO_2 control over respiration, indicated this fish was solely dependent upon aquatic respiration when aquatic PCO_2 was below 10 torr or above 25–30 torr, but that within 10 and 25 torr CO_2, gill ventilation occurred. This finding has been often cited as evidence for the role of CO_2 in controlling the respiration of a lower vertebrate, and Willmer's figure found its way into many reviews (Carter, 1957, figures 3 and 4; Johansen, 1970, figure 15; Shelton *et al.*, 1986, figure 20A) and texts (Gordon *et al.*, 1982, figure 5-12). Unfortunately, Willmer's conclusions about the role of low CO_2 have not been verified in subsequent respiratory control studies with erythrinids or with other air-breathing fishes (Chapter 6).

Subsequent studies on erythrinid air breathing include that of Lüling (1964b) who observed the natural history and air-breathing behavior of *H. unitaeniatus* and suggested a possible accessory respiratory function of the skin and gills in hypoxia and during terrestrial excursions. In essentially confirming the conclusions of Carter and Beadle (1931), Weibezahn (1967) found differences in the ABO gas contents of *H. unitaeniatus* that had just taken air as opposed to those that had been maintained below the water surface in O_2 deficient water.

Several studies of erythrinid air-breathing physiology, metabolism, and blood respiratory properties were undertaken during the 1976 Amazon Expedition of the *Alpha Helix*. Examination of metabolism and respiratory partitioning (Stevens and Holeton, 1978b) revealed that the total $\dot{V}O_2$ of *Hoplerythrinus* was higher than that of *Erythrinus* but that about 40% of the respiration of each species was aerial. Randall *et al.* (1978b) found that air exposure increased the blood PCO_2 and the aerial R of *Hoplerythrinus.* In erythrinids and lebiasinids (see following) and several other families, inspiration precedes expiration during each air-breathing cycle, and studies of *Hoplerythrinus* by Kramer (1978a) led to the important discovery that new air first flows into the anterior chamber of the gas bladder and enters the posterior chamber simultaneous with the exit of old air via the pneumatic duct, which minimizes mixing.

Lebiasinidae. Lebiasinids and erythrinids are quite similar in body plan and have the same basic structure in their ABO design (Graham *et al.*, 1978b, and Chapter 3). Lebiasinids occur in freshwaters of South America and also extend into Central America. There are six genera and over 50 species. These fishes are carnivorous and most are less than 20 cm in total length. Included in this family is *Copella arnoldi,* the species that jumps from water to spawn on leaves (Table 1.1). Rowntree (1903) diagrammed the gas bladder structure of *Lebiasina* and remarked that its cellular structure was similar to that of *Erythrinus.* Air-breathing is known to occur in two lebiasinids, *Lebiasina bimaculata* and *Piabucina festae* (Graham *et al.*, 1977, 1978b). These fishes are continuous, but not obligatory, aquatic air breathers and may be capable of overland movement.

There are 16–17 species of *Lebiasina* and *Piabucina* including five undescribed species (Weitzman, pers. comm.). The air-breathing status of most of these fishes is still unestablished. For example, both *P. panamensis* and *P. erythrnoides* do not breathe air, but whether or not three other species *(P. boruca, P. aureoguttata,* and *P. elongata)* do is unknown. *L. bimaculata* has a trabeculated gas bladder (Cuvier and Valenciennes, 1846, plate 588), but the condition of this organ in other species is not known.

As seen in Figure 2.3, the Lebiasinidae and Erythrinidae are not as closely related as might be surmised on the basis of their similarities. Weitzman (1964) hypothesized that the stock giving rise to the Erythrinidae split off from the main characin line much earlier than did that leading to the Lebiasinidae, and that the similarity of the ABOs in these two families is the result of convergent evolution. However, reconsidering the problem in the light of recent methodological improvements in phylogenetic studies, he now believes *(in litt.)* that there is not enough information to draw such conclusions. It is virtually certain that air breathing evolved independently in both families.

The presence of air-breathers and non-air-breathers within the same genus is an exceptional feature that distinguishes the Lebiasinidae from all other air-breathing fish families (Graham *et al.*, 1977, 1978b) and may even suggest the reverse evolution of this trait. It is possible, within the Lebiasinidae, to make comparisons among closely related species for the effects of air breathing. Comparison of *P. festae* and *P. panamensis* suggests that the gas bladder's function in increasing buoyancy of the non-air breathing *P. panamensis* in hypoxic water could have initially led to selection for ABO specialization in this organ (Graham *et al.*, 1978b). Similarly, the evolutionary transition to air

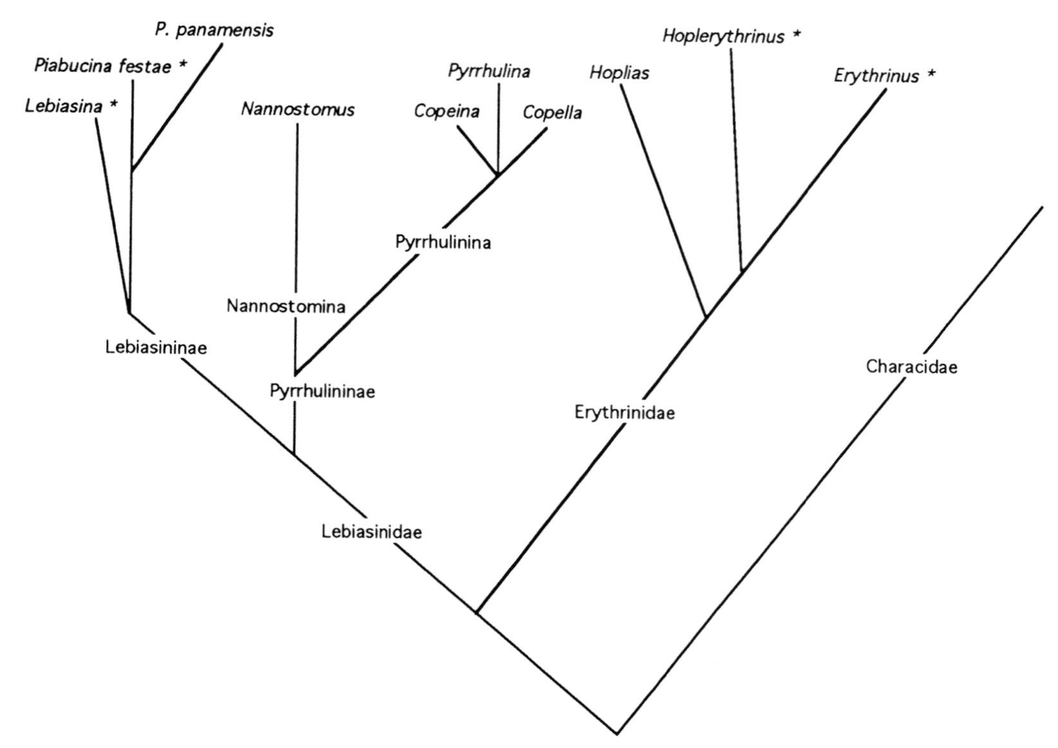

FIGURE 2.3 Postulated cladistic relationships for the Lebiasinidae, Erythrinidae, and Characidae, based on characteristics used in taxonomic classification, and showing the probable independent evolution of air breathing in four genera (asterisks) of two families, as well as the possible evolutionary reversal of this condition in *Piabucina panamensis.* Information contained in this figure is based on Weitzman (1964 and pers. comm.) and Vari (1995).

breathing in *P. festae,* while apparently elevating its metabolic rate, has resulted in concommitant losses in its capacity for aquatic ventilation and O_2 uptake in hypoxic water (Chapter 5).

Siluriformes

Pangasiidae. Pangasiids occur in Southeast Asian brackish and freshwaters from Pakistan to Borneo. Based on the recent revision of this family by Roberts and Vidthayanon (1991), there are 21 species of pangasiids occurring in two genera (*Pangasius* [19 sp], *Helicophagus* [2]). These fishes are toothless plant eaters. Some grow to a very large size (over 2 m) and several have become important for aquaculture.

Air breathing has been observed in a few species of *Pangasius.* It was documented for *P. sutchi* (now *hypophthalmus*) by Browman and Kramer (1985) who determined that the air-gulping frequency of this continuous air breather increased with aquatic hypoxia. The latter observation was also confirmed for *P. pangasius* by Thakur *et al.* (1987). Browman and Kramer concluded that *P. hypophthalmus* is an obligatory air breather, but the data varied: Some fish survived only 50 minutes without air; others endured up to seven days (one up to 18 days). These workers observed that air-breathing pangasiids are unique among the Siluriformes in their use of a modified gas bladder as an ABO. A fundamental question arising from this is, in view of the diverse air-breathing organs found in this order, whether this organ's function is primitive or reacquired (Browman and Kramer, 1985; Zheng and Liu, 1988).

Pangasiid specializations for air breathing were first noted by Taylor (1831) who described the highly septated (= cellular) gas bladder of *Pimelodus* (= *Pangasius*) *pangasius.* The taxonomic description of *Pangasius buchanani* (now *P. pangasius*) by Day (1877) also detailed its cellular gas bladder. Kawamoto *et al.* (1972) provided detailed illustrations of this structure and its vascular pattern in *P. nasutus (conchophilus)* living in the Mekong Delta but did not comment on the structure of this organ in two other species (*P. macronemas* and *P. larnaudi*) from the same region. Other descriptions are found in Thakur *et al.* (1987), and Jionghua *et al.* (1988).

Browman and Kramer (1985) reported that the ABO structure of *P. sutchi (hypophthalmus)* was different from that described for *P. pangasius* by Day (1877) in having neither the double-walled anterior chamber described by Day nor the paired, narrow extensions of the second chamber that Day observed to continue to the level of the anal fin. Thakur *et al.* (1987) described and illustrated the ABO of *P. pangasius* but did not evaluate its structure relative to Day's earlier description or the points raised by Browman and Kramer. Roberts and Vidthyanon (1991) illustrate the variation found in pangasiid gas-bladder morphology. They confirm the gas bladder description for *P. hypophthalmus* given by Browman and Kramer (1985) as well as the double chamber in *P. pangasius.* These workers, however, question Browman and Kramer's conclusion that the gas bladder of *P. hypophthalmus* has a respiratory function. Their observations of fresh specimens indicated that although heavily trabeculated, the gas bladder is not particularly vascularized. They suggest the alimentary canal as the ABO site.

Clariidae. Walking catfishes *(Clarias)* are among the best known air-breathing fishes. One of the earliest descriptions of a fish ABO was that of Geoffroy St. Hilaire (1802a) for *Silurus* (= *Clarias*) *anguillaris.* Clariids are widely distributed, extending throughout Africa and across southern Asia to the Phillipines. Fossil clariids date back to the Pliocene (David, 1935). *Clarias batrachus,* the most studied species, occurs throughout Asia. Albino specimens of this species were sent all over the world by the aquarium trade and are now established in many areas including Guam and southern Florida (Lachner *et al.,* 1970; Courtenay and Stauffer, 1984).

There are about 10 genera and close to 75 species of clariids. These come in a variety of sizes (lengths range from 10 to 200 cm) and weigh as much as 60 kg. They are carnivores but will feed opportunistically and scavenge almost anything. Greenwood (1961) reviews the evolutionary pathway taken by air breathing in this family. Numerous species in at least three genera, *Clarias, Heterobranchus,* and *Dinotopterus,* appear, on the basis of ABO structure, to be proficient air breathers. The genus *Clarias* is subdivided into three subgenera: *C. Heterobranchoides, C. Clarias,* and *C. Allanbenchelys* (Table 2.2, Figure 2.4). Most studies of air breathing have been done with species in the *C. Heterobranchoides* group; however, all species of *Clarias* appear to possess the ABO and have at least a minimal air-breathing capability. In addition, there are four

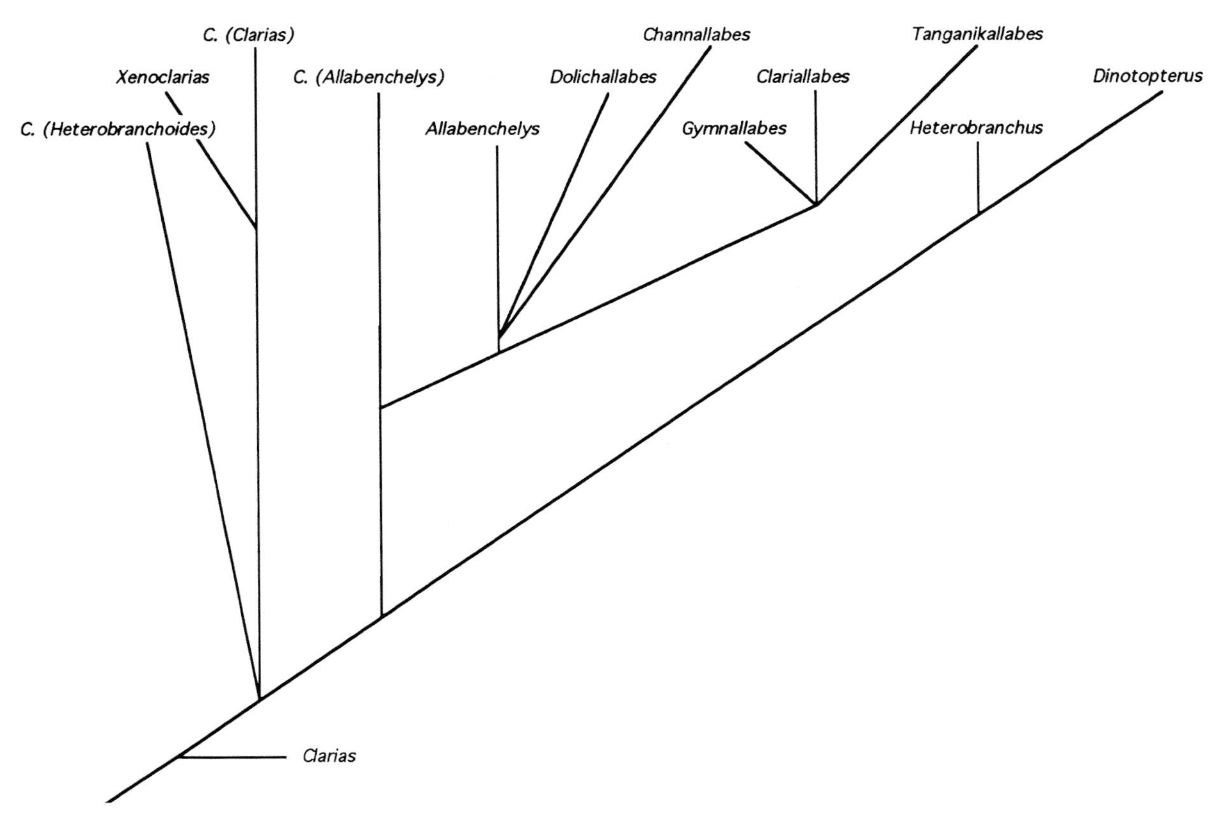

FIGURE 2.4 Tentative cladistic arrangement for the clariid catfish genera and sub-genera derived from the genus *Clarias* and illustrating the groups which have undergone regressive ABO development (see text for details).

known species of *Heterobranchus* all of which have an ABO (David, 1935) and Greenwood (1961) lists four of 13 species of *Dinotopterus* in which the ABO appears sufficiently developed to permit aerial respiration. There have, however, been no studies of air breathing in either *Heterobranchus* or *Dinotopterus* or in the seven other genera derived from *Clarias* in which ABO components are generally atrophied (David, 1935; Marlier, 1938; Greenwood, 1961).

The clariid ABO is composed of paired suprabranchial chambers located in the dorsal-posterior part of the branchial cavity and nearly filled by respiratory trees (arborescent organs), extensions from the upper parts of branchial arches 2 and 4. The opening of each chamber is valved by fan-shaped extensions of gill lamella on the upper end of each branchial arch. The upper surfaces of these, the floor and inner walls of the chamber, and the arborescent surfaces are covered by a modified gill epithelium that functions for aerial respiration. Das (1927) describes the natural history of *C. batrachus* as well as the ontogenetic development of its ABO. There have been numerous descriptions of the gross and fine structure of the ABO in *Clarias* (Moussa, 1956, 1957; Munshi, 1961, 1976; Singh *et al.*, 1982a; Chapter 3).

Clarias is known to make nocturnal terrestrial excursions. According to Smith (1945), *C. lazera* crawls across land

> to search for better living or feeding conditions or perhaps to escape enemies.

Vaillant (1895) reported that this fish left water at night to feed on millet, and Inger (1957) found a preserved fish in the Chicago Natural History Museum that had its stomach filled with a similar grain. Species of *Clarias* are also able to survive exposure in mud and can burrow into moist, sandy substrates to await the return of the rains (Day, 1877; Das, 1927; Donnelly, 1973, 1978). Observations that *Clarias* suddenly appears in formerly dry pools at the start of the rainy season gave rise to the myth that they fell from the sky with the rain (Das, 1927). Not all substrates are suitable for burrowing and some populations of *Clarias* that become isolated in drying pools are subject to dehydration and intense avian predation and, depending on the vagaries of nature, have a slim chance of surviving until the rainy season arrives. Papers by Donnelly (1973, 1978) detail this process and aspects of fish behavior including the effect of a thick mud gruel on altering the clariid air-ventilation sequence. An especially detailed account of this by Bruton (1979) also describes several interesting aspects of behavior including the group activity of "dredging" the center of a shallow pool to make the water deeper.

The air breathing of *Clarias* has been the subject of diverse study. Loftus (1979) documented synchronous air breathing in Florida *C. batrachus.* Most members of this genus air breathe on a regular basis, typically at low frequencies in normoxia. Insofar as obligatory air breathing is concerned, there are conflicting data for some species. The most thorough studies with *C. batrachus, C. lazera, C. gariepinus,* and *C. macrocephalus* indicate they are not obligatory air breathers (Hora, 1935; Magid, 1971; Donnelly, 1973; Jordan, 1976; Bevan and Kramer, 1987a). Still, there are reports of obligatory air breathing in *C. batrachus, C. lazera* and *C. mossambicus* (Das, 1927; Moussa, 1956; Greenwood, 1961; Singh and Hughes, 1971). Johnston *et al.* (1983) described *Clarias mossambicus* as an obligatory air breather based only on their finding that two small fish survived less than 52 hours in normoxic water without access to air.

Intra- and interspecific variations in the extent of ABO development in *Clarias* may explain the differences in air-breathing capacity found by various investigators. For example, it is surprising that *C. mossambicus* is described as an obligatory air breather in view of Trewavas' (1962) observation that it has a reduced ABO relative to other species (Chapter 3). *Clarias maclareni* also has a diminutive ABO (Greenwood, 1961; Trewavas, 1962) and population differences in the degree of development of the arborescent organs in *C. dumerili* have been documented (Greenwood, 1961). As the following discussion of the anabantoids demonstrates, environmental conditions at the time of ABO development may determine organ size, growth, and importance to respiration in later life.

In several species of *Clarias*, respiratory partitioning has been examined in relation to body size and aquatic hypoxia. Babiker (1979) reported that small *C. lazera* obtained 15% of their $\dot{V}O_2$ aerially but larger fish were dependent on air to a greater extent (40–50% of total $\dot{V}O_2$). The latter conflicts with the finding of Ar and Zacks (1989) that 1–2 kg *C. lazera* hardly used air breathing in normoxic water and only used it for about half their total $\dot{V}O_2$ in moderate hypoxia (70 torr). Ar and Zacks also reported correlations between hypoxia and air-breathing frequency and aerial $\dot{V}O_2$. This differs from data of Johnston *et al.* (1983) who found that the ABO of *C. mossambicus* accounted for 25% of its $\dot{V}O_2$ in normoxic water and 70% in hypoxia, even though fish in both tests had about the same air-gulping frequency. These workers showed that following acclimation to aquatic hypoxia for 27 days, *C. mossambicus* was able to sustain a higher aquatic $\dot{V}O_2$ in hypoxic water which in turn lowered the percentage contribution to total $\dot{V}O_2$ of its ABO. The hypoxia acclimated group, however, breathed air more frequently in hypoxia than did control groups, suggesting that the ratio of ABO ventilation to extraction is not precisely regulated in this species (Johnston *et al.*, 1983). No effect of hypoxia acclimation was seen on the myotomal-muscle mitochondrial volume or capillary density in *C. mossambicus* indicating that, in terms of compensation for environmental O_2 limitation, air-breathing adaptations are independent of tissue specializations (Johnston *et al.*, 1983).

Singh and Hughes (1971) found that in normoxic water *C. batrachus* obtained 58% of its O_2 from air but also had a low aerial R. The skin of this species functions for up to 17% of its total aquatic $\dot{V}O_2$. Singh and Hughes (1971) also found that in progressive hypoxia *C. batrachus* becomes an O_2 conformer and, even though it was air breathing and had a reduced $\dot{V}O_2$, its aerial R was elevated suggesting a restricted gill perfusion.

Jordan (1976) reached several different conclusions about the air breathing of Florida *C. batrachus.* Her data showed this fish to be neither an obligatory air breather nor an O_2 conformer. Rather, she found it regulated $\dot{V}O_2$ down to a PO_2 of 40 torr, when air breathing frequency increased dramatically. Jordan (1976) also concluded that *C. batrachus* uses less aerial O_2 in normoxic water (range 5 to 50% of total $\dot{V}O_2$) than was reported by Singh and Hughes (1971) but that the fraction of total $\dot{V}O_2$ taken through the skin of this fish in air was greater (50%) than earlier reported. Like Singh and Hughes, Jordan (1976) did find that air-exposed fish reduced their total $\dot{V}O_2$ and she measured a threefold increase in the $\dot{V}O_2$ of active fish in air. Her work also showed that the air-breathing frequency of *C. batrachus* is greater during nocturnal hours and that it increases in aquatic hypoxia. Finally, she interpreted the finding that smaller fish surfaced for air more frequently than did larger fish to indicate a greater dependence upon air breathing. This is at odds with Babiker's (1979) findings and needs further examination with data on the scaling of both ABO–O_2 utilization and respiratory partitioning. It would be surprising if smaller *C. batrachus* are more in need of air than larger specimens in view of the general observation, made repeatedly in the air-breathing fish literature, that air-breathing dependency in a species usually increases with size (Chapter 5).

Bevan and Kramer (1987a) found that *C. macrocephalus* increased air-breathing and gill ventilation frequency in progressively hypoxic water down to a PO_2 of about 40 torr, a value similar to the critical O_2 tension shown by Jordan (1976). Below this tension, air-breathing frequency continued to increase whereas gill ventilation was sharply reduced, suggesting disuse of the gills. A study of air breathing in relation to depth (Bevan and Kramer, 1987b) showed that specimens of *C. macrocephalus* exposed to hypoxia (40 torr) in deep (235 cm) tanks breathed air less often and ventilated their gills to a greater extent than in shallower (60 cm) tanks also at 40 torr PO_2. These workers point out that the time and energy spent surfacing are not trivial, and their data suggest that fish in deeper water somehow handle their air breaths in a more effective manner that reduces air-breathing frequency. A separate study (Bevan and Kramer, 1987b), however, revealed no effect of surfacing distance on the growth rate of young fish. The finding (Kulakkattolickal and Kramer, 1988) that air breathing in hypoxic water increased the resistance of *C. macrocephalus* to the toxic extract of *Croton* seeds demonstrates the role of air breathing in "isolating" a fish from its aquatic environment and illustrates the role of air breathing in habitats where toxic extracts might accumulate on a seasonal basis, a result of soil leaching, or the decomposition of organic material in stagnant waters.

Clariid evolution. The Clariidae also illustrates what can befall the ABO when evolutionary radiation of a lineage proceeds into environments where selection pressure for air breathing is reduced. Clariid taxonomy is based in part on the prominence of suprabranchial chamber structural components, including the size of gill fans and arborescent organs and the effect of suprabranchial chamber volume on the size and shape of the skull-roof bones (David, 1935; Greenwood, 1961). As detailed in Figure 2.4, there are three air-breathing clariid genera (*Heterobranchus, Dinotopterus, Clarias*), all of which have the various component ABO parts. *Clarias* in turn contains three subgenera (*Heterobranchoides, Clarias, Allanbenchelys*) all with different degrees of ABO development. The four species of *Heterobranchus* all have suprabranchial organs that are comparable to, or slightly more developed than, those of most species of *C. Heterobranchoides* (David, 1935; Marlier, 1938). The organs are less prominent in *C. Clarias* and *C. Allabenchelys* and appear sufficiently developed for air breathing in only four species of *Dinotopterus.* The structures are either extremely reduced or absent in the several genera derived from the former two (*Xenoclarias, Allabenchelys, Clariallabes, Gymnallabes,* and *Taganikallabes,* Figure 2.4). Species in the latter three genera (known as eelcats) are small and anguilliform with small or absent paired fins. Many of these have taken up life in swift streams where they burrow in rubble. Others burrow into mud and have vestigial eyes, while some have radiated into open lakes where they live at considerable depth.

Another genus that has undergone changes in ABO structure is *Dinotopterus,* a group of 13 species with close affinities to *C. Heterobranchoides* (Figure 2.4). Species of *Dinotopterus* occur in both Lakes Malawi and Tanganyika where they have radiated into the open waters and depths (to 70 m) and adopted feeding habits ranging from planktivory to piscivory. Greenwood (1961) found that, relative to *Clarias,* the suprabranchial chambers in nine species of *Dinotopterus* are reduced as are the arborescent organs, which are even absent in some. Living at depth precludes air breathing and, based on the large relative gill size of *Dinotopterus* compared to *Clarias,* suggests that most species of *Dinotopterus* have returned to a greater dependence upon aquatic respiration (Greenwood, 1961).

Greenwood (1961) describes the pattern seen for the clariid ABO (Figure 2.4) as regressive organ development. He notes that the regressive clariids are largely confined into a narrow range of distribution in central Africa (5 N to 15 S) where they have diversified into unfilled niches (e.g., flowing waters and the depths and open waters of Lakes Malawi and Tanganyika). In contrast, the fully air-breathing species with their independence of a variety of habitat conditions, have radiated across both Africa and Asia.

Heteropneustidae. This family contains two species, *Heteropneustes fossilis* and *H. microps,* both non-obligatory, aquatic air breathers. These fishes reach a maximum length of 20–25 cm, are carnivorous benthic feeders, and possess venomous spines. They are distributed across southern India from Pakistan to Thailand, and they frequent still water habitats in both fresh and brackish waters and are known to estivate in mud when their habitats become dry (Day, 1877; Das, 1927). The early literature contains reference to a third species, *H. singio,* but only two are now recognized. Most air-breathing investigations have been carried out on *H. fossilis* which occurs in India and Burma.

A complete account of the respiratory morphology of *H. fossilis* is contained in Munshi (1993). Some of the modern literature still refers to this group by the older family and generic names (Saccobranchidae, *Saccobranchus*) which vividly described its ABO (Day, 1877; Das, 1927; Hora, 1935; Bertin, 1958). The organ consists of an expanded pharyngeal chamber located behind and above the gills. This space connects in turn with a pair of horn-like air sacs that extend posteriorly through the myotomal muscle on either side of the neural spines to nearly as far as the tail. The pharyngeal chamber and the air sacs are covered by a fairly continuous sheet of biserial gill lamellae laid down in the form of respiratory islets and paths (Munshi, 1962a, 1976; and Chapter 3). As in *Clarias,* gill fans, formed by the uppermost part of each branchial arch, control the inhalant and exhalant ports of the ABO. Details of ABO development are found in Singh *et al.* (1981) and early observations of circulatory specializations (see Chapter 4) related to air breathing were made by Hyrtl (1854b) and Burne (1896). The gills of *Heteropneustes* are well developed, and Munshi and Singh (1968) concluded that this species was better suited for sustaining aquatic respiration than other endemic Indian air-breathers *(Anabas, Clarias,* or *Channa).* The skin of *H. fossilis* has been shown to function as an accessory respiratory surface in water and probably also in air (Hughes and Singh, 1971).

Concerning air breathing, Hughes and Singh (1971) found that large *H. fossilis* are not obligatory air breathers, and that, in normoxic water, this fish "shows little inclination for air breathing." In spite of this observation, these workers reported that this fish uses its ABO for about 40% of its total $\dot{V}O_2$ in normoxic water and about 60% in hypoxia. As in other species, the aerial R of *H. fossilis* is low but increases with greater rates of air breathing. Hughes and Singh (1971) also found that *H. fossilis* held in hypoxic water, forcibly submerged in normoxic water, or exposed to air, all responded by dropping total $\dot{V}O_2$, which may be adaptive for survival in the absence of water. Congruent with this is the discovery by Saha and Ratha (1987, 1989, and unpublished) that *H. fossilis* has significant concentrations of the ornithine-urea cycle enzymes in its liver and kidney and thus may be able to switch to ureotelism when exposed to air during the dry season.

Aspects of the energetic costs of surfacing for air were examined by Aruachalam *et al.* (1976) who found that *H. fossilis* living in water deeper than 40 cm and fish that were starved both tended to breathe air less frequently. Fish living in deeper water (60 cm) also consumed more food than fish at shallower depths but had lower metabolic conversion efficiencies owing to the apparently higher metabolic costs of swimming greater distances for air, having to ventilate water to a greater extent, and having to digest a greater quantity of food (Aruachalam *et al.,* 1976). Sheel and Singh (1981) found that the scaling of metabolic rate with body mass went through an inflection (to a lower slope) at the body size where this fish makes the ontogenetic transition to air breathing. Concerning social aspects of air breathing by *H. fossilis,* Hora (1935, page 5) observed:

> in some places shoals of them rise together for aerial respiration.

Aspredinidae. Banjo catfishes occur in the tropical regions of South America where they frequent both brackish and freshwater habitats. This family contains 13 genera and at least 46 species all of which lack scales, are highly cryptic, and usually nocturnally active. Most species are less than 10 cm long and all are scavengers but also take live prey. The first observation of air breathing in this family was reported for an Amazon species of *Bunocephalus* which gulped air and floated at the surface of hypoxic water (Kramer *et al.*, 1978). Kramer and McClure (1982) subsequently observed air breathing in small (1–5 g) *B. amaurus*. Kramer *et al.* (unpublished observations) describe these fishes as facultative air breathers that hold air in the mouth.

Trichomycteridae (Pygidiidae). The pencil—or parasitic—catfishes, as they are commonly known, occur in freshwaters throughout South America and up into Panama and Costa Rica. This diverse family has 27 genera and at least 175 species most of which reach a maximum length of less than 15 cm. These fishes lack scales, are bottom dwellers, and often burrow into their habitat substrate. Except for the parasitic forms, these fishes feed on small benthic invertebrates.

Air breathing has been documented in only two species, *Pygidium striatum* (Gee, 1976) and *Eremophilus mutisii* (Garzon and del Castillo, 1986; Cala, 1987b; Cala *et al.*, 1990). Both species are facultative air breathers and use their stomach as an ABO.

Callichthyidae. The callichthyids are one of two armored catfish families (see following) so named because of their covering of bony plated scales. Included in this family are eight genera and about 110 species ranging throughout the freshwaters of South America and into Panama. These fishes are omnivorous feeders and, while most reach a maximum length of 7 to 10 cm, species of *Hoplosternum* (Figure 2.5) grow to 20 cm.

Air breathing has been documented in five genera (*Brochis*, *Callichthys*, *Corydoras*, *Hoplosternum*, and *Dianema*, Table 2.1) but may be present in every genus and species in this family (Kramer and McClure, 1980). These fish use a section of their intestine as an ABO (Jobert, 1877; Carter and Beadle, 1931; Dorn, 1983). Depending upon species, from 50 to 80% of the length of the intestinal lumen is covered by a layer of thin, vascularized respiratory epithelium (Carter and Beadle, 1931; Huebner and Chee, 1978). Callichthyids breathe air continuously, and Jobert (1877, for *C. callichthys*) first observed that new air is gulped at about the same time, or just before, old air is released from the vent. Although Carter and Beadle (1931) did not confirm the anal-exhalation mechanism (also see Carter, 1957), it has since been verified in *Callichthys*, *Corydoras* (Kramer and McClure, 1980), *Hoplosternum* (Gee, 1976), and *Brochis* (Gee and Graham, 1978).

Callichthyids are known to survive in mud following the evaporation of water (Day, 1868) and have been reported to make terrestrial excursions (Beebe, 1945; Lüling, 1973). With one exception (*Hoplosternum littorale*, Carter and Beadle [1931]) the callichthyids that have been examined (*Corydoras aeneus*, *Brochis splendens*, and *H. thoracatum*) are not obligatory air breathers. Although Carter and Beadle (1931) found *H. littorale* to be an obligatory air breather, they did not remark about the respiratory dependence of another species they studied, *H. pectoralis*.

Callichthyids often swim in small groups and synchronous air breathing has been reported for *H. thoracatum* and *Corydoras aeneus* (Kramer and Graham, 1976; Kramer and McClure, 1980). Although they are largely bottom feeders, callichthyids do often swim in midwater; and in some circumstances, the contribution of the ABO to buoyancy takes on more impor-

FIGURE 2.5 *Hoplosternum thoracatum* (10 cm, Panama).

tance than its role in respiration (Gee and Graham, 1978). Callichthyids are highly amenable to laboratory study and Kramer and McClure (1981) used *Corydoras aeneus* to demonstrate how air-breathing costs, in the form of the time and energy required to swim to the surface for air, were affected by water depth and O_2.

The combined effects of competition for a limited food supply and aquatic O_2 tension on air breathing were examined by Kramer and Braun (1983). They showed that a group of *C. aeneus* in fairly well oxygenated water (72 torr or greater) markedly reduced their air-breathing frequency over the period required to consume a limited food supply. This was in contrast to fish in hypoxia (44 and 24 torr) which, because they could make less use of aquatic O_2, could not reduce air-breathing frequency to the same extent.

Loricariidae. Loricariids, another family of armored catfish, are bottom-dwelling species that have dorso-ventrally compressed bodies (Figure 2.6) and a ventral mouth with broad, rough lips specialized for grazing on periphyton, the slimy layer of algae and bacteria that commonly coats submerged objects (Power, 1984). Species range in maximum length from 4 to 40 cm. This family, which occurs throughout the South American tropics and into Panama and Costa Rica, is highly diversified and its taxonomy is not completely known. There are about 70 genera and from 450 to 600 species (Isbrucker, 1980). Owing to the uncertain taxonomy of the group, the specific identity of some loricariids used in air-breathing studies remains unknown (Table 2.1). The names of some genera have also changed, and it is likely that some species were not correctly identified. Nevertheless, at least 10 genera and nearly 20 species in this family have been described as air breathers (Table 2.2).

Early reports of air breathing in this family (Cuvier and Valenciennes, 1840; Jobert, 1878) implicated the intestine as an ABO. Carter and Beadle (1931), however, established that the stomach is an ABO and described its histology and vascular anatomy and respiratory gas contents. Subsequent studies have verified this for other loricariids (Gradwell, 1971; Gee, 1976); however, given the diversity of this group, other ABO types may yet be discovered. Air breathing is done by rapidly swimming to the surface and gulping air. Old air is expired via the gill openings or mouth just prior to, or during, surfacing (Gradwell, 1971). For the most part loricariids appear to be facultative air breathers (Gradwell, 1971; Gee, 1976; Graham, 1983). Other air-breathing patterns may, however, be present. Carter and Beadle (1931), for example, reported always finding air in the stomach of *Ancistrus anisitsi*, and *Hypostomus* will occasionally dart to the surface for air in conjunction with feeding or swimming (Graham and Baird, 1982).

Variability in the O_2 threshold triggering air breathing, and in the capacities of different loricariid species for air breathing, has been demonstrated. Gee (1976) compared the air-breathing responses to hypoxia of six Panamanian loricariids, finding one (*Leptoancistrus*) was not an air breather, that two (*Chaetostoma* and *Sturisoma*) could do so marginally, and three (*Loricaria* [= *Rineloricaria*] *Ancistrus*, and *Plecostomus*

FIGURE 2.6 *Ancistrus spinosus* (25 cm, Panama).

[= Hypostomus]) could sustain this action for longer periods. Among the latter, *R. uracantha* does not endure long periods of forced facultative air breathing as well as do *Hypostomus* and *Ancistrus* (Graham, 1985). Similarly, Carter and Beadle (1931) described another species, *Loricaria typus,* as a non-air breather. When given time to acclimate to hypoxia and air breathing, *Hypostomus* and *Ancistrus* (Gee, 1976; Graham, 1983, 1985) and *Pterogoplichthys* (Weber *et al.*, 1979; Val *et al.*, 1990) undergo physiological adjustments that increase Hb–O_2 affinity and improve air-breathing efficiency (Chapter 7). *Ancistrus chagresi (spinosus),* for example, increases its ABO volume and extracts more O_2 from each breath, thereby reducing its air-breathing frequency. It also sustains a higher $\dot{V}O_2$ in hypoxic water (Graham, 1983). At least two species, *A. spinosus* and *H. plecostomus,* perform synchronous air breathing (Kramer and Graham, 1976; Graham and Baird, 1982).

Gymnotiformes

This order contains six families of freshwater electrogenic fishes which occur throughout tropical South America and Central America. Known commonly as knifefishes, gymnotids are carnivores (worms, insect larvae, small fish), using their electrogenic capacities for detecting and, in the case of *Electrophorus,* stunning prey. They swim with a straight body, using their anal and pectoral fins for propulsion. Air breathing is known to occur in three families, Hypopomidae, Gymnotidae, and the Electrophoridae.

Hypopomidae. This family has four genera and about 12 species. Air breathing has been reported for three species in the genus *Hypopomus, H. brevirostris* (Carter and Beadle, 1931), an unidentified species (Kramer *et al.*, 1978), and *H. occidentalis* (see following). There are about four species of *Hypopomus* all of which reach a maximum length of near 20 cm.

Carter and Beadle observed that *Hypopomus* holds inspired air in its mouth and, because minimal vascularity was found in the walls of the buccal and pharyngeal chambers, concluded that the gills were the site of aerial respiration. Observations of gill morphology further suggested to Carter and Beadle (1931) that the large ratio of secondary lamellae breadth (i.e., height) to thickness in *H. brevirostris* was a specialization for air breathing; however, this ratio proved to have minimal predictive value for air-breathing fishes and has not been applied in subsequent studies.

Carter and Beadle (1931) did not remark on the air-breathing capacities of another species they studied, *Hypopomus occidentalis.* Initial observations with this species by D.L. Kramer and me in Panama failed to demonstrate facultative air-breathing, though it did exhibit ASR in hypoxia and in a subsequent survey Kramer (1983a) listed it as a non-air breather. A few years later, however, I did see surface gulping by *H. occidentalis* that had been kept in a bucket of extremely hypoxic water. These fish gulped such a large volume of air that they were seemingly helpless to do anything but float vertically at the water surface while air was held, which is similar to the description for *H. brevirostris* (Carter and Beadle, 1931). Hagedorn (1988) also reports unpublished observations confirming air breathing by this species in Panama. Beebe (1945) lists *Hypopomus* among the species he encountered "traveling through the jungle a considerable distance from water" which is surprising in view of its delicate and soft body, relatively small paired fin structure, and dependence on electrodetection. Finally, Saul (1975) observed that this fish would burrow in soft mud.

Gymnotidae. This family contains one genus, *Gymnotus,* of which there are three species. These fishes reach a maximum length of 60 cm and feed at night, finding their prey by electrolocation. Air breathing was documented for *G. carapo* (Liem *et al.*, 1984). The vascularized posterior chamber of its gas bladder serves the continuous, non-obligatory air breathing of *G. carapo.* To facilitate ABO ventilation, the pneumatic duct of this species is enlarged. The ventilatory sequence begins with inspiration which is driven by a buccal force-pump that hyperinflates the ABO with fresh air. The ensuing exhalation occurs mainly through pressure equilibration and leads to the voiding of a mixture of old and new air. *Gymnotus carapo* is a proficient bimodal breather that, depending upon aquatic O_2 tension, can use more or less aerial O_2 to maintain its $\dot{V}O_2$. Moreover, its capacity for aquatic respiration is sufficient for survival without access to air, even down to an aquatic PO_2 of 50 torr. But, its total $\dot{V}O_2$ is reduced by about 25% when air access is denied.

Electrophoridae. This family has a monotypic species *Electrophorus electricus,* the electric eel, which is restricted to tropical South America. Reaching a length of over 1 m, *Electrophorus* has a powerful electric discharge that is used for defense and for stunning its prey which include fishes, insects, and crustaceans. Studies of *Electrophorus* led to important eighteenth century discoveries about electricity (Wu, 1984). The ability of this species to breathe air (which none of the other strong electric fishes do) was significant because it facilitated the transport of *Electrophorus* to Europe and its endurance of the rigors of captivity and experimentation.

E. electricus is an obligatory air breather, using its

mouth as an ABO. There have been numerous accounts of its ABO structure (Valentin, 1842; Hunter, 1861; Evans, 1929; Böker, 1933; Carter, 1935; Richter, 1935). The floor and roof of the mouth and the branchial chamber arches and walls are covered by a vascularized epithelium. Also, vascular papillae extend from the roof and floor of the mouth into the buccal space where they interdigitate to form a high surface-area vascular labyrinth in the midst of the inspired air. By holding air in its mouth *Electrophorus* is one of very few air breathers (others include *Hypopomus, Pseudapocryptes,* the Synbranchidae) that cannot ventilate the gills while air breathing. Even so, the gills of this fish are small and seldom ventilated during the periods when air is not held, gills therefore play a minimal role in gas exchange (Johansen *et al.*, 1968b).

Over two hundred years ago Garden (1775) reported that *Electrophorus* rises to the surface every four or five minutes to breathe air. More recent studies (summarized in Chapter 5) indicate that air gulps occur from as frequently as every minute to perhaps only three to four times per hour, depending on factors such as its activity state, and water temperature and O_2 content (Farber and Rahn, 1970). Farber and Rahn (1970) also found that *E. electricus* obtains 78% of its $\dot{V}O_2$ from air but releases 81% of its respiratory CO_2 to water, primarily through the skin. Paradoxically, and in spite of its high aquatic gas exchange ratio, *E. electricus* also appears to have chronically elevated arterial and venous CO_2 tensions and a correspondingly high plasma bicarbonate level, perhaps owing to limited gill area and perfusion (Johansen *et al.*, 1968b; Garey and Rahn, 1970; Rahn and Garey, 1973; and Chapter 7).

Salmoniformes

The salmoniforms are a diverse and widely distributed group (Nelson, 1994). Air breathing is limited to only three families, Umbridae, Lepidogalaxiidae, and Galaxiidae.

Umbridae. The mudminnows occur in freshwaters of the Northern Hemisphere. There are three genera and at least six species. Air breathing is known for two genera, *Umbra* and *Dallia.*

Species of *Umbra* occur in both North America *(U. limi* and *U. pygmaea)* and Europe *(U. lacustris* and *U. krameri).* These fishes live in creeks, ponds, and lakes, rarely grow to longer than 15 cm, and are carnivores. Umbrids tolerate environmental extremes ranging from drying mud in summer to overwintering under ice.

Studies of the anatomy and physiology of the *Umbra* gas bladder convinced Rauther (1914) of its aerial-respiratory function. Additional work on various aspects of respiration, including blood circulation pattern and ABO gas content were done by Geyer and Mann (1939a,b on *U. lacustris*) and Black (1945, *U. limi*). Kramer *(in litt.)* has observed air breathing in *U. pygmaea.* Gee (1980, 1981) determined that air breathing by *U. limi* is continuous, but not obligatory, and subject to control by disturbing factors such as the activity of potential predators. Also, in normoxic water at least, buoyancy control rather than O_2 access can be a stronger driving force for regular air gulping.

While it seems generally true that air-breathing fishes have unlimited access to air, *Umbra* is at least one exception because it survives hypoxia under ice by "air breathing" in gas pockets often containing low percentages of O_2 (Klinger *et al.*, 1982; Magnuson *et al.*, 1983, 1985). Gee (1981) found that air-gulping *U. limi* did not respond adversely to N_2 when this gas was substituted for air. While such a condition could not be endured indefinitely, this response contrasts markedly with the immediate effect on air-breathing frequency and tidal volume reported for other species given a low O_2 air phase (e.g., *Pantodon,* Schwartz, 1969; *Monopterus,* Lomholt and Johansen, 1974; Graham *et al.*, 1995) and may be adaptive in view of the variable O_2 content of some under-ice air spaces.

Blackfish of the genus *Dallia* are similar, ecologically speaking, to *Umbra* except that they occupy freshwater habitats in Arctic North America and Siberia. These fish rarely exceed a length of 10 cm and are carnivorous. Observations by Crawford (1971, 1974) verified air breathing in the Alaska blackfish *(Dallia pectoralis),* demonstrating that a section of its esophagus functions as an ABO. The respiratory region of the esophagus, which is separated from the stomach by a sphincter, is lined by an epithelium containing a dense concentration of capillaries. The physostomous gas bladder of *D. pectoralis* has *retia mirabilia* for gas secretion but lacks respiratory epithelium.

Lepidogalaxiidae. There is one species in this family, the salamander fish *Lepidogalaxias salamandroides* (Allen and Berra, 1989). This small (7 cm) fish inhabits an acidic (pH < 5) creek tributary of the Shannon River in the heathlands of southwestern Australia. Its diet includes aquatic crustaceans, midge larvae, and terrestrial invertebrates. During dry periods of the year the salamander fish estivates under leaves or by burrowing into mud; with the onset of rain, fish suddenly appear in previously dry pools.

Merrick and Schmida (1984) describe *L. salamandriodes* as the only Australian fish that truly estivates (i.e., has a reduced metabolism, uses stored fat, and becomes ureogenic), and recent studies of the natural history and physiology by several investigators have contributed much new information. Works by Pusey (1986, 1989a,b, 1990) detailed estivation metabolism and nitrogen balance and examined dry season survivorship. Observations by Berra and coworkers (Berra and Allen, 1989; Berra *et al.*, 1989) focused on behavioral aspects of burrowing and functional morphology. These workers refuted earlier assertions that the gas bladder of *L. salamandroides* functions as an ABO for fish in burrows. They found the gas bladder of this fish to lack respiratory surfaces and concluded that its minimal O_2 requirements while estivating could be met by diffusion across a mucus-coated skin. A primarily cutaneous respiration was confirmed by Martin *et al.* (1993) who also showed that the salamander fish has a high aerial respiratory exchange ratio. Observations by these and other workers further suggested that seasonal estivation requires a period of induction.

Galaxiidae. Galaxiids are freshwater and diadromus and occur in Australia, New Zealand, and southernmost Africa and South America. There are eight genera and about 49 species. The New Zealand mudfishes, *Neochanna apoda, N. burrowsius,* and *N. diversus* (Subfamily Galaxinae) are small (8 cm) carnivorous fishes that occur in temporary streams. Detailed natural history observations on *N. burrowsius* were made by Eldon (1979a,b,c) and Eldon *et al.* (1978) determined how seasonal drought affected survivorship. In hypoxic water *N. burrowsius* is a facultative air breather and does this by floating at the surface while holding air in its mouth or by lying at the water's edge with its head emergent (Eldon, 1979a,b). While in water and during estivation, *N. burrowsius* uses its skin to sustain nearly half of its $\dot{V}O_2$ (Meredith *et al.*, 1982). Experiments by Wells *et al.* (1984) revealed little effect of three weeks' air exposure on the blood respiratory properties of this species and Ling and Dean (1995) found that emergent *N. diversus* have a lowered aerobic metabolism.

Similar capabilities for estivation and air breathing are likely present in other galaxiids. McDowall (1978, figure 6:12, page 65), for example, illustrates another New Zealand species, *Galaxias brevipinnus,* climbing out of water along a spillway. He suggests that this fish, which occupies fast-flowing streams, may emerge to avoid turbulent flow. In times of low water, it burrows and is presumed to overwinter in mud. *G. brevipinnus* is also widely distributed in Australia where it is known to climb water falls (Merrick and Schmida, 1984). Emergent behavior has also been observed for *Galaxias fasciatus* (A.S. Meredith *in litt.* to D.L. Kramer) and *Galaxias cleaveri* is thought to estivate (Merrick and Schmida, 1984). Estivation is thought also to occur in *Galaxiella pusilla,* and *G. munda* (Backhouse and Vanner, 1978; Allen, 1982; Merrick and Schmida, 1984).

Superorder Paracanthopterygii

Gobiesociformes

Gobiesocidae. There are 36 genera and about 114 species of gobiesociform fishes contained in two families, Gobiesocidae and Alabetidae. The latter is a small family (one genus, four species) restricted to Australia and Tasmania. The Gobiesocidae (clingfishes) therefore accounts for nearly all fish in this order, and these occur primarily in marine, intertidal habitats throughout tropical and temperate zones of the world's oceans. (Four freshwater species of *Gobiesox* occur in Central America.)

Clingfishes are dorsoventrally flattened and have their pelvic fins and adjacent tissues modified into a ventral sucker for attachment to substrate in habitats where there is forceful water movement. Most species are small (12 cm or much less), and their diets consist mainly of small invertebrates. Clingfish are unique among teleosts in having only three branchial arches (most fishes have four or more) and a venous circulation featuring accessory cardinal sinuses. It is not known if these differences are environmental adaptations, a consequence of changes in body morphology, or both.

Air breathing has not been examined in most clingfishes, but amphibious air breathing seems likely for many species owing to their air-exposure during low tide. Kramer (1983a), for example, observed a weak and inconsistent air-breathing response by a freshwater *Gobiesox* exposed to hypoxic water. By contrast, the intertidal *Arcos rhodospilus* would breathe air while emerged but did not surface to gulp air in hypoxic water. For *G. maeandricus,* a species not commonly found out of water, Martin (1993) measured a relatively high (0.7) aerial gas-exchange ratio.

The capacity of *Lepadogaster bimaculatus* to endure long exposures to air was noted by Simroth (1891, page 81). More recent studies have documented this for *Sicyases sanguinius* (Ebeling *et al.*, 1970; Gordon *et al.*, 1970), *Phallerodiscus funebris, Gobiesox pinniger, Tomicodon humeralis, T. boehlkii,* and *T. petersi* (Eger, 1971). Eger (1971) investigated the exposure tolerances of four Gulf of California clingfishes, finding that they could endure up to 93 hours air exposure and could tolerate dehydration up to as much as a 60% loss of body mass, which exceeds the water-loss capacity of most amphibians. Eger also demonstrated specific differences in desiccation resistance relating to littoral distribution. *Tomicodon humeralis* which occurs in the higher intertidal areas has the greatest tolerance to air exposure and dehydration.

The gills and adjacent epithelium and the scaleless skin all probably function for aerial respiration in most clingfish. *Sicyases sanguineus,* a particularly large (30 cm) species found in Chile, respires aerially using its branchial surfaces (Vargas and Concha, 1957; Gordon *et al.*, 1970) and a patch of skin below the jaw (Ebeling *et al.*, 1970). Marusic *et al.* (1981) also suggested that O_2 might be absorbed from air swallowed by emergent fish because the gut of such fish had become engorged with blood. When kept moist, *S. sanguineus* can survive air exposure for up to three days; however, fishes in which the gill surfaces were covered with algal paste (in order to determine their relative contribution to air breathing) lasted only 12 hours in air (Vargas and Concha, 1957). Most clingfish seem to simply tolerate low-tide exposure by respiring aerially and taking cover under rocks or algal mats to reduce desiccation. However, *S. sanguineus* is an active, high intertidal forager during low tides, and observations of its distribution and feeding behavior (Paine and Palmer, 1978) suggest that its emergence from water may facilitate digestion.

Superorder Acanthopterygii

Acanthopterygians are the largest and most diverse major grouping of fishes containing 22 suborders, 150 families, 1367 genera, and nearly 7800 species. As will be documented here, the air-breathing acanthopterygians are diverse, occupy a variety of habitats, and range in their dependence on air breathing from facultative to continuous, while some are amphibious.

Cyprinodontiformes

This is a large order (13 families, 120 genera, and about 845 species) which is broadly distributed (tropical and temperate freshwaters and oceans) and includes a few fishes that routinely emerge from water. Among these are the flyingfishes (Exocoetidae, discussed in Chapter 1), two families (the halfbeaks [Hemiramphidae] and needlefishes [Belonidae]) with active species that often jump into the air, and some species of live-bearers (Poeciliidae) which emerge briefly (Baylis, 1982; Lefebvre and Spahn, 1987). There are no studies related to air breathing or aerial exposure in these fishes. On the other hand, there is more documentation of the terrestrial behavior of a few species in two other cyprinidont families, the rivulines (Aplocheilidae), and the killifishes (Cyprinodontidae), which will now be discussed.

Aplocheilidae. About 15 genera and 210 species are in this family, most of which occur in freshwater and are small (8 cm) carnivorous feeders (small fishes and insects). There is one marine species, *Rivulus marmoratus,* which is also a self-fertilizing hermaphrodite that occurs in the shallow waters and along the mud banks of mangrove forests in southern Florida and the Caribbean (Huehner *et al.*, 1985). Amphibious activities related to feeding, reproduction, avoidance of predators, and avoidance of adverse aquatic conditions have been documented for at least five species of *Rivulus (beniensis, brunneus, hartii, limoncochae, marmoratus).*

Because of their jumping ability, rivulines occur in, and have colonized, many habitats that contain few other fishes such as small hill streams, tributaries of larger streams, small pools, and potholes. In many such habitats, rivulines are annuals because seasonal drying may regularly eliminate them. Some rivulines leave hard-shelled eggs in an arrested metabolic state (and thus capable of surviving the exigencies of the impending dry season) to complete development when the rains start. Lüling (1971) observed that in the late wet season, *Rivulus beniensis,* an annual species of the Peruvian Andes, begins to jump repeatedly across the moist land in order to find the deepest temporary pools. According to Lüling (1971) this behavior at least assures the start of the next generation by concentrating the reproductive population in microhabitats that will remain wet the longest and where fewer predators may be present.

On the other hand, some populations living in seasonally marginal habitats may be able to "wait out" the dry season in residual pools. Saul (1975) found that *R. limoncochae* survived long droughts by burrowing into mud. Seghers and Nielsen (1982) reported a behavior similar to *R. beniensis* for populations of *R. hartii* located in the ephemeral, high elevation streams of Trinidad. They observed that, toward the end of the dry season, these fishes congregated in the deepest pools of the habitat where they endured extreme conditions of crowding, hypoxia, and even the complete disappearance of water. In the latter case, fish survived under piles of moist leaves until the onset of the wet season.

Emergence for food has been observed for *R. hartii* (Seghers, 1978) and *R. marmoratus* (Huehner *et al.*, 1985). *R. marmoratus* will also leave water to avoid aggressive encounters as will *R. brunneus* to avoid predators (Abel *et al.*, 1987). The prolonged tolerance of rivulines for air exposure and their ability to sustain jumping suggest they are capable of aerial respiration; however, there are few data. These species also adopt ASR in experimental hypoxia (Kramer and McClure, 1982; Kramer, 1983a) but do not necessarily emerge from water. Abel *et al.* (1987) determined that *R. marmoratus* will emerge from water containing high concentrations of hydrogen sulfide and that it is capable of respiring aerially. Grizzle and Thiyagarajah (1987) found that the dorsal head skin of this species contains epidermal capillaries that likely enhance aerial respiration in shallow water and during emergence. *R. marmoratus* reportedly lays its eggs on land and will stalk and consume insect prey while out of water (Abel *et al.*, 1987). Huehner *et al.* (1985), who did not observe fish consume prey out of water, did determine that individual fish would stay emerged for up to 7 days, at which time they returned frequently to water. *R. marmoratus* moves by sinusoidal movements through the leaf litter and can also use tail flips to cover larger distances.

Cyprinodontidae. Killifishes are diverse (29 genera and 268 species) and broadly dispersed in marine, fresh, and brackish waters. These fishes are small (8 cm) and feed mainly on invertebrates. Mast (1915) observed that *Fundulus majalis* trapped in sand pools on a falling tide would exit the seaside of the pool and jump in directed fashion across the sand towards open water. Terrestrial behavior and suncompass orientation have been shown for *F. notti* by Goodyear (1970), who also observed that this fish would emerge on to land when approached by potential predators (e.g., *Micropterus*). Baylis (1982) similarly observed that *Lucania parva* (and the poeciliid *Gambusia affinis*) could avoid the predatory attacks of the bluegill *(Lepomis macrochirus)* by leaping onto lily pads. Although this behavior seemed successful in the short term, Baylis' observations revealed a number of lily pads that had circular holes or scars, suggesting that the bluegill countered the avoidance behavior with a subsurface "buttonholing" attack on the emergent fish using its shadow as a target. Emergent behavior has also been seen in other species *(F. grandis, F. similis, F. confluentus,* V. Springer, *in litt.)* and Kushlan (1973) suggested that emergence by *F. confluentus* in some habitats may be related to egg laying. Nothing is known about the air-breathing capacity of these fishes.

Scorpaeniformes

In this very large order (20 families, 269 genera, and about 1160 families), accounts of terrestrial excursions and air breathing are limited to littoral fishes in the sculpin family Cottidae.

Cottidae. There are 70 genera and about 300 species of sculpins distributed in marine and freshwater habitats throughout the Northern Hemisphere and in New Zealand. The highest diversity of this carnivorous group is in the North Pacific, and it is for the intertidal genus *Clinocottus* (maximum length about 15 cm) living along the California coast that both various stimuli for emergence and a capacity for air breathing have been documented. Wright and Raymond (1978) observed that *C. recalvus* inhabiting hypoxic tidepools within a dimly lighted cave either crawled out of water completely, or far enough to expose their heads to air. Fish living in pools outside the cave, however, were not observed to emerge, even though these pools also had low O_2 tensions.

Wright and Raymond (1978) concluded that emergence was likely triggered by aquatic hypoxia, but that other stimuli such as the high ammonia concentration of cave pools, dim light, and the absence of aerial predators may have had synergistic effects. Their respirometry studies showed that *C. recalvus* in water and exposed to air had the same $\dot{V}O_2$, although aerial $\dot{V}CO_2$ was low. This species was also shown to spring from water when approached by predatory fish. Emergence was also observed for *C. globiceps* from British Columbia (Wright and Raymond, 1978). I have observed partial and full emergence of *C. analis* in hypoxic bowls of seawater and studies of this species by Martin (1991, 1993) have shown it is capable of aerial exposures of at least 24 hours during which the gills, buccopharyngeal surfaces, and skin are utilized for high rates of aerial $\dot{V}O_2$ and $\dot{V}CO_2$. A survey of several air-exposed cottids (some of which do not naturally emerge from water) has shown the general presence of a high aerial respiratory exchange ratio (R) (Martin, 1993, 1995). Davenport and Woolmington (1981) found that *Taurulus (Cottus) bubalis* will emerge from hypoxic tidepools when the PO_2 declines to 6 torr. Emergence and air breathing in response to tidepool hypoxia are likely for other littoral-zone cottids (Martin, 1995).

Perciformes

Suborder Zoarcoidei

Stichaeidae. The pricklebacks are a large marine family containing 31 genera and about 60 species. These fishes are carnivores and most species are small, but some grow to 30 cm length. Most occur in the North Pacific with a few in the North Atlantic. Three North Pacific genera, *Xiphister (Epigeichthys), Anoplarchus,* and *Cebidichthys,* occur intertidally and often become exposed to air at low tide (Barton, 1985; Horn and Gibson, 1988; Martin, 1993). Schmidt (1942) suggested that the gill structure of another species *Chirolophis polyactocephalus* indicated it was adapted for air exposure. Buckley and Horn (1980) reported that air exposure of both *Cebidichthys violaceus* and *Xiphister mucosus* resulted in a decreased plasma pH, and increases in both plasma PO_2 and blood hematocrit. The blood of small *C. violaceus* also exhibited a smaller Bohr effect and a larger buffering capacity (Chapter 7) than did the blood of either larger *C. violaceus* or *Xiphister.* Horn and Riegle (1981) compared the desiccation tolerances of *Xiphister, Anoplarchus,* and *Cebidichthys* and found them all to be incapable of regulating water-loss rate which occurred passively during air exposure. These workers did, however, find that *C. violaceus,* which occurs higher in the intertidal zone than the other stichaeids, has a longer emersion endurance because it is larger and thus has both a higher total body water content and a lower body surface area to volume ratio. Daxboeck and Heming (1982) demonstrated that *Xiphister atropurpureus* can respire aerially, probably using its skin and gills for aerial O_2 consumption. The $\dot{V}O_2$ of this species in air equaled its rate in water; however, the aerial respiratory exchange ratio (R) determined by these workers was below 0.4. In contrast, subsequent studies by Martin (1993) indicate that *X. atropurpureus* and several other stichaeids have high aerial R values (Chapter 5).

Pholididae. Members of this family, termed gunnels, are similar to pricklebacks in terms of morphology, body size, and diet. They are also mostly found in the North Pacific with a few species in the North Atlantic. There are four genera and 13 species in this family. At least three genera, *Xererpes, Pholis* (three species), and *Apodichthys* occur intertidally and are subject to tidal emersion (Barton, 1985; Horn and Gibson, 1988; Martin, 1993). Horn and Riegle (1981) found that the air-exposure tolerance of *Xererpes fucorum* is generally similar to that of the intertidal stichaeids (see preceding) with which it commonly occurs. Martin (1993) measured an aerial R of 0.86 for this species.

Suborder Blennioidei

Tripterygiidae. A primarily tropical and subtropical marine family, the carnivorous tripterygiids number about 19 genera with from 95 to 110 species. They occur in shallow water and intertidally and rarely exceed a length of 12 cm. One species, *Helcogramma medium* was shown by Innes and Wells (1985) to crawl from hypoxic tidepools and breathe air. These workers reported that this fish lacks a specialized ABO and has no capacity to increase its aerial $\dot{V}O_2$ in response to enforced activity.

Labrisomidae. Members of this family are carnivorous and entirely marine. There are 16 genera and about 41 species distributed in primarily shallow water and littoral zone habitats of the new world tropics. Amphibious behavior is known for only two species: *Mnierpes macrocephalus,* which occurs along the Pacific coast of Central America, and *Dialommus fuscus,* which is endemic to the Galapagos Islands (Graham 1970, 1973). These fishes do not exceed 15 cm length and are highly similar. Both occur on rocky shores and *Mnierpes* (Figure 2.7) is known to be active during daylight hours. Experiments show this species will emerge from hypoxic water and will also leave water when approached by aquatic predators (Graham, 1970). *Mnierpes* has, relative to other amphibious species, a limited tolerance of air exposure; however, its aerial $\dot{V}O_2$ is similar to its $\dot{V}O_2$ in water and, it can also expel CO_2 aerially with a high R. Its skin and gills each function for about 50% of its total aerial $\dot{V}O_2$ (Graham, 1973). Both *Mnierpes* and *Dialommus* have eyes that are specialized for aerial vision through the presence of flattened corneas (Figure 2.8) which minimize aerial light refraction while not compromising normal aquatic vision (Graham and Rosenblatt, 1970; Graham, 1971; Herald and Herald, 1973; Stevens and Parsons, 1980).

Blenniidae. Blennies are a diverse, highly derived family consisting of about 53 genera and 300 species. Most are less than 15 cm in total length and commonly occur in littoral habitats of the tropics and temperate zones, subsisting on algae and small invertebrates. The amphibious behavior of blennies has been long known. Simroth (1891) observed that *Salarias quadricornis*

> lies happily in the sun, on the cliffs projecting above the surf where it is moistened by the spray of the waves.

Counting *Salarias,* amphibious behavior is known for at least seven genera: *Andamia, Alticus, Istiblennius, Entomacrodus, Coryphoblennius,* and *Blennius.* Field observations of the latter two in the Atlantic, Adriatic, and Mediterranean have documented amphibious behavior in several species: *C. galerita, B. dalmatinus, B. pholis, B. trigloides, B. montagui, B. pavo, B. sphinx, B. gattorugine,* and *B. erythrocephalus* (Soljan, 1932; Heymer, 1982; Laming *et al.*, 1982; Louisy, 1987). The other five genera listed are tropical, and these display the greatest degree of terrestrial behavior (Hora, 1932; Magnus, 1963, 1966; Mohamed, 1980; Zander, 1983; Graham *et al.*, 1985, Louisy, 1987).

Tropical amphibious blennies feed, court, defend territories, and guard their nests in the high littoral zone. They are commonly active out of water and usually rest at night exposed to air (Soljan, 1932; Abel, 1973; Zander, 1983; Graham *et al.*, 1985). Many of them feed on the filamentous algae and associated fauna of the high-energy littoral zone; Soljan (1932) described the highly specialized feeding of *B. galerita* on the appendages of *Chthalamus* barnacles. Many of these

FIGURE 2.7 Two *Mnierpes macrocephalus* in the littoral zone of Pacific Panama. (Fish on left holds a piece of food in its jaws; specimens about 6 to 10 cm.)

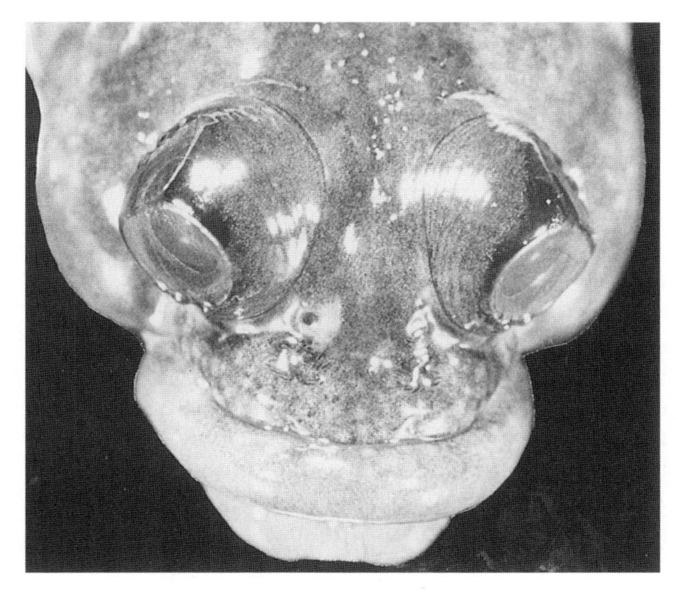

FIGURE 2.8 Flat corneal surfaces of *M. macrocephalus*. (Reprinted with permission from *Science* 169:586–588. Graham and Rosenblatt, 1970. American Association for the Advancement of Science.)

fishes also have well-developed abilities for terrestrial navigation and for orientation to water (Graham *et al.*, 1985).

All of the amphibious blennies have some capacity for air breathing even though most lack morphological specializations for this (Hora, 1932; Graham, 1976a; Laming *et al.*, 1982; Bridges, 1988). Their tolerance of air exposure is brief relative to other amphibious groups, and, because the blenny habitat is usually well oxygenated by wave action, they infrequently encounter hypoxia. These fishes are thus more likely to breathe air while amphibious (and probably respond to aquatic hypoxia by emergence) and would be less likely to gulp air while in water (Graham *et al.*, 1985; Bridges, 1988). The conclusion of Rao and Hora (1938) that *Andamia heteroptera* is an obligatory air breather can be questioned in view of the unspecified aquatic O_2 conditions in their tests. Their observations raise the possibility that, with emergence as an alternative, amphibious blennies in general may have a reduced capacity to withstand moderate aquatic hypoxia.

Comparative studies of *Alticus, Blennius,* and *Coryphoblennius* by Zander (1967, 1972a,b, 1974, 1983) documented progressive changes in functions related to the transition from aquatic to amphibious life. These include altered eye structure to favor aerial vision, reduction of the lateral line, modification of the paired and median fins to favor terrestrial locomotion, and the development of cutaneous blood vessels for greater aerial gas exchange (Zander and Bartsch, 1972; Louisy, 1987). Recent works with *Alticus kirki* described its microhabitat salinity, temperature, and immersion versus emersion preferences (Brown *et al.*, 1991) and compared its rates of aerial and aquatic respiration (Brown *et al.*, 1992). This fish can make use of cutaneous respiration; however, it is one of few marine species apparently having a lower total $\dot{V}O_2$ in air than in water (Chapter 5). Even when its aerial $\dot{V}O_2$ is increased twofold by forced activity, *A. kirki* is still able to maintain a relatively high aerial R (Martin and Lighton, 1989).

The actual number of amphibious species that occur among various tropical blenny genera is not presently known because details are lacking about the behavior of many species. There are, for example, from eight to 12 species of *Salarias,* all unstudied. There are also four to five species of *Andamia* living in the Indo-Pacific region, but no work with this genus has been done since the early study by Rao and Hora (1938) describing aerial respiration and adaptations for terrestriality in *A. heteroptera* and *A. reyi.* Shen *et al.* (1986) reported the amphibious behavior of *A. tetradactylus* (and *A. reyi*) in Taiwan. There are between 22 and 24 species of *Entomacrodus* living on tropical reefs around the world, but amphibious behavior has been documented for only four species, *E. vermiculatus* (Abel, 1973), *E. nigricans, E. chiostictus,* and *E. marmoratus* (Graham *et al.*, 1985; D. Strasburg, *in litt.*). Although there have been a number of reports about the ecology, behavior, morphology, and aerial gas exchange of *Alticus kirki* and *A. saliens* in the Red Sea (Magnus, 1963, 1966; Klausewitz, 1964; Zander, 1972a,b; Martin and Lighton, 1989; Brown *et al.*, 1991, 1992), the remaining seven to nine species of *Alticus* remain unstudied. Similarly, there are few accounts of terrestrial behavior for the approximately 25–30 species of *Istiblennius.*

Comprehensive studies of air-breathing physiology have been carried out on *Blennius pholis,* the intertidal shanny which occurs in North Atlantic and Mediterranean tidepools and often exposed in rock crevices or under wrack during low tide (Qasim, 1957). Although not normally active on land, this fish will emerge from tidepools which have a low O_2 content (Davenport and Woolmington, 1981). Daniel (ms) found that *B. pholis* could survive in moist air for up to five days, that it could carry on aerial O_2 uptake at a rate comparable to its $\dot{V}O_2$ in water, and that it was also proficient at removing CO_2 to air. Nonnote and Kirsch (1978) reported that *B. pholis* could absorb O_2 through its skin while in air or water. Laming *et al.* (1982) discovered that emerged shannys gulped air into the esophagus where numerous capillaries are positioned close to the surface of that organ. Emergence causes a brief bradycardia followed by a resumption of normal heart rate. Davenport and Sayer (1986) reported that although reduced, the nitrogen excretion of this fish was not interrupted by air exposure and that ammonia and not urea is the main excretory product, with most of the loss occurring into skin mucus. Pelster *et al.* (1988) determined that erythrocyte carbonic anhydrase plays an important role in the high rate of aerial CO_2 release by *B. pholis* (Chapter 5).

Suborder Gobioidei

Eleotrididae. This euryhaline family, called gudgeons or sleepers, occurs worldwide in the tropics, most usually in still waters and amongst thick vegetation. Eleotrids number about 40 genera and at least 150 species. They are omnivorous and rarely exceed a maximum length of 30 cm. The habitat of many eleotrids naturally exposes them to hypoxia, and many species are adept at ASR (Kramer, 1983a). Gee (1986) showed that an increase in physoclistous gas bladder volume improved the ASR efficacy of four Australian eleotrids by lowering the energetic cost of remaining at the surface and altering the incident angle between the body and the surface to optimize surface ventilation.

Although probably present in a number of eleotrids, air breathing has not been described for many species. Kramer (1983a), for example, did not observe air breathing by any of four Panamanian eleotrids he exposed to hypoxia. Similarly, a survey of five Australian eleotrids (Gee and Gee, 1991) revealed that while these fish resorted to ASR in hypoxia, none gulped air. *Oxyeleotris marmorata* is known to use ASR in hypoxic water, and Fenwick and Lam (1988) reported it also air breathes by holding air in its mouth. These workers compared *Oxyeleotris* and other species to determine how relative gill dependency (the *a priori* assumption was that this would be less in air breathers) affected calcium-ion flux rates. They found a direct correspondence between calcium flux and gill use with the calcium ion loss rate of *O. marmorata* being less than that of a non-air-breathing cichlid, *Oreochromis mossambicus,* but greater than that of the amphibious air-breathing *Periophthalmodon schlosseri* (Gobiidae). Nothing, however, is known about the ABO morphology or air-breathing behavior in *O. marmorata,* and in contrast to previous assertions, observations in my laboratory, while confirming ASR, have not shown this species to breathe air.

Facultative aquatic air breathing has been documented for *Dormitator latifrons* (Figure 2.9A; Todd, 1973). Exposure of this fish to hypoxia induces facultative air breathing in which the skin on the frontal head surface serves as the ABO. This surface is gradually (over 10–20 days) transformed into a thick capillary bed that measures about 10% of the total body surface area (Todd, 1973, figure 1; and Figure 2.9B). The physoclistous gas bladder of *Dormitator,* while not involved in respiration, is vital to its ability for facultative air breathing because a positive buoyancy must

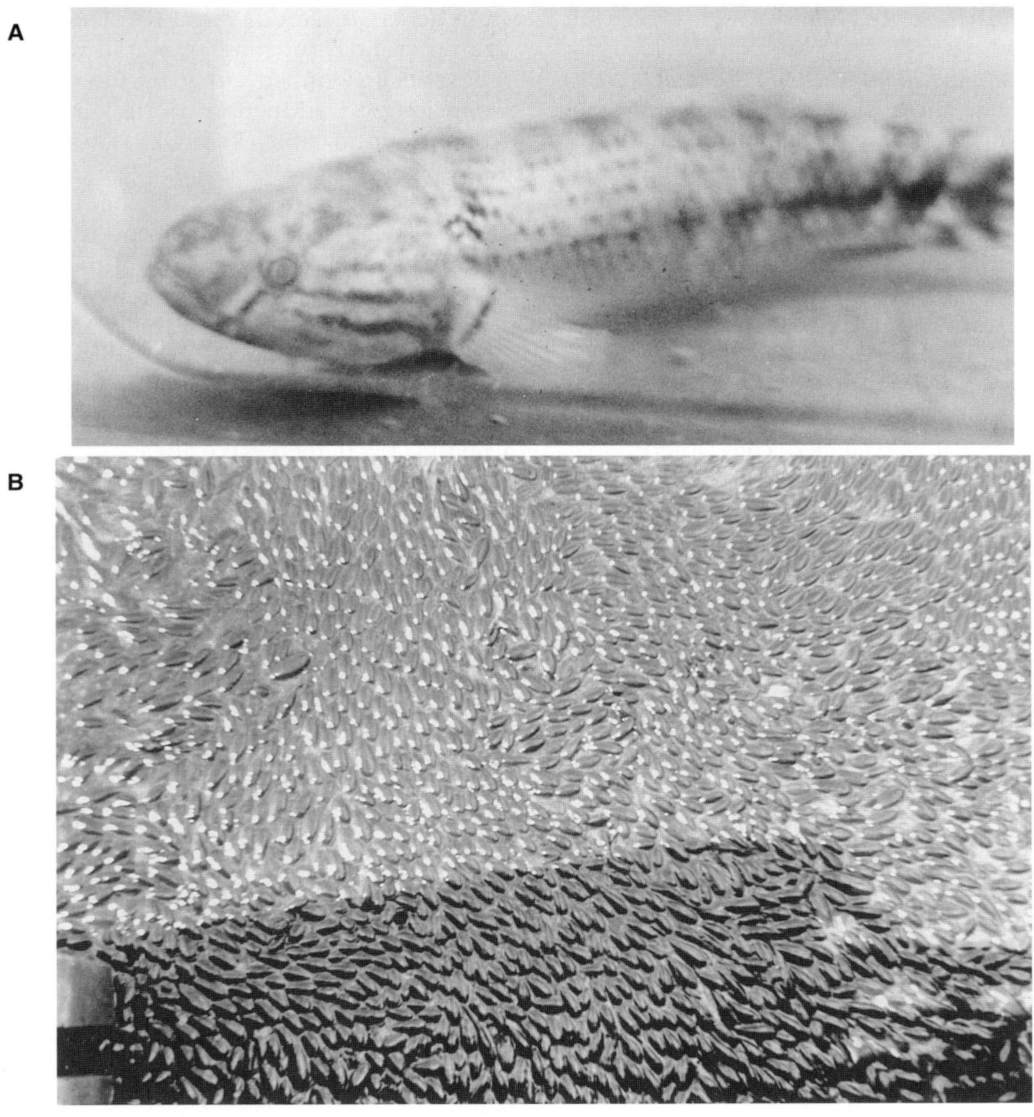

FIGURE 2.9 **A,** *Dormiator latifrons,* about 25 cm, Panama. **B,** Overhead view of a large number of *Dormitator* aggregated at a dam, apparently driven out of their backwater habitats in response to heavy use of insecticide sprays. Each fish is positively buoyant and its aerial-respiratory frontal skin patch is exposed to air. Picture was taken on Farfan Road, Panama, September, 1972.

be achieved in order to elevate the vascularized frontal skin into air (Figure 2.9B). Still, and paradoxically, O_2 is needed to inflate the gas bladder, and in hypoxic water this is best supplied by air and by ASR. Thus, in the early phases of air breathing (about six hours are needed to inflate the gas bladder) *Dormitator* will, using submerged objects or the shallow bottom, attempt to prop itself far enough up in the water to expose the top of its head to air (Todd, 1973; and personal observations). Other changes accompanying the transition to air breathing by *Dormitator* include increases in both its blood hematocrit and hemoglobin concentration (Todd, 1972; Graham, 1985; Chapter 7).

Gobiidae. Gobies are found mainly in marine and brackish waters, but frequently occur in freshwater. They are most common in the shallow waters of tropical and subtropical regions. With as many as 200 genera and about 1500 species, the gobies are the largest family of marine fishes and probably the most abundant benthic species in tropical littoral communities. Gobies are mostly carnivorous, and all of them have their pelvic fins modified to form an adhesive, or sucking disc, that functions for support and for substrate attachment in turbulent habitats. Most species are less than 20 cm long at maximum size. There are a number of air-breathing gobies including species that do this aquatically as well as during terrestrial sojourns.

Mudskippers and their allies (Subfamily Oxudercinae). Most notable among the air-breathing gobies are the mudskippers; colorful and highly active inhabitants of rocky shores, mangrove swamps, and tidal flats throughout the Indo-Pacific Ocean and west Africa. Mudskippers are the best known and most studied of the amphibious air-breathing fishes and the many and diverse aspects of their biology have been reviewed by Macnae (1968a,b) and by Clayton (1993). An early account of mudskipper natural history and taxonomy is found in Cuvier and Valenciennes (1837). These workers differentiated eight species of *Periophthalmus* and six species of *Boleophthalmus.* They interpreted mudskipper eye, fin, and branchial structure in relation to amphibious behavior, pointed out that life in the mudflats and the capacity to burrow reduced predation by large fish and birds, and noted that *Periophthalmus* was more amphibious than *Boleophthalmus.* All species are active land dwellers that spend considerable time exposed to air where they feed on a range of prey, as well as some plant materials, and spend sufficient time out of water to be frequent victims of mosquitos. Most species are marine; however, they also occur in both fresh and brackish waters. Mudskippers are easily spotted by their conspicuous amphibious antics, their prominent and turreted eyes, and a high and usually brightly colored first dorsal fin. Although capable of jumping and skipping across the mud, these fish can also crawl in a characteristic "crutching" mode in which the body is held stiff and pulled along by the alternate anterior extension and adduction of the jointed, highly dextrous pectoral fins (Van Dijk, 1959; Harris, 1960; Klausewitz, 1967, 1968a,b).

A taxonomic revision of the mudskippers by Murdy (1989) will prove enormously valuable to all future studies (Figure 2.10). Mudskippers are a diverse group and their taxonomy had not been uniformly known, particularly by biologists who have mainly studied their air-breathing and terrestrial behavior. At one time or another as many as 100 mudskipper species have been named and, as reviewed by Clayton (1993), there have always been problems with mudskipper-species identification owing to the similarities and broad geographic distributions of these fishes. Eggert's work (1929a,b, 1935), for example, enumerated 17 species and, reflecting the broad distribution and similarities of the group, 44 subspecies. One impediment to standard identification is that the behavior patterns of the species can vary in different habitats. In one area, a species may occur in the mangroves; whereas, in another region it may live on a rocky shore (Macnae, 1968b). The probable misidentification of mudskipper species and genera has likely contributed to some of the discordant findings and conclusions about their air-breathing and related aspects of their physiology as well as other elements of their biology (Clayton, 1993).

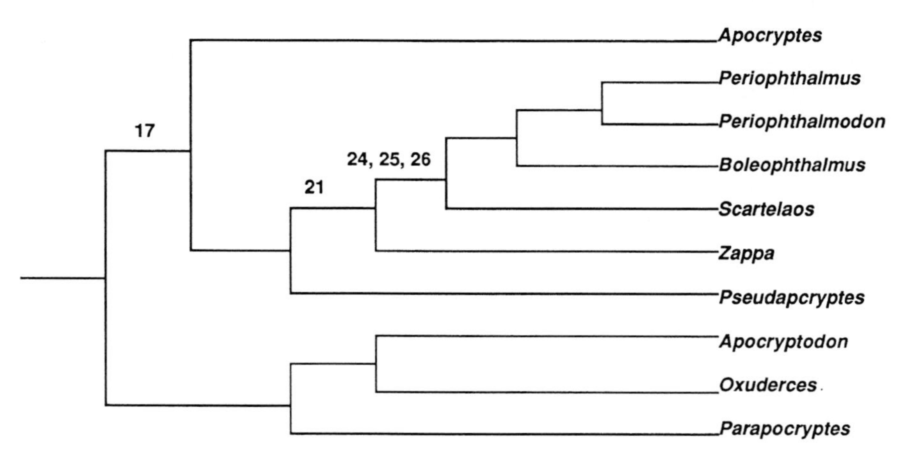

FIGURE 2.10 Murdy's (1989) cladogram showing the evolutionary relationships among the 10 genera of the Oxudercinae. A total of 39 characters were used. The five characters indicated all relate to air breathing and terrestrial life. Character 17 indicates the presence of some type of auxiliary buccopharyngeal or opercular epithelium or a cutaneous respiratory capacity. Number 21 separates five genera capable of short periods of survival out of water. Characters 24 (truly amphibious behavior), 25 (dorsal fins separated), and 26 (dermal eye cups) are specializations for life on land and separate the four mudskipper genera *(Periophthalmus, Periophthalmodon, Boleophthalmus,* and *Scartelaos)* from the other oxudercines.

Murdy's (1989) work on mudskipper taxonomy also presents keys to species identification and geographical distribution patterns and thus should minimize problems of species identifications in future investigations. The mudskippers are in the gobiid subfamily Oxudercinae, a group consisting of ten genera: *Apocryptes, Zappa, Pseudapocryptes, Apocryptodon, Parapocryptes, Oxuderces, Scartelaos, Boleophthalmus, Periophthalmus, Periophthalmodon.* Figure 2.10 shows Murdy's cladogram for these genera. Of the 39 cladistic characters used to delineate the oxudercines, the position of five of these relating directly to air breathing and terrestriality are shown in Figure 2.10. Character 17, for example, separates seven genera having some type of modified respiratory epithelium. Character 21 distinguished five genera capable of surviving short periods out of water whereas characters 24–26, which relate directly to life out of water distinguish the four genera that are usually referred to as the mudskippers. The number of species enumerated for each mudskipper genus by Murdy (1989) are: *Periophthalmodon* (3), *Boleophthalmus* (5), *Scartelaos* (4), and *Periophthalmus* (12). Table 2.3 lists these species, along with the other known air-breathing oxudercines. The important question from a comparative standpoint is how do the species names in Table 2.3 relate to the species identifications used in all previous works on mudskippers. Parenthetical notations in the text of both Murdy (1989) and Clayton (1993) align the earlier and current nomenclature for the various species. Table 2.4 matches Murdy's specific names with species names published by workers investigating aspects of mudskipper air-breathing or terrestrial life. My survey of the literature shows that studies have been reported for no less than 12 different species of *Periophthalmus,* including an unnamed species, 4 species of *Boleophthalmus,* 2 species of *Periophthalmodon,* and 2 species of *Scartelaos* (Table 2.4). While I will refer to the species names used by the original workers in this text, readers can refer to Table 2.4 for the now-accepted taxonomic nomenclature.

Studies with mudskippers have touched on all facets of their respiratory and amphibious capabilities. Hickson (1889) observed that emergent mudskippers frequently had their tails in water and suggested that this was for respiration. Hickson (1889, page 31) wrote that the mudskipper had progressed to the point of having:

> transferred the chief point of its respiratory functions from its gills to its tail.

Experiments by Haddon (1889), in which mudskipper tails were covered with gold sizing, appeared to confirm the respiratory function; coated fish survived a shorter time in air. Other observations have ranged from consideration of whether or not mudskippers are obligatory air breathers (Day, 1868, 1877; Sowerby, 1923) to pectoral fin structure in relation to terrestrial locomotion (Harris, 1960), specializations for aerial vision (Karsten, 1923) and to various aspects of their aerial respiration and metabolism (Willem and Boelaert, 1937; Hillman and Withers, 1987). Mudskipper territoriality, terrestrial orientation, and feeding behavior have all been examined (Stebbins and Kalk, 1961; Macnae, 1968a,b; Nursall, 1981; Sponder and Lauder, 1981; and Gordon *et al.*, 1985).

Studies of metabolic adaptations for air exposure include Harms' (1935) idea, rejected almost immediately, that the thyroid gland played a key role in unlocking latent evolutionary adaptations for terrestrial life present in all water-dwelling fishes. Harms (1935, and earlier experiments) claimed that if various species of *Periophthalmus* (and *Blennius)* were exposed to water containing thyroxin or if this compound was injected into them, they reacted by becoming "adapted to land," they left water for greater periods, their skin thickened and their pectoral fins became longer. Finally, according to Harms, the injected thyroxin

TABLE 2.3 Recognized Species of Mudskippers and Other Air-Breathing Species in the Subfamily Oxudercinae[a]

Periophthalmus (12 species, all likely amphibious air breathers)		
chrysospilos	*kalolo*	*novemradiatus*
malaccensis	*novaguineaensis*	*waltoni*
modestus	*weberi*	*barbarus*
gracilis	*argentilineatus*	*minutus*
Periophthalmodon (3 species, all probably amphibious air breathers)		
freycineti		
schlosseri		
septemraidatus		
Boleophthalmus (all 5 species likely amphibious air breathers)		
boddarti		
caeruleomaculatus		
dussumieri		
pectinirostris		
birdsongi		
Scartelaos (amphibous behavior seen in these fishes)		
histophorus		
tenuis		
cantoris		
gigas		
Pseudapocryptes lanceolatus (amphibious and also gulps air)		
Apocryptes bata (amphibious and also gulps air)		

[a]Based on the taxonomic revison of Murdy (1989), with notes on amphibious behavior or air breathing.

TABLE 2.4 Genera and Species of Mudskipper from Different Localities That Have Been Subjected to Experimental or Behavioral Observations Related to Air Breathing or Other Aspects of Terrestriality[a]

Genus and species	Locations	Reference	Species name (Murdy, 1989)
Periophthalmus			
P. koelreuteri	Nigeria	Harris, 1960	*kalolo*
	India	Natarajan and Rajulu, 1983	*kalolo*
	Indonesia	Schöttle, 1931	*kalolo*
	Sri Lanka	Karsten, 1923	*kalolo*[b]
P. sobrinus	East Africa	Stebbins and Kalk, 1961	*argentilineatus*
	Madagascar	Gordon *et al.*, 1969	*argentilineatus*
		Teal and Carey, 1967	*argentilineatus*[b]
		Brillet, 1975, 1976, 1986	*argentilineatus*[b]
	Mozambique	Macnae and Kalk, 1962	*argentilineatus*[b]
P. barbarus	India	Lele and Kulkarni, 1939	[c]
	[d]	Hillman and Withers, 1987	*argentilineatus*
P. pearsei	India	Singh and Munshi, 1969	*novemradiatus*
P. cantonensis	Japan	Morii, 1979	*modestus*
		Iwata *et al.*, 1981	*modestus*
		Tamura and Moriyama, 1976	*modestus*
	Hong Kong	Gordon *et al.*, 1985	*modestus*
	India	Pearse, 1932	*modestus*[b]
P. chrysospilos	Singapore	Lee *et al.*, 1987	*chrysospilos*
		Low *et al.*, 1988, 1990	*chrysospilos*
P. schlosseri	Singapore	Siau and Ip, 1987	[e]
	Indonesia	Schöttle, 1931	[e]
	India	Venkateswarlu, 1966	*Pn. schlosseri*[b]
P. gracilis	Queensland	Gregory, 1977	*gracilis*
		Nursall, 1981	*gracilis*
P. expeditionium	Queensland	Gregory, 1977	*noveaguineaensis*
		Nursall, 1981	*novaeguineaensis*
		Milward, 1974	*novaeguineaensis*[b]
P. vulgaris	Queensland	Nursall, 1981	*argentilineatus*
		Dall and Milward, 1969	*argentilineatus*
		Milward, 1974	*argentilineatus*[b]
	India	Singh and Munshi, 1968	[f]
	Indonesia	Schöttle, 1931	[g]
	Malaysia	Macintosh, 1979	*argentilineatus*[b]
P. dipus	Indonesia	Schöttle, 1931	*argentilineatus*
P. sp.	Queensland	Nursall, 1981	*minutus*
Periophthalmodon			
Pn. australis	Queensland	Bandurski *et al.*, 1968	*freycineti*
		Garey, 1962	*freycineti*
Pn. schlosseri	Malaysia	Fenwick and Lam, 1988	*freycineti*
	Singapore	Low *et al.*, 1988, 1990	*schlosseri*
	Queensland	Nursall, 1981	*freycineti*
		Milward, 1974	*freycineti*[b]
	Andaman Is.	Yadav *et al.*, 1990	*P. minutus*
Boleophthalmus			
B. chinensis	Japan	Tamura *et al.*, 1976	*pectinirostris*
B. pectinirostris	Japan	Morii, 1979	*pectinirostris*
B. boddarti	Singapore	Siau and Ip, 1987, Low *et al.*, 1990	*boddarti*
	India	Biswas *et al.*, 1981	*boddarti*
	Kuwait	Clayton and Vaughan, 1986, 1988	*dussumieri*[b]
		Hughes and Al-Kadhomiy, 1986	*dussumieri*

(continues)

TABLE 2.4 *(continued)*

Genus and species	Locations	Reference	Species name (Murdy, 1989)
B. dentatus	India	Hoda and Akhtar, 1985	*dussumieri*[b]
		Soni and George, 1986	*dussumieri*[b]
Scartelaos			
S. histophorus	Queensland	Gregory, 1977	*histophorus*
S. viridis	Singapore	Macnae, 1968b	*histophorus*
	Kuwait	Tytler and Vaughn, 1983	*tenuis*

[a]Right-hand column indicates correct species name (synonym) based on Murdy's (1989) taxonomic revision, data for geographical distributions, and information provided by Clayton (1993).
[b]Synonymy designated by Clayton, 1993.
[c]*P. barbarus* not found in India; three species found there are *chrysospilos, novemradiatus,* and *argentilineatus.*
[d]No locality data.
[e]Murdy regards *P. schlosseri* as synonymous with *Pn. schlosseri.* Assuming workers did identify the genus *Periophthalmus* correctly, the only species listed ranging through both areas are *argentilineatus, chrysospilos,* and possibly *minutus.*
[f]*P. vulgaris* does not occur in India; the three possible species are listed in [c] (above).
[g]*P. vulgaris* does not occur in Indonesia. Likely species in this locality are listed in [e] (above).

stimulated the fishes' own thyroid gland to take over and sustain the process of terrestrial adaptation.

Concerning the adaptation of respiratory surfaces for air breathing, Schöttle's (1931) classic study of mudskipper comparative respiratory morphometrics and respiration first identified the trend for a reduction in gill area with increased terrestriality. She further found that the loss of gill area in *Periophthalmus* and *Boleophthalmus* was paralleled by the proliferation of large areas of heavily vascularized epithelium, especially in the pharynx. Whereas buccal capillaries in *Gobius* were diffuse and buried below the epithelium, Schöttle (1931) observed that these were more numerous and penetrated the epithelium in *Boleophthalmus.* In *Periophthalmus schlosseri,* dense capillary beds are actually within the epithelium where they formed furrows that increased surface area.

Numerous morphological studies have confirmed Schöttle's early findings for mudskippers. Singh and Munshi (1969) found *Periophthalmus vulgaris* to have a reduced number of relatively thick secondary gill lamellae, while both the buccopharyngeal epithelium and branchiostegal membrane of this fish are vascularized. Tamura *et al.* (1976) added functional data showing differences in the branchial and cutaneous aerial respiratory capacities of *Boleophthalmus* and *Periophthalmus.* The gills of *B. chinensis* play a greater role in its aerial respiration than does its skin (59 versus 43%); whereas, in *P. cantonensis* the skin and not the gills is the site of most of the aerial O_2 uptake (76 versus 27%). Tamura and colleagues concluded that the gills are phased out in the transition to terrestrial air breathing. Related to this, Low *et al.* (1988, 1990) demonstrated a direct correlation between the degree of terrestriality and gill modification among species of *Boleophthalmus, Periophthalmus,* and *Periophthalmodon.* All of these observations suggest that both the problems of gill collapse and desiccation during air exposure are precluded for mudskippers by the development of a respiratory surface better suited for aerial gas exchange and likely useful for aquatic respiration as well (Graham *et al.*, 1987; and Chapter 3).

A series of papers by Gordon and co-workers (Gordon, 1970; Gordon *et al.*, 1965, 1968, 1969, 1978, 1985) focused on physiological and metabolic aspects of air exposure. Harms (1935) and other early workers had experimented with mudskipper survival times in air. Gordon *et al.* (1968) determined that *P. sobrinus* lasted up to 37 hours in air of 70 to 80% relative humidity, passively losing water at a rate of 1.8% of initial body mass/h to a lethal point of about a 22% reduction of initial body mass. As with other species, mudskipper survival times in air were reduced by direct sunlight. There is no support for either Hickson's (1889) conjecture or Haddon's (1889) report that mudskippers survive in air by keeping their tails immersed for aquatic respiration; however, the mudskipper's vascularized skin enables cutaneous respiration both in water and air (Chapter 3). Schöttle (1931) noted the greatest number of vascularized papillae along the tail of *Boleophthalmus* and dense concentrations of capillaries between the body scales of *Periophthalmus* and *Periophthalmodon.* For *Periophthalmus sobrinus,* Teal and Carey (1967) found that cutaneous respiration accounted for 50 to 60% of total aerial and about 40% of total aquatic respiration. Tamura *et al.* (1976) measured similar proportions of skin $\dot{V}O_2$ in *P. cantonensis* (48% in water, 76% in air) and *B. chinensis* (36% water, 43% air).

A common observation for mudskippers is that

they spend most of their time out of water (Stebbins and Kalk, 1961; Nursall, 1981), and some species have been demonstrated to actually prefer air to water (Gordon *et al.*, 1985). To determine how the aerobic metabolism of mudskippers is affected by long-term exposure to air, indices of metabolic stability such as aerial $\dot{V}O_2$ and aerobic scope, heart rate, and metabolite levels have been investigated in a number of species (see Chapters 5 and 8). When kept in air *P. sobrinus* is able to maintain its $\dot{V}O_2$ at the same level as in water, does not alter its heart rate, and does not accumulate lactate (Gordon *et al.*, 1968). Data on the aerobic and anaerobic capacities of mudskippers in air and water also have a bearing on this. Teal and Carey (1967) found that *P. sobrinus* reduced its $\dot{V}O_2$ in hypoxic air and, following the return to normoxia, consumed O_2 at a higher rate to repay an O_2 debt. Bandurski and colleagues (1968) showed that *Periophthalmodon australis* accumulates lactate in its brain and muscle tissue during vigorous activity and when exposed to anoxia. However, Gordon *et al.* (1969) found no lactate accumulation in air-exposed, resting *P. sobrinus,* and in contrast to the above, it was subsequently reported (Gordon, 1978) that *P. vulgaris* had an unusually short tolerance of anoxia and produced very small amounts of lactate in response to this treatment. Recent experiments (Chew *et al.*, 1990; Ip *et al.*, 1991) indicate that hypoxia exposure depresses metabolic activity and thus does not result in elevated lactate. This suggests that lactate formed during vigorous activity (i.e., struggling at the start of an experiment) may be a complicating factor in some of the earlier hypoxia studies (Clayton, 1993).

Hillman and Withers (1987) found that during forced terrestrial activity, *Periophthalmus barbarus* could increase its aerial $\dot{V}O_2$ by a factor of 3.1 over resting $\dot{V}O_2$. This metabolic scope is in the range observed for most ectotherms, including fishes in water and urodele amphibians. It is also similar to the $\dot{V}O_2$ differences observed by Teal and Carey (1967) between resting and struggling *P. sobrinus.*

Accounts differ regarding the need of emergent mudskippers to carry water in their branchial chambers (Graham, 1976a). Stebbins and Kalk (1961) and Macnae (1968a,b) reported that *P. cantonensis* needed to carry water in its chamber. Emerged fish were observed to regularly engage in a head-rolling behavior thought to have the effect of sloshing water throughout the chamber, as well as to periodically return to the water's edge to replenish the volume, particularly after feeding. By contrast, obligatory water retention was not found for *P. sobrinus* (Gordon *et al.*, 1969), and both feeding and cineradiographic studies with *P. koelreuteri* showed it also did not hold branchial water (Sponder and Lauder, 1981).

Although no mudskippers have been reported to be obligatory air breathers in normoxic water (Biswas *et al.*, 1979 [sometimes cited as Niva *et al.*]), some species respond to submersion as though aquatic life is alien to them. Garey (1962) found that when he submerged *Periophthalmodon australis* its heart rate actually decreased. This is the only known occurrence of a "diving bradycardia" in fishes and was noted again for this species during tests with anoxia tolerance (Bandurski *et al.*, 1968). Bradycardia in water correlates well with the behavior of some mudskippers which pass high tides in burrows that are partially hypoxic (Garey, 1962; Macnae, 1968a,b). Nevertheless, we do not know how mudskippers endure long exposure to hypoxia, such as during high tide or extended periods of egg guarding (Gordon *et al.*, 1985). Part of the answer may be that mudskipper burrows are not especially hypoxic (Clayton, 1993). It is also possible that mudskipper behavioral preferences and air-breathing physiology are subject to change with conditions in its microhabitat. Thus, a fish in a flooded burrow (or one containing hypoxic air) might reduce its $\dot{V}O_2$ (see preceding references), periodically emerge to obtain O_2, or both. Alternatively, Hora (1933) reported digging up estivating *Periophthalmus* and *Pseudapocryptes* (see following) from a dry burrow.

Separate studies with *Periophthalmus cantonensis* suggest that behavioral preferences, do in fact, change with environmental conditions. Using a large respirometer that allowed fish to move freely between air and relatively deep water, Tamura *et al.* (1976), found that *P. cantonensis* (and *B. chinensis*) consumed nearly 70% of its total $\dot{V}O_2$ aquatically, and by implication (no behavioral data were reported) spent most of its time submerged, or at least partially submerged, in water. By contrast, Gordon *et al.* (1985) used behavioral choice experiments to determine that *P. cantonensis* preferred complete air exposure on mud to immersion in shallow dishes of water.

Finally, there are seriously divergent findings regarding the nitrogen metabolism of emerged mudskippers (reviewed by Clayton, 1993, and see Chapter 8). Gordon *et al.* (1969, 1978) reported that both *P. sobrinus* and *P. cantonensis* switched from ammonotelism to ureotelism while emersed, and stored urea was excreted following the return to water. Morii *et al.* (1978, 1979), however, did not confirm this for *P. cantonensis,* finding instead that this species and *B. pectinorostris* both stored ammonia while in air. Gregory (1977) further determined that the activities of the ornithine-urea cycle enzymes in *Scartelaos histophorus, P. expeditionium,* and *P. gracilis* were not sufficient for significant ureotelism. In line with this, results for *P. cantonensis* indicate that ammonia concentrated by emersed fish is detoxified by conversion to free amino acids and stored in body tissues (Iwata *et al.*, 1981).

Two other air-breathing oxudercines are *Pseudapocryptes lanceolatus* and *Apocryptes bata* which reside on mud flats in India. From accounts written for these fishes (Das, 1930, 1933, 1934; Hora, 1935), both species appear to be briefly amphibious and usually maintain water contact by moving with the tide. If they become exposed to air, they can survive by burrowing into the mud and Hora (1933) reports finding estivating *Pseudapocryptes.* In hypoxic water *Pseudapocryptes* gulps air at the surface and holds it against the vascular regions on its inner opercular walls (Das, 1930, 1934). Experiments by Das (1934) reporting *Pseudapocryptes* to be an obligatory air breather, remain equivocal because of uncertain aquatic O_2 tensions (Chapter 5).

Other air-breathing gobies. Several other gobies also breathe air. Todd and Ebeling (1966) showed that when exposed to progressively hypoxic water, *Gillichthys mirabilis* first consumes the O_2 contained in its physoclistous gas bladder to briefly maintain its metabolism before shifting to facultative air breathing. In hypoxic water this fish regularly surfaces for air gulps, which it holds in its mouth, or it may emerge entirely (Todd and Ebeling, 1966; Todd, 1971). In addition, young and mid-sized *G. mirabilis* hold air bubbles in their mouths while floating at the water surface, presumably for purposes of ASR. This species has also been observed to climb out of water onto seaweed. Todd (1968) showed that emerged *G. mirabilis* could orientate to water and additionally suggested that its amphibious exposure may be a mechanism for shedding ectoparasitic trematodes. Two other species, *G. seta* and *Quietula guaymasiae,* have similar, but apparently less well-developed, aerial-respiratory capacities (Todd and Ebeling, 1966).

Brief amphibious excursions were observed for the Indo-Pacific intertidal goby *Kelloggella cardinalis* by Larson (1983). Along the Pacific coast of Panama, the mud flat species, *Gobionellus saggitula,* can been seen feeding above the water line during low tides, but there is no information about its air breathing (Todd, 1976). Kramer's (1983a) survey of the hypoxia responses of Panamanian fishes showed that the freshwater gobies *Awaous transandeanus* and *Evorthodus lyricus* would occasionally gulp air while conducting ASR in hypoxic water and more data are needed.

A survey (Gee and Gee, 1991) of 14 Australian gobies that used ASR in hypoxia revealed that nine of these *(Favonigobius exquisitus, Pseudogobius olorum, Arenigobius bifrenatus, A. frenatus, Cryptocentroides cristatus, Mugilogobius paludis, M. stigmaticus, M.* sp. 9, and *Chlamydogobius* sp. nov.) would also gulp air on a regular basis. Whereas respirometry is needed to confirm O_2 consumption from these gulps, histological studies by Gee and Gee (1995) suggest O_2 consumption. They found capillary rich patches in the buccal areas contacting the bubble held in *Cryptocentroides, Mugilogobius, Chlamydogobius,* and *Arenigobius* and that, in most cases, the development of these vascular patches was enhanced in hypoxic water. These workers also also noted that *Mugilogobius, Chlamydogobius,* and *Arenigobius* have capillaries close to the body surface in regions exposed to air during ASR. It is thus likely that air gulps taken by these gobies facilitate ASR (Burggren, 1982; Gee, 1986), and contribute to aerial respiration via the buccal epithelium and skin (Gee and Gee, 1991, 1995). Merrick and Schmida (1984) have also suggested amphibious aerial respiration for *Chlamydogobius eremius,* which occurs in the desert of central Australia and has been seen to remain emerged from warm artesian spring pools for long periods.

Gobioididae. This family of eel-shaped gobies numbers about 19 species in eight genera. These carnivorous fishes are small but quite long and slender, and some species lack scales. Gobioidids are found worldwide in tropical habitats, occurring mostly in shallow marine, brackish, and freshwaters, usually burrowing in soft silt. Hora (1935) remarked that the behavior of *Taenioides rubicundus* is much like that of *Apocryptes* and *Pseudapocryptes* (previous) and his figure 2 shows several *Taenioides* floating at the water surface of a bucket of foul water, having inflated their branchial chambers with air.

Suborder Mastacembeloidei

Mastacembelidae. This family, called the ditch— or spiny—eels (the aquarium trade refers to them as tire tread eels), occurs throughout tropical Africa and southern Asia. There are about 60 species and several genera; *Macrotremus* and *Mastcembelus* are most frequently mentioned in connection with life out of water. Like the clariids, the mastacembelids of tropical Africa have become greatly diversified (Lowe-McConnell, 1987). These eel-shaped fishes are predators, and various species range in maximum length from 30 cm to 1 m. Mastacembelids live in both fresh and brackish waters and occur in thick vegetation or burrow into sand or mud. They have an elongated snout and tubular nostrils which are left projecting when they burrow. Little is known about the air breathing capacity or behavior of these fishes. In a 1979 letter to D.L. Kramer, Isaac Isbrucker reported seeing a captive *Mastacembelus erythrotaenia* emerge from water and move about the terrestrial part of a large tank. The capacity of *Macrognathus (Rhynchobdella)* and *Mastacembelus* to endure long-term confinement in the mud of dry stream beds and ponds in

India was documented by Day (1877). He had at first thought the fish to be moribund but found they revived when placed in water and concluded that they were estivating. Day's early work suggested two species of *Mastacembelus, (pancalus* and *armatus)* and *Macrognathus aculeatus* were obligatory air breathers. However, this was not verified in better controlled experiments by Ghosh (1934) and Hora (1935).

By all accounts, including my observations of specimens in the British Museum, mastacembelids lack an ABO (Hora, 1935; Job, 1941; Sufi, 1956). They have been seen to gulp at the surface; however, Hora (1935) states that inspired air is passed through the gills and released immediately. Such observations could be interpreted to mean that, rather than air breathing, some mastacembelids have been seen to occasionally gulp air while conducting ASR in hypoxia. Experiments by Kramer and McClure (1982), however, showed that *Mastacembelus circumcinctus* and *Macrognathus aculeatus* did not gulp air or perform ASR very effectively in hypoxic water and subsequently died. On the other hand, several species in this family can respire aerially when emerged and appear to tolerate periodic habitat drying, possibly by depressing their metabolism and estivating in the mud. Mittal and Munshi (1971) found that the skin of *Mastacembelus pancalus* is relatively thin and contains numerous epithelial capillaries and thus may be suited for aerial gas exchange (Chapter 3).

Suborder Anabantiodei

The anabantoids are called "labyrinth fishes" because every species in the entire suborder possesses paired suprabranchial chambers that function for air breathing (Das, 1927; Bader, 1937; Peters, 1978). There are about 70 species of anabantoids placed in 16 genera and five different families. The families Anabantidae and Belontiidae are diverse and widely distributed. The other three, Helostomatidae, Osphronemidae, and Luciocephalidae, are monospecific and have narrow distributions within southern Asia. As with some other air-breathing species (e.g., *Channa* and *Clarias*), the hardiness of these fishes and their value as a human food source has resulted in extended distributions (e.g., *Anabas testudineus, Osphronemus gouramy*).

Early discussions of the form and possible function of the anabantoid labyrinth organ are contained in Peters (1846, 1853), Zograff (1886, 1888), and Henninger (1907). The organs (= suprabranchial chambers) are positioned behind and above the gills with the bones of the skull and operculum forming the roof and walls, and the muscles of the jaw composing the floor. Valves at chamber apertures largely isolate them from the adjacent buccal and pharyngeal cavities which means, that because the entire volume of inspired air is held within the suprabranchial chamber, aerial respiration is unaffected by processes such as feeding and aquatic ventilation (Liem, 1987). Each chamber houses a bony labyrinth, a complex structure with a large surface area. Both the labyrinth surface and the inner chamber wall are lined with a respiratory epithelium.

Anabantoids have a variety of air-breathing specializations, from the presence of a divided ventral aorta that separates afferent blood into systemic and ABO streams (Olson *et al.*, 1986, and Chapter 4) to a well developed ABO and a specialized air-ventilatory mode (Peters, 1978). Peters (1978) found that these fishes use either a monophasic or biphasic air ventilation pattern. In the monophasic pattern, old air is displaced through the ABO and out the opercula by the inward flow of the newly inspired breath. In biphasic breathing, the ABO is first flushed by a reversed stream of water (entering via the opercula) and then refilled with freshly inspired air. Combining both cineradiographic and electromyographic data and taking into account factors such as water depth, body posture, and branchial muscle actions, Liem (1987) embellished Peters' ventilatory classification scheme to triphasic (= monophasic) and quadruphasic (= biphasic). Both these workers were of the opinion that, because water is not needed to flush the ABO in the triphasic respiratory mode, anabantoids using this pattern would be more efficient terrestrial air breathers and thus more likely to undergo terrestrial excursions (Peters, 1978; Liem, 1987).

All anabantoids are continuous aquatic air breathers, and this process appears obligatory for most species. However (see Chapter 3), suprabranchial organ development varies among families and species in a manner that affects air breathing and several genera appear to be non-obligatory air breathers (Pinter, 1986). An observation made for the anabantoids thus far examined is that they begin gulping air well before their ABO is fully formed (Peters, 1978; Prasad and Prasad, 1985). Also, Bader (1937) made the important discovery that variation in ABO development was related to the ambient conditions experienced by young fish during the 20 or so days after hatching and while ABO development is taking place. Bader found that *Macropodus* kept in aerated water without access to air did not drown and that fish treated in this manner failed to develop a complete ABO.

Anabantidae. This family, known as the climbing gouramis, contains three genera, *Anabas* (1 species), *Ctenopoma* (at least 25 species) and *Sandelia* (2). Air-breathing has been well-studied in *A. testudineus* (= *scandens*) which occurs from Pakistan, throughout India and southeast Asia to the Phillipines and Taiwan. This species is found in all types of habitats, even brackish waters, and is primarily carnivorous; although, it will eat rice. It reaches a maximum length of about 25 cm and is the most primitive anabantoid, having both the largest suprabranchial organs and the greatest labyrinth area relative to body mass of all the anabantoids. *Anabas* (Figure 2.11) is an obligatory air-breather (Hughes and Singh, 1970a,b). Air breathing commences about 20 days after hatching, but before the ABO is fully formed (Singh and Mishra, 1980), and leads to a reduction of the mass exponent for aquatic $\dot{V}O_2$ by about 40% (Mishra and Singh, 1979).

Anabas can endure air exposure for long periods provided it remains moist. When trapped in pools that become dry, this fish will dig into the mud and estivate for as long as the entire dry season (Das, 1927). In some situations, *Anabas* will move across land apparently in search of water, its terrestrial locomotion being assisted by spinous projections on the operculum and pectoral fins. Overland migrations thus occur through tall grass and take place at night or after rainfall (Das, 1927; Liem, 1987), with distances up to 180 m being covered by daybreak (Pinter, 1986). *Anabas* ventilates its ABO triphasically (i.e., without using the flushing actions of water) and thus has a better tolerance of air exposure than other anabantoids which, with the exception of *Ctenopoma* and *Helostoma* (see following), are quadruphasic ventilators (Liem, 1987). Das (1927) reported seeing emerged fish searching for earthworms and also engaging in "male-male interactions and courtship." Liem (1987) did not observe feeding on the part of emerged *Anabas;* however, his laboratory studies did confirm that while emergence could not be initiated by aquatic hypoxia, factors such as starvation and habitat crowding (inter- and intraspecific interactions, including agonistic behavior by *Oreochromis*) did increase terrestriality.

The name "climbing perch" derives from early reports by local people that *A. testudinus* in India climbed palmyra trees (*Borassus flabellifer*) in order to suck their juice. Missionaries and other reliable sources reported fish crawling along the ground after rains (Daldorff, 1797), hanging from trees, or living in water-filled slits in the bark (Mitchell, 1864). However, to quote from Olson *et al.* (1986),

> [Its] reputation for terrestrial exploits into the branches of trees has far exceeded its physical capabilities.

There are no direct observations of tree climbing by *Anabas,* and Das (1927) suggests that, because these trees occur along the edges of the water and are partly submerged in the wet season, fish may frequent the branches and become isolated when the water recedes, or that they might be captured by kites or crows and deposited in the trees for later consumption.

Both *Ctenopoma* and *Sandelia* occur in Africa. Known generally as African climbing perch, the numerous species of *Ctenopoma* are all carnivorous. These fishes range in maximum length from 8 to 20 cm and occur throughout the tropical parts of the continent (Pinter, 1986). Some species live in hypoxic habitats where air breathing would be required for survival, but others do not, and it seems that few *Ctenopoma* are obligatory air breathers (Pinter, 1986). Liem's (1987) observation that *Ctenopoma* air breathes facultatively (and triphasically) is consistent with the

FIGURE 2.11 *Anabas testudineus,* about 15 cm, India.

ability of at least two species, *C. kingsleyae* and *C. multispinnis,* to endure confinement in mud for dry periods and migrate overland (Pinter, 1986).

The two species of *Sandelia, S. capensis* and *S. bainsi,* occur in southern Africa where they grow to a maximum length of about 18 cm and are carnivorous (Pinter, 1986). Both species have highly reduced labyrinth surfaces (Barnard, 1943). Peters (1978) is of the opinion that the surface area of these structures is barely sufficient to support the minimal rate of aerial respiration. Both Liem (1963a) and Beadle (1981) interpret the reduced dependence upon air breathing in *Sandelia* to be the result of anabantid radiation into habitats where the risk of exposure to hypoxia is less.

Belontiidae. Many belontiids are popular aquarium fish. The more than 45 species in this family are divided among three subfamilies and placed in 12 genera (Table 2.1). These genera, their common names (if any), number of species, and maximum body length are as follows: *Belontia* (comb tail gourami or paradise fish, 2 species, 13 cm), *Macropodus* (paradise fish, 3 species, 8 cm), *Colisa* (various designations of gourami, 4 species, 5–12 cm species range), *Trichogaster* (various designations of gourami, 4 species, 4–15 cm species range), *Parasphaerichthys* (1 species, *ocellatus,* 3 cm), *Sphaerichthyes* (various designations of gourami including chocolate, 3 species, to 6 cm), *Betta* (various designations of fighting fish, 16 species, 6–11 cm), *Ctenops* (1 species, *nobilis,* 10 cm), *Mulpulutta* (spotted pointed tail gourami, 1 species *kretseri,* 9 cm), *Parosphromenus* (various designations of gourami, 4 species, 3–4 cm), *Pseudophromenus* (2 species, *cupanus* and *dayi,* 7 cm), and *Trichopsis* (talking or croaking gouramis, 3 species, 7 cm). According to Pinter (1986) the belontiid diet varies broadly. *Betta* and *Colisa* are carnivorous, feeding on fish, shrimp, and insects at and above the water surface; one Indian species of *Colisa* also reportedly feeds in a manner similar to the archerfish *(Toxotes). Belontia, Trichopsis,* and *Macropodus* are omnivorous, whereas *Trichogaster* is primarily herbivorus.

Belontiids range from India, throughout southern Asia and into Korea. They occupy swift streams, brackish lagoons, potholes, and ditches and extend into areas with dense aquatic or overhanging vegetation where, because hypoxia is common, air breathing is advantageous (Das, 1927; Pinter, 1986). Some species expand into new areas with seasonal floods, and Das (1927) speculated that species of *Betta* have extended their distribution to above waterfalls by making short overland migrations. Field investigations by Liem (1987) did not uncover any evidence of terrestrial behavior by *B. splendens* and in laboratory studies *B. splendens, Macropodus opercularis,* and *Trichogaster trichopterus* (all quadruphasic breathers) did not emerge from water even when exposed to high temperature, hypoxia, starvation, and overcrowding.

Knowledge of belontiid air breathing is incomplete, being based on only a few studies and information gleaned from the aquarium literature. Pinter (1986) suggests that, while all belontiids are air breathers, most, including *Belontia, Mulpulutta,* and *Parosphromenus,* are non-obligatory. The low air-breathing frequency of *Colisa lalia* in normoxia (Wolf and Kramer, 1987) suggests it is also not an obligatory air breather. On the other hand, air breathing by both *Trichogaster trichopterus* (Burggren, 1979) and *T. pectoralis* (Natarajan and Rajulu, 1982) is obligatory. Inevitably, discrepancies exist in the designation of certain belontiids as either obligatory or non-obligatory. *Colisa fasciata,* for example, was described by Ojha *et al.* (1977, as *C. fasciatus*) as a non-obligatory air breather, while Mustafa and Mubarak (1980) report it to be obligatory, a status also implied in the data of Prasad and Singh (1984, as *C. fasciatus*). This could reflect habitat influences on ABO development (Bader, 1937) or may be due to either the incorrect identification of a species (several Indian belontiids that were at one time placed in the genus *Trichopsis* are now in the genus *Colisa*) or inadequate experimental design.

Prasad (1988) examined the ontogeny of respiratory partitioning in *C. fasciata* and found that air breathing begins at about 30 mg body mass (12 mm length), which is before its ABO is fully developed, and that juvenile fish used air for about 36% of their total $\dot{V}O_2$. Investigations with *Trichogaster trichopterus* (Burggren, 1979; Burggren and Haswell, 1979) reveal that the fraction of total $\dot{V}O_2$ acquired via the ABO (42% in normoxic water) increased in hypoxia and that the partitioning of CO_2 loss between air and water in this species also depends on aquatic conditions and is independent of that for O_2. *T. trichopterus* tolerates air exposure for 6–7 hours while maintaining its $\dot{V}O_2$ at near control rates by hyperventilating. The skin of air-exposed fish also accounts for between 10 and 12% of total $\dot{V}O_2$ (Burggren, 1979). Based on experiments with carbonic anhydrase and its inhibitor acetazolamide, Burggren and Haswell (1979) concluded that the branchial derivation of the ABO in *T. trichopterus* endows it (and presumably all anabantoids) with a large concentration of carbonic anhydrase to facilitate aerial CO_2 release (Chapter 5).

Energetic aspects of air breathing related to aquaculture were examined by Ponniah and Pandian (1977) who determined the combined effects of ration size, air-breathing frequency, and the requisite swimming distance for air on the growth of *Macropodus (Polyacanthus) cupanus.* Fish (0.5 g, 3.4 cm) fed the maximum ration (174 cal/g day) and kept in normoxic water (15 cm deep) surfaced for air 1383 times/day (a distance of 415 m); whereas, starved fish surfaced for air 439 times/day (132 m). Fish on the maximal ration, but in hypoxia (81 torr), made 2165 surfacings/day (650 m). Based on conversion-efficiency estimates and swimming costs for surfacing, it was determined that the optimal ration size of *M. cupanus* was 84 cal/g day. Bevan and Kramer (1986) also found that depth had a negative effect on the air-breathing frequency of *Colisa chuna.* Additional studies with *M. cupanus* (Jacob *et al.*, 1982) show that its air breathing permits it to absorb less amounts of toxic pesticides than non-air-breathing fishes making it more effective in the biological control of insects.

The effect of predatory danger on air breathing was investigated by Wolf and Kramer (1987) who found that in the presence of a predator (the air-breathing *Channa micropeltes*), *Colisa lalia* breathed air less frequently and spent more time within the protective cover of a clump of artificial plants. In experiments using a Green Heron *(Butorides striatus)* as a predator, Kramer *et al.* (1983) found that the air-breathing frequencies of *Trichogaster trichopterus* and *Macropodus opercularis* were reduced when the bird was permitted to forage in the tank and that fish avoided surfacing for air in areas near the bird. Species of both *Trichogaster* and *Colisa* have been shown to be synchronous air breathers (Forselius, 1957; Kramer and Graham, 1976; Fitzpatrick, unpublished ms). One air-breathing-related behavior that remains relatively uninvestigated is sound production. In talking gouramis *(Trichopsis)* audible tones emanating from the ABO play a role in establishing male territories and dominance. Lastly, as was observed by Pinter (1986), most belontiid activities, including territorial defense, ritualized behavior, and courtship displays are always subject to interruption by the need for a fish to "take an air breath."

Helostomatidae. This monotypic family contains the kissing gourami, *Helostoma temmincki,* which occurs throughout southeast Asia and reaches a maximum length of 30 cm. The common name of this fish stems from conspecific "face to face" behavioral displays seen in aquaria. These actions are accentuated by its enlarged lips, a specialization for algal grazing. Another feeding specialization is closely spaced gill rakers which act as a sieve for plankton feeding (Liem, 1967a).

There are few studies on the air breathing of this species. Kramer *et al.* (1983) found that in response to the Green Heron, *H. temmincki* reduced its air-breathing frequency and generally moved to areas away from the bird. Liem (1987) reported that this species was capable of either quadruphasic or triphasic air ventilation; however, neither laboratory nor field observations have ever indicated amphibious behavior.

Osphronemidae. The only species in this family is the giant gourami, *Osphronemus goramy,* a herbivore which grows to a maximum length of about 60 cm. A native of Sumatra, this species has been introduced into tropical habitats around the world because of its tasty flesh. Air breathing by this species was first documented by Day (1877); however, there is little additional information. Natarajan (1980) examined several blood-respiratory properties (Chapter 7) of this species (as *O. olfax*—a specific name reflecting the early idea that the suprabranchial chamber was an olfactory device) in relation to air breathing. He also reported that *O. goramy* is an obligate air breather that takes air at intervals of from 5 to 25 minutes, depending upon aquatic O_2 level, and normally obtains 69% of its O_2 requirement aerially.

Luciocephalidae. The single member of this family is the pikehead gourami, *Luciocephalus pulcher,* a carnivorous species that grows to about 18 cm and occurs from the Malay Peninsula to Sumatra. This fish is a quadruphasic air ventilator (Liem, 1987), and its ABO differs from other anabantoids in lacking a pharyngeal opening and having only an opercular opening (Lauder and Liem, 1981, 1983a,b). This difference may relate to the narrow head shape of *L. pulcher.* According to Liem (1987), this morphology has little effect on ABO ventilation; although, as revealed by X-ray cinematography, the pikehead differs from other anabantoids in its ability to hold air both in its ABO and pharyngeal chamber.

There have been no studies of gas exchange in this species, but conflicting impressions exist regarding its capacity for air breathing. Liem (1967b) concluded that the organ had a respiratory function and remarked on the dense vascularization of its lamellae. He also suggested its possible role in sound amplification. Pinter (1986), however, concluded that the suprabranchial organ of *L. pulcher* functions primarily for sound detection and, in contrast to Liem (1967b), suggested its respiratory function could at most be facultative because the organ is structurally much simpler than in other anabantoids, has less surface area, and is poorly vascularized.

Channiformes

Channidae. The snakeheads are all placed in the genus *Channa* (formerly *Ophiocephalus*), and there are at least 12 species. Channids are predaceous and feed mainly on other fishes. Some species reach a maximum length of 20 cm, others about 1.2 m. *Channa* occurs in freshwater throughout tropical Africa, across southern Asia to Korea and Japan, and into southeast Asia. Like the other Asian air-breathing fishes *(Clarias, Anabas)*, snakeheads have been transported widely by humans and now occur on many of the tropical Pacific Island chains, including Hawaii. Live *Channa* are imported or cultured for food by many expatriate Chinese communities.

Channids are valued for food, and their importance in aquaculture has spurred interest in all aspects of their physiology including air breathing. All species breathe air continuously, and this is apparently obligatory for them at their normal tropical and subtropical habitat temperatures; however, both intra- and interspecific variability exists for this (Hora, 1935). *Channa* will also burrow in soft mud to survive temporary drought, and some species will migrate in search of new habitats (Day, 1877; Uchida and Fujimoto, 1933). The tenacity of these fishes for life, due in part to their air-breathing capacity, is revealed by the account of Hamilton Buchanan (1822):

> In China, species of the genus *Ophicephalus* are often carried alive in pails of water, and slices are cut for sale as wanted, the fish selling dear while it retains life, while what remains after death is considered of little value.

Paired suprabranchial chambers located behind and slightly above the gills function as the ABO of *Channa* (Das, 1927; Bader, 1937; Dehadrai, 1962b; Munshi 1962b; Singh *et al.*, 1988). Das (1927) gives a general account of the natural history of Indian *Channa* and describes the morphology and development of the ABO. Included in his analysis is a thoughtful contrast of the degree of ABO development in species that obtain different maximum body sizes and occur in habitats requiring different levels of air-breathing proficiency. *C. striatus,* which attains the largest size, has the most extensive ABO surface area, larger, for example, than that of *C. gachua,* one of the smallest species, which has a relatively smooth-surfaced ABO (Das, 1927).

The suprabranchial chamber of *Channa* forms in the dorsal and medial recesses in the skull and is bordered laterally by the opercular bones. The chambers have respiratory epithelium on their walls as well as on an interior labyrinth structure that has a large surface area. The volume of air inspired by a fish usually exceeds that of the suprabranchial chamber and thus overflows into the dorsal buccopharyngeal cavity which has respiratory epithelium on most exposed surfaces (Munshi, 1962b; Liem, 1984). Thus, unlike all anabantoids except *Luciocephalus,* there is free exchange of air between the suprabranchial and buccopharyngeal chambers with the probability of some respiratory exchange occurring within the latter (Liem, 1963a, 1984; Lauder and Liem, 1981).

Analyses of the air-ventilation pattern of four species of *Channa (punctatus, striatus, gachua, marulius)* by Liem (1984) shows that it is quadruphasic with both body inclination and the flooding of the ABO from behind (via opercula) being integral aspects of exhalation. Thus, as in anabantoids, a need for water limits the terrestrial capabilities of these fishes (Liem, 1984, 1987). Air inhalation is less dependent on a large buccal force-pump pressure and is a rather passive process compared to some other air breathers (Liem, 1984). Air enters the mouth as the buccal cavity expands, the mouth then closes and the buccal cavity is slightly compressed which forces the air back past the gills where it pops up into the ABO, displacing water in the process.

The channid circulatory system is specialized through the presence of two ventral aortae (Ishimatsu *et al.*, 1979; Ishimatsu and Itazawa, 1983a, 1993; also Chapter 4), as in some anabantoids. The two ventral aortae, in conjunction with separate venous returns from the systemic and ABO circulations, assure a small degree of isolation of these two flows to increase both air-breathing and O_2 transport efficiency (Ishimatsu and Itazawa, 1983a).

Studies of *Channa* air-breathing physiology cover a broad spectrum. Anatomy, physiology, and cardiorespiratory function during air breathing are reviewed by Ishimatsu and Itazawa (1993). The ontogenetic development of the ABO and air breathing in two commercially important Indian species *(C. striatus* and *C. punctatus)* was examined by R.P. Singh *et al.* (1982) and B.R. Singh *et al.* (1982b, 1986, 1988). Young *Channa* respire aquatically until about 20 days after hatching (a body length of 11–12 mm; mass 7–10 mg) when they begin bimodal respiration. The transition to air breathing in young fish reduces the exponent relating mass and total $\dot{V}O_2$.

Experiments with *Channa striatus* (Pandian and Vivekanandan, 1976; Vivekanandan, 1976, 1977) examined interrelationships between food availability, hypoxia, temperature, body size, and water depth on assimilation efficiency and growth. Both feeding and hypoxia increased daily air-breathing rate; however, a greater food ration was needed for fish dwelling in deeper (i.e., longer swimming distance to air) tanks. Starvation, by contrast markedly lowered the air-breathing frequency of *C. striatus* (from 900 to 300/day) and starved fish tended to "hang" at the surface, apparently reducing the energetic costs of surfacing for air.

Bimodal gas exchange experiments reveal that all species of *Channa* thus far examined typically obtain from 60 to 85% of their $\dot{V}O_2$ aerially, with this fraction being affected by body mass, activity, temperature, and aquatic PO_2 and PCO_2 (Ojha *et al.*, 1978; Rama Samy and Reddy, 1978; Itazawa and Ishimatsu, 1981; Yu and Woo, 1985; Glass *et al.*, 1986). Aquatic hypoxia, for example, increased the air-breathing frequency but otherwise did not affect the hunting behavior of *Channa micropeltes* (Wolf and Kramer, 1987). Yu and Woo (1987a,b) compared the short- and long-term effects of aquatic hypoxia on the respiration and metabolism of *C. maculata.* Even with its capacity to breathe air, exposure of this species to 10–30 torr hypoxia for one to two hours caused reductions in heart rate, blood pressure, urine output, and hematocrit. After three days, however, complete compensation to hypoxia was seen in all these parameters.

Studies of ventilatory behavior, and the effects of air exposure on acid-base physiology were carried out by Ishimatsu and Itazawa, (1981, 1983b) on *C. argus.* In air, this fish had higher ventilation rates, but a lower ABO–PO_2 and a higher ABO–PCO_2 than did fish in water. Artificial ventilation of the ABO with an air stream, however, eliminated these differences and also relieved hypercapnic acidosis. *Channa gachua* exposed to air for up to 10 hours was found to mobilize the bulk of its liver glycogen stores and transfer these to the heart and body muscle for possible use as an anaerobic energy source (Ramaswamy, 1983). Also, upon its return to water, *C. gachua* excreted a large volume of urea suggesting it becomes ureotelic in air (Ramaswamy and Reddy, 1983). This warrants further study, particularly in light of the recent finding (Saha and Ratha, 1989) that *C. punctatus* lacks the complete set of ornithine-urea cycle enzymes (Chapter 8).

Other investigations connected with aerial respiration have considered the problems caused for the aquaculture of these fishes by insecticide poisoning and parasite infestations (Natarajan, 1981, 1984a; Ravindranath *et al.*, 1985). Finally, the way in which mouth brooding (e.g., by *C. orientalis*) affects air breathing remains uninvestigated.

Synbranchiformes

Synbranchidae. Commonly called swamp eels, the synbranchids are widely distributed in tropical fresh and brackish waters, frequenting ponds, swamps, forest streams, rivers, mats of floating vegetation, and even small ditches and temporary pools. Also, one species, *Ophisternon infernale* is a blind cave dweller. There are fifteen species of swamp eels placed in four genera: *Synbranchus* (2 species), *Ophisternon* (6), *Monopterus* (6), and *Macrotrema* (1). Another synbranchid genus frequently mentioned in the literature is *Amphipnous.* Rosen and Greenwood (1976) have placed *Amphipnous* in synonymy with *Monopterus;* however, as discussed by Nelson (1994), some workers still regard this as a valid genus and member of a separate family, the Amphipnoidae. The latter interpretation is based largely on the presence of pharyngeal pouches for air breathing in only three (formerly *Amphipnous*) species (*M. cuchia, M. fossorius,* and *M. indicus* types 1 and 2).

Because of their amphibious capability, some synbranchids have, like *Rivulus,* become established in habitats beyond the range of most other fishes. In South America and Panama, for example, synbranchids are one of few fishes occurring above major waterfalls (Breder, 1927; Lüling, 1958) and *Synbranchus* (Figure 2.12) has become an ecologically important predator on tadpoles in some high elevation habitats (Zaret and Rand, 1971). Synbranchids are protogynous hermaphrodites, a characteristic enhancing their ability to colonize newly formed habitats or support populations subsisting in severe habitats (Liem, 1963b; Breder and Rosen, 1966).

Adult swamp eels lack paired fins and either lack or have very small scales. They reach a maximum length of about 1.5 m and are voracious, nocturnal predators. *Synbranchus* in Panama feeds at night by methodically searching the shallows and burrows for anything edible. Southeast-Asian *Monopterus* lay in mud burrows near the water line and spring upon terrestrial crabs that wander by, but are, in turn, subject to capture by native fishermen using tethered crabs as bait (Shih,

FIGURE 2.12 *Synbranchus marmoratus* at the surface of a mud burrow. (Photo by L. Ford, Scripps Institution of Oceanography.)

1940). Liem (1987) also reports seeing *Monopterus* feed while out of water.

Synbranchid habitats are subject to hypoxia, and it is probable that most species are capable of air breathing. These fishes also air breathe during amphibious excursions (*S. marmoratus* has been observed out of water on river banks covered by dense vegetation and at night [Lüling, 1969, 1973, 1980; Kramer *et al.*, 1978; Graham, pers. obs.]) and while seasonally confined in humid burrows (Das, 1927; Johansen, 1966; Bicudo and Johansen, 1979). Day (1868) dug estivating *Monopterus cuchia* out of the mud in India. Wu and Kung (1940) reported that *M. albus* tolerates seasonal draining of rice patties and that it can survive several days in market bins without water. Lüling (1973) diagrammed the burrow system of South American *Synbranchus marmoratus* and reported that as the dry season progressed, fish burrowed deeper to remain in contact with the water table. Similar observations for *M. albus* were made by Liem (1987). Experiments with *S. marmoratus* show it can endure six to nine months' exposure in drying laboratory burrows (Bicudo and Johansen, 1979; Graham, unpublished). Fish in these burrows have a reduced metabolic rate, yet they move along the tubes and try to escape if uncovered (Liem, 1987; Graham, pers. obs.).

The air-breathing capacity of synbranchids has been long known. Taylor (1831) described the paired pharyngeal pouches of *Amphipnous cuchia,* concluding that, because the gills were so reduced, the pouches had to function for accessory air breathing. Both Müller (1842) and Hyrtl (1858) examined various details of synbranchid ABO structure and blood circulation in relation to air breathing. Volz' (1906) suggestion that *Monopterus* was capable of intestinal respiration was not substantiated (Wu and Kung, 1940; Liem, 1961, 1967c).

Synbranchids respire aerially using a vascular epithelium that lines the mouth, pharynx, and branchial chambers (Wu and Liu, 1940, 1943; Liem, 1987); the gills also function for air breathing (Carter and Beadle, 1931; Johansen, 1966). Paired pharyngeal air sacs, first described by Taylor (1831), are present in *M. cuchia, M. fossorius,* and *M. indicus* (Rosen and Greenwood, 1976; Liem, 1987; Munshi *et al.*, 1989). The skin has also been shown to function for both aerial and aquatic gas exchange (Liem, 1961, 1981; Lomholt and Johansen, 1976; Heisler, 1982; Graham *et al.*, 1987) as well as for aquatic ion regulation (Stiffler *et al.*, 1986).

The relationship between air breathing and the modification of synbranchid gills was first noted by Taylor (1831). Carter and Beadle (1931) reported a large ratio of gill-fold breadth to thickness in *S. marmoratus* as a specialization for air breathing, but subsequent work has not verified these observations, partly because it is not exactly clear what these authors' meant by "breadth" and "thickness" (Johansen, 1966). Liem (1987, figure 8) chronicles the range of gill development in this family. In *Macrotrema* the gills are fully functional for aquatic respiration. Those of *Synbranchus* are also fully developed (Taylor, 1914; Johansen, 1966) but modified in a manner that assures their function in both air and water (i.e., the hemibranchs have a staggered, asymmetrical origin along the arch and the primary lamellar supports are thick and long). By contrast, the gill filaments of most *Monopterus* have been changed into a respiratory mucosa with most species having few lamella on the first three gill arches and none on arch 4 (Wu and Liu, 1940, 1943; Rosen and Greenwood, 1976; Liem, 1987; Munshi *et al.*, 1989; and Chapter 3). It is not surprising that air breathing seems obligatory for all *Monopterus.* Both *Ophisternon* and *Synbranchus,* however, have a full complement of gills (Rosen and Greenwood, 1976) and air breathing by *S. marmoratus* is facultative (Johansen, 1966; Graham and Baird, 1984).

Swamp eels exposed to air breathe by periodically ventilating their mouths or by holding their mouths open for extended periods (Figure 2.12). In aquatic air breathing the gulped air is held in the mouth, and the inspired volume is large enough to inflate the buccal, pharyngeal, and branchial cavities. The breaths taken are so large that the head of the fish can float (illustrated in Johansen, 1968; Lüling, 1973) unless the body can be levered against a fulcrum (such as the burrow wall) to force its submergence. In shallow water, *Monopterus* often holds its inflated head vertically and motionless just below the surface. This posture, while facilitating air ventilation, also enables the fish to keep its head motionless near the surface and may favor surprise attack on unwary prey that swim past. To prevent the loss of air from the branchial chamber and to prevent mud and other debris from entering the chamber during burrowing, the synbranchid opercular opening (except in *Macrotrema)* is reduced to a single, common ventral slit (Rosen and Greenwood, 1976).

The complete spectrum of air-breathing adaptation is seen in the Synbranchidae, ranging from the apparently non-air-breathing, marine species *Macrotrema caligans* to the facultative air-breathing *Synbranchus* (and probably *Ophisternon* as well) to the obligatory-breathing species of *Monopterus* (Liem, 1987). There are no data for either *Ophisternon* or *Macrotrema,* and most studies have been conducted on *Monopterus* and *S. marmoratus.* Nevertheless, our knowledge of synbranchid air breathing is broad, with studies to date having examined respiratory control and the interaction between air breathing, gill ventilation, and cardiac frequency (Johansen, 1966; Lomholt and Johansen, 1974; Graham and Baird, 1984; Roberts and Graham, 1985; Graham *et al.*, 1995a); the effect of air breathing on blood Hb–O_2 affinity and blood acid-base status (Johansen *et al.*, 1978b; Bicudo and Johansen, 1979; Heisler, 1977, 1982), the role of the elevated synbranchid blood Hb concentration in O_2 storage (Lomholt and Johansen, 1976; Graham, 1985); and the importance of body size in determining the transition to air breathing, the partitioning of aerial and aquatic respiration, and the use of the skin as a respiratory surface (Singh and Thakur, 1979; Graham *et al.*, 1987). Liem (1987) has investigated the role of physical and behavioral-ecological factors in stimulating the terrestrial behavior of *Monopterus,* finding that overcrowding and interspecific interactions (i.e., competition with and aggression by the non-air-breathing species, *Nandus* and *Oreochromis*) triggered more frequent emersion, whereas hypoxia, decreased water level, and increased water temperature had little effect on this behavior.

In tests similar to those done with *Lepidosiren,* Cunningham (1932) reported that male *S. marmoratus,* which also guard nestlings, are capable of exuding aerially obtained O_2 across their skin and into nest water. He also stated that, as a specialization for nest care, males had a greater skin capillary density than did females. There have been no follow-up studies on either of these observations. Larval synbranchids use their vascularized pectoral fins (which are shed in later life) and skin for accessory aquatic gas exchange (Liem, 1981). The swamp eel heart is displaced posteriorly in the body giving these fish an extremely long ventral aorta (Rosen and Greenwood, 1976; and Chapter 4). There have been no ideas forwarded regarding the functionality of this, although my observations suggest it serves both the augmentation of venous return during head-up positioning and the avoidance of cardiac compression during air breathing (which greatly stretches the branchial region including the normal heart position) or during burrowing. Synbranchids also have an enlarged urinary bladder which appears to have a role in water retention during burrow confinement (Liem, unpubl. obs.; Stiffler *et al.*, 1986).

SUMMARY AND OVERVIEW

This overview focuses on air-breathing fish diversity and on existing uncertainties about air breathing in some species. A synthesis of the other information contained in this chapter, found in Chapter 9, treats the presence, type, and degree of development of air breathing in different fish groups in relation to environmental and ecological factors and variables such as species distribution, body size, and behavior.

The Present Estimate of Air-Breathing Fish Diversity

Based on information in this chapter, the number of known air-breathing species is now estimated to be 374 (Table 2.1). These occur in 125 genera and are distributed among 49 families spanning 17 orders of the Osteichthyes. For most of these fishes, evidence for air breathing exists in the form of the presence of an ABO and air-ventilatory behavior or the frequent display of natural emergence behavior. Evidence for some species is, however, not yet conclusive and more work is needed. Included in this category are some of the clariid genera and subgenera (Table 2.2), *Galaxiella, Chirocentrus,* a number of the air-gulping gobies, the cottids, and others. Also, claims of air breathing in *Oxyeleotris* are not substantiated.

The first comprehensive list of air breathers by Carter and Beadle (1931) totaled 28 genera. Most successive counts have increased this. Leiner (1938) listed 41 genera. Other counts were: Bertin (1958), 43; Saxena (1963), 34; Gans (1970), 59. All of these estimates lie well below the current list of 125 genera. The current tally thus reflects the considerable expansion in our knowledge of air-breathing fishes over the past 65 years. It also reflects the expanded definition of the group used here, which, as detailed in Chapter 1, encompasses the amphibious species. Even with the amphibious species the estimate of 374 species is conservative because in cases where several species in a genus are probably amphibious air breathers (e.g., the tropical blennies, Table 2.1), only the documented observations were counted.

In spite of our progress, the present list of air-breathing fishes is incomplete; there is insufficient detail for many families and genera known to contain air breathers. Based on the present state of knowledge, I expect another 100 species of air-breathing fishes will be documented in the next half-century. Most discoveries will come from within families where this adaptation is now known but where details about various species are lacking. The Cobitididae, Pangasiidae, Aspredinidae, Callicthyidae, Loricariidae, Gobiesocidae, Tripterygiidae, Labrisomidae, Blenniidae, Eleotridae, Gobiidae, and the Clariidae will be important in this respect.

Future work must also investigate other families

where air breathing may be present. There are certain groups that, in view of their habitat and diversity could be expected to contain air breathers but these have yet to be discovered. There are numerous species that have either been suggested or reported to be air breathers, however, without documentation. These will now be discussed. Future inquiries into these groups will doubtlessly lead to new discovery.

Families and Species in Which Air Breathing May Be Present

Conspicuously absent from the ranks of known air-breathing fishes are any species from the diverse (330 species) teleostean infradivison Clupeomorpha. Most members of this group, which includes the herring, use a well-developed physostomous gas bladder for buoyancy and sound reception. Surface air gulping is also known to occur; however, air breathing and structural modification of the gas bladder for aerial gas exchange has never been shown for any member of this group (although de Beaufort's [1909] description of *Chirocentrus* has been so interpreted by some workers).

One family that surprisingly lacks air breathing is the Cyprinidae. Rivaling the Gobiidae in total species diversity (over 2000), the cyprinids have species (e.g., minnows and carp) occupying habitats much like those frequented by the air-breathing loaches. Although titled "Air breathing abilities of the common carp," the study by Nakamura (1994) merely documents the capacity of chilled fish to respire cutaneously during their transportation to market. Various poeciliids have also been observed to emerge briefly from water (Baylis, 1982; Lefebvre and Spahn, 1987); however, the terrestrial capacity of these fishes remains unexamined. There are accounts of terrestrial excursions by the four-eyed fishes (Anablepidae). *Anableps anableps* and *A. microlepis* have been observed to crawl out of water and rest on the mud banks (Rauther, 1910), but there are few additional details.

The present analysis suggests that the following species should be investigated for air breathing. First, *Novumbra hubbsi* (Umbridae) is suspected because air breathing is present in both the other genera *(Dallia* and *Umbra)* of this family. More study is also needed for *Cichlasoma bimaculatum* (Cichlidae). Lowe-McConnell (1969) suggested that this species may be able to breathe air because it survives for a long time out of water, and her dissections showed it to have a well-vascularized, air-filled stomach. Finally, as reviewed by Kramer (1978b,c) a number of species spawn either in extremely shallow water or while emerged. Included among these are three air-breathing genera: *Fundulus, Galaxias,* and *Neochanna.* Best known among the terrestrial spawners are the grunion (*Leuresthes tenuius* and *L. sardina,* Atherinidae), which tolerate exposures of several minutes in duration. Other emergent spawners include another atherinid, *Hubbesia gilberti* (the females emerge to shed eggs at high tide); the characins, *Brycon* and *Copella;* the puffer (*Fugu,* Tetraodontidae); the smelt (*Hypomesius,* Osmeridae); and the capelin (*Mallotus,* Osmeridae, only the western North Atlantic population does this and adults die after spawning). The responses of *L. tenuis* to aerial exposure include bradycardia (Garey, 1962), an increase in muscle anaerobiosis (Scholander *et al.*, 1962), and probably minimal rates of aerial respiration, although the latter remains unstudied. Little else is known about the emergent responses of these fishes.

Species for Which Reports of Air Breathing Remain Unsubstantiated

My literature review fails to support claims of air breathing made for various species by the following authors:

Carter and Beadle (1931): *Doras* (Doradidae)

Hora (1933, 1935): *Amblyceps* (Amblycipitidae), *Mystus gulio* (Bagridae), *Pisodonophis boro* (Ophichthidae)

Leiner (1938): *Macrones* (Bagridae), *Sternarchus* (Gymnotidae), *Nemachilus* (Cobitididae), *Otocinclus* (Loricariidae), *Amblyopus, Trypauchen* (Gobiiformes), *Tetrodon* (sic), *Diodon, Triacanthus, Monacanthus* (Acanthopterygii)

Hora and Law (1942): *Amblyceps, Olyra*

Carter (1957): *Doras, Amblyceps*

Bertin (1958): *Girardinus, Lebias, Orestias* (Cyprinodontidae)

Saxena (1963): *Achirus* (Soleidae); *Amblyceps, Rita* (Bagridae), *Doras*

Gans (1970): *Girardinus, Lebias, Orestias, Doras, Neogobius* (?)

Dehadrai and Tripathi (1976): *Doras*

Farmer *et al.* (1979): *Rhamphichthys* (Rhamphichthyidae), *Eigenmania, Sternopygus* (Sternopygidae), *Parahemiodon* (?), *Spatuloricaria* (Loricariidae?)

This is discussed for each group.

Doradidae

Since the paper by Jobert (1878), the South American freshwater genus *Doras* has been included in most lists of air-breathing fishes. Jobert, however, did not make any field observations and did not observe air breathing in *Doras.* He did report that

Doras could survive out of water longer than *Hypostomus,* and his observations of fish in market bins indicated they could survive long periods out of water, even in direct sunlight. His anatomical observations indicated that *Doras* had gas in its intestine (it was not stated how fish were handled prior to dissection); however, the region of the organ where this gas was found had numerous villi and did not seem particularly specialized for respiration. Jobert reported that while some characteristic capillary tufts (houppes sanquines) were present, the intestine of *Doras* seemed much less specialized for air breathing than that of either *Callichthys* or *Hypostomus* for which he made detailed observations.

My observations of doradids in the British Museum did not suggest extreme gut specialization for air breathing. Nevertheless, it seems likely that some doradids are air breathers, which can be easily tested because several genera are marketed by commercial aquarium traders. Sterba (1963) reports that "supplemental internal respiration is wide spread" in this family and lists *Trachycorystes striatulus* as an air breather.

Amblyceps, Mystus, Macrones, Rita, and *Pisodonophis*

These five Indian genera are described as air breathers in various reports. Hora (1935) lists *Amblyceps,* a resident of torrential highland streams, as a facultative air breather that gulps air when exposed to air during periods of low water. Although widely regarded as an air breather with an ABO (Singh *et al.,* 1989a), the literature provides neither documented accounts of the behavior nor a description of the organ. Also, my observations of *Amblyceps* in the British Museum did not indicate adaptations for air breathing.

Various species of *Mystus, Macrones,* and *Rita* (Bagridae) have been consistently referred to as air breathers by numerous investigators (preceding references, also see Ramaswamy and Reddy, 1977, 1983). There are no published details regarding the behavior, ABO, or physiology of these fishes.

Day's (1877) observations on the air ventilation of *Pisodonophis boro* during air exposure have been frequently cited as evidence for amphibious air breathing, which is not the case. Although occurring in shallow rice paddies, *P. boro* is not generally described as an air breather (Subramanian, 1984).

Leiner's (1938) List of Air-Breathing Fishes

The origin of the diverse and rather extensive list of questionable air breathers presented by Leiner (1938) remains uncertain. His list includes several tropical marine genera including the puffers *Tetrodon* (probably *Tetradon)* and *Diodon,* which inflate themselves with air when pulled from water, the filefish *Monacanthus* and triple spine *(Triacanthus),* for which there are no supporting data. Liener's claims for air breathing in the cobitid *Nemachilus,* the gymnotid *Sternarchus,* and the gobiid *Amblyopus* (a synonym for *Taenioides* [Gobioididae, see preceding text]), while reasonable, are all unconfirmed.

Leiner (1938) also listed the eel goby *Trypauchen vagina* (Gobiidae) as an air breather, and while he presented no data, this warrants further discussion. Harms (1935) had also implied the presence of air breathing in *T. vagina* but gave no data. Schöttle (1931) indicated that this fish resides in burrows containing anoxic water and is unlikely to leave water. Her analyses showed that the skin of *T. vagina* contained capillaries and could have an accessory respiratory function. If *Trypauchen vagina* can breathe air, it would, like *Taenioides,* probably do so by surfacing long enough to gulp air and then retreat to its burrow. One characteristic of the Trypauchenidae is the presence of pouched cavities on each side of the body just above the operculum (Nelson, 1994). The function of these structures is unknown; however, they are not connected to the opercular cavity.

Cyprinodontidae

No documentation is known for the various cyprinodontid genera listed as air breathers by Bertin (1958) and Gans (1970). Gans (1970) reported that *Orestias* uses a branchial diverticulum as an ABO; however, there are no supporting data.

Other Species

There are three final observations. First, there are no reports of air breathing in *Olyra* or *Neogobius,* both listed previously, or in *Glyptothorax telchitta,* a species listed as air breathing in other early works. Second, the study by Breder (1941) showed that both *Achirus* and *Chiloscyllium* would resort to ASR in hypoxia. Saxena (1963) apparently concluded this was air breathing. Lastly, of the 17 Amazonian genera listed by Farmer *et al.* (1979) as air breathers, the five listed above remain unconfirmed.

CHAPTER

3

Respiratory Organs

The illustration of the swim-bladder in fishes is a good one, because it shows us clearly the highly important fact that an organ originally constructed for one purpose, namely, flotation, may be converted into one for a widely different purpose, namely respiration. . . . All physiologists admit that the swim-bladder is homologous, or "ideally similar" in position and structure with the lungs of the higher vertebrate animals: hence there is no reason to doubt that the swim-bladder has actually been converted into lungs, or an organ used exclusively for respiration.

C. Darwin, *The Origin of Species*, 1859

INTRODUCTION

The structure of fish air-breathing organs (ABO), their comparative morphometrics, air-ventilation mechanics, tidal volumes, and the timing and sequence of air inhalation and expiration, are the subjects of this chapter. Also considered here are gill and skin morphology in relation to air breathing; in many species these organs serve an aerial and aquatic-respiratory function, either for CO_2 release, auxiliary O_2 uptake, or both.

HISTORICAL

Descriptions of fish ABOs began in the early 19th century. One of the first to be done was for *Clarias (Silurus anguillaris)*, by Geoffroy St. Hilaire (1802a). Taylor (1831) described ABO structure and breathing patterns in *Heteropneustes (Silurus), Channa, Pangasius (Pimelodus), Monopterus,* and also *Clarias (Macropteronotus).* The *Histoire Naturelle des Poissons,* the massive work of Cuvier and Valenciennes published between 1828 and 1849, contains comparative observations on the ABO structure of anabantoids, *Channa,* and several other fishes including the clariids, gars, *Amia, Heterotis,* and the characins. Owen (1841) and Quekett (1844) compared the structure and vascularization of the lungs and gas bladders of several fishes. The Austrian anatomist Joseph Hyrtl detailed aspects of ABO structure and circulation for no less than nine genera: (*Lepidosiren,* 1845 [often cited as 1843-44]; *Lepisosteus,* 1852a; *Channa,* 1853; *Heteropneustes,* 1854a; *Heterotis,* 1854b; *Gymnarchus,* 1856; *Amphipnous* and *Monopterus,* 1858; *Polyacanthus* [= *Trichopsis*], 1863) and *Polypterus* (1870). Early work on the cobitidid respiratory intestine was done by Erman (1808) and Leydig (1853). By the time Rauther (1910) compiled the first comparative survey of ABO morphology, the major features of most organs had been described. Subsequent ABO accounts were written by Carter and Beadle (1931), Marlier (1938), Leiner (1938), Carter (1957), Bertin (1958), and Johansen (1970). Comparative descriptions of species within limited geographic regions are found in Das (1927), Hora (1935), Dorn (1983), Munshi (1985), Dutta and Munshi (1985), Munshi and Hughes (1992), and Val and de Almeida-Val (1995).

An early controversy surrounding ABO function stemmed from the views of several workers (Cuvier and Valenciennes, 1831, 1840, 1846; Owen, 1846;

Günther, 1880) that the curious organs found in clariids, anabantoids, channids, and other fishes were in fact water reservoirs that kept gills moist when fish were air-exposed or perhaps filtered the water when the fish was trapped in a thick gruel. This idea was successfully challenged by demonstrations that the organs of fish exposed to air did not contain water and that the organs of fish in water contained gas (Day, 1868; Das, 1927). The role of these organs in aerial respiration was further confirmed by the obligatory air breathing of several species (Taylor, 1831; Boake, 1865). Finally, the finding that many of these organs had a venous drainage was also an impediment to functional interpretation because of the prevailing and rigid view of the time that veins "always" contained "dark blood," while only arteries carried "light blood" (Chapter 4).

ANATOMY AND MORPHOMETRICS

Analyses of ABO structure have progressed from the gross descriptions to ultrastructure and morphometrics. Transmission and scanning electron microscopy (TEM, SEM) have revealed many structural details. Morphometric analyses, principally by G.M. Hughes, J.S.D. Munshi, and their numerous colleagues, have yielded valuable comparative data on ABO and gill surface areas (Munshi 1962a,b; Hughes 1972, 1976; Hughes and Munshi, 1968; Hughes and Morgan, 1973; Munshi *et al.*, 1978).

Scaling

Interspecific ABO comparisons are usually done using scaling equations. These describe the size of an organ parameter, such as surface area, in relation to body mass (M) by a power equation:

$$Y = aM^b \qquad (3.1)$$

where a is the intercept and b is the scaling or mass coefficient. In the log form, Equation 3.1 describes a straight line with intercept a and slope b:

$$\log Y = \log a + b \log M \qquad (3.2)$$

Diffusing Capacity

Combination of gross and fine scale morphometrics enables estimation of morphological diffusing capacity (D_t), an important index of, assuming that neither ventilation nor perfusion are limiting, a tissue's capacity to exchange gases (Weibel, 1970). This is calculated by:

$$D_t = \mathrm{K}A / t \qquad (3.3)$$

where K (in the case of O_2) is the Krogh constant for O_2 tissue permeation (except where otherwise noted, the standard value of $1.5 \cdot 10^{-4}$ ml O_2/min/torr/cm^2/µm can be assumed in all subsequent discussions and calculations), A is the respiratory surface area, and t is the tissue diffusion thickness (i.e., air or water to blood distance). The units of D are: ml/min/µm/cm^2/torr.

Estimates of total organ diffusing capacity (D_L), while more common in morphometric analyses of mammalian tissue, have also been attempted for a few air-breathing fish species (Maina and Maloiy, 1985; Munshi *et al.*, 1989). These require estimation (from TEM) of the separate, but in-series, diffusing capacities (or O_2 conductances) of all surfaces in the gas diffusion path; the tissue (t), the plasma (p), and the red cell (r):

$$D_L = D_t + D_p + D_r \qquad (3.4)$$

where each conductance term is as in Equation 3.3. For the calculation of D_p, the product of a plasma permeation coefficient ($\mathrm{K_p}$) and an average surface-area expression for endothelial (S_e) and red cells (S_r) is divided by the diffusion distance t

$$D_p = \mathrm{K_p}(S_e + S_r/2)\,(1/t) \qquad (3.5)$$

Estimation of D_r is from the volume of capillary blood surrounding the erythrocytes (V_c) and the eyrthocytic conductance term (θO_2)

$$D_r = (V_c)\,(\theta O_2) \qquad (3.6)$$

where θO_2 contains terms for intraerythrocyte O_2 diffusion, facilitated diffusion by oxyhemoglobin, and the binding-reaction velocity of hemoglobin and O_2. The utility of such fine-scale diffusion analyses is questionable because none of the various constants have been determined for fish red cells, and iterations are usually done using mammalian values with correction factors for temperature, red cell number, and surface area, etc. (Yamaguchi *et al.*, 1987).

Development

As will be amplified throughout this chapter, ABO ontogeny has been the subject of numerous studies. Important work by B.R. Singh and co-workers in India documented the contribution of embryonic germ layers to ABO tissue differentiation in a number of families. Their work shows that, in one way or another, the gills or rudimentary branchial tissues contribute to the formation of many ABO structures developing in or adjacent to the head (Singh *et al.*, 1989b, 1990; Singh, 1993).

TYPES OF AIR-BREATHING ORGANS

Previous ABO classification schemes have been based on organ position and usually list from four to six types (Carter, 1957; Bertin, 1958). However, this approach over emphasizes supposed functional differences attributable to variations in ABO location and thus diverts attention from the evolutionary constraints and developmental limitations influencing ABO location, derivation, structure, and function.

Synopsis of ABO Origin

Consideration of historical and evolutionary factors affecting fish air breathing makes it possible to view ABO diversity in relation to phylogeny and to understand why these organs take the form and occur where they do. This perspective is based on the tenet that air breathing is an ancient piscine trait and that a primitive lung served as a primary aerial-exchange surface. This view also holds that, over the course of fish evolution, selective factors have acted on the lung and gradually altered its form and function with the result being the non-respiratory, and frequently physoclistous, organ of modern actinopterygians. Selective forces have also affected the evolution and adaptive radiation of fishes, with one result being the independent acquisition of air breathing in some advanced teleostean lineages. However, many of the lineages arrived at this point of requiring air breathing without the requisite functional plasticity in their specialized gas bladder. Novel ABO structures were thus invented.

Several structural and developmental limitations affect the formation of a new ABO or the modification of existing tissues for this function. The principal requirements for an ABO are the acquistion, retention, and absorption of aerial O_2. Few structures other than the mouth and jaws can be used to capture air. In terms of holding inspired air, spacious cavities or compliant spaces with a large blood supply and potentially modifiable vascular surfaces either already occur or can be developed in the mouth, branchial region, or digestive track. Thus, and although evolving independently, the form taken by the ABOs of these modern fishes is characterized by either evolutionary parallelism or convergence driven by limitations of structure, space, and surface area.

ABO Diversity

Given this evolutionary and structural-limitation approach, only a few main ABO groups need be defined (Figure 3.1; Table 3.1). First, there are the lungs and respiratory gas bladders of the more primitive groups including some lower teleosts. Then, there are the ABOs of the more advanced teleosts where, in the absence of a modifiable gas bladder, new structures were recruited including:

i. Organs formed in the head region:
   –Buccal, pharyngeal, branchial, and opercular chambers, or their surfaces
   –Pharyngeal and branchial pouches, the gills, and gill derivatives, and
ii. Organs formed along the digestive tube including the esophagus, the pneumatic duct, stomach, and intestine.

Finally, the skin serves an auxiliary aerial-respiratory function in many species.

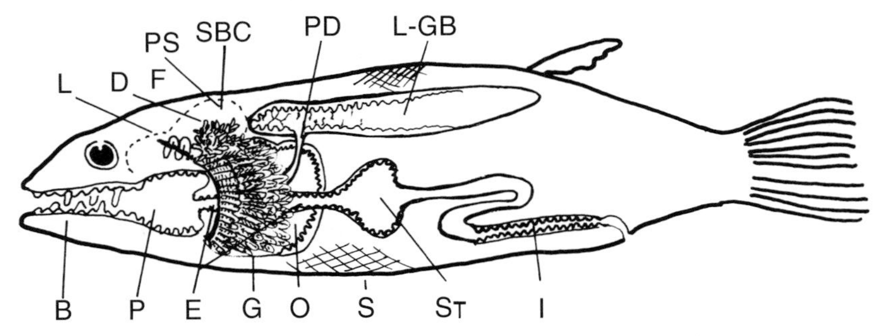

FIGURE 3.1 Generalized air-breathing fish *"Aerorespirichthys"* illustrating the ABOs presently known, including: Modified epithelial surfaces in the buccal (B), pharyngeal (P), esophageal (E), and opercular (O) chambers, as well as the gills (G), skin (S), stomach (ST), and intestine (I). Modified spaces include, the suprabranchial chamber (SBC), or pharyngeal sacs (PS). Modified chambers include the pneumatic duct (PD), and the lung or respiratory gas bladder (L-GB). Projections into these spaces include the labyrinth (L), dendrites (D), and gill fans (F).

TABLE 3.1 Classification of Fish Air-Breathing Organs

A. Lungs and Respiratory Gas Bladders of the More Primitive Bony Fishes

Lungs

Protopterus, Lepidosiren, Neoceratodus, Polypterus, Erpetoichthys

Respiratory Gas Bladders

Lepisosteus, Atractosteus, Amia, Arapaima, Heterotis, Pantodon, Gymnarchus, Notopterus, Papyrocranus, Xenomystus, Megalops, Phractolaemus, Erythrinus, Hoplerythrinus, Lebiasina, Piabucina, Pangasius, Gymnotus, Umbra

B. Air-Breathing Organs of the More Advanced Teleosts

Organs in the Head Region

Buccal and Pharyngeal Epithelial Surfaces

Electrophorus, Hypopomus, Sicyases, Alticus, Mnierpes, Entomacrodus, Periophthalmus, Periophthalmodon, Boleophthalmus, Scartelaos, Gillichthys, Channa, Monopterus, Synbranchus, Ophisternon, Blennius, Gobius

Branchial and Opercular Epithelial Surfaces

Mudskippers, *Pseudapocryptes*, Synbranchidae

Pouches Formed Adjacent to the Pharynx

Channa, Monopterus

Branchial Diverticulae

Heteropneustes, Clarias, Anabantoids

Gills

Hypopomus, Mnierpes, Synbranchus, Mastacembelus

Organs Located along the Digestive Tube

Pneumatic Duct

Anguilla

Esophagus

Dallia

Stomach

Loricariidae, Trichomycteridae

Intestine

Cobitididae, Callichthyidae

C. Skin

Erpetoichthys, Anguilla, Misgurnus, Clarias, Heteropneustes, Electrophorus, Neochanna, Xiphister, Sicyases, Mnierpes, Alticus, Coryphoblennius, Blennius, Periophthalmus, Boleophthalmus, Dormitator, Mastacembelus, Macrognathus, Monopterus, Synbranchus, Ophisternon

This simplified ABO classification scheme (Table 3.1) groups all the primitive fishes possessing lungs or respiratory gas bladders. Among the modern advanced fishes, the classification can be elaborated, but species found in the various ABO categories are diverse and not closely related. Thus, even though air breathing has evolved numerous times and independently, the location of aerial exchange sites has remained largely under the conservative influences of structures "predisposed" for air gulping and sites in the body where gas storage and the requisite vascularization could be developed.

It is also apparent in Table 3.1 that some taxa make use of more than one ABO type. In some cases, this results from the large area and continuity of the respiratory surfaces, even across chamber boundaries. It is nearly impossible to empirically discriminate between respiration taking place simultaneously in the gills and in the adjacent branchial, pharyngeal, or opercular epithelia. Inspection of Table 3.1 also indicates that the need for, and development of, a specialized aerial-exchange surface depends heavily on whether a species is an amphibious or aquatic air breather. Proficient amphibious air breathers in many cases do so without an elaborate ABO.

The morphology of fish ABOs will now be described. To emphasize the phylogenetic relationships and the evolutionary succession of various ABO types, the lung and gas bladder are examined first.

LUNGS AND RESPIRATORY GAS BLADDERS

Definitions

Throughout this book the term "gas bladder" is used to describe the gas-filled organ present in many fishes that serves a hydrostatic and other purposes, including air breathing. Two other terms, "swim bladder" and "air bladder" are avoided because of their reduced accuracy in describing either this organ's function or its contents. First, fish do not "swim" with the bladder (although its buoyant properties do aid swimming by providing static lift). Second, the bladder seldom contains "air," that is, a gas content closely approximating that of atmospheric air (Chapter 1).

For clarity, lung and gas bladder differences are detailed before discussing their evolutionary relationship. Vertebrate lungs share most of the following characteristics:

i. Embryonic origin as a small outpocketing from the ventral wall of the alimentary canal that persists and gives rise to ventrally positioned (i.e., closer to the ventral body wall) and paired organs
ii. The presence of a valvular glottis in the floor of the alimentary tract that guards the entrance to the lung
iii. The presence of a pulmonary circulation (i.e., afferent and efferent vessels leading more or less directly from the heart to the lungs and returning, Chapter 4).

Gas bladders by contrast:

i. Have an embryonic origin from the side or dorsal aspect of the alimentary canal, occur higher in the body (for vertical stability in water), and are not paired (although the bilobed structure of some reflects a primitively paired state)

ii. Do not always have a glottis and may or may not retain an open pneumatic duct (i.e., physostomous versus physoclistous, Chapter 2)
iii. In most cases, receive blood in parallel with the systemic circulation and thus lack a specialized circulatory loop functionally equivalent to a pulmonary circulation (Chapter 4).

Origin and Phylogeny of the Vertebrate Lung

The homology of the vertebrate lung and gas bladder was established by the British comparative anatomist Richard Owen (1846). As indicated by this chapter's epigraph, Darwin (1859) held that the piscine gas bladder had, through natural selection, given rise to the lung. Liem (1988) has reviewed the different and even opposite views that have been held as to which of these organs is ancestral. This is an important consideration because of the underlying question of whether initial selective forces for buoyancy, for respiration, or both led to the appearance of the primitive organ. Lift from a bolus of inspired air particularly a ventrally positioned one, would have facilitated shallow-water movement by densely armored Paleozoic fishes (Eaton, 1960; Schaeffer, 1965; Schmalhausen, 1968; Wells and Dorr, 1985). Alternatively, interesting essays by Morris (1885, 1892)

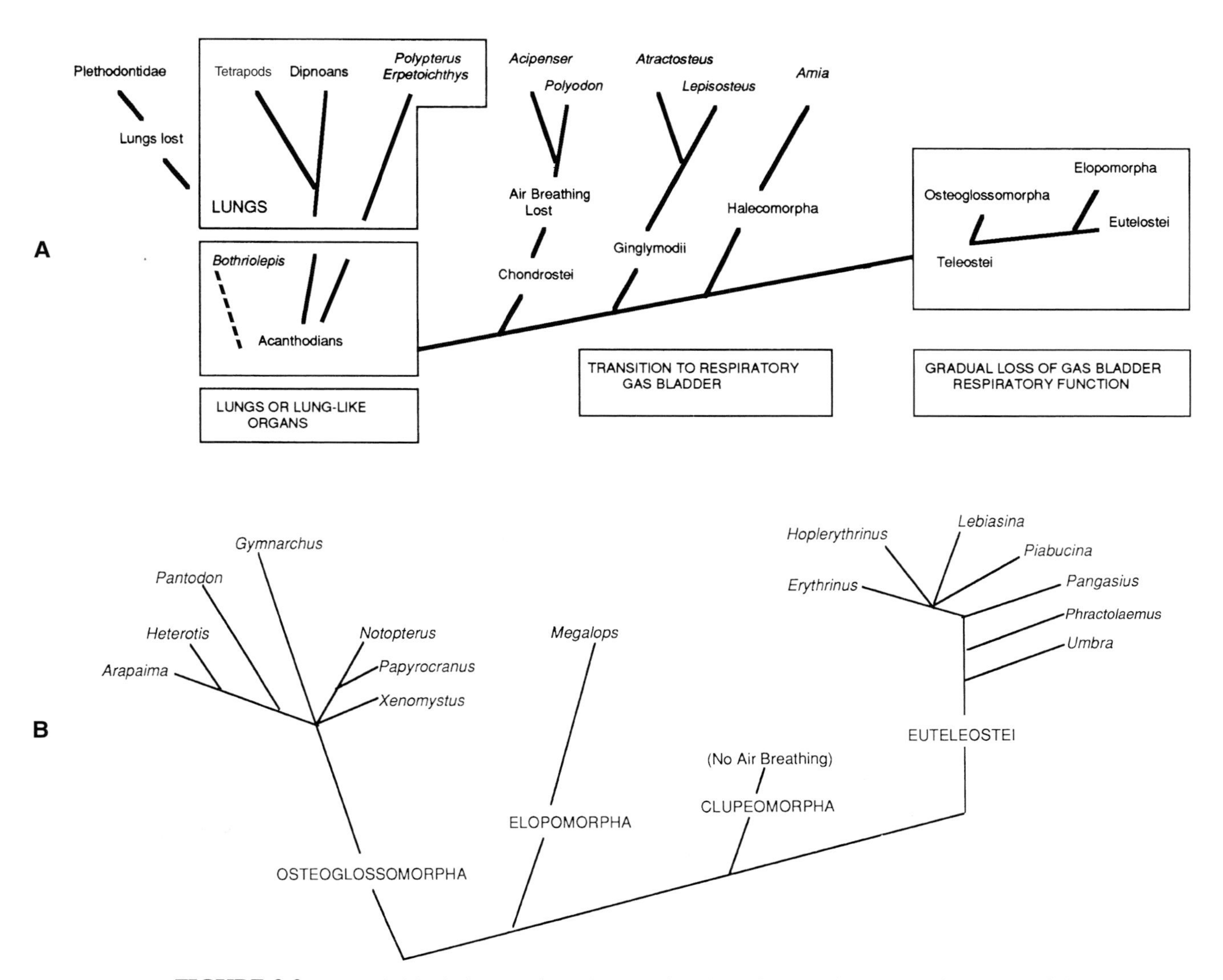

FIGURE 3.2 **A,** Probable phylogeny of vertebrate air breathing showing the origin and evolution of the lung and other ABOs among early fishes, the evolutionary development of the lung in tetrapods and various fishes, the gradual transition among fishes, from lung to gas bladder, and the loss of air breathing and lungs or lung-like structures from diverse groups. **B,** Distribution of the respiratory gas bladder among the four teleostean Infradivisions with a listing of the lung or gas-bladder air-breathing genera.

argue credibly that initial selection for this organ had to have occurred in conjunction with a more vital function such as auxiliary respiration. The basis for Morris' argument is the general "hit or miss" presence, among extant fishes, of either gas bladder-related buoyancy or sound-reception functions (i.e., about 50% lack these two gas-bladder specializations and many species have lost the organ [and see Quekett, 1844]).

It is now generally agreed that a respiratory lung, perhaps appearing first in early jawed vertebrates, was the ancestral organ (Romer, 1966; Liem, 1988). This origin and the different evolutionary routes taken by the vertebrate lung and related structures are depicted in Figure 3.2. Schaeffer (1965) postulated that air breathing might have been present in the acanthodians, and Denison (1941) described lung-like organs in the antiarch placoderm *Bothriolepis*. (Although *Bothriolepis* is not on the main vertebrate evolutionary line, Denison's observations stimulated a controversy [Myers, 1942; Stensio, 1948] that remains unresolved [Wells and Dorr, 1985]. Because there are numerous fossils of *Bothriolepis* and many are in a good preservation state, it would be useful to re-examine Denison's proposition using tomography or related modern techniques.)

Figure 3.2 shows that in the line leading to tetrapods, the primitive lung was gradually modified into the organs found in species ranging from lungfishes to mammals. By contrast, the fate of the primitive lung was different among the actinopterygians, where, according to a theory first advanced by Sagemehl (1885), a gradual transition into a non-respiratory gas bladder occurred. Two extant Chondrosteans (sturgeons and the paddlefish) are not air breathers. (In the case of the sturgeons, which are primarily benthic, the gas bladder is highly reduced.) Among the tetrapods, the plethodontids and a few other salamanders are the only known forms to have apparently undergone lung loss (Piiper *et al.*, 1976).

The Transition from Lung to Gas Bladder

Comparative anatomical and embryological evidence (reviewed in classic works by Goodrich [1930] and Romer [1964]) and phylogenetic relationships (Lauder and Liem, 1983a,b) offer general support for the "lung to gas bladder" hypothesis of Sagemehl. Briefly, this theory holds that the evolutionary transition from paired, ventral lungs to a dorsal single gas bladder occurred gradually, and very likely in stages similar to conditions seen in various extant species, as illustrated in Figure 3.3. The polypterid arrangement (ventral, paired lungs together with a complete right and left pulmonary circulation) would thus be archetypical. Intermediate stages might have been similar

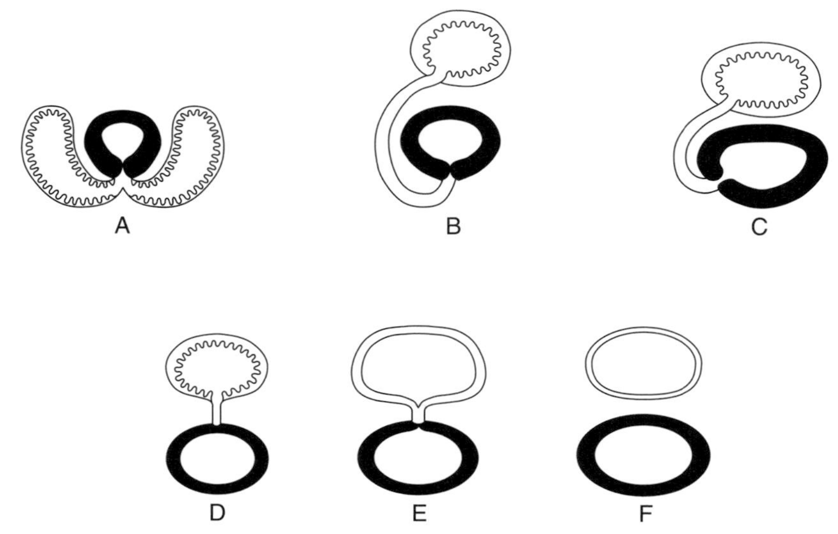

FIGURE 3.3 Possible intermediate morphological stages in the evolutionary transition from a ventral lung (A) to a dorsal, non-respiratory gas bladder. Alimentary canal in each stage represented by thick black wall; respiratory surface by convoluted line. **A,** Ventral lungs. **B,** Dorsal lung with a ventral glottis and elongated pneumatic duct. **C,** Gut rotation. **D,** Dorsal pneumatic duct. **E,** Configuration as in D but with loss of respiratory function. **F,** Physoclistous gas bladder.

to *Neoceratodus*, where a monopneumonous condition reflects a primary function in buoyancy rather than respiration. However, even though the lung is dorsal, it connects to a ventral glottis via an elongated pneumatic duct. Also, its left pulmonary artery passes around the alimentary canal in order to reach the lung. *Amia* may also represent an intermediate stage. It retains a pulmonary circulation but has an unpaired, dorsal respiratory bladder, a dorsal glottis, and a short pneumatic duct. The final phases of the transition are represented in the teleosts, where the pneumatic duct lies closer to the gas bladder (a result of alimentary canal rotation, see following) or is absent entirely. Also, and reflecting the increased importance of the organ's non-respiratory functions, the pulmonary circulation is absent.

While never completely resolved, objections to the Sagemehl theory (most notably, how to account for discrepancies in lung-vagal innervation and the presence of bilaterally symmetrical pulmonary circulation in *Amia*), have not interfered with its general acceptance. Additional comparative and developmental data have also proved important (Kerr, 1907, 1908; Ballantyne, 1927; Liem, 1988). Proponents of the theory sought to embellish it, and Dean (1895) presented his now widely recognized figure (Figure 3.4) illustrating likely intermediate stages. Although this figure is found in many texts, it has not illuminated this fas-

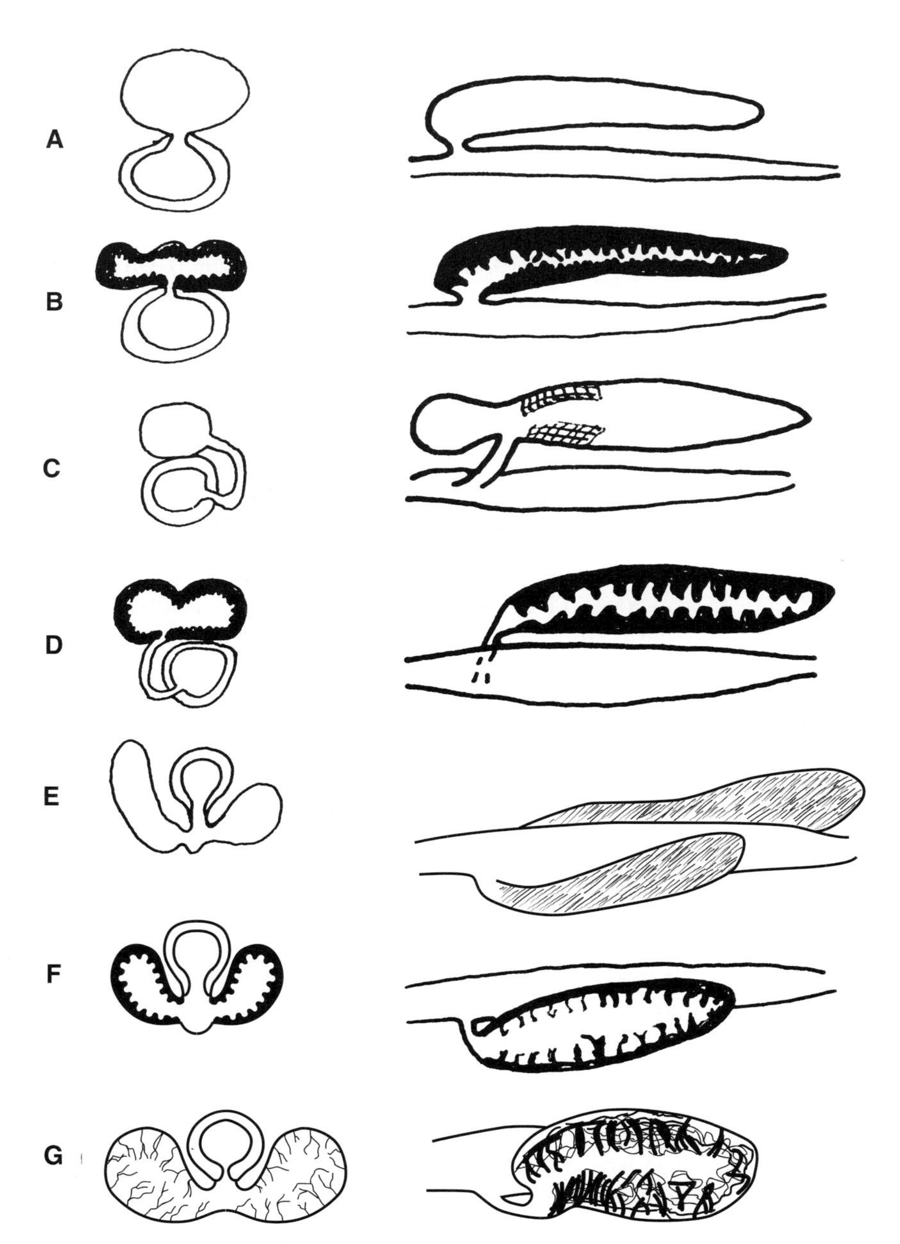

FIGURE 3.4 Modified version of Bashford Dean's (1895) original figure showing the evolutionary progression of the relative position between the gut and lung or gas bladder in fishes and tetrapods. Cross sections depict relationships as seen from head-end of the body. Morphological conditions are: **A,** Physotomous, non-air-breathing: Sturgeon and many teleosts. **B,** *Lepisosteus* and *Amia*. **C,** *Erythrinus* (note origin of pneumatic duct from the left side of gut). **D,** *Neoceratodus* (note that the pneumatic duct originates ventrally but ascends on the right side of the gut, Dean had originally illustrated this on the left side). **E,** *Polypterus* and *Erpetoichthys* (note that Dean did not illustrate furrows or smaller size of the left lung). **F,** *Lepidosiren* and *Protopterus*. **G,** Tetrapods.

cinating aspect of organ evolution as much as could have been hoped. This is because subsequent users often failed to note that Dean selected stages to show the possible scheme for pneumatic duct position and not the actual phylogeny of the transition (i.e., the examples are not in phylogenetic order). Also, most later renditions failed to note that both clockwise and counter-clockwise alimentary tract rotations had occurred in the various groups.

Occurrence of Lungs and Respiratory Gas Bladders

Table 3.1 lists 24 bony fish genera (47 species) known to breathe air using a lung or a respiratory gas bladder and, as seen in Figure 3.2, these are broadly distributed throughout the bony fish lineages, from the primitive dipnoans and actinopterygians to the euteleosts. The lungfishes (*Neoceratodus*, *Lepidosiren*, *Protopterus*) and the polypterids (*Polypterus* and *Erpetoichthys*) all possess lungs. Respiratory bladders are present in both *Amia* (Halecomorpha) and the gars (*Lepisosteus* and *Astracosteus*, Ginglyostoma). These organs occur high in the body, are bilobed, and differ from those of more primitive groups in having a dorsally originating pneumatic duct. *Amia* also differs from gars in retaining a pulmonary circulation (Chapter 4).

As discussed, the dependence of teleosts on alternative air-breathing structures is strongly correlated with their adaptive radiation and with evolutionary canalization of gas bladder structure (e.g., a physoclistous organ in *Channa*, the anabantoids, and cobitidids) or its enclosure by bone (e.g., many catfishes) and functions such as buoyancy and sound reception that are at cross purposes with cyclic air ventilation for aerial gas exchange. It is thus not surprising to find, among teleosts, that a respiratory gas bladder has a scattered appearance. Although well represented in the most primitive Osteoglossomorpha, the respiratory gas bladder is absent in the Clupeomorpha, and relatively rare in both the Elopomorpha and among the Euteleostei, occurring mainly in the ostariophysan lineages of the latter (Figure 3.2). Among these fishes, respiratory gas bladders differ greatly in complexity suggesting the organ's respiratory capacity may represent the last vestige of a primitive organ trait in some groups or in other groups, the "reacquisition" of aerial respiration by this organ.

Comparative Lung and Gas Bladder Structure: A Systematic Survey

Fish lungs and respiratory gas bladders are now compared for structure, morphometrics, and air-ventilation mechanics. General morphology is considered first, and diagrams of ABO structure for each group (Figure 3.5) serve as reference. Information presented here is from the original literature, from general treatments of fish lungs and respiratory gas bladders (Owen, 1846; Rowntree, 1903; de Beaufort, 1909; Ballantyne, 1927; Leiner, 1938; Fange, 1976, 1983; Hughes and Weibel, 1976, 1978; Wood and Lenfant, 1976; Liem, 1988, 1989a) and my study of specimens in the British Natural History Museum. As an aside, it is noteworthy that use by early workers of the term "cellular" to describe the complex varieties of compartmentalization, pneumatization, and surface-adding trabeculations of these organs slightly predated the widespread adoption of this term to connote the fundamental unit of tissue organization.

Lungfishes

The lungs of lungfish received considerable early study by Owen (1841), Quekett (1844), Hyrtl (1845), Günther (1871), and Parker (1892). Poll (1962) compared structures in the different species of *Protopterus* with that of *Lepidosiren* and Grigg (1965) provided additional details for *Neoceratodus.* Works by Klika and Lelek (1967), Hughes and Weibel (1976, 1978) and Maina and Malioy (1985) examined lung morphometrics.

Neoceratodus has only a single lung. In the four species of *Protopterus* and in *Lepidosiren,* the lungs are paired although fused anteriorly. In all species, the organs are long, fill most of the dorsal coelom, and obtain blood flow via a paired pulmonary circulation (Chapter 4). Lung connection with the glottis (located on the ventral pharynx) is via a long pneumatic duct (Figures 3.3–3.5) that contains layers of smooth muscle.

The lungs of these fishes contain an elaborate network of septa (Figure 3.6) that subdivide the lumen into respiratory compartments, or alveoli. Septa are formed by connective tissue and contain a layer of smooth muscle. They are covered by a respiratory surface consisting of a thin epithelium over a capillary bed. A layer of smooth muscle is also found in the lung wall itself (Hughes and Weibel, 1976).

The overall structure and degree of septation found in the lungs of *Protopterus* and *Lepidosiren* are highly similar. These lungs also have a much larger surface area (i.e., more and smaller compartments) than does the lung of *Neoceratodus*. As previously noted, exceptions for *Neoceratodus* are consonant with reduced requirements for aerial respiration compared to either the African or South American lungfishes. This fish is a facultative air breather, and its single lung is important for buoyancy (Grigg, 1965).

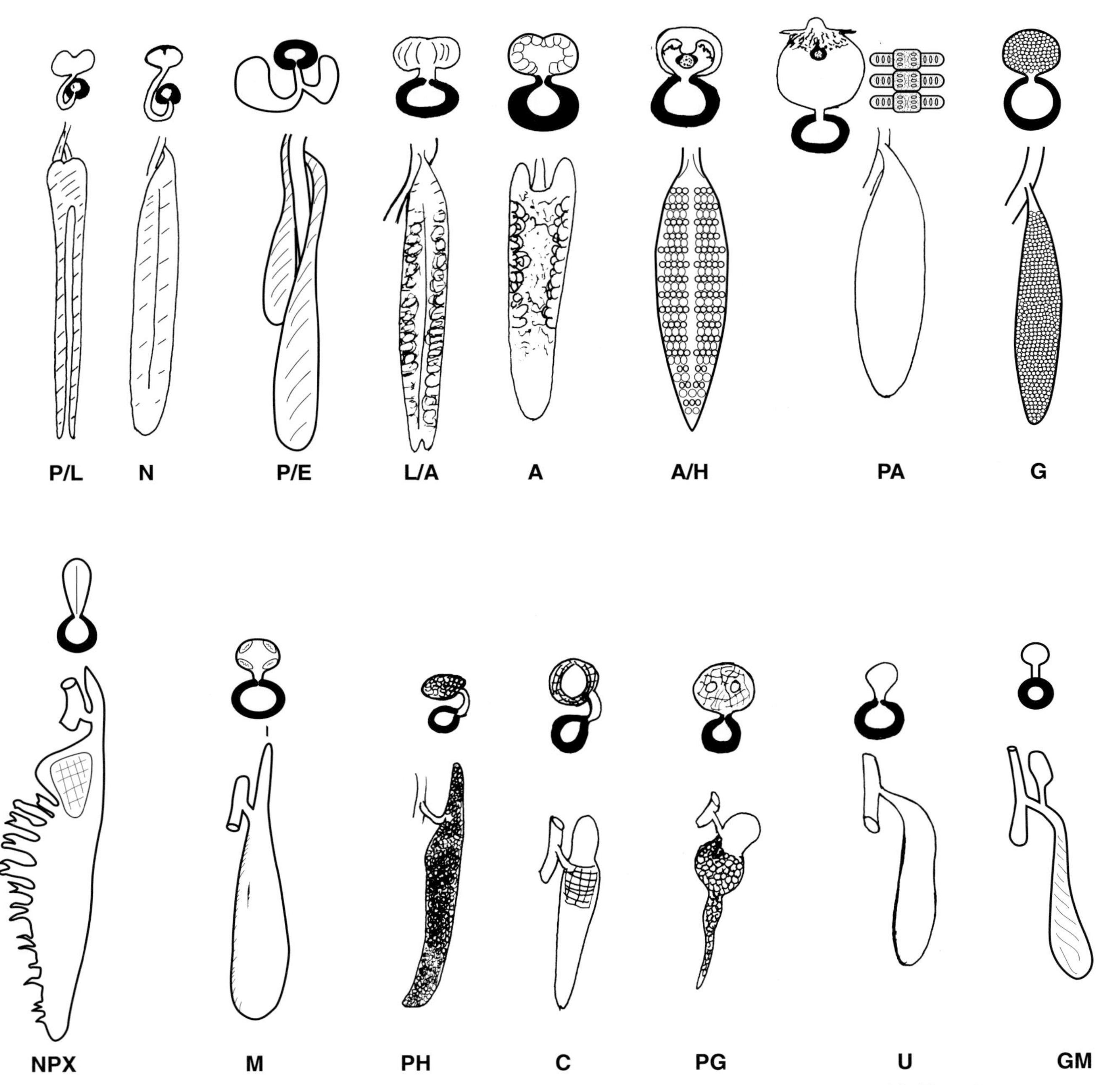

FIGURE 3.5 Profile and anterior transverse-section views of the lungs and respiratory gas bladders of the air-breathing fishes. Abbreviations: (P/L) *Protopterus* and *Lepidosiren;* (N) *Neoceratodus;* (P/E) *Polypterus* and *Erpetoichthys;* (L/A) *Lepisosteus* and *Atractosteus;* (A) *Amia;* (A/H) *Arapaima* and *Heterotis* (note kidney suspension); (Pa) *Pantodon* (suspended kidney and fenestrated vertebrae also shown); (G) *Gymnarchus;* (NPX) *Notopterus, Papyrocranus,* and *Xenomystus;* (M) *Megalops;* (PH) *Phractolaemus;* (C) Characoids; (PG) *Pangasius;* (U) Umbra; and (GM) *Gymnotus.* (Note that, beginning with *Pantodon,* most profile projections are lateral.)

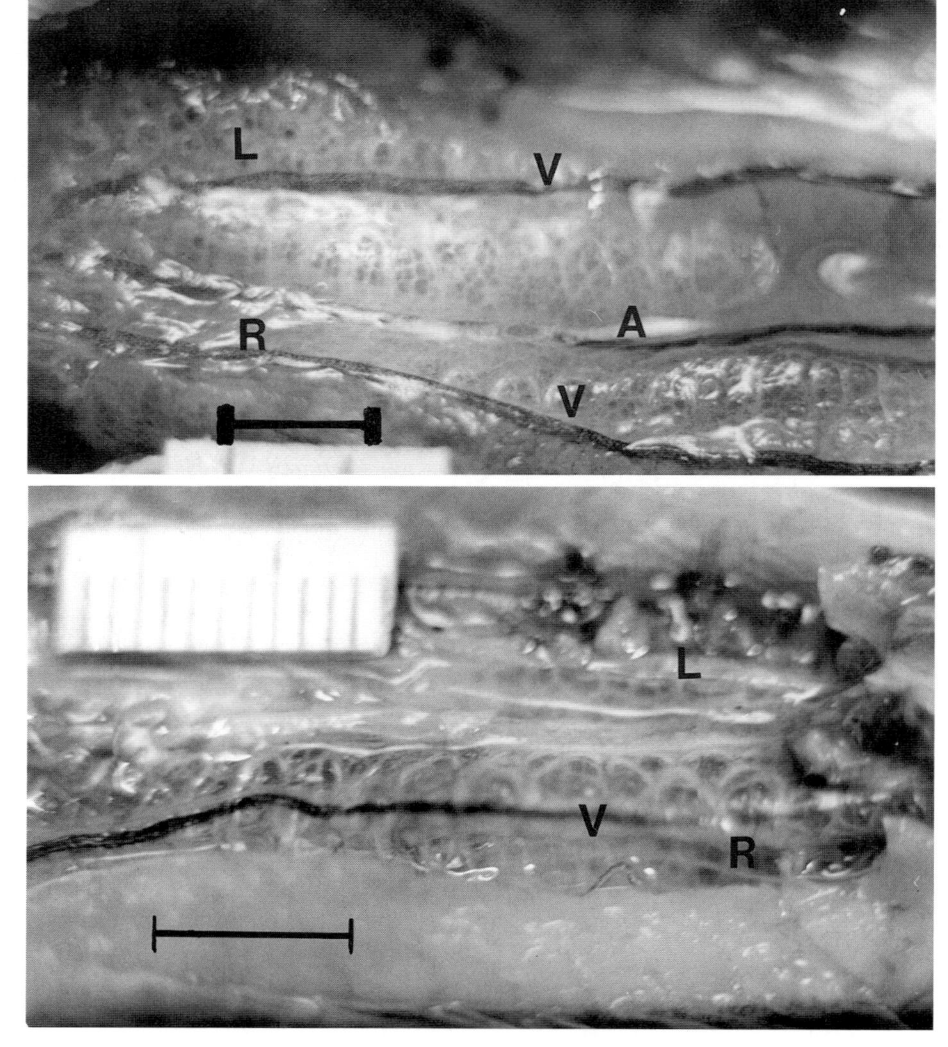

FIGURE 3.6 Ventral views of the anterior *(top)* and posterior *(lower)* sections of the lung of a 28 cm preserved *Protopterus* sp. Head direction is to the left. Note that both right (R) and left (L) lung lobes are evident at both positions and that there are regional differences in alveolar size. A, pulmonary artery; V, pulmonary vein. Scale = 7.0 mm.

Polypterus and *Erpetoichthys*

The paired, ventrally originating lungs of these fishes were noted by Geoffrey Saint Hilaire (1802b) in his description of the genus *Polypterus*. Early work with these organs include their circulation (Müller, 1842, 1844; Hyrtl, 1870) and accounts of gross and fine structure for *Erpetoichthys* (Rauther, 1922; Purser, 1926) and for several species of *Polypterus* (Leydig, 1854; Gérard, 1931; Ballantyne, 1927; Daget, 1950; Klika and Lelek, 1967). Poll and Deswattines (1967) reviewed the literature and conducted a comparative survey of lung structure and circulation in nine species of *Polypterus*. In addition, the work of J.S. Budgett (1901, 1902), whose extensive field studies on *Polypterus* provided many details about its ontogeny and lung development, are considered a fundamental contribution to solidifying ideas about lung–gas bladder phylogeny (Kerr, 1907, 1908).

The lungs of *Polypterus* and *Erpetoichthys* arise from a ventral glottis and pass dorso-posteriorly around the alimentary canal into the coelom. In both genera, the glottis, a muscle-ridged slit, opens into the right lung; the left lung being connected to the right by a separate opening. In *Polypterus*, each lung has about the same diameter; however, the length of the left lung is limited by the stomach and posteriorly the right lung occupies a mid-dorsal position and expands greatly in some species (Figure 3.7; Ballantyne, 1927;

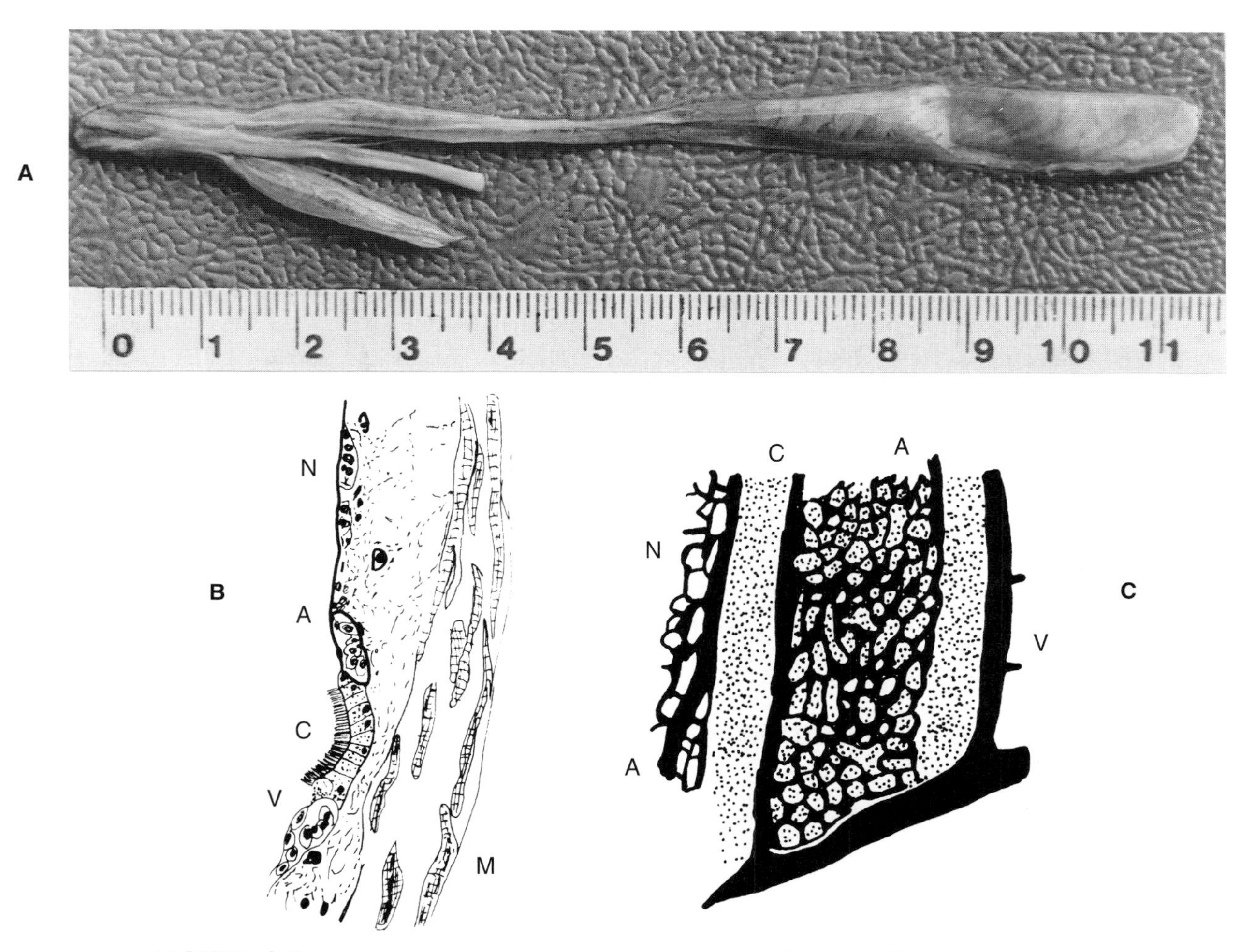

FIGURE 3.7 **A,** Dorsal view of the paired lungs of a 20 cm *Polypterus.* The longer right lung also expands posteriorly and furrows can be seen in the posterior lobe. Central tube is the esophagus. **B,** Transverse section of the lung wall (×305) and, **C,** frontal view of the lung wall (×125) showing the capillary network (N), and ciliated furrow (C). Other abbreviations: A, artery; V, vein, N, capillary network; M, striated muscle; C, ciliated furrow. (**B** and **C** redrawn from Purser (1926) by permission of the Royal Society of Edinburgh from *Transactions of the Royal Society of Edinburgh,* volume 54, pp. 767–784.)

Daget, 1950; Poll and Deswattines, 1967). The lungs of *Erpetoichthys* are similar; the left lung is curtailed by the stomach, and both tubes are relatively thin owing to the slender body of this fish.

The thin-walled lungs of these fishes are nearly transparent (Figure 3.7). The inner surface lacks all septation and is relatively smooth except for a highly distinct, macroscopic pattern of longitudinal striations, termed furrows (Figure 3.7). Furrows are shallow grooves in the lung epithelium containing rich concentrations of ciliated, granular, and mucous-producing cells. The respiratory epithelium is formed by a layer of flat cells lining luminal surfaces between furrows. The capillary network largely parallels the furrows. Both Leydig, and Poll and Deswattines reported that capillaries lie within the epithelial cells, but this is not apparent in the figures of Klika and Lelek (1967). Blood flow to and from lung capillaries is via segmental arteries and veins traversing the organs at regular intervals.

The most extensive development of the respiratory and furrowed epithelium occurs in the central region of the right lung and near the level of the stomach in the left lung. Two layers of striated muscle line the lung walls, becoming thicker in the posterior reaches, and function in compressive exhalation (Brainerd *et al.,* 1989).

Lepisosteus

Early studies of the gar respiratory gas bladder include the descriptions of inner-wall septation (Van der Hoeven, 1841) and organ structure and circulation

(Quekett, 1844; Hyrtl, 1852a,b) and respiratory behavior (Wilder, 1875). Detailed structural accounts are contained in Potter's (1927) landmark study while more recent works (Crawford, 1971; Rahn *et al.*, 1971) provide comparative and morphometric data.

This large and bilobed gas bladder (Figure 3.8) fills the gar's dorsal coelom. It is connected to the alimentary canal by a short, vertical pneumatic duct that extends from the dorsally positioned pharyngeal glottis to the anterior end of the gas bladder. The duct's walls are thickened by the presence of oblique muscular bands that regulate diameter.

Viewed in cross-section (Figure 3.8), the gas bladder is oval with an open lumen occupying about the central third of its area. Series of paired trabeculae subdivide the lateral sections of the organ into a sequence of about 30 compartments. Each compartment opens onto the central lumen, and its walls are formed by the interior bladder walls and by thin sheets of connective tissue that radiate from the trabeculae to the inner

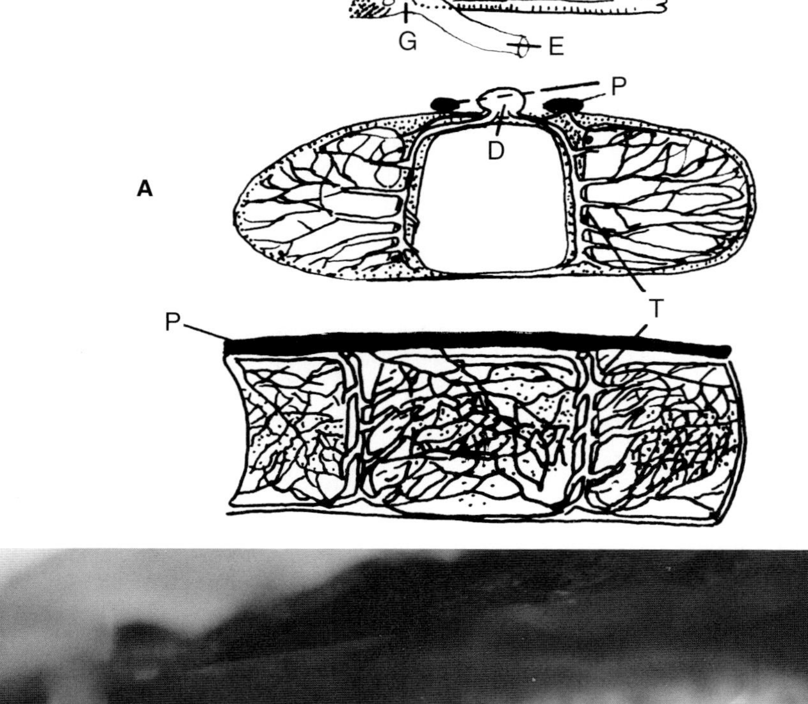

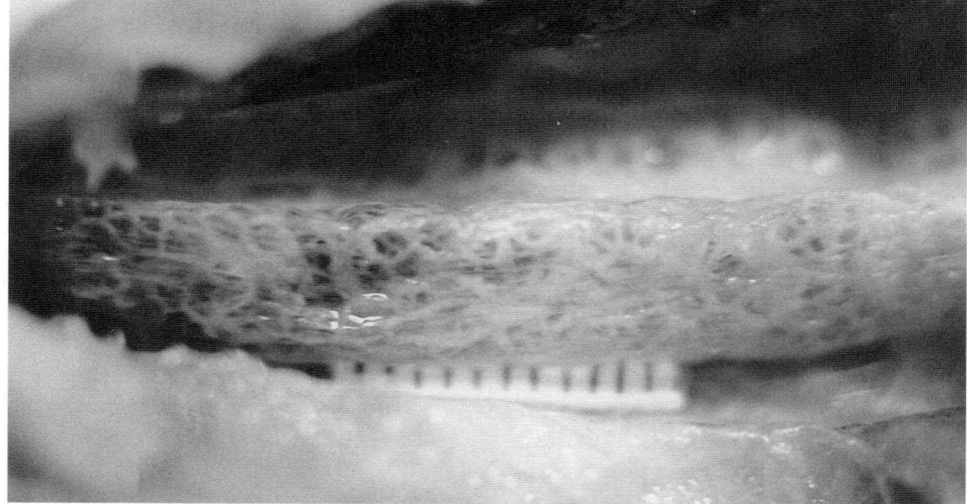

FIGURE 3.8 **A,** Redrawing of Potter's original illustration of the respiratory gas bladder of *Lepisosteus*. *Top*. Ventral view showing glottis (G), and esophagus (E). *Middle*. Gas-bladder cross-section showing the central lumen and the side pockets formed by trabeculae (T). Vascular inflow is via the dorsal aorta (D) while drainage is into the right post cardinal vein (P). *Lower*. Saggital section showing vascularization of the organ wall, the post cardinal vein, and trabeculae. (Redrawn from *Journal of Experimental Zoology*. Copyright 1927. Potter. Reprinted by permission of Wiley-Liss, Inc., a subsidiary of John Wiley & Sons, Inc.) **B,** Photograph shows the anterior region of the ABO in a 34 cm preserved *L. platostomus*. Scale = 12 mm.

walls. These sheets and the inner wall contain numerous adjacent capillaries covered by an epithelium. The gas-bladder walls and the septal trabeculae both contain bands of striated muscle that enable compression of the organ, probably for buoyancy regulation and exhalation.

Amia

Vascularized cells on the roof and lateral walls of the gas bladder of *Amia* (Figure 3.9) were described by Cuvier and Valenciennes (1846, pages 408–409) who concluded that this structure was involved in respiration. Additional anatomical details are contained in Wilder (1875, 1877), Ballantyne (1927), Crawford (1971), and Milsom and Jones (1985).

The large, wide, and highly vascularized gas bladder occupies nearly the entire breadth and length of the dorsal coelom. The organ is bilobed in the front but tapers towards the caudal end. The pneumatic duct is short and leads from the mid-dorsal esophagus to the anterior gas bladder. This ABO does not have the auditory extensions found in many other species (see following).

Comparative aspects: *Amia* and gars. Crawford (1971) contrasted the respiratory structures in gas bladders of comparably sized *Amia* and *Lepisosteus*. Features common to both organs are dorsal position, a short, muscular pneumatic duct, and a dorsal glottis. Differences between the two include glottis position (on the pharynx in gar, on the posterior esophagus in *Amia*) and the degree of muscular development within the organs (in *Amia* the gas bladder is encircled by a thick, striated muscle layer [Rahn *et al.*, 1971], while muscles in the wall of the gar's organ are less developed). Crawford (1971) also noted that *Amia* has less alveolar development than gar. He reported that alveoli in *Amia* are larger (only six or seven alveolar compartments occur along the organ versus 30 or more in gar) and, unlike in gar, they are not uniformly distributed throughout the organ. Alveoli are in fact poorly developed in the posterior reaches of the *Amia* ABO, even though the walls in this region are well vascularized (Figure 3.9). *Amia* also possesses a greater diversity of epithelial cell types than does the gar. Despite these structural differences, Crawford (1971) found that the ABOs of *Amia* and gar did not differ appreciably in either capillary and epithelial cell arrangements or blood-gas barrier thicknesses.

Arapaima and *Heterotis*

The respiratory gas bladders of *Arapaima* and *Heterotis* are remarkably similar and distinct from other species. Cuvier and Valenciennes (1846, page 439) wrote about this structure,

> . . . mais qu'un organe curieux, comme un poumon d'oiseau, existe le long de la colonne vertébrale, et que l'intérior ressemble à un gâteau de miel.

Other early ABO descriptions include a detailed study and illustration for *Heterotis* (Hyrtl, 1854b) and a brief description for *Arapaima* (*Sudis*) by Jobert (1878). Additional detail was added by Lüling (1964a) and others. Especially notable contributions by Poll and Nysten (1962) and Dorn (1968) include comparative structural and morphometric data for both genera in comparison to other Osteoglossidae. More recent studies (Greenwood and Liem, 1984; Liem, 1989a) have focused on details of ABO ventilation. Comparative morphological, histochemical, and ultrastructural observations for *Arapaima* and *Hoplerythrinus* are contained in Cruz-Landem and Cruz-Höfling (1979) and in Cruz-Höfling *et al.* (1981).

In both *Arapaima* and *Heterotis*, the gas bladder is wide and extends the full length of the coelom. Its lateral and dorsal walls are fused to the body wall, and a thin ridge of connective tissue extending along the organ's dorsal wall subdivides its roof. The richest concentration of respiratory epithelium occurs in the

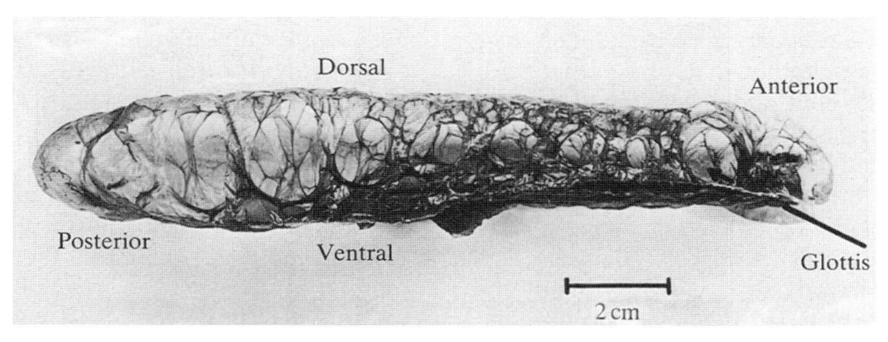

FIGURE 3.9 Respiratory gas bladder of *Amia*. (Milsom and Jones, 1985. Reprinted by permission of the Company of Biologists, Ltd. from the *Journal of Experimental Biology*.)

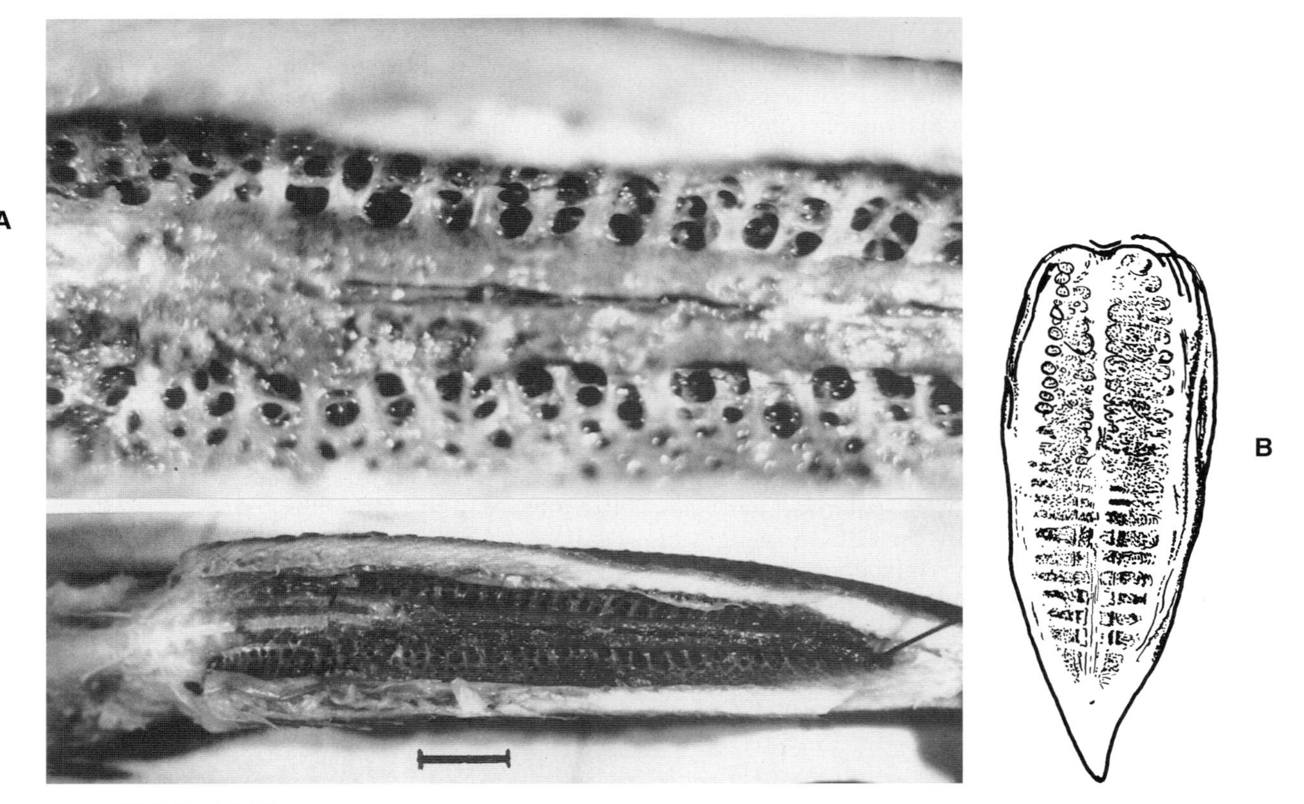

FIGURE 3.10 **A,** *Upper.* Detail for the ABO inner roof of *Arapaima* showing trabecular structure and the paired kidneys. *Lower.* Ventral view showing how the respiratory gas bladder fills the coelomic cavity (scale = 1 cm). **B,** Redrawing of Hyrtl's (1854a) original illustration of the inner gas bladder wall of *Heterotis.*

organ's dorsal and lateral walls. Here the spongy tissues are elaborated within space formed between the roof and lateral walls and supporting trabeculae that extend laterally from the mid-dorsal ridge (Figure 3.10A and B). Each trabeculation gives off lateral branches that in turn branch, making an intricate meshwork of vascular tissue.

Dorn (1968) observed that the ventral gas bladder wall is thin and compliant but highly vascularized. She also noted slight differences in the fine structure of the respiratory chambers in *Heterotis* and *Arapaima*. Greenwood and Liem (1984) report that no pneumatic duct is present in these fishes. Rather, in its anterior reaches, the ventral ABO wall is continuous with the roof of the esophagus, and contact between the two is via a longitudinal slit surrounded by sphincter muscles.

A consequence of the gas bladder's increased size is that the kidney, suspended by the dorsal mesentery, is contained within the latter two-thirds of the ABO (Figure 3.10A). This is obviously a space-conserving mechanism; however, it may have functional implications for gas exchange and transport, for blood-pressure regulation, and for kidney biochemistry (Chapters 4 and 8).

Pantodon

The ABO of *Pantodon* has similarities with *Arapaima* and *Heterotis.* Nysten (1962) reported on the structure of this organ, and a companion paper (Poll and Nysten, 1962) compared it with the respiratory gas bladders of *Arapaima* and *Heterotis* and some of the non-air-breathing osteoglossomorphs.

In *Pantodon*, the respiratory gas bladder fills the dorsal coelom and extends nearly the full length of that cavity. As in *Arapaima* and *Heterotis*, the lateral and dorsal walls of the gas bladder are fused to the body wall while the compliant ventral wall rests on the viscera, and the kidney is supported within the gas-bladder lumen by the dorsal mesentery.

The respiratory epithelium of this gas bladder is found primarily on its superior lateral and dorsal surfaces (Figure 3.11). Due to its intimate association with the dorsal body wall, the dorsal surface of the gas bladder completely contours the vertebral centra and parapophyses as well as proximal ends of the ribs and the intervertebral recesses. A respiratory epithelium covers all of these surfaces and becomes particularly well developed within each intervertebral pocket.

A

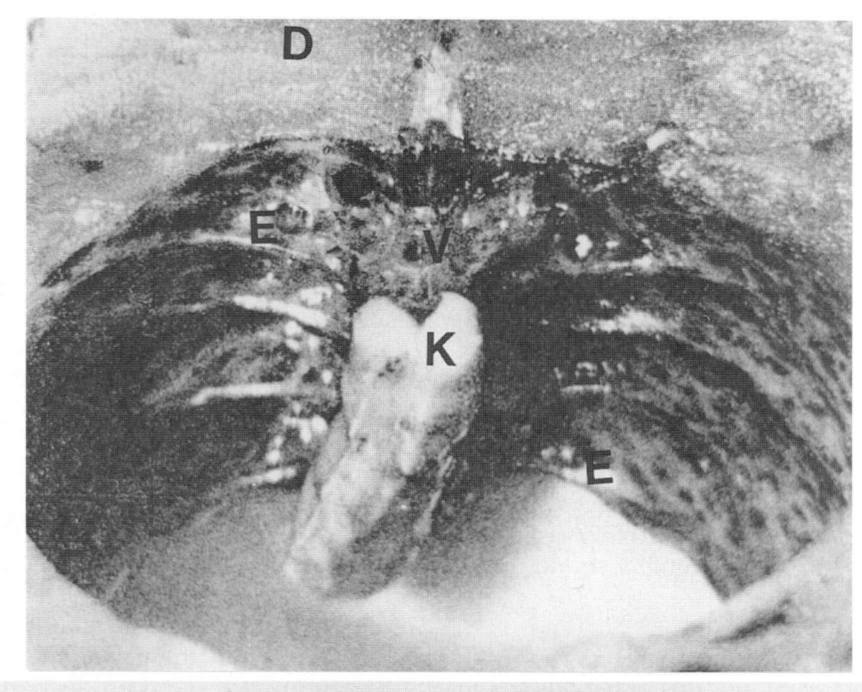

B

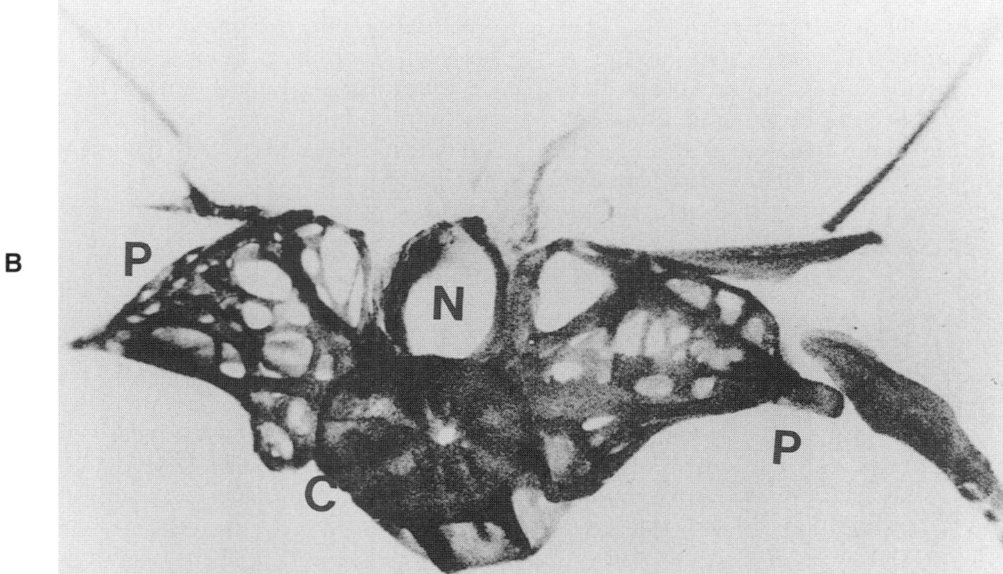

FIGURE 3.11 The original illustrations of *Pantodon* by Nysten (1962, Plates II and IX). **A,** Saggital body section showing the respiratory epithelium (E) lining the dorsal surface of the gas bladder and filling the intervertebral spaces. K, kidney; V, vertebrae, D, dorsal body wall. **B,** End view of vertebrae number 5 showing the large degree of fenestration in its body and parapophyses. C, centrum; N, neural tube; P, parapophyses. (Reprinted with permission of Musée royal de l'Afrique centrale from *Les Annales du Musée royal de l'Afrique centrale, Tervuren [Belgique].*)

Moreover, the vertebral bones themselves are highly pneumatized and extensions of the gas bladder pass through numerous and closely spaced fenestrae to link with a complex air labyrinth system that extends deep into bone and near the spinal cord (Figure 3.11). Nysten (1962) viewed the primary function of this modification to be for density reduction, thus enabling *Pantodon* to float at the water surface. Her suggestion that these surfaces may also have a respiratory function (thus enormously increasing the organ's conductance) is probably correct but needs to be examined by histological studies.

The gas bladder in *Pantodon* connects to the underlying esophagus by a short, vertical pneumatic duct. This duct contains two lateral, longitudinal bands of muscle that, as interpreted by Nysten (1962), alter the duct's diameter and its connection with the esophagus. These muscles then act in harmony with others to appropriately direct esophageal contents (air or food). At the level where the pneumatic duct attaches, paired anterior lobes of the gas bladder extend forward to the proximity of the skull to likely augment sound transmission.

Gymnarchus

This species differs from all other osteoglossids in having an especially well-developed respiratory gas bladder surface (Figure 3.12) and also having a "quasi-pulmonary" circulation (Chapter 4). The first ABO description for this species was by Professor Förg (1853) in a letter communicated to the French Academy by G. Duvernoy. Hyrtl (1856) made cursory observations of ABO and pneumatic duct structure in relation to blood circulation. Assheton (1907, based on material collected by J.S. Budgett) and Ballantyne (1927) detailed aspects of ABO ontogeny, showing it to be initially bilobed with a right-side projection to the otic vesicles.

The otic vesicles lose their connection with the gas bladder in adult *Gymnarchus*, and the respiratory gas bladder, still bilobed in its anterior end, extends the full length of the coelomic cavity. The organ's large central cavity is surrounded by air chambers that are in turn subdivided into numerous alveoli. These completely fill the organ and have a uniform density and size throughout. Gas-bladder connection to the alimentary tract is via a short pneumatic duct that emerges from just left of the esophageal midline and enters the anterior end of the gas bladder. There are no detailed or more recent accounts of the structure or morphometrics of this ABO other than its air-ventilation mechanics (Liem, 1989a).

Notopterus, Papyrocranus, and *Xenomystus*

The unique shape of the gas bladder in these fishes (Figure 3.5) reflects both their exotic body form (Figure 2.2) as well as the organ's functions (in addition to air breathing) for buoyancy control, sound reception, and sound production. The organ has both an anterior otic extension and posterior projections between the pterygiophores and is divided medially by a vertical septum (Figure 3.13). The short, wide pneumatic duct proceeds vertically from the dorsal left side of the esophagus to the left-ventral side of the gas bladder.

In *Notopterus*, a single layer of epithelium and inter-

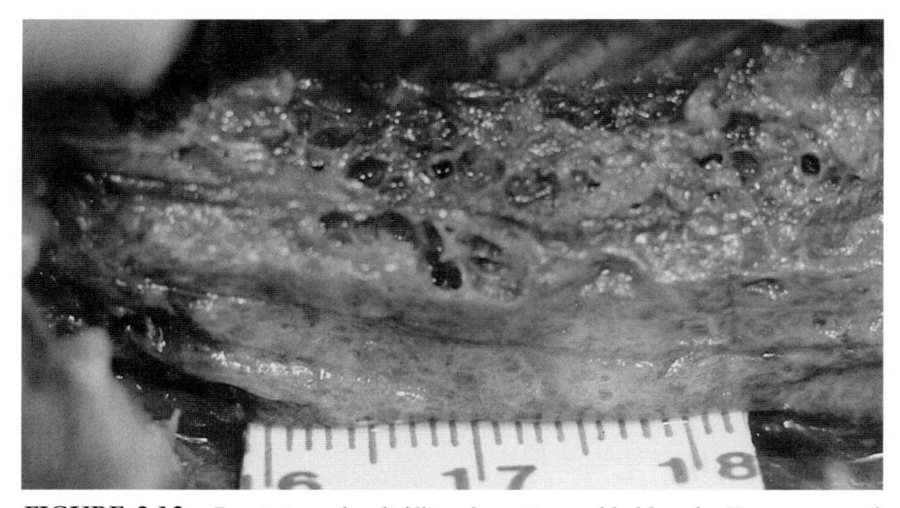

FIGURE 3.12 Respiratory alveoli filling the entire gas bladder of a 55 cm preserved *Gymnarchus niloticus*. Scale in cm. (Note that alveoli at top of photo are sliced open whereas those just above the ruler are intact.)

FIGURE 3.13 View of the gas bladder of *Notopterus* sectioned near respiratory region just posterior of the pectoral fins (left). Picture taken in a market in Bhagalpur, India (1974).

spersed mucous cells form the respiratory surface which appears to extend throughout the organ (Dehadrai, 1962a). The underlying ABO wall is richly vascularized and contains a layer of striated muscle. Liem (1988, 1989a) showed SEM details of the ABO epithelium in *N. chitala* and indicated this occurs in the anterior part of the main gas bladder chamber. He also described the respiratory mechanics of this species. Based on structural similarities with the gas bladder of *Notopterus*, Greenwood (1963) concluded the organs in both *Xenomystus* and *Papyrocranus* are also respiratory.

Megalops

The respiratory gas bladder of *M. cyprinoides* was illustrated by de Beaufort (1909) and that of *M. atlanticus* by Babcock (1951). In both species, the aerial-respiratory surface consists of four strips of spongy, lung-like tissue lying along the dorsal, ventral, and lateral walls of the posterior gas bladder (Figure 3.14). These extend posteriorly to about 80% of the organ's length. Histological sections (Babcock, 1951) show these spongy strips to be capillaries covered by a thin epithelium.

A short, wide, and heavily walled pneumatic duct connects the esophagus to the gas bladder near the junction between the anterior and posterior chambers. The anterior gas bladder chamber is a thin tubular projection that contacts the otic capsule to augment sound transmission. With the exception of its alveolar ridges, the posterior gas bladder chamber is smooth.

The total surface area of the alveolar strips does not appear particularly large. Histological cross sections do, however, give the impression that the spongy layers are not entirely embedded in the gas-bladder wall. While verification is required, this implies that the organ's effective gas-exchange area is much greater than suggested by just the luminal surface.

The ABOs of the two *Megalops* species have not been systematically compared, but their illustrations do not lend support to Liem's (1989a) suggestion that the respiratory surface in *M. cyprinoides* is more developed than in *M. atlanticus*. Observations by S.F. Hildebrand (contained in a letter to Babcock [1951, pages 51–55]) indicate, moreover, that respiratory surfaces are equally developed in both small (30 cm) and large (180 cm) Atlantic tarpon. This dispels the impression (Liem, 1989a) that a respiratory gas bladder is functional in young, swamp-dwelling fish, but non-functional in open-water inhabiting adults. (Furthermore, adult fish are known to exhibit what is termed a periodic rolling behavior, most likely for air ventilation [Graham, 1976a].)

Phractolaemus

The respiratory gas bladder of *P. ansorgei* was described by Thys van Audenaerde (1959, 1961). It is large, extending throughout the coelom, actually increasing in diameter near the tail, and is completely filled with alveolar vessicles. There are no morphometric data on this organ; however, small respiratory chambers occur throughout and the total respiratory surface development rivals that of *Gymnarchus* (Figure 3.15). The relatively long pneumatic duct exits the upper left side of the esophagus and enters the anterior left side of the ABO at about 10% of its total length.

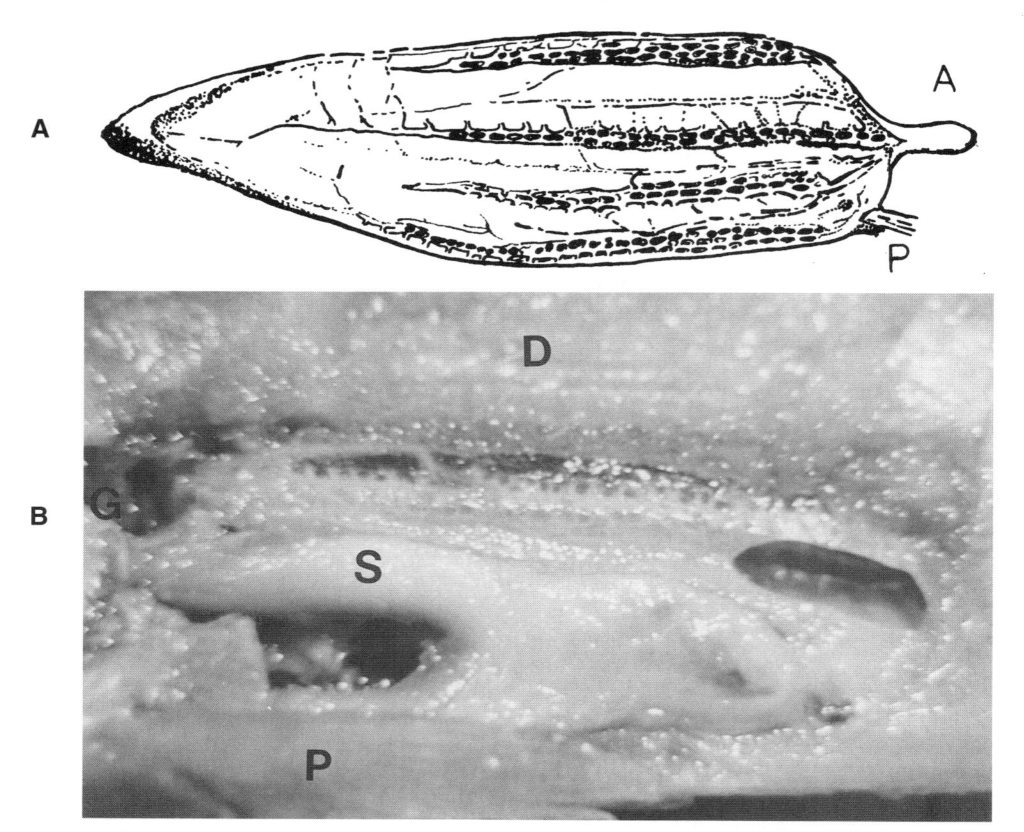

FIGURE 3.14 **A,** Redrawing of de Beaufort's (1909) illustration of the respiratory gas bladder of *Megalops cyprinoides.* Organ has been opened on its right side and slightly everted to reveal the four vascular strips. Also apparent are the pneumatic duct (P) and the organ's anterior auditory projection (A). **B,** Left-side view of the gas bladder and one vascular strip in a 13 cm preserved *M. atlanticus.* D, dorsal muscle; G, posterior edge of gills, S, stomach, P, pectoral fin.

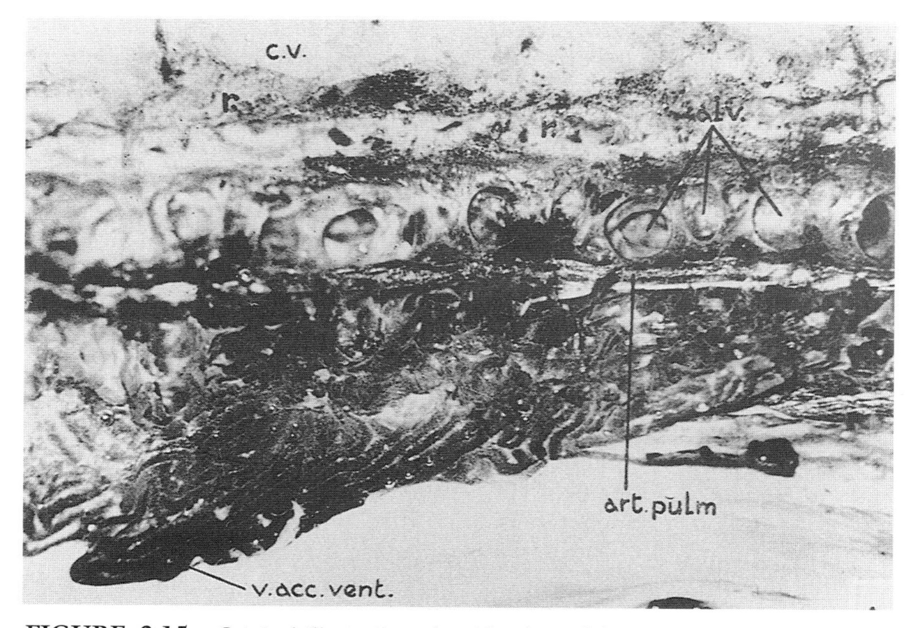

FIGURE 3.15 Original illustration of a side view of the respiratory gas bladder in *Phractolaemus* from Thys Van Den Audenaerde (1961, Plate VIII b). C.V., vertebral column; V. Acc. Vent., ventral collecting veins; Art. Pulm., pulmonary artery; Alv., alveoli; r, kidney. (Reprinted with permission of Musée royal de l'Afrique centrale from *Les Annales du Musée royal de l'Afrique centrale, Tervuren [Belgique].*)

Erythrinidae and Lebiasinidae

In these two characin families similar respiratory gas bladders occur in four genera (*Erythrinus, Hoplerythrinus, Lebiasina,* and *Piabucina*). Accounts of "cellular" gas bladders in these fishes date from Cuvier and Valenciennes (1846) who illustrated (plate 588) many details of the organ. Jobert (1878) made reference to the complex ABO structure of *Erythrinus*, and Rowntree (1903) illustrated the organ in *Lebiasina*. For *Hoplerythrinus*, Carter and Beadle (1931) provided a complete organ description, showing its gross features, the cross-section detail of its cellular structure, and illustrating capillary infiltration into the wall of each cell. Böker (1933) also illustrated the ABO of *Hoplerythrinus* and detailed the pneumatic duct structure. Graham *et al.* (1977, 1978b) described the gas-bladder anatomy of *Piabucina* and compared the respiratory surface structures of *Piabucina, Hoplerythrinus*, and *Erythrinus*. Histological and ultrastructural observations were made for *Hoplerythrinus* by Cruz-Landim and Cruz-Höfling (1979) and Cruz-Höfling *et al.* (1981).

The long, dorsal gas bladders in these fishes consist of anterior and posterior chambers connected through a narrow duct (Figure 3.16). The anterior chamber contacts the Weberian ossicles with its dorso-anterior surface and additionally acts for gas storage during ABO ventilation (follows). Modifications for aerial respiration occur in the anterior third of the organ's posterior chamber. Viewed from the outside, this region is slightly expanded, profusely vascularized, and spongy in appearance. The inner surface of this area is subdivided into a series of compartments, formed by trabeculae, that attach to the dorsal and ventral ligaments and wrap around the organ's inner surface (Figure 3.16). Supported on these trabeculations are layered arrays of circular and longitudinal filaments. These are linked by sheets of connective tissue and form several layers of spongy cells on the chamber's inner wall. The thin septal wall of each compartment is lined with capillaries.

Except for the dorsal and ventral longitudinal tendons, the more caudal regions of the posterior chamber are smooth. The pneumatic duct has a large diameter and is somewhat elongated, extending from the left side of the esophagus to the left-anterior part of the posterior chamber.

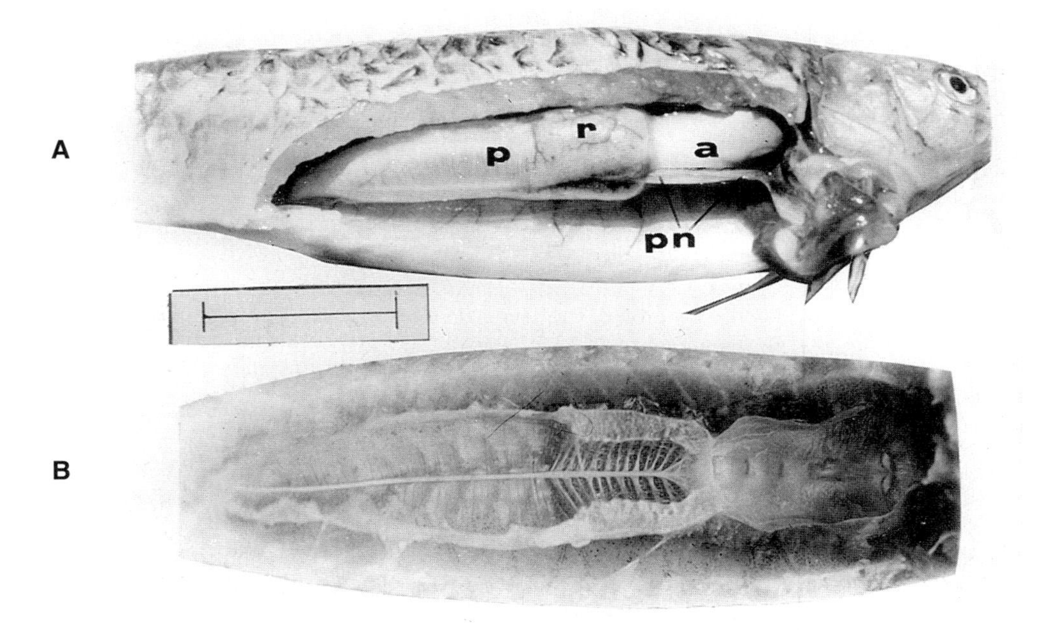

FIGURE 3.16 **A,** Lateral external view of the gas bladder in a 13.9 cm, 79 g *Piabucina festae* showing the anterior (a) and posterior (p) chambers, the respiratory region (r) in the posterior chamber, and the pneumatic duct (pn) running below the anterior chamber; scale = 2.7 cm. **B,** Ventral view of the gas bladder of a preserved *P. festae* (118 mm). Organ has been cut open along its horizontal midline to expose its inner walls and roof. Two bright areas in the front of the anterior chamber (right) are Weberian ossicles that connect to the chamber; scale = 1.6 cm. (From Graham *et al.*, 1977. Reprinted with permission of Springer-Verlag New York, Inc. from the *Journal of Comparative Physiology*, volume 122, pp. 295-310.)

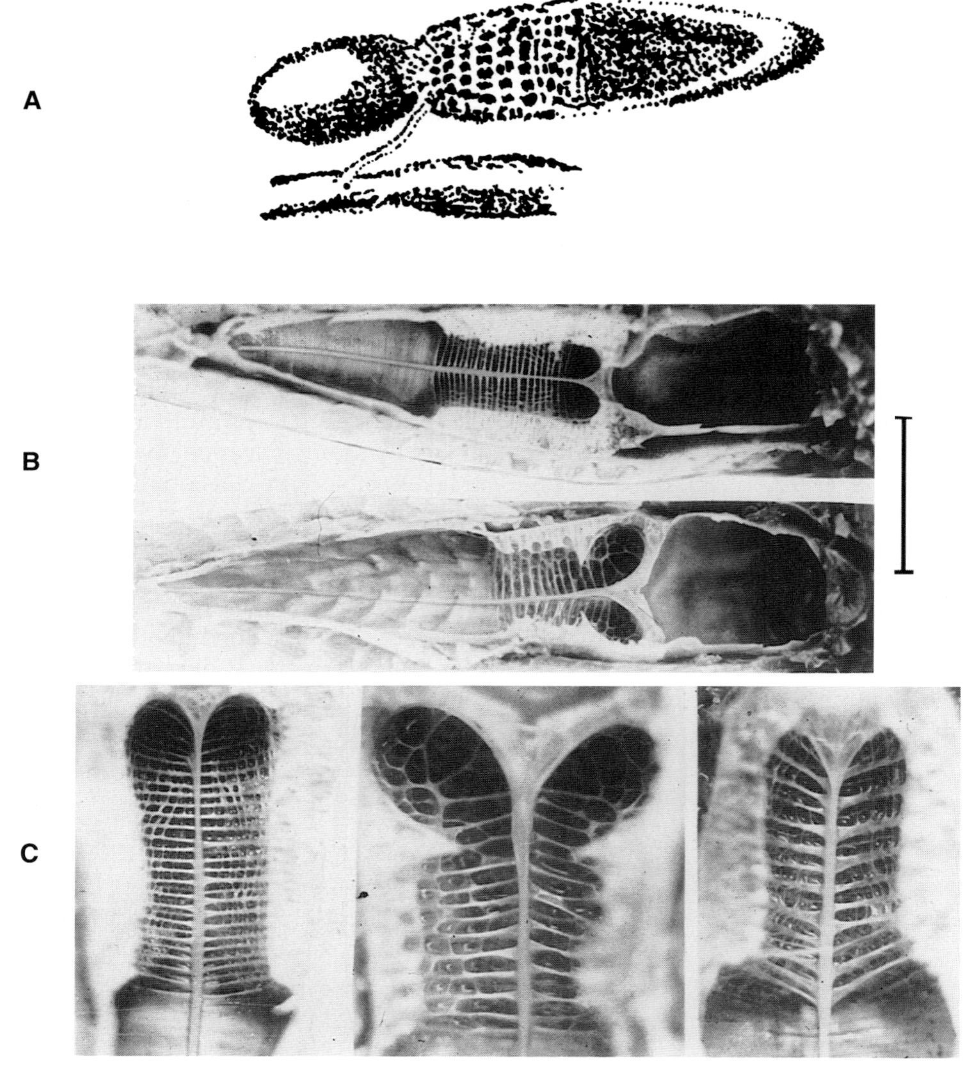

FIGURE 3.17 **A,** Redrawing of respiratory gas bladder of *Lebiasina*. (From Rowntree, 1903.) **B,** Ventral views of the gas bladder of *Hoplerythrinus unitaeniatus* (118 mm); and, just below, *E. erythrinus* (135 mm; scale 1.6 cm). The organs were cut open along their horizontal midline to expose the inner walls and roof. **C,** Ventral views of the respiratory sections in *Hoplerythrinus, Erythrinus,* and *Piabucina festae;* scale = 0.65 cm. (From Graham *et al.*, 1978. Reprinted with permission of The University of Chicago Press.)

Figure 3.17 shows generic differences in respiratory surface structure. Compared to *Piabucina*, the transverse trabeculae in *Hoplerythrinus* are more regularly spaced and twice as numerous, making for smaller cells. In *Erythrinus*, the front part of the respiratory region is wider and nearly bisected laterally by an extended medial band. Slightly posterior to this, the chamber lumen is constricted by an expanded trabecular ring. The functional significance of these differences is unknown. There are no recent illustrations of the organ of *Lebiasina*.

Pangasius

A respiratory gas bladder has been described in several species of *Pangasius*. Brief accounts of this organ are found in Taylor (1831), Owen (1846), and Day (1877). Modern descriptions are contained in works by Browman and Kramer (1985), Thakur *et al.* (1987), Zheng and Liu (1988), and Jionghua *et al.* (1988). As illustrated by Roberts and Vidthayanon (1991), there is considerable variation in the gas-bladder structure of different *Pangasius* species and these workers have expressed doubt about the respiratory

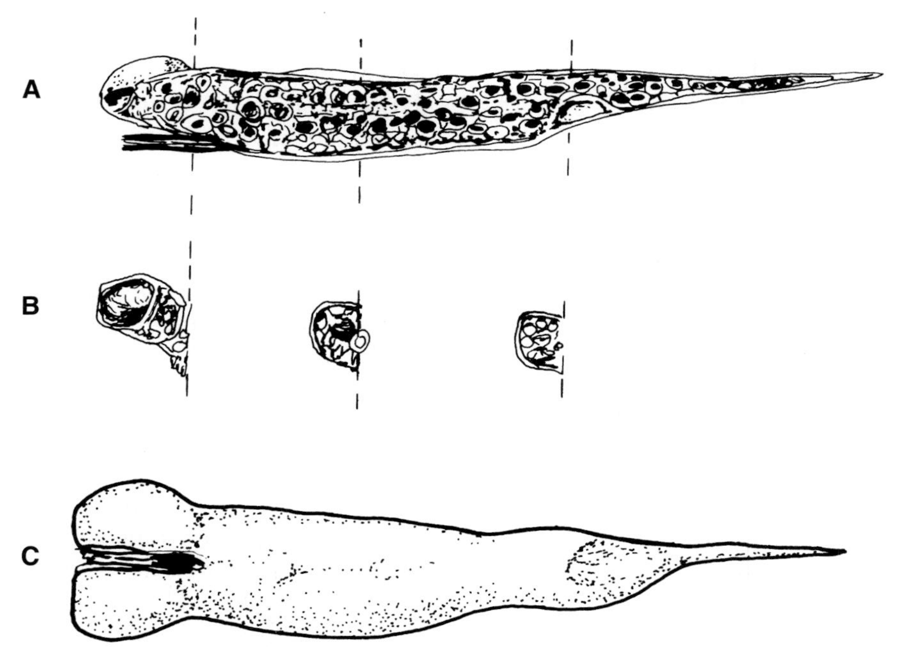

FIGURE 3.18 **A,** Sagittal mid-section of the respiratory gas bladder of *Pangasius nassutus* showing its cellular structure. **B,** Transverse sections of the organ at points corresponding to vertical lines. **C,** Ventral view showing the bilobed anterior section and the position of the pneumatic duct. (Redrawn from Kawamoto *et al.*, 1972.)

function of this organ (Chapter 2). For *P. sutchi*, however, the detailed accounting of this organ's ontogeny and ultrastructural detail given by Liu (1993) leaves little doubt as to its respiratory function.

Browman and Kramer (1985) describe the gas bladder of *P. sutchi* (*hypophthalmus*) as a single chamber that is wide and bilobed anteriorly but gradually tapers to a point at the end of the coelom (Figure 3.18). The anterior chamber has smooth walls and an open medial space. The balance of the organ, about 80%, is entirely trabeculated except for paired cylindrical passages connecting the medial space of the anterior chamber. The pneumatic duct rises from the roof of the esophagus and enters the front of the gas bladder. An SEM of the respiratory epithelium was illustrated by Liem (1988) and ultrastructural details are presented in Liu (1993).

Gymnotus

The aerial respiratory function of the posterior gas bladder chamber was described for *G. carapo* by Liem *et al.* (1984). The organ is housed within pleural ribs in the upper coelom and is considerably longer in *Gymnotus* than other members of the Gymnotidae. The long pneumatic duct extends from the dorsal esophagus to the anterior-ventral edge of the second chamber. Near its connection with the esophagus, the duct is thick and muscular and contains a sphincter valve. At about one-half the distance between the esophagus and the posterior chamber, a small branch extends anterodorsally from the pneumatic duct and connects with the anterior gas bladder (Figure 3.5).

Liem *et al.* (1984) describe the inner wall of the posterior chamber as "highly vascularized," but no other details are known. My study of preserved specimens confirmed the Liem *et al.* account and showed the inner wall of the posterior chamber to be smooth.

Umbra

All species of *Umbra* possess a respiratory gas bladder (Rauther, 1914; Geyer and Mann, 1939a,b; Black, 1945; Jasinski, 1965). The organ is expanded dorsally but is thin walled and lacks septations (Figure 3.5). General organ structure and wall histology were shown by Rauther (1914). Jasinski (1965) observed that in *U. krameri* the organ's vascular lining was uniformly distributed over about 80% of the surface and had an average capillary density of about 9750/linear mm of wall. A large diameter, but short, pneumatic duct runs vertically from the dorsal wall of the esophagus directly into the lower-anterior margin of the gas bladder. In addition to a respiratory epithelium, the gas bladder of *Umbra* has a *rete* and a gas gland (structures associated with counter-current O_2 secretion).

Comparative Morphometrics and Functional Morphology of Fish Lungs and Respiratory Gas Bladders

Morphometrics

My attempt to compile morphometric data for fish lungs and respiratory gas bladders reveals that little comparative data actually exist (Table 3.2). Estimates of total ABO volume (range 2–8% of body mass) and tidal volume are available for only five of the 24 genera having these types of ABOs. There are few estimates of ABO surface area and these range considerably, from less than 1 to 14% of body mass. Measurements of organ blood-air diffusion distances are more numerous (although many of the values in Table 3.2 are the result of my having measured published photos and drawings of organ sections) and show an overall range of from less than 1 to about 2 μm.

Several studies point out differences in the degree of septation and relative vascularization of the ABO. Klicka and Lelek (1967) found that polypterid and dipnoan lungs have about the same diffusion barriers; however, the inner walls of polypterid lungs are smooth and have much less total surface. Crawford (1971) also noted that gars have more, and smaller, alveolar compartments than does an *Amia* of the same body size; however, both organs have about the same diffusion distances.

Detailed morphometric analyses have only been conducted on the lungs of *Lepidosiren* (Hughes and Weibel, 1976, 1978) and *Protopterus aethiopicus* (Maina and Maloiy, 1985). Comparison of these data (Table 3.2) leads to the unsatisfactory conclusion that the lung of *Protopterus* has about half the diffusion (mean harmonic thickness) distance as that of *Lepidosiren* but over 15 times the total surface area (< 1 versus 14 cm^2/g, Table 3.2), and thus about a 41 times greater D_t. While specific differences are expected, the surface area estimate for *Protopterus* by Maina and Maloiy (1985) is exceptionally high (i.e., greater than varanid

TABLE 3.2 Comparative Morphometrics of Fish Lungs and Respiratory Gas Bladders

Species	Organ volume (% mass)	Tidal volume (% organ volume)[a]	Organ surface area (cm^2g)	Diffusion barrier[b] (μm)	D_t	D_L	Surfactant lining present[20,21,22]
Protopterus aethiopicus	1.7[5]		14	0.37	0.163	0.002	X
Protopterus aethiopicus	2.9[6]	80					
P. annectens				0.5–1[7]			
P. amphibius	3–6[27]						
Lepidosiren paradoxa[4]	2.3		0.85	0.86	0.004	0.003	X
Neoceratodus forsteri							X
Polypterus senegalus	2.94–3.67[24]	40[19]					
Polypterus sp.				0.5[7,11b]			X
Erpetoichthys calabaricus				1–2[10b]			
Lepisosteus	8.3[1]	50		<1[18b]			
Lepisosteus	8–10[2]	40	0.38-0.69[3]				
L. oculatus		25±3(n)[25] 35±1(h)					
Amia calva	8.0	15–34[26] 25[28]		<1[18b]			X
Arapaima gigas	10[16]	70[16]		0.5[14b]			
Notopterus chitala				<4[23b]			
Megalops cyprinoides				<1[15b]			
Hoplerythrinus unitaeniatus	3–8	19–130[12]		0.5–2[9]			X
E. erythrinus				0.1–2[13b]			
Piabucina festae[8]		14					
Umbra pygmaea				0.5[17]			

[a]Unless otherwise stated.
[b]Denotes estimates from illustrations.
[1]Potter, 1927, *L. osseus* and *L. platystomus*, 600 g; [2]Rahn *et al.*, 1971, *L. osseus*, 600–1730 g; [3]Rahn *et al.*, 1971, *L. osseus*, mass and area ($n = 2$) 650 g–250 cm^2; 1730 g–1200 cm^2; [4]Hughes and Weibel, 1976; [5]Maina and Maloiy, 1985; [6]Jesse *et al.*, 1967 (species either *aethiopicus* or *dolloi*, 43–66 g); [7]Klicka and Lelek, 1967; [8]Graham *et al.*, 1977; [9]Cruz-Landem and Cruz-Höflung, 1979; [10]Purser, 1926; [11]Poll and Deswattines, 1967; [12]Kramer, 1978a; [13]Carter and Beadle, 1931; [14]Dorn, 1968, 1983; [15]de Beaufort, 1909; [16]Randall *et al.*, 1978a; [17]Rauther, 1914; [18]Crawford, 1971; [19]Magid *et al.*, 1970; [20]Phleger and Saunders, 1978; [21]Liem, 1988; [22]Hughes and Weibel, 1978; [23]Dehadrai, 1962a; [24]Babiker, 1984a (240–380 g); [25]Smatresk and Cameron, 1982a (values are ml/kg ± SE in normoxia and hypoxia); [26]Hedrick and Jones, 1993; [27]Lomholt, 1993; [28]Deyst and Liem, 1985.

lizards and approaching mammalian levels), and the authors claim for a much greater area in *Protopterus* versus *Lepidosiren* is neither supported by visual comparisons nor by earlier quantitative statements regarding differences in lung compartmentalization (Robertson, 1913; Poll, 1962). Clearly, more and better data are needed to resolve this discrepancy and to learn more about the comparative morphometrics of lungs and respiratory gas bladders.

Epithelial Cell Classification

Differences in epithelial cell morphology have been noted for various organs (Klicka and Lelek, 1967; Crawford, 1971; Cruz-Höfling *et al.*, 1981; Liem, 1988). Regardless, it is fair to state that the respiratory epithelia in most fish lungs and respiratory gas bladders are basically similar. In cases where tissues have been examined closely, the cells resemble the type I epithelial cells forming most of the exchange surfaces present in the tetrapod lung (Hughes, 1973a, 1978; Hughes *et al.*, 1973a; Liem, 1988). These fish respiratory organs also have rows of ciliated epithelia (e.g., the furrows in polypterids) as well as tufted, mucous-producing cells that are considered homologous with the type II cells of the tetrapod lung (Hughes and Weibel, 1976; Liem, 1988). Thus, in all organs examined to date, cells functionally similar to (if not in all cases homologous with) cell types I and II are found.

Surfactants

Occurrence. For the relatively few organs examined thus far, evidence (from EM and washings) exists for the presence of organ-lining substances with surfactant-like properties associated primarily with type II cells (Hughes, 1973a; Hughes *et al.*, 1973a; Pattle, 1976; Cruz-Höfling *et al.*, 1981; Liem, 1988). In further similarity with the tetrapod lung, ontogenetic changes from type II to type I cells have been noted in gars (Henderson, 1975). Also, lungfish and polypterid lungs have been shown to have laminated osmophilic bodies structurally similar to the cytosomes (lipid surfactant-containing bodies) of tetrapod lungs (Hughes *et al.*, 1973a; Hughes and Weibel, 1978; Liem, 1989a).

In tetrapods and particularly mammals, pulmonary surfactants play an important role in preventing the surface-tension induced collapse of the minute lung alveoli (Pattle, 1976; Phleger and Saunders, 1978; Daniels *et al.*, 1995). There are several reasons why the respiratory lungs and gas bladders of fishes would require a surfactant. Many of these organs have a complex gas exchanging surface that may be subject to collapse. In addition, most of these organs are routinely deflated (i.e., opposing walls come into contact) during expiration. Smits *et al.* (1994) demonstrated that surfactants in the ABOs of *Erpetoichthys*, *Polypterus*, and *Lepisosteus* have an "antiglue" function and permit the reinflation of the ABO. This was shown by measuring a significant increase in the pressure required to inflate the organ once the surfactant had been washed away.

Structure and adaptive significance. Comparative studies have shown that the surfactants of the primitive actinoptergyians and lungfish have greater amounts of cholesterol but lower amounts of disaturated phosholipids than do those of reptiles and mammals (Orgeig and Daniels, 1995). These differences correlate with the respiratory complexity of the ABO and with the greater surface tension reducing requirements of the mammalian lung (which correspondingly has a greater quantity of disaturated phospholipids, Daniels *et al.*, 1995). Comparisons of the lung surfactants of *Neoceratodus*, *Protopterus*, and *Lepidosiren* show that the former, which has a less complex lung, has a correspondingly greater quantity of cholesterol and fewer disaturated phosphates. In fact, the cholesterol and disaturated phosphate profiles of both *Lepidosiren* and *Protopterus* more closely resemble those of tetrapods. These three lungfishes also differ in the types of phospholipid making up their surfactant: *Neoceratodus* has a lower percentage of phosphatidylcholine but a greater amount of sphingomyelin than the other two lungfish genera (Daniels *et al.*, 1995; Orgeig and Daniels, 1995). The functional significance of these and other phospholipid profile differences in relation to lung structure is unknown.

Surfactants probably have other functions and may have first evolved in primitive air breathers for reasons other than their effect on surface tension. Functions suggested for the lung and respiratory gas bladder surfactants of fishes include: Prevention of epithelial dehydration, facilitation of foreign substance removal, prevention of tissue oxidation in the presence of high O_2 tensions, the mechanical cushioning of the epithelium during the pressure surges and wall distensions taking place during buccal force ventilation, and anti-edema (Hughes and Weibel, 1978; Liem, 1988, 1989a; Daniels *et al.*, 1995).

Air-Ventilation Mechanics

The buccal pump. Ventilation studies using X-ray motion pictures, pressure recordings, and electromyograms (emgs) all demonstrate that, with two exceptions, fishes use a buccal force pump to inject air into their lungs and respiratory gas bladders (Liem, 1988). Although all genera have not been examined, the functionality of the buccal force pump has been shown for lungfishes, *Amia*, *Lepisosteus*, *Notopterus*, *Arapaima*,

Gymnotus, *Pangasius*, *Gymnarchus*, and *Hoplerythrinus* (McMahon, 1969; Kramer, 1978a; Liem, 1988).

McMahon (1969) described a two-step buccal pump lung ventilatory mechanism for *Protopterus aethiopicus*. This begins as the fish contacts the water surface and emerses its mouth. The first step, buccal and branchial expansion, then occurs and, as the mouth opens, fresh air fills the cavity. Lung gas is then exhaled into the buccal chamber space and, because the mouth remains open, the stale air becomes equilibrated with ambient air. Because the inflated lung has an elevated pressure, lung emptying is driven by elastic recoil. The contraction of smooth muscle layers within the organ's wall may also contribute to exhalation as might hydrostatic forces. The latter contribution depends upon fish body angle relative to the water surface. In water as deep or deeper than body length, a fish will be nearly vertical. In shallow water it will raise its head by abruptly curving its body. The second step, air inhalation, involves a series of actions, jaw closure, elevation of the buccal floor, and sealing the operculae, all of which compress the enclosed air and force it into the lung.

Liem (1988, 1989a) described the buccal ventilatory pump associated with the lungs and gas bladders of most actinopterygians as having four phases:

1. The transfer phase: Exhalant gas is passed out of the ABO and into the buccal chamber. In most instances this is actively driven by the combined actions of elastic recoil and the compression of muscles in the ABO wall or its septa, together with the suction generated by rapid expansion of the buccal floor. Both buoyant forces and the hydrostatic pressure gradient in an inclined fish can also assist gas transfer.
2. The expulsion phase: Air is eliminated from the mouth or operculae, either by passive (hydrostatic or buoyancy mechanism) or active forces (rapid raising of the mouth floor and jaw closure).
3. The intake phase: The action of expansive muscles associated with the jaw and operculae rapidly lowers the mouth-floor and aspirates air.
4. The compression phase: Both jaw closure and mouth floor elevation force new air into the ABO.

Variations on the buccal-pump theme. Cineradiographic and fluoroscopic analyses by Brainerd (1994) confirmed the buccal ventilatory phases of lungfish and actinopterygians described by McMahon (1969) and Liem (1988, 1989a). Brainerd (1994), moreover, extended her comparisons to include several amphibians and concluded that differences between the four-stroke ventilatory mechanism of actinopterygians and the two-stroke mechanism of the sarcopterygians (lungfish as well as frogs and salamanders, and most tetrapods) indicate the probable independent origin of each of these pumps from different types of pre-existing pumps and not divergence from the respiratory pump of a common ancestor. Brainerd suggests that the two-stroke buccal pump evolved from the gill irrigation pump whereas the four-stroke pump originated from a combination of suction-feeding and coughing actions.

It does not appear that Brainerd (1994) takes issue with the hypothesis of a common air-breathing ancestor for the actinopterygians and sarcopterygians. The question becomes whether or not the differences she has detailed for the two- and four-stroke ventilatory mechanisms warrant the postulate that, while conserving air-breathing by a lung or lung-like organ, one of these lineages entirely restructured its ventilatory mechanics. In my opinion, we need to search for additional evidence documenting autapomorphy in actinopterygian and sarcopterygian buccal ventilatory pumps. As Brainerd (1994) points out, hers is not the most parsimonious conclusion and there are several exceptions that complicate her interpretation. For example, not all of the lung or gas bladder ventilating actinopterygians employ the four-stroke pattern (*Hoplerythrinus*, *Gymnotus*). Similarly, not all amphibians use a two-stroke lung-ventilation pattern (*Amphiuma*, *Xenopus*).

Aspiratory ventilation. The exceptions to buccal force ventilation occur in the polypterids (*Polypterus* and *Erpetoichthys*) which use "recoil aspiration." Brainerd *et al.* (1989) showed that exhalation by these fishes is driven by contraction of the lung wall which also deforms the body wall and establishes a negative pleuroperitoneal pressure of between -4 to -6 torr. The lungs are then filled by aspiration; the source of this suction being the post-expiration recoil of the ganoid scale-reinforced skin and body wall. In his analyses of the lungs of *Erpetoichthys*, Purser (1926) had also concluded that lung filling had to occur by aspiration. Purser wrote,

> Movement of the body wall in and out is therefore very much restricted and the result of the forcible expiration is to put such an amount of tension upon the body wall that it expands the lungs automatically when the muscles are relaxed again, much as in the Chelonia.

Whereas both *Erpetoichthys* and *Polypterus* have been observed to inhale through their mouths, the presence of a negative-pressure pulmonary aspiratory mechanism adds considerable credibility to the observations made by Budgett (1903) and Magid (1966) that *Polypterus* could rapidly inhale air through its spiracles. Claims for suctional lung filling by aestivating

Protopterus (Lomholt *et al.*, 1975) have not been substantiated (DeLaney and Fishman, 1977). Similar claims of aspiratory filling for *Arapaima* (Farrell and Randall, 1978) were also rejected (Greenwood and Liem, 1984).

Factors Contributing to Specific Ventilatory Differences

A common feature shared by all air-breathing fishes except the polypterids is use of the buccal force pump to fill the ABO. However, there is a cadre of species-specific differences that can affect the timing and number of steps preceding the buccal forcing of air into the ABO (Chapter 6).

Morphology. Included among morphological factors are: Skull size and snout length. Branchial chamber capacity and its vertical rise above the glottis. Position of the glottis. Lung or gas bladder morphology including position, volume, tidal volume, and the presence of a muscular lining. Buoyancy and body flexibility and, the presence of thick scales that may preclude hydrostatic compression of the body or augment the aspiration of air.

The composite effect of these variables can be illustrated by comparing air breathing in a lungfish and *Amia*. An African or South American lungfish resting among weeds in shallow water can leisurely approach the water surface and take as long as six seconds (see following) to empty and fill its lungs. These animals are negatively buoyant and can, while contacting the substrate, bend their bodies vertically in order to extend the mouth to the surface. An *Amia*, while likely to occur in a similar habitat setting, has neither the body flexibility nor fin dexterity of a lungfish. *Amia* is also more nearly neutrally buoyant and thus less often in contact with a supporting substrate when air breathing. It cannot elevate its mouth above the water surface for as long. It would be important to compare the breathing pattern of the Australian *Neocertadous* because it has a somewhat stiffer body than the other lungfishes and it occurs in more open water and has facultative breathing and buoyancy requirements similar to those of *Amia* (Chapter 5).

Ventilatory time. The time required for air ventilation can vary among and within species. An undisturbed lungfish may take two to six seconds to complete its ventilation (Brainerd, 1994). Other ventilatory times reported by Brainerd (1994) are: *Polypterus senegalus*, 350 ms; *Lepisosteus oculatus*, 400 ms; *Amia calva*, 850 ms. Fishes disturbed by potential predators can ventilate in much less than one second (Kramer, 1978a) and some reduce time at the surface by completing the transfer and expulsion phases during ascent, by making a very rapid (300 ms) intake, and then carrying out compression during descent (Kramer and Graham, 1976; Liem, 1988).

Ventilation volume. We lack information about the degree to which fish can completely empty or turnover their lung-respiratory gas bladder contents with each ventilation. Liem's X-ray data and other measures (tidal volume and buoyancy) suggest that most lungs and respiratory gas bladders have a post-exhalation residual gas volume. Owing to the airway dead space separating the ABO and ambient air, there may also be a tendency for some groups to re-inhale stale air (Brainerd, 1994). Because of the important bearing these variables have on respiratory gas-exchange efficacy, more data are needed.

Anatomical work to date has revealed the general presence of muscle layers in the walls and septa of these organs, and this clearly implicates the potential use of an active gas transfer phase under certain conditions. However, the quantities of muscle differ and some organs have smooth rather than striated fibers (or a mixture of both) which likely affects the potential rate of organ compression.

The expiration–inspiration sequence. Another unresolved question related to this concerns the priority of expiration and inspiration in organ ventilation (Gans, 1970; Kramer, 1978a; Liem, 1989a). From the description of buccal-force pump mechanics, it would be expected that fish normally exhale prior to inhaling and most do. However, this is definitely not the case for *Pantodon*, the four characin genera, *Gymnotus*, *Umbra*, and perhaps others, which inhale new air before expiring the old. Also, whereas Liem (1989a) reports that *Notopterus chitala* exhales before inhaling, Kramer *(in litt.)* reported the opposite sequence in films made for both *N. chitala* and *Xenomystus nigri*.

We do not understand the significance of this reversed order. Kramer's (1978a) X-ray analysis of air ventilation in *Hoplerythrinus* showed that inspired air was first stored in the anterior chamber and then passed to the posterior (respiratory) chamber after a bolus of old gas was ejected into the pneumatic duct. This indicates that the mixing of old and new air, although taking place, is minimized (Graham *et al.*, 1977, 1978b). *Gymnotus*, however, mixes the inspired and resident gases prior to expiration (Liem *et al.*, 1984). It may be that inspiration before expiration minimizes time at the water surface and thus serves an anti-predator function. Alternatively, it may be that muscles in the chamber wall cannot compress the organ sufficiently to cause ventilation or that expiration briefly compromises the organ's buoyancy and sound detection functions. The relatively long pneumatic duct in some of these genera may also limit gas exhalation rate, particularly without the large exhalation pressure. Prior inspiration may thus pressurize the organ to permit passive exhalation (Liem *et al.*, 1984).

ABOS OF THE HIGHER TELEOSTS

The balance of this chapter focuses on the ABOs and aerial respiratory surfaces of the higher euteleosts. For comparative purposes ABOs occurring in similar regions of the body are described together.

ABOs of the Head Region

Epithelial Surfaces That Function as ABOs

Specialized air-breathing respiratory epithelia on surfaces of the buccal, pharyngeal, branchial, and opercular chambers have been reported for no less than 16 air-breathing genera (Table 3.1). Most fishes possessing this feature are either amphibious or hold air in their mouths while air breathing. Modifications for air breathing range from increased buccopharyngeal vascularization (Oglialiro, 1947) to elaboration of ABO surface area through expanded diverticulae or pouches (Das, 1934). It is common for the respiratory epithelium to have a contiguous distribution across several adjoining chamber surfaces in some species.

The following account of the chamber surfaces modified for aerial respiration is confined to genera for which some structural detail is known: *Electrophorus*, several gobies (mudskippers, *Gillichthys* and *Pseudapocryptes*), *Channa*, and *Monopterus*. Most species in this category (Tables 2.1 and 3.1) have not been studied.

Electrophorus

Structure. A description of the ABO of *E. electricus* (formerly *Gymnotus electricus*) was found in the unpublished works of the 19th century biologist John Hunter (Evans, 1929). The ABO of this fish was one of the last to be described; however, four highly similar papers appeared in the span of six years (Evans, 1929; Böker, 1933; Carter, 1935; and Richter, 1935). Works by Evans, Böker, and Carter made similar interpretations of histological sections of the vascular epithelium. Böker published a photo of the entire ABO and Carter provided an especially detailed analysis of its circulatory pattern, as did Richter. Dorn (1972, 1983) has also provided details on the fine structure of this ABO.

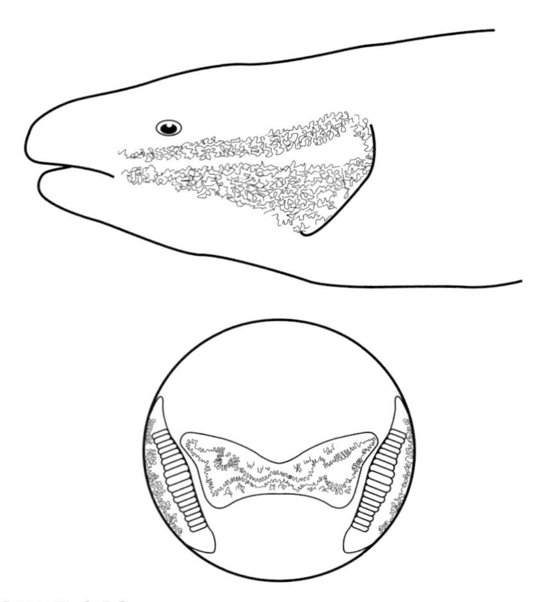

FIGURE 3.19 Side and transverse views of the buccopharyngeal epithelium of *E. electrophorus*. Transverse section shows relatively reduced gills, presence of a respiratory epithelium in the opercular chamber, and the interdigitation of epithelial tufts on the upper and lower jaws. (Modified from Carter, 1935. Printed with permission of the Journal of the Linnean Society of London.)

The *Electrophorus* ABO consists of a vascular epithelium that:

i. Occupies nearly the entire upper and lower surfaces of the mouth (i.e., from the jaw to the pharyngeal teeth)
ii. Occurs on each branchial arch and proliferates in the dorsal part of the branchial chamber, and
iii. Extends to the opercular wall (Figure 3.19).

Respiratory tissue in the mouth occurs mainly in tufts that are arranged in longitudinal rows and give the appearance of a patch of cauliflower. The tufts are supported on elevated connective-tissue platforms, and their placement is such that, when the mouth is closed around an air breath, the longitudinal rows and individual tufts interdigitate for maximum exposure of the vascular surface (Figure 3.20).

The epithelial tufts are formed on a base of soft cartilage covered by a layer of fibrous tissue containing numerous blood vessels and sinuses. The covering layer of vascular epithelium is thin and capillaries appear along the entire surface of most areas, and the blood–air diffusion distance is 1–2 μm. In some areas, this distance is even less because emergence of capillaries into the lumen gives rise to a corrugated surface.

Morphometrics. The estimated total ABO surface areas of five electric eels (mass 0.3–5 kg) was propor-

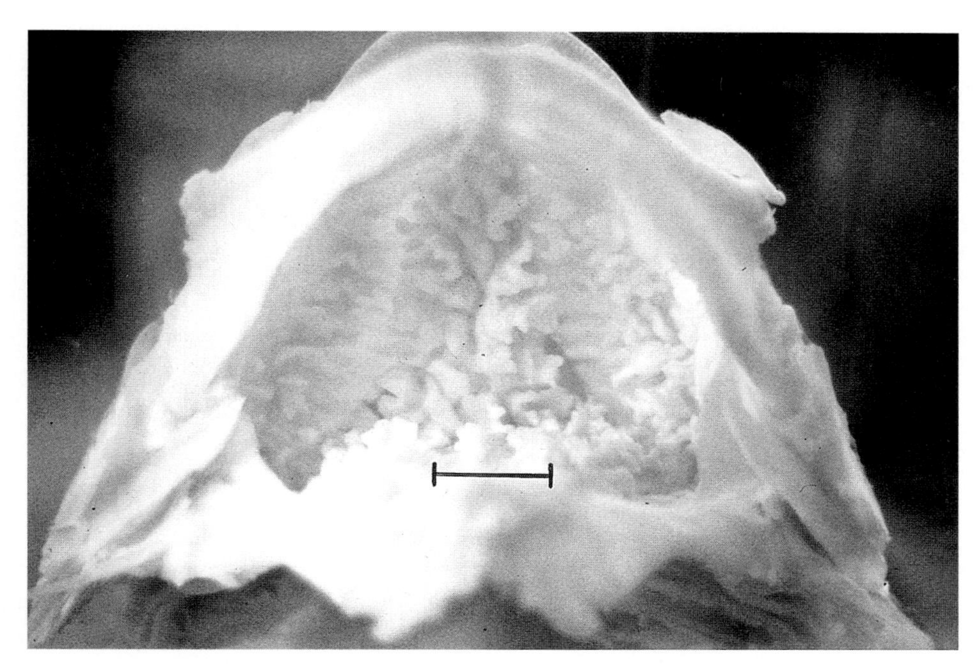

FIGURE 3.20 Dorsal view of the lower jaw epithelium of *Electrophorus;* scale = 1 cm.

tional to body size and ranged between 65 and 372 cm^2 (Johansen *et al.*, 1968b). Expressed in relation to mass, average ABO surface area was 0.16 cm^2/g (range 0.07–0.22) or 14.1% (12.7–15.5) of body surface area.

Air-ventilation mechanics. *Electrophorus* inspires through its mouth and exhales via its opercular slits. Replicate observations (Farber and Rahn, 1970) showed the average inspired ABO volume in a 2.8 kg fish was 35 ml (0.0125 ml/g) and that the ABO was completely emptied during exhalation.

Mudskippers

Cuvier and Valenciennes (1837, page 179) speculated that the reduced size of the branchial chamber opening of mudskippers, ". . . Les orifices ètroits de leurs ouïes," permitted them to live out of water and the role of branchial chamber specializations has been amply demonstrated.

Structure. Amphibious, mudskippers breathe air by opening their mouths or by filling the entire expanse (buccal, pharyngeal, branchial, and opercular cavities) with air. All of these surfaces, together with the gills and the skin, are considered to be functional for mudskipper aerial-respiration (Schöttle, 1931; Graham, 1976a; Biswas *et al.*, 1981 [cited in some places as Niva *et al.*]).

Structural modifications favoring air breathing have been identified in *Periophthalmus vulgaris* (Singh and Munshi, 1969), *Boleophthalmus boddarti* (Schöttle, 1931; Biswas *et al.*, 1981), and *Periophthalmodon schlosseri* (Yadav *et al.*, 1990) and are likely present in all mudskippers. These include (Figure 3.21):

i. A membranous curtain extending between the ceratobranchial segment of the first gill arch and the hyomandibular bone that reduces the size of the first branchial cleft and forms a small inhalant aperture into the branchial and opercular chambers
ii. A vascularized branchiostegal membrane that effectively surrounds the gills to control ABO volume and exhalation, and
iii. An increased volume in both the pharynx (slight) and particularly the opercular chambers (i.e., both infra- and suprabranchial recesses are present).

Respiratory surfaces and morphometrics. Schöttle (1931) differentiated various mudskippers on the basis of the degree of capillary penetration into the epithelial surfaces of the buccopharynx; capillaries were predominately sub-epithelial in *P. koelreuteri,* occurred both below and within the epithelium in *B. boddarti,* and were almost exclusively intra-epithelial in *P. schlosseri.*

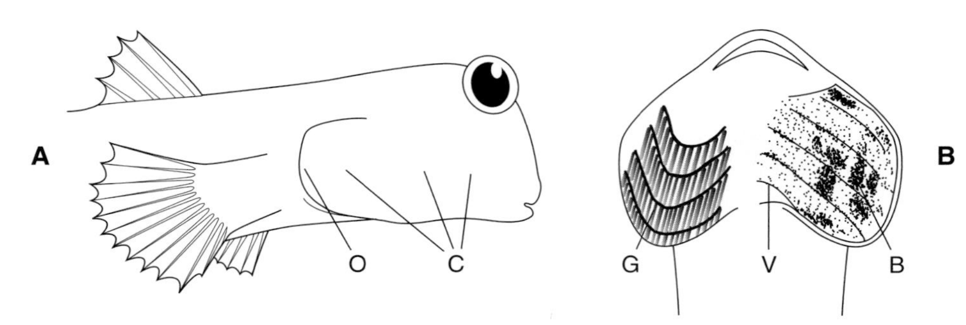

FIGURE 3.21 **A,** Side view of *Periophthalmus* showing the expanded buccal, branchial, and opercular chambers (C) and opercular opening (O). **B,** Ventral view of branchial chamber with operculum lifted revealing the branchiostegal membrane (B) that wraps around the gills (G) and controls ABO volume through a ventro-medial valve (V). (Modified from Singh and Munshi, 1968.)

TABLE 3.3 Comparative ABO Morphometrics among the "Higher" Teleosts

Species	Scaling equation total area (cm^2)		A_{100}		Diffusion distance	Diffusing capacity	
	a	*b*	(cm^2)	(cm^2/g)	(μm)	D_t (ml O_2/min/torr/kg)	D_L
Clarias batrachus[1]							
SBC	0.522	0.736	15.48	0.155	—	0.0420	
gill fans	0.131	0.707	3.40	0.034	—	0.0005	
arborescent organ	0.600	0.840	28.72	0.287	0.55	0.0773	
C. mossambicus[2]							
SBC	0.430	0.687	10.17	0.102	0.313	0.050	
gill fans	—	—	—	—	—		
arborescent organ	0.177	0.754	5.70	0.047	0.287	0.070	
Heteropneustes fossilis[3]							
air sac	1.469	0.662	30.98	0.310	1.605	0.0288	
Anabas testudineus[4]							
SBC	0.554	0.574	7.79	0.078	0.210	0.0539	
labyrinth organ	0.807	0.779	29.17	0.292	0.210	0.2286	
Channa punctatus[7]							
SBC	1.591	0.696	39.23	0.392	0.780	0.0753	
C. striatus[1]							
SBC	1.894	0.543	23.09	0.231	1.359	0.0254	
C. gachua[1]							
SBC	0.860	0.678	19.52	0.195	0.56	0.0524	
Boleophthalmus boddarti[6]							
opercular chamber	0.523	0.544	6.40	0.064	1.22	0.0080	
opercular chamber and gills	3.363	0.682	77.75	—	—		
Pseudapocryptes lanceolatus[1]							
opercular chamber	0.334	0.738	9.99	0.100	—		
Monopterus cuchia[5]							
air sac (<60 g)	0.178	0.714	4.77	0.048	0.44	0.0165	
air sac (>60 g)	0.0173	1.19	4.15	0.041	—		
M. cuchia (200 g)[8]							
air sac and buccopharyngeal epithelium	—	—	10.00	0.100	0.72	0.0209	0.00165
with added mucus and plasma diffusion layers	—	—	10.00	0.100	2.41	0.0062	

[1]Munshi, 1985; [2]Maina and Maloiy, 1986; [3]Hughes *et al.*, 1974b; [4]Hughes *et al.*, 1973b; [5]Hughes *et al.*, 1974a; [6]Biswas *et al.*, 1981; [7]Hakim *et al.*, 1978; [8]Munshi *et al.*, 1989.

Singh and Munshi (1969) reported that the respiratory epithelium of *P. vulgaris* lines the mouth, the expanded pharyngeal and opercular chambers, as well as the branchiostegal membrane. They reported this epithelium to be particularly well developed within the opercular pouches where it was highly vascular (with blood-air diffusion distances from 1 to 3.2 μm) and folded into alveolar-like spaces that also contained strips of muscle, apparently for expelling air during ventilation.

Biswas *et al.* (1981) describe the opercular-pouch expansion in *B. boddarti* as "balloon-like." They determined the scaling of the opercular pouch and gill areas for this species (Table 3.3) and combined these to estimate ABO surface area. The estimated mass coefficient for gill and opercular chamber surface area is $M^{0.682}$, and the total area of these two surfaces in a 100 g fish is 77.8 cm^2, with the opercular chamber only accounting for 8.2% of the area (Table 3.3). This, however, underestimates total air-breathing surface area because the buccal, pharyngeal, and cutaneous surfaces are not included. Biswas *et al.* also measured an opercular tissue diffusion distance of 1.22 μm (similar to that of *P. vulgaris*) and estimated an opercular-chamber D value of 0.008 (Table 3.3) for a 100 g fish.

Air-ventilation mechanics. There are no emg studies of mudskipper air ventilation, but Singh and Munshi (1969) did describe the actions of numerous muscles believed to be involved in the air ventilation of *P. vulgaris*. By their account, expansion of the mouth fills the buccal and pharyngeal chambers with air which is then, by contraction of the buccopharynx, squeezed through the inhalent aperture into the branchial-opercular chambers. Several workers have suggested that air is held under a slight pressure, but this has not been measured. Also, the amount of water held in the chamber is debated and may vary among species (Chapter 2).

Exhalation occurs by contraction of the opercular chambers through actions of the fibers of at least five different muscles inserting on the operculum. Contraction of the branchiostegal membrane (which is invested with strips of the superior hyoideus muscle) is also important for exhalation. This membrane wraps around the gills, is fused ventrally, and has a flap valve on its ventro-lateral surface. This valve is located just inside the opercular opening and functions as the exhalant aperture for the ABO (Figure 3.21).

Pseudapocryptes and *Apocryptes*

Structure. *Pseudapocryptes lanceolatus* air breathes during brief terrestrial excursions and when exposed to hypoxia; in doing the latter, it often floats at the water surface (Das, 1934). A vascularized epithelium lining the mouth, branchial chamber, and opercular chambers forms the ABO of this species (Schöttle, 1931; Das, 1934; Yadav and Singh, 1989). Similar to mudskippers, the buccal chamber and particularly the inner opercular wall in *Pseudapocryptes* and *Apocryptes* are richly vascularized (Figure 3.22). Also, the bran-

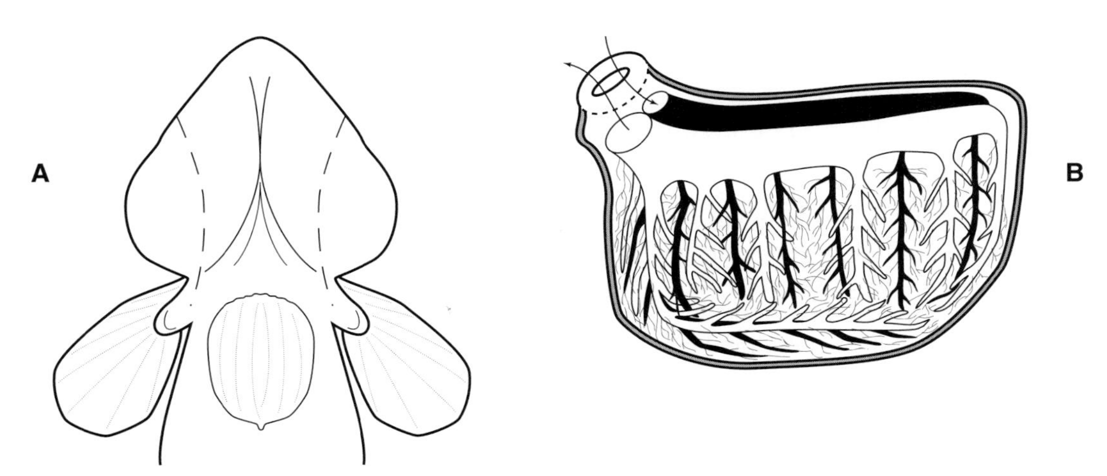

FIGURE 3.22 **A,** Ventral view of *Pseudapocryptes lanceolatus* showing pectoral and pelvic fins, the ventral convergence of the operculae, and the expanded opercular pouches (dashed lines). **B,** Medial view of the respiratory surface of the right operculum showing the progression of the opercular artery (black) and vein (white) along the dorsal margin and their several vertical branches. Arrows indicate blood flow. (Adapted from Das, 1934. Printed with permission of The Royal Society of London from Proceedings B., volume 115, pp. 422–431, figures 2b and 8.)

chiostegal membrane of this fish surrounds its gills and is an effective opercular seal, being both thick and endowed with numerous mucous cells.

Respiratory surfaces and morphometrics. Data for the allometric scaling of the opercular chamber of *Apocryptes* (Munshi, 1985) are contained in Table 3.3. In a 100 g fish, this organ has a surface area of about 10 cm^2. However, other surfaces doubtlessly contributing to aerial respiration remain unquantified. There are no estimates of ABO–tissue diffusion distance although Schöttle (1931) pointed out that capillaries occur within and commonly on the surface of this epithelium.

Air-ventilation mechanics. It is unknown what atmospheric pressures prevail when air is held; however, the opercular chamber of *P. lanceolatus* is highly distensible and its walls become bulged during air breathing. To prevent air loss, the opercular margins are firmly adducted to the body and sealed by mucus exuded from the branchiostegal membrane. With its operculae restricted, air ventilation by this species becomes tidal, and Das (1934) observed that with a sealed operculae and flexible opercular bones, air ventilation is similar to the action of a bellows.

Gillichthys

Gillichthys mirabilis gulps air at the surface and is also amphibious. The ABO of this species is formed by a triangular patch of respiratory epithelium in the roof of its mouth (Todd and Ebeling, 1966). Although vascular surfaces also occur on the tongue, the buccal roof provides the greatest area of contact with gulped air. Todd and Ebeling estimated the capillary density in this region to be about 6.5 vessels within each linear mm. These vessels lie close to the surface and are surrounded by numerous mucous cells.

Observations by Todd and Ebeling indicated that within 3 to 5 minutes of the induction of air breathing (by emersion, aquatic hypoxia, or hypercapnia), capillaries in the buccal roof patch and on the tongue of this fish begin to redden and that these become fully engorged with blood after 10–20 minutes of air breathing. As expected from these observations, histological sections from fish induced to breathe air showed increases in capillary distension compared to non-air-breathing controls. The increased perfusion actually bulged the vessels into the lumen, forming a corrugated surface. Air breathing also brought notable changes in mucous cell activity, apparently to lessen desiccation of the air-exposed surface (Todd and Ebeling, 1966). Extensive surface vascularization is not found in either the pharyngeal or opercular chambers.

Cryptocentroides, Mugilogobius, Chlamydogobius, and *Arenigobius*

Species in these four Australian goby genera hold gas bubbles in their mouths to facilitate ASR (defined in Chapter 1). Histological examinations of buccal chamber areas contacting these bubbles reveal a concentration of capillaries just below the epithelium at distances ranging from 9 to 31 μm from the surface (Gee and Gee, 1995). Areas showing increased capillarity include the roof of the mouth encompassing the anterior area under the eyes, the sides of the buccal chamber, and the tongue.

Similar to the findings for *Gillichthys* by Todd and Ebeling (1966), Gee and Gee (1995) observed that hypoxia exposure triggered an increased engorgement of these vessels in *Cryptocentroides cristatus, Arenigobius bifrenatus,* and *Chlamydogobius* sp. In *Mugilogobius palidus* these vessels appear fully engorged with blood at all times, however, hypoxia exposure did reveal additional vessels located within 10 μm of the surface.

Pouches Formed Adjacent to the Pharynx

Channidae

Structure. Early descriptions of the snakehead ABO are found in Taylor (1831), Cuvier and Valenciennes (1831, text and plate 206), and Hyrtl (1853). Later accounts are found in Rauther (1910), Senna (1924), Das (1927), Bader (1937), Marlier (1938), Munshi (1962b), and Liem (1980, 1984). Ishimatsu and Itazawa (1993) provide an integrated review of channid-ABO morphology, circulation, and physiology. Microvascular casting studies have yielded new details about microcirculation in the ABO (Olson *et al.,* 1994) and gills (Munshi *et al.,* 1994) of this genus.

The snakehead ABO consists of a pair of suprabranchial chambers (SBC) situated in the roof of the pharynx, above the gills, and adjacent to the skull (Figure 3.23). Bones of the skull form the organ's roof and median wall. The SBC is somewhat divided into anterior and posterior compartments by a shelf-like outgrowth of the hyomandibular bone. Air mixes easily between the anterior and posterior compartments, and the SBC is in open communication with the pharynx through three openings (Figure 3.24).

Respiratory surfaces. A respiratory epithelium lines the SBC, the bony projections (also termed nodules, dendrites or even labyrinthine organs by some workers, although not to be confused with the true

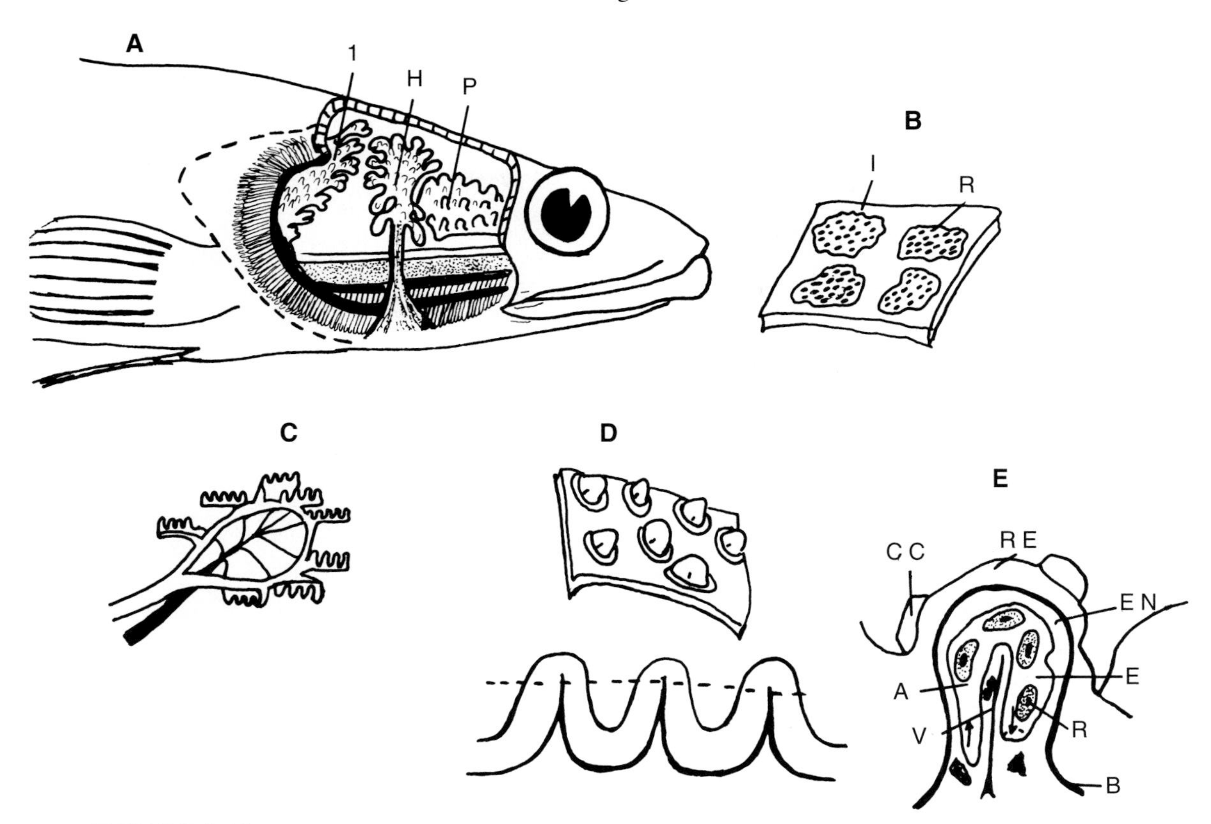

FIGURE 3.23 **A,** Lateral view of *Channa* showing the aerial-respiratory dendrites on the parasphenoid (P), hyomandibular (H), and first gill arch (1) filling the SBC. Upper body wall cut away to show how SBC extends vertically from the roof of the skull to the stippled area which marks the floor of the pharynx. Dashed line marks the perimeter of the operculum. **B,** Section of the SBC wall showing the distribution of four respiratory islets (I) and their numerous rosettes (R). **C,** Afferent (open) and efferent (dark) circulation to papillae within a rosette. **D,** Epithelial surface view of respiratory papillae and a side view of in-series papillae along one vessel (dashed line shows the level of penetration into the lumen). **E,** Papillar cross-section. RE, epithelial surface; CC, cap cells; B, basement membrane; EN, endothelial cells (cut away to show lumen); A, afferent and E, efferent sides of the loop (arrows show flow direction); V, central valve; R, red cell. (Adapted from Hughes and Munshi, 1973a; 1986. Used with permission of G.M. Hughes and J.S.D. Munshi.)

labyrinth apparatus of the anabantoids, see following) present on the hyomandibular and parasphenoid bones, the epibranchial section of gill arch 1, and the roof of the buccopharynx (Figure 3.23A; Hughes and Munshi, 1986; Munshi *et al.*, 1994). The respiratory tissue takes the form of vascular islets (i.e., patches surrounded by non-vascular lanes) which are in turn composed of numerous clumps called vascular rosettes. Each rosette is supplied by its own afferent and efferent blood vessel and is also spatially isolated from other rosettes by tissue clefts that, by forming micro-air pockets, increase rosette surface area (Figure 3.23B and C).

Individual rosettes consist of from 25 to as many as 100 vascular papillae which are vertical capillary loops bulging into the ABO's lumen (Figure 3.23D). Microvascular casts suggest this wave-shaped structure increases blood and air contact for efficient gas exchange (Munshi, 1985; Munshi *et al.*, 1994). Light and TEM sections show the papillae are lined with endothelial cells, have lumen diameters approximating red cell thickness, and contain unicellular valves presumed to influence blood flow (Figure 3.23E). Hughes and Munshi (1986) report regional ABO differences in both islet density and papillar ultrastructure. Buccopharyngeal papillae, for example, are less dense and also occur in cup-like receptacles and can be retracted into a depression, a useful expedient during feeding.

Comparative ABO morphometrics. Analyses to date have focused mainly on the dimensions of the SBC and not on the total ABO exchange area of the SBC and other aerial-respiratory surfaces. Consider-

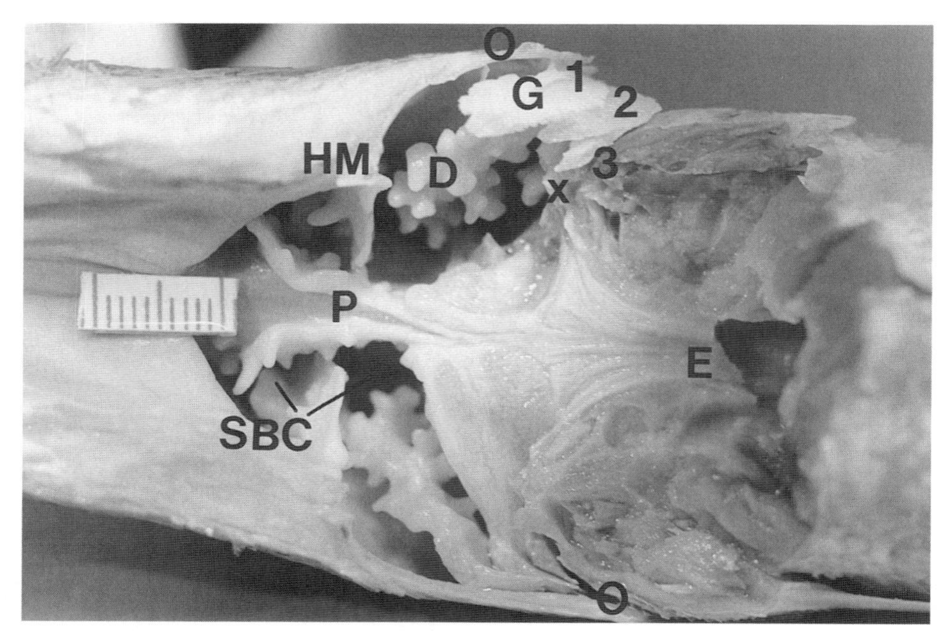

FIGURE 3.24 Ventral view of the head and ABO of a 30 cm preserved *Channa micropeltes*. Head has been sectioned at the level of the esophagus (E) and the roof of the mouth to show openings into the suprabranchial chamber (SBC). O, operculum; G, gill arches 1, 2, and 3; D, dendrites on gill arch 1. Note that dendrites occur on the hyomandibular bone (HM), the parasphenoid bone (P), and on gill arch 2 (x).

able variation exists in SBC–morphometric parameters for three *Channa* species (Table 3.3). Figure 3.24 shows that the Malaysian species *C. micropeltes* differs from other species in having respiratory nodules on the surface of branchial arch 2.

Das (1927) correlated body size with ABO development in *C. gaucha* (smaller species and smaller SBC) and *C. striatus*. This is supported by the relative organ volumes of these two fishes (Table 3.3). However, *C. punctatus*, which also attains a large body size, does not fit this pattern (Hakim *et al.*, 1978). Das (1927) also described the SBC of *C. gaucha* as being relatively smooth, but this is not consistent with its apparently larger mass exponent for surface area (Table 3.3). Nor is it consistent with the report (Hughes and Munshi, 1973a; Singh *et al.*, 1988) that respiratory papillae on the hyomandibular and first gill arch surfaces in *C. striatus* are more complex and that papillae cover a much greater area, including both the parasphenoid and the anterior region of the pharynx.

The air-blood diffusion distance in the snakehead SBC contains three cell layers (epithelial, basal, endothelial). The thicknesses of these layers in two species are as follows: *C. striatus* (1.36 μm total membrane thickness), epithelial layer (1.2), basal lamina (0.06), and endothelial cell (0.09); *C. punctatus* (0.78 μm total thickness) epithelial (0.47), basal (0.06), endothelial (0.26). Hughes and Munshi (1973a, 1986) also observed that the respiratory papillae in *C. striatus* are partially covered by epithelial cap cells which add to diffusion distance. Table 3.3 reveals the strong effect of interspecific differences in SBC area and diffusion thickness on values of *D*.

Air-ventilation mechanics. Based on anatomical observations (Munshi, 1962b, 1976; Singh *et al.*, 1988), the conventional view of SBC ventilation mechanics was that sets of muscles (levator arcus branchialum = levatores externi) extending from the skull to each branchial arch (and thus crossing the compliant posterior and lateral walls of the SBC) were active agents in compressing the organ to cause exhalation and that inhalation by elastic recoil was the consequence of subsequent muscular relaxation. However, this interpretation must be abandoned in the light of Liem's (1980, 1984) convincing emg and X-ray cineradiographic analyses revealing that the SBC is not compressed (i.e., organ pressure did not rise) during exhalation and that air ventilation has striking parallels with the normal teleost coughing reflex.

Liem (1984) was able to show that air-ventilation in four *Channa* species (*punctatus*, *striatus*, *gaucha*, *marulius*) resembles that of the anabantoids (see following) in being driven by a reversed water flow. He describes three phases of air ventilation (exhalation, aspiration, compression) and states that the discrete actions of no less than 11 muscle pairs are required. Exhalation occurs first and is driven by sequential expansion of the buccal and then the opercular cavities which, as the operculae are elevated, causes a reversed water inflow and displaces ABO air into the mouth. The emerged mouth is then opened to release air. Body inclination angle (about 40°, although determined largely by water depth and fish length) during ventilation determines the extent that the ABO is completely flushed of all residual gas (Ishimatsu and Itazawa, 1981, 1993). New air is then aspirated by an action similar to inertial suction employed for feeding; the jaw is quickly adducted and then reopened to aspirate a volume of fresh air sufficient to fill the buccopharynx. The chamber containing water and the newly acquired air is then compressed by a rising buccal floor and adduction of the opercular walls. This action forces air back towards the gills where it then pops up into the SBC and displaces water which is ejected from the operculae.

Liem (1984) also observed that the relatively open access between the pharynx and the SBC allows air in-flow to be a rather passive process (i.e., largely a function of its buoyancy) compared to other air breathers. Normal inspired air volumes are sufficient to overflow the SBC and thus contact the respiratory surfaces located in the roof of the pharynx and mouth. Thus, even with air in its ABO, *Channa* is still able to ventilate its gills normally. The relatively open contact between the SBC and the pharynx, however, has the effect of dislodging respiratory air during inertial suction feeding.

ABO ontogeny. *C. striatus* begins air breathing about 20 days after hatching (12 mm length) and well before the ABO is fully developed (Singh *et al.*, 1988). Expanded knowledge of this ABO has led to rejection of the initial claim (Munshi, 1962b) that its respiratory epithelium was derived from gill secondary lamellae (Hughes and Munshi, 1973a). Rather, R.P. Singh *et al.* (1982) and B.R. Singh *et al.* (1988) showed this epithelium is differentiated from rudimentary gill tissues very early in ontogeny and before lamellae appear. These workers report that the SBC takes form within the interstices of the pharyngeal roof rather than as a pharyngeal outpocketing.

The embryonic gill mass (i.e., the rudiments of gill arch 5 [and 4 in some species] and integumental ectoderm) give rise to the SBC respiratory epithelium. With its opening into the pharynx, the SBC's respiratory epithelium extends to cover this area as well as the dendritic plates of gill arch 1 and the hyomandibula. Singh *et al.* (1988) report specific differences in the timing of epithelial succession and papillar development; in *C. striatus*, succession is completed when fish reach about 2.5 cm length, and papilla take form at about 5 cm. Respiratory nodules (well-developed papillar surfaces that form on the bony projections) do not form completely in this species until 10–12 cm and then continue to grow with increased body size (at least up to 24 cm).

Monopterus and the Synbranchidae

Structure. With the probable exception of *Macrotrema,* all synbranchids air breathe by inflating their mouths and pharyngeal chambers with large air gulps (Figure 3.25A). Inspired air is thus in contact with respiratory epithelium lining the mouth, pharynx, and the branchial and opercular cavities, and gills, as well as, in the three synbranchid species that possess them, the paired respiratory sacs. The synbranchid opercular opening (Figure 3.25B) is reduced to a single, common ventral slit (Rosen and Greenwood, 1976) which likely enables control of inspired air and prevents debris from entering the chamber during burrowing.

Respiratory air sacs. Paired lung-like air sacs (Figure 3.25A) occur above the gills and on each side of the head in *Monopterus cuchia*, *M. fossorius*, and *M. indicus* (Taylor, 1831; Hyrtl, 1858; Das, 1927; Wu and Kung, 1940; Munshi and Singh, 1968; Rosen and Greenwood, 1976; Munshi *et al.*, 1989, 1990). Each sac connects to the pharyngeal cavity by a single, unvalved aperture that serves both inhalation and exhalation. The inner wall of the sac has large folds and about two-thirds of its surface is covered by a respiratory epithelium (Munshi *et al.*, 1989).

Respiratory surfaces. The vascular epithelium of the air sacs and the buccopharynx of *Monopterus* is highly convergent with that in *Channa*. Vascular islets and non-vascular lanes are present, and the islets in turn contain rosettes and individual papillae (Figure 3.25C–F). The papillar surfaces are smooth with microvilli on their margins. As in *Channa*, no pillar cells are present. The papillae are isolated around separate micro-air pockets (that ensure maximum surface area) and also take the form of small endothelial-cell domes, complete with valves, that bulge into the lumen of the ABO (Hughes and Munshi, 1973a; Munshi *et al.*, 1989, 1990; Singh *et al.*, 1991).

Although formerly regarded as intra-epithelial capillary endings, papillae in the ABO of *Monopterus* have, as in *Channa*, been recently interpreted (Munshi, 1985; Munshi *et al.*, 1989, 1990) as surface contacting sections of elongated capillaries that traverse the ABO

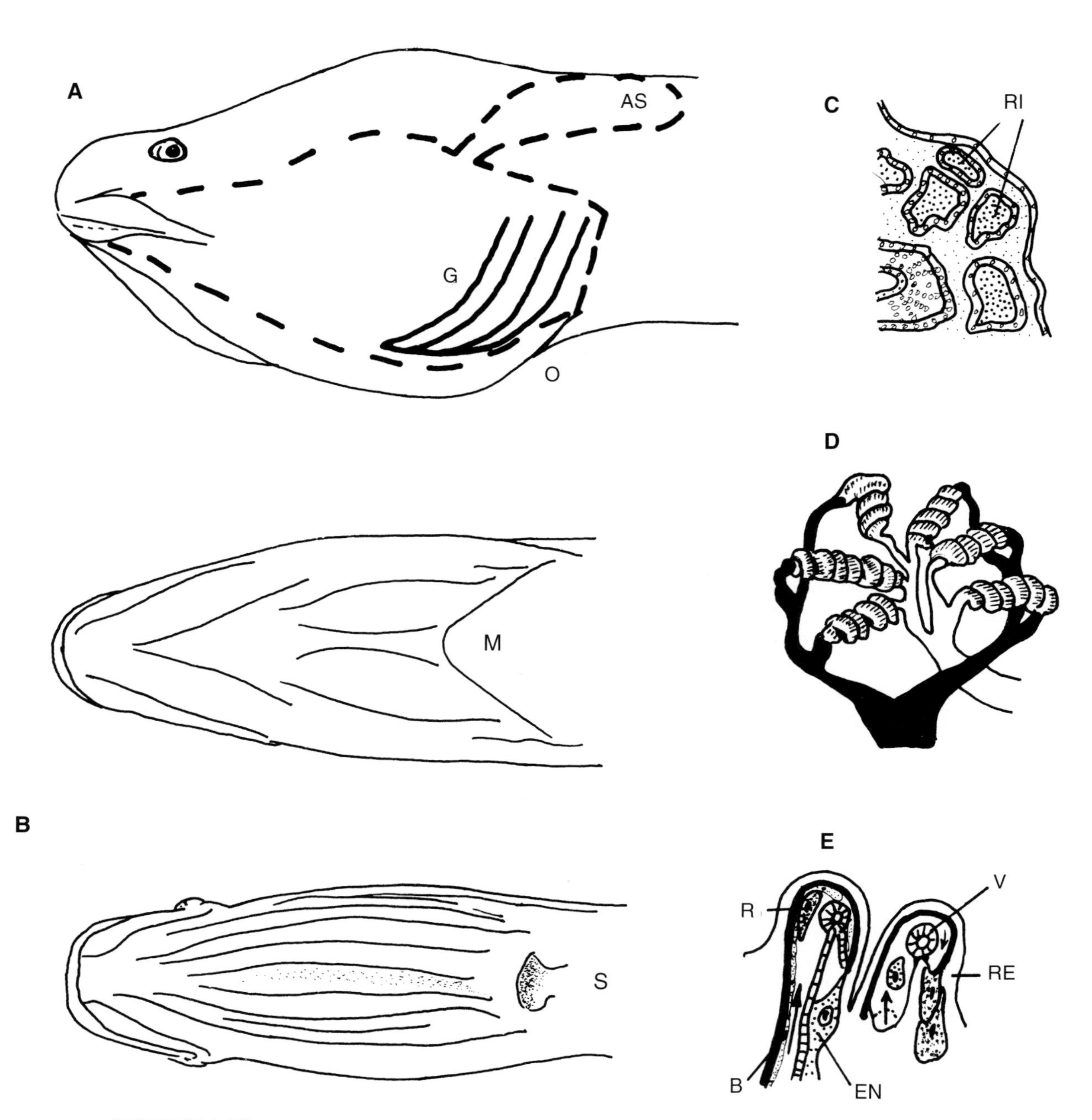

FIGURE 3.25 **A,** Side view of *Monopterus* with an inflated ABO chamber showing the relative positions of the air sac (AS), gills (G), and opercular opening (O). Dashed line shows perimeter of the fully inflated ABO. **B,** Ventral views of *Monopterus* (M) and *Synbranchus* (S) showing the longitudinal folds that expand with ABO inflation, and the ventral position and reduced size of the opercular valve. **C,** Respiratory islet (RI) clusters in the air sac wall. **D,** Afferent (open) and efferent (solid) circulation to the spiral respiratory papillae. **E,** Oblique section through two papillae. RE, respiratory epithelium; B, basement membrane; EN, endothelial lining; V, papillar valve; R, red cell. **F,** SEM of spiral respiratory papillae (x 1,950). Arrows show endothelial valve position. (**A** and **B** adapted from Rosen and Greenwood, 1976; **C** and **E** adapted from Hughes and Munshi, 1973a, with permission of G.M. Hughes and J.S.D. Munshi; **D** and **F** copied from Munshi *et al.*, 1990, by permission of Wiley-Liss, Inc., a subsidiary of John Wiley & Sons, Inc., from the *Journal of Morphology*, volume 203, pp. 181–201.)

(continues)

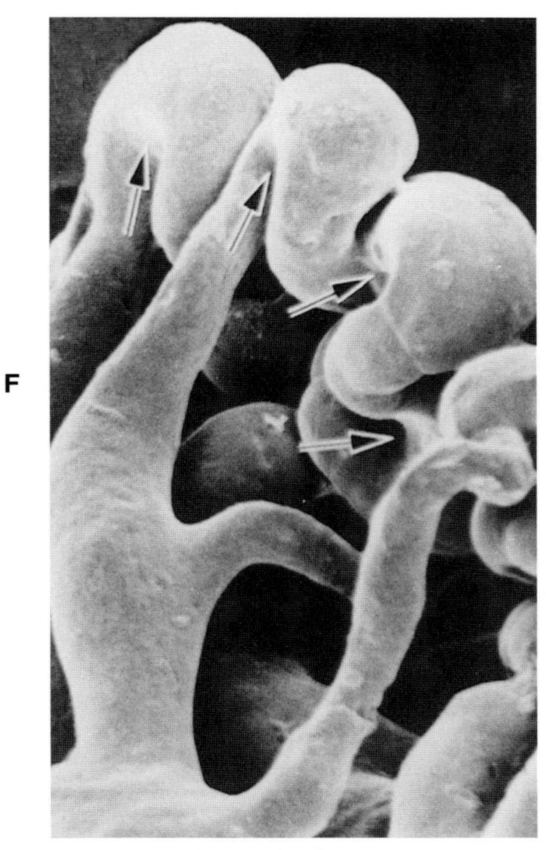

FIGURE 3.25 *(continued)*

surface in a spiraling or wave-shaped manner (Figure 3.25D–F). The non-vascular surfaces are composed of microridged epithelial cells and both mucous and sensory cells.

Morphometrics. Taylor (1831) observed that the air sacs of *M. cuchia* were relatively small. Hughes *et al.* (1974a) confirmed this by estimating a mean organ surface area of only 5.056 mm^2/g. The mean tissue diffusion thickness measured for this fish is 0.44 μm and there is a major shift in air-sac allometry with growth. Below a mass of 60 g, sac–area scales in proportion to $M^{0.714}$; whereas, in larger fish, area is proportional to $M^{1.19}$.

Using an estimated air-sac surface area of 48.4 cm^2/kg and a diffusion barrier thickness of 0.44 μm, Hughes *et al.* (1974a) calculated the air-sac D of a 100 g *M. cuchia* to be 0.0165 (Table 3.3). Munshi *et al.* (1989) examined the morphometrics of all the aerial respiratory surfaces (air sacs and buccopharyngeal epithelium) and found this area to be twice that of the respiratory sacs alone; whereas, total sac area in a 100 g fish is about 5.0 cm^2, total ABO area is about 10 cm^2 (Table 3.3). These workers determined that only about two-thirds of the air-sac surface was covered by a respiratory epithelium, and they measured air–blood diffusion distances greater than had been previously estimated (i.e., 0.72 μm versus 0.44 μm). Based on these values, total ABO D_L was estimated to be 0.0208 (Munshi *et al.*, 1989, table 2, also see Table 3.3).

TEM sections additionally suggested to Munshi *et al.* (1989) that, when the layers of mucus (0.72 μm), plasma (0.19 μm), and erythrocytes were added, the functional ABO diffusion thickness might be as great as 2.41 μm which would vastly reduce total organ D; however, both further discussion and analysis of this point become somewhat obscure in the Munshi *et al.* (1989) paper. For example, I estimate (Table 3.3) that, with an increase in diffusion distance to 2.41 μm, ABO D would fall to 0.006. This is not as low as the D = 0.00165 estimated by Munshi *et al.* (their tables 2 and 3), nor is it clear how these authors arrived at their number. They remark (page 463) that even with a doubling of ABO surface area, a nearly six-fold increase in diffusion distance results in over a 10 fold increase in D. However, as is apparent from Equation 3.3 and Table 3.3, these changes in A and t should result in an approximately 60% reduction in D. Finally, it is also not clear if, or how, Munshi *et al.* (1989) applied a correction for their estimated one-third reduction in functional air-sac respiratory area.

Munshi *et al.* (1989) conclude that the combined ABO surfaces of *Monopterus* are considerably less effective for aerial O_2 transfer than are those of other air-breathing fishes; a conclusion seemingly supported by the general view of this species as sluggish due to its high ratio of diffusion pressure ($\dot{V}O_2/D$) and interspecific comparisons of D. However, because the methods used for these various calculations of D are unclear and because I cannot duplicate some of their numbers, the validity of the Munshi *et al.* conclusions remains open to question. I submit that because its total ABO surface area (Table 3.3), its degree of papillary development, and vascular density are all similar to those of most other air-breathing fishes, the efficacy of the ABO in *Monopterus* should compare more favorably. In addition, the vastly atrophied gills of this species (which make it an obligatory air breather), and its demonstrated capacity to respire both aerially and aquatically using its modified ABO tissues (Lomholt and Johansen, 1974), both suggest a greater relative efficiency for gas exchange through this respiratory surface than implied by Munshi *et al.* (1989).

Comparative aspects: *Monopterus* and the other synbranchids. No quantitative ABO data exist for *Synbranchus, Ophisternon,* or the species of *Monopterus*

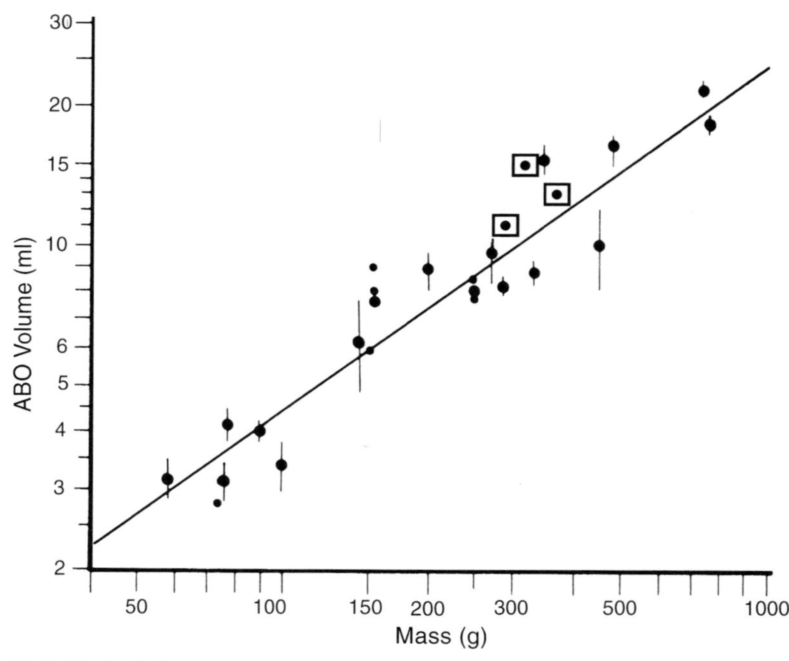

FIGURE 3.26 Relationship between ABO volume and body mass for *Synbranchus marmoratus*. Regression equation: ABO volume = $0.150M^{0.737}$, r = 0.958, N = 17; slope confidence limits ± 0.088 (Data from Graham and Baird, 1984, by permission of the Company of Biologists, Ltd., from the *Journal of Experimental Biology*.) Squares with dots are three ABO volume values for *Monopterus cuchia* (Lomholt and Johansen, 1974.)

lacking respiratory sacs, all of which breathe air using modified respiratory epithelial surfaces in their buccopharynx, gill arches, and opercular chamber. Graham and Baird (1984) determined the scaling of ABO volume (inspired air) in *S. marmoratus* (Figure 3.26). They found that breath-to-breath volumes did not vary markedly among individual fish and confirmed earlier findings (Johansen, 1966; Lomholt and Johansen, 1976) that the ABO is completely emptied by exhalation. Figure 3.26 also contains ABO volume data obtained by Lomholt and Johansen (1974) for three specimens of *M. cuchia* in hypoxic water. Even though *M. cuchia* has respiratory sacs, its ABO volume is highly similar to that of *Synbranchus*.

Air-ventilation mechanics. In all air-breathing synbranchids, inspired air volume is sufficient to float the head (Johansen, 1968; Lüling, 1973) unless the body can be used to forcibly retract it into a burrow. In shallow water, both *Synbranchus* and *Monopterus* often hold their inflated heads vertically and motionless just below the surface, a behavior which may be part of a stealthy feeding tactic. As in *Channa*, feeding requires these fish to expel their air breaths; however, my observations show that this can be done rapidly in conjunction with suction feeding.

Air ventilation by *Monopterus* and *Synbranchus* was detailed by Liem (1980, 1987) and does not appear different in species with or without air sacs. Exhalation occurs prior to inspiration, may take place either at, or below, the water surface, and can be either active (through compressive forces generated by the geniohyoideus posterior and adductor mandibulae) or passive (without muscular activity; a result of hydrostatic pressure and gas buoyancy). Singh *et al.* (1984) state that ABO exhalation also involves actions of the superior hyohyoideus muscle which arises on the branchiostegal rays and traverses the dorso-lateral wall of the sac enroute to the body myotomes. With the mouth at the surface, inhalation is initiated by an abrupt lowering of the buccopharyngeal floor. This action is brought about by contraction of both the hypaxial and sternohyoideus muscles which causes a 90° depression of the hyoid. Inhalation is often associated with an audible hiss (Taylor, 1831). Once air fills the chamber, the mouth is closed by the adductor mandibulae.

Whereas the muscular actions during air breathing by most teleosts can be functionally linked to those associated with feeding (triphasic) or coughing (quadruphasic), Liem (1980, 1987) observed that synbranchid muscular actions during air inspiration and expiration are not derived from these actions; synbranchids differ from other teleosts in their exclusive dependence upon ventrally positioned muscles (geniohyoideus, sternohyoideus, and hypaxials) for ventilation. This, together with the highly derived ventral opercular valve, leaves little doubt that strong selection for optimal ABO function has taken place.

ABO ontogeny. Because the gills of *M. cuchia* undergo very limited development, recruitment of the buccopharyngeal epithelium in both aquatic and aerial respiration must occur almost immediately after hatching and before the respiratory tissues fully develop. *M. cuchia* commences air breathing at about 3.5 cm (Singh *et al.*, 1984; Singh, 1988) which is before its buccopharyngeal epithelium becomes differentiated into respiratory islets and lanes (5 cm) and well before the size (15 cm) when the respiratory air sacs become functional.

As was shown for *Channa*, the air sacs of *M. cuchia* develop within the wall of the pharynx rather than as an outgrowth of that chamber. The buccal and pharyngeal respiratory epithelium of this species is formed by modified tissue in gill arch 1 and, because differentiation from larval gill arches occurs early, this tissue does not assume a lamellar appearance and lacks pillar cells. These mature gill characteristics are also lacking in the air sacs which develop from the gill mass.

It is clear that ontogenetic development in *Monopterus* is programmed from the very outset to manufacture air sacs and not gills. In view of the constant need for functional intermediates, we have little insight, as to how the course of natural selection enabled gill development to become re-tooled on the ontogenetic assembly line.

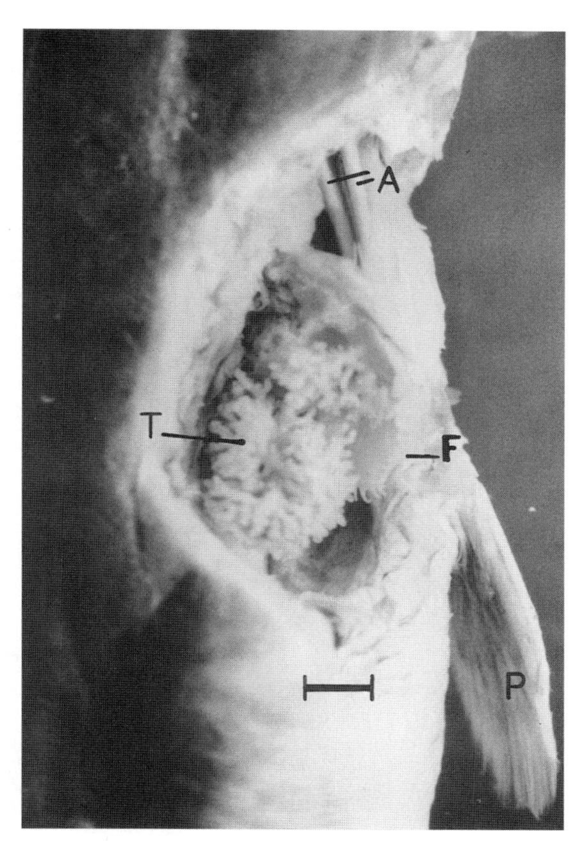

FIGURE 3.27 Dorsal view of the opened right SBC of a 35 cm *Clarias gariepinus* showing the two respiratory trees (T) and gill fans (F). A, gill arches; P, pectoral fin. Scale = 0.5 cm.

ABOs of the Branchial Region

Fishes making up this group include those with branchial and opercular surfaces modified for air breathing (discussed previously), those that air breathe using structures derived from their gills, branchial chamber, or both, and those in which the gills themselves function for aerial respiration.

ABOs Derived from Gills and a Modified Branchial Chamber

Clariidae

Rauther (1910) summarized the 19th century works on clariid ABO structure. Two early descriptions were by Geoffroy St. Hilaire (1802a, *Silurus anguillaris*) and Taylor (1831, *Macropteronotus magur*). Important later contributions were on the natural history and ABO ontogeny of *C. magur* (*batrachus*) (Das, 1927), ABO dissections and histological studies for *C. lazera* (Moussa, 1956), and detailed morphological studies on *C. batrachus* by Munshi (1961). The following ABO description is based on these accounts and the additional comparative morphometric data contained in Hughes and Munshi (1973a), Munshi (1985), and Maina and Maloiy (1986). Details are known only for species of *Clarias*; limited comparative data for other genera are found in David (1935, *Heterobranchus*) and Greenwood (1961, *Dinotopterus*).

Structure. The gas exchanging surfaces of the ABO include the lining of the SBC, the arborescent organs, and the gill fans (Figure 3.27). All of these structures form within, or as part of, paired SBCs situated above the gills in the dorsal part of the skull. The SBCs have a vascular epithelial lining and are spacious, extending posteriorly beyond the opercular chamber and pectoral girdle to border the coelomic mesenteries. Each SBC is partially subdivided into anterior and posterior compartments. Top and side

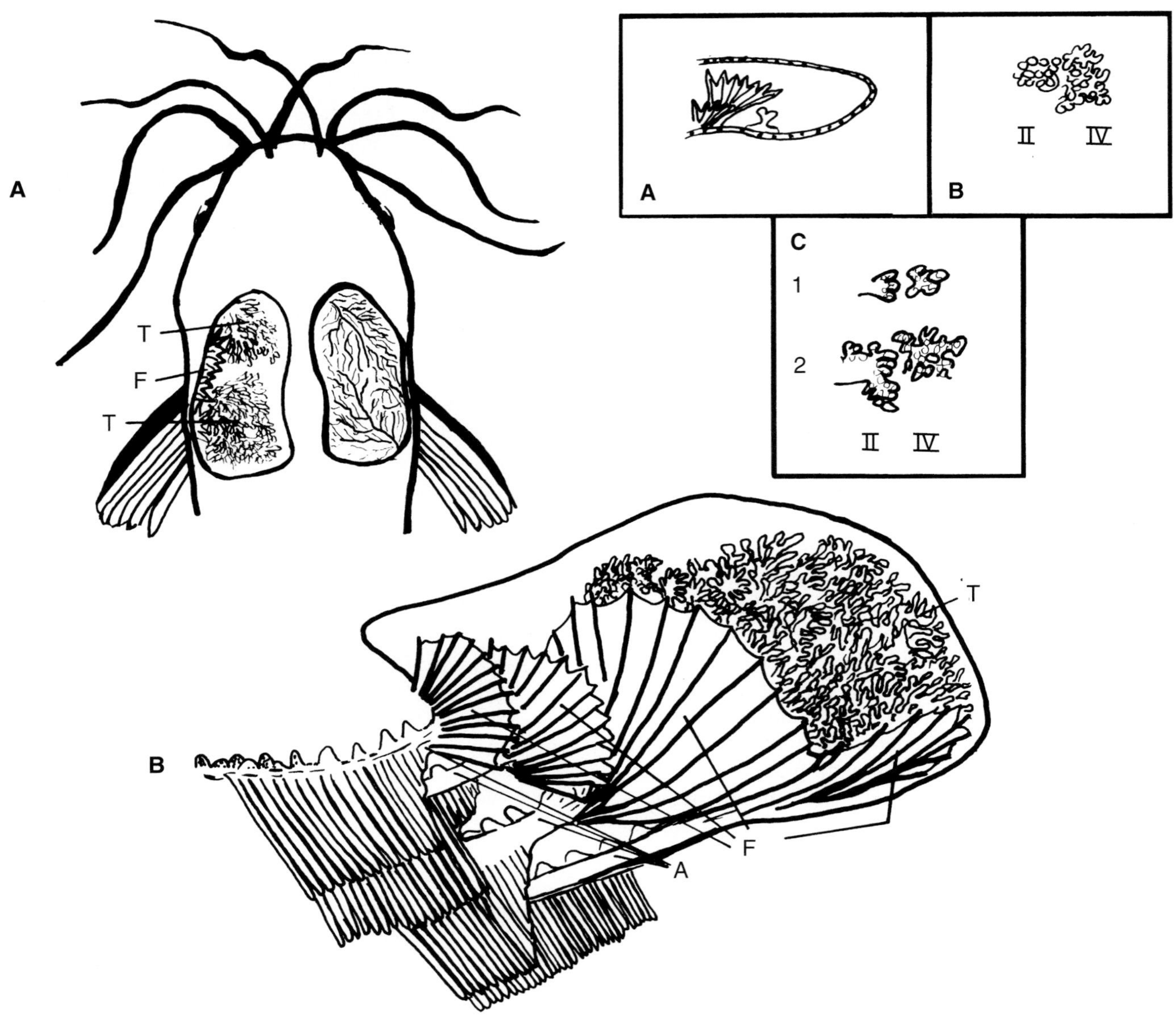

FIGURE 3.28 **A,** Dorsal view of the paired SBCs of *Clarias.* Chamber on right shows the vascularity of the SBC roof. Left chamber shows, after roof removal, the respiratory trees (T) and fans (F). **B,** Side view of the left SBC showing gill arches (A), fans, and trees. *Boxes;* (A) Lateral diagrammatic view of the left SBC of *Dinotopterus jacksoni* showing the arch II fan and the vestigial knob-shaped tree. (B) Dorsal view of the highly reduced respiratory trees (II and IV) on the left side of a 151 mm *C. walkeri.* (C) Dorsal views of the atrophied trees (II and IV) on the left side of 110 mm (1) and 222 (2) mm specimens of *C. maclareni.*

views (Figure 3.28) show each SBC is kidney shaped and considerably narrowed anteriorly. The anterior roof of the chamber is formed by skull bones while the anterior floor consists of the branchial arches. Posteriorly, the chamber lies below the dorsal body wall and rests against the esophagus, the coelomic mesentery and visceral organs, including lobes of the kidney and liver, and part of the gas bladder. The posterior dorsal wall of the SBC is diagonally wrapped by the sheet-like cucullaris muscle which partially compresses it during exhalation.

The inner wall of the SBC is bordered by the skull and vertebral column; whereas, the outer wall is made by the four gill fans. Gill fans form near the top of each arch and consist of fused epibranchial filaments. These become incorporated into the lateral wall of the SBC and separate the chamber from the opercular cavity (Figure 3.27).

Inside the SBC are the anterior and posterior arborescent organs (also termed dendrites or respiratory trees)–knobby ended structures covered by a respiratory membrane and having a large surface area. Being confined to the anterior part of the SBC, the anterior tree is usually smaller than the posterior tree which, in *C. lazera*, *C. batrachus*, and *C. gariepinus* (Figure 3.27), nearly fills the more spacious posterior chamber. Dendritic organ size varies greatly among species (Figure 3.28 and Greenwood, 1961).

The importance of gill fans in aerial respiration derives from their well-developed gas-exchanging surface (composed of fused gill lamellae) and their functions in ABO ventilation and air retention. The anterior edge of gill fan 1 forms the barrier between the air chamber and the opercular cavity. Fan 1 and the larger fan 2 spread posteriorly and dorsally to border the anterior arborescent organ. A notch between fans 2 and 3 serves as the inhalent aperture for air into the SBC. Fan 3, about six times the area of fan 1, overlaps considerably with fans 1 and 2 and thus covers most of the posterior arborescent organ. Fan 4, which is small and ribbon-like, extends laterally and caudally toward the pectoral girdle. A small notch between fans 3 and 4 serves as the exhalation aperture.

Respiratory surfaces. In light of their intimate anatomical relationships, it is not surprising that the respiratory surfaces of the arborescent organs, the SBC, and the gill-fan are formed by a modified gill epithelium. This was documented by Munshi (1961) who discovered pillar cells were the structural entities of the vascularized membranes. Munshi also illustrated the manner in which, during ontogeny, the gill lamellae forming on these surfaces undergo a fusion, shortening, and a unidimensional growth pattern which juxtaposes secondary lamellae that would have otherwise ended up on opposite sides of the primary lamella (Figure 3.29).

The respiratory regions on these membranes thus occur in long islets (also termed pleats, alveoli, vascular tongues, rosettes, islands, and papillae), separated from one another by non-vascular lanes. Owing to their origin from serial lamellae, parallel sets of islets wind in various directions across the surface of the membrane, creating a pattern similar to a brain coral. Interspecific differences in this design were noted by Greenwood (1961) and, not surprisingly, magnification of the islets reveals a parallel series of respiratory tracks (i.e., rows of modified secondary lamellae), each with a transecting capillary flow. Large numbers of mucous secreting cells occur in amongst all these surfaces.

Comparative morphometrics. Table 3.3 shows comparative morphometric data for *C. batrachus* and *C. mossambicus*. Both organs have similar diffusion distances; however, the areas in *C. mossambicus* are less than *C. batrachus*. The functional significance of this difference remains uninvestigated and is surprising in view of the conclusion (Maina and Malioy, 1986), that the general structure of the ABO in *C. mossambicus* closely resembled that described for *C. batrachus* by Munshi (1961). Moussa (1956) observed some small histologic differences in the suprabranchial chamber wall tissues and vascularization in *C. lazera* as compared to what was described for *C. batrachus* by Das (1927).

Preliminary data of Greenwood (1961) show that species of *Clarias* have a relatively larger SBC volume but a lower relative gill area than species of *Dinotopterus*. Ar and Zacks (1989) found that *C. lazera* (1.4–2 kg) had inspired air volumes of from 50 to 80 ml (mean 0.04 ml/g) and that the entire ABO gas volume

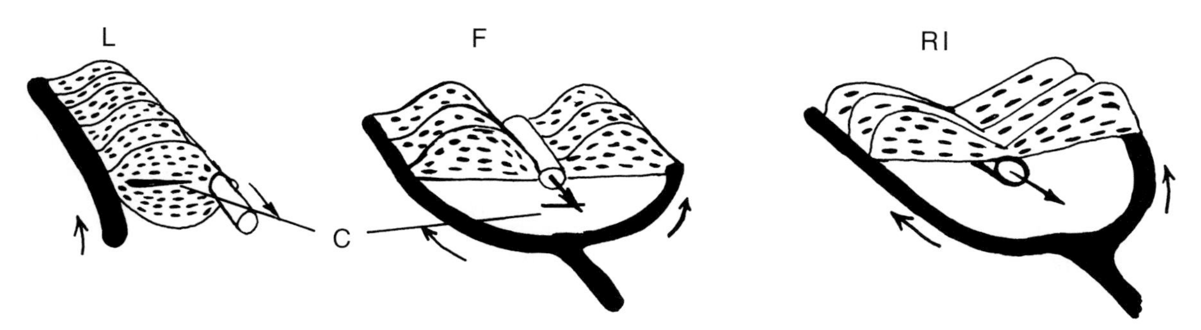

FIGURE 3.29 Probable stages in the evolutionary development of the clariid ABO showing the ontogenetic transition from a row of gill lamellae (L) to the condition in a gill fan (F) and then to a respiratory islet (RI). Note the loss of the cartilage support (C) and the deposition of the afferent (solid) and efferent (open) arteries in the underlying connective tissue. Arrows indicate blood flow. (Adapted from Hughes and Munshi, 1973a, with permission of G.M. Hughes and J.S.D. Munshi.)

was ejected during exhalation. Extrapolating from this, the largest known *Clarias* (60 kg) would likely eject a 2.4 liter bubble!

Air-ventilation mechanics. Clever observations using methylene blue enabled Moussa (1957) to determine that the SBCs of *C. lazera* do not normally come in contact with water. Clariid air exhalation precedes inhalation and may occur during ascent. Exhalation is initiated by contraction of the pharyngeal and suprabranchial chambers (cucullaris muscle). During this process, the mouth and inhalent ABO aperture (between fans 2 and 3) are kept closed to force old air into (via the exhalent aperture) the opercular chamber from where bubbles are released. Following this, gulped air expands the buccal and pharyngeal cavities and displaces branchial arch position sufficiently to open the inhalent aperture and fill the ABO. Subsequent compression of the mouth and pharynx squeezes air into the ABO while concurrent relaxation of the cucullaris muscle establishes a partial vacuum within the SBC to facilitate air inflow (Moussa, 1956; Singh, 1976; Vandewalle and Chardon, 1991). This sequence contains several characteristics of Liem's (1987) triphasic air-breathing pattern. However, Liem (also see Ar and Zacks, 1989) further observed that, depending upon both water depth and quality, *Clarias* may also adopt a quadruphasic pattern in which, during exhalation, the SBC is flushed by a reversed flow of water from the opercular chamber. (Details for both triphasic and quadruphasic ventilation are found in the section on the Anabantoidei.) Related to this, Donnelly (1973) observed that *C. gariepinus* trapped in thick mud made use of a simpler ventilatory pattern in which air was first taken into the mouth, followed by the ejection of presumably old air from the operculae.

ABO ontogeny. Singh *et al.* (1982a) report that the first air breaths by *C. mossambicus* are taken at age 18–20 days (11–12 mm length). Their developmental studies showed that in 8–9 mm fish both the gill lamellae and the modified lamellar lining of the SBC have been formed. They also showed that the embryonic gill mass, an aggregation of ectoderm, mesenchyme (embryonic connective tissue that gives rise to much of the mesodermal tissues), and rudiments of embryonic arch 5, all combine to form the lining of the SBC, while secondary lamellae migrating from the first four arches form its epithelial surface. Singh *et al.* (1982a) found both the anterior and posterior arborescent organs to be formed in 8 mm fish, which contrasts with observations (Das, 1927) that the anterior respiratory tree had yet to form in 22 mm *C. batrachus*. Singh *et al.* (1982a) have challenged the conventional view that respiratory trees form within, and grow out from, the proximal epibranchial segments of gill arches 2 and 4. They interpreted serial histological sections of young fishes to indicate that trees form within the embryonic gill mass and migrate forward during development to attach to arches 2 and 4. Inspection of their figures, however, did not entirely convince me of this. Moreover, their discussion does not resolve uncertainties about–assuming their interpretation is correct–how the cartilagenous core and afferent and efferent vessels within the respiratory tree would be differentiated.

Heteropneustidae

Structure. A brief description of the ABO and its blood circulation pattern in *H. fossilis* is found in Taylor (1831, *Silurus singio*). More detailed accounts are contained in Das (1927) and Munshi (1962a). Olson *et al.* (1990) examined ABO anatomy and circulatory specialization as revealed by SEMs of microvascular casts. Combined aspects of ABO morphology and physiology have been reviewed in Munshi (1993).

The ABO of this fish is similar in many ways to that of the clariids with the major differences being the absence of respiratory trees and the presence of paired tubular, sac-like chambers extending from the SBC back into the body. These tubes are embedded in the epaxial myotomes and extend to as much as one-half body length (Figures 3.30A–C). As in *Clarias*, paired SBCs form above the gills and next to the cranium, and gill fans form at the top of each branchial arch. Gill fan number 1 becomes embedded in the anterior wall of the SBC, fans 2 and 3 guard the inhalant aperture, and fan 4 regulates the exhalant opening. Fans 3 and 4 also partially subdivide the SBC into anterior and posterior parts and thus slightly occlude the entrance to the air sacs which emanate from the posterior chamber. As in *Clarias*, the walls of the posterior SBC and the sacs are ensheathed by the cucullaris muscle which functions in ABO ventilation. Moussa (1956) stated that a single notch between fans 2 and 3 served as both an inspiration and expiration port in *H. fossilis*. This, however, is at variance with most other accounts identifying separate entry and exit ports for air, as in *Clarias* (Munshi, 1961; Singh *et al.*, 1981).

Respiratory surfaces. The configuration of the respiratory epithelium covering the SBCs, respiratory tubes, and gill fans confirm its derivation from gill tissue. There are, however, slight regional differences in the appearance of the respiratory epithelium (Hughes and Munshi, 1978). The main feature of the SBC and gill-fan epithelium is the alternating presence of

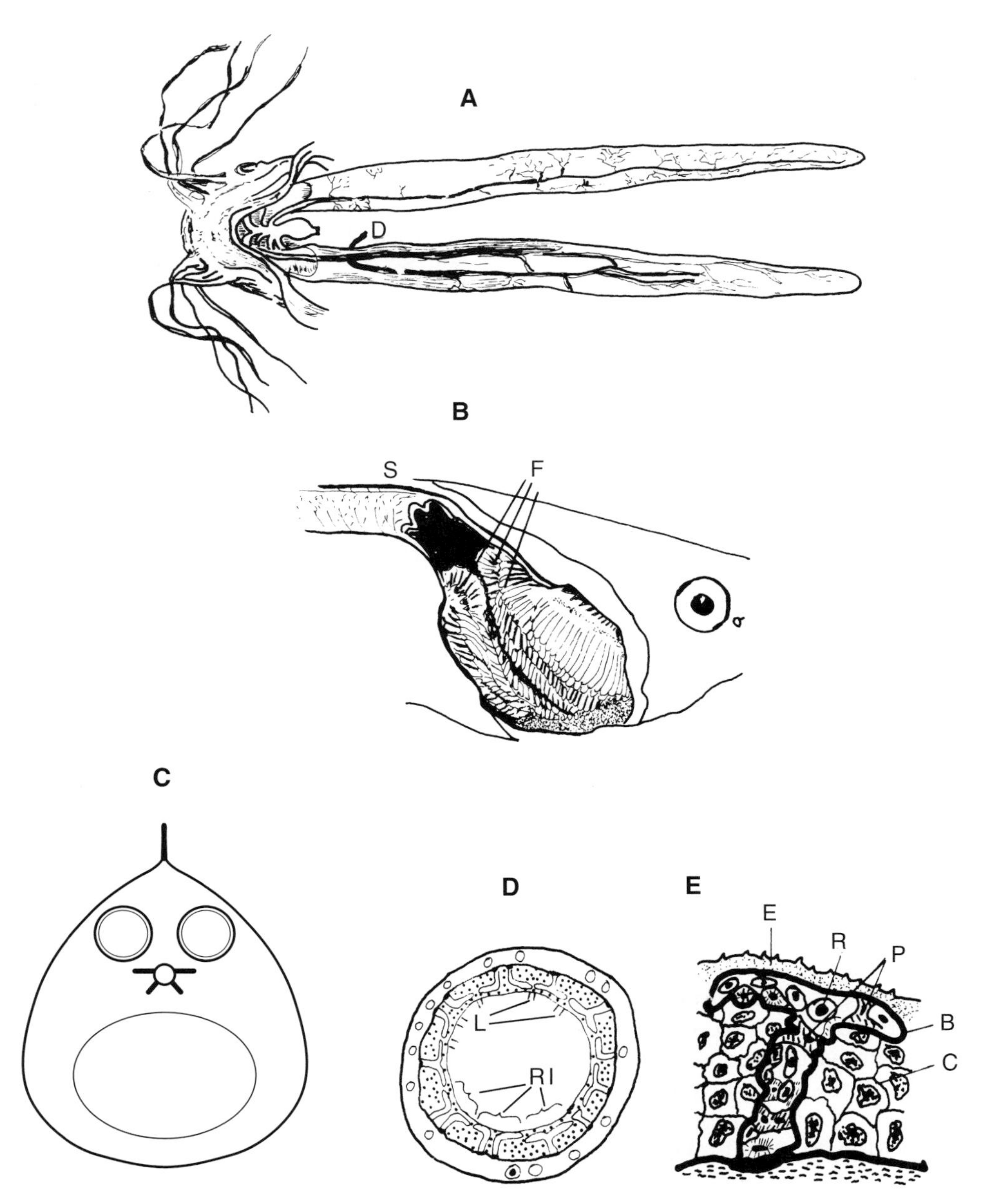

FIGURE 3.30 **A,** Modified drawing of Hyrtl's (1854b) original illustration of the ventral view of the SBC and air sacs of *Heteropneustes fossilis*. Ventral surface of the left air sac shows path of the afferent artery of gill arch number 4 from the heart. Right sac shows dorsal path of afferent number 1 and the path of efferent drainage to the dorsal aorta (D) (See Chapter 4 for detail). **B,** Side view showing relationship of gill fans (F) to opening of the air sac. **C,** Transverse body section showing air-sac position in relation to the vertebral column and coelomic cavity. **D,** Transverse section of an air sac showing respiratory islets (RI) and non-respiratory lanes (L). **E,** Section through a respiratory islet showing the covering respiratory epithelium (E), the vascular channel surrounded by a cuboidal epithelium (C), and basement membrane (B), and the capillary held open by pillar cells (P and diagonal lines). R, red cells. (**D-E** redrawn from Hughes and Munshi, 1973a, with permission of G.M. Hughes and J.S.D. Munshi.)

slightly elevated (0.1 × 0.02 mm) respiratory islets and lanes of flattened cells (Figures 3.30D and E). The lanes, which are composed of micro-ridged cells, border the islets and are all interconnected via branches. The islets are variously sized and occur in long tracks or whorls. SEMs reveal the basic double row arrangement of the islets, reflective of their derivation from secondary lamellae. However, the islet lamellae are smaller than those on the gills and have numerous discontinuous blood spaces which probably lower effective exchange area (Olson *et al.*, 1990).

A highly similar pattern of islets and lanes is also seen in the air sacs; however, the embryonic derivation of these structures is distinct from those in the SBC (see following). Four distinctive longitudinal ridges develop along the inner wall of the air sacs (Figure 3.30D). TEMs of the ABO epithelium (Hughes and Munshi, 1978) show the lanes are non-vascular. The islets, however, are essentially blood channels which, to increase surface contact, often bifurcate near the epithelial surface of the lumen. These channels are held open by pillar cells, another diagnostic feature of gill derivation. The mean air–blood diffusion path across the islet, discounting a mucous layer, is estimated to be about 1.7 μm. This consists of the epithelial lining (1.4 μm), the basement membrane (0.05 μm), and the pillar cell lining (0.2 μm). Mucous cells are freely distributed in the ABO surface and function to prevent desiccation (Singh *et al.*, 1974).

Morphometrics. Respiratory sac-scaling and morphometric data are contained in Table 3.3. No ontogenetic changes occur in ABO morphology. Sac area in a 100 g fish is nearly 31 cm^2 and *D* is near values for other species (Hughes *et al.*, 1974b). Sac area, however, does not constitute the entire ABO surface area of this fish, and more data are needed. Also, sac surface area estimates were based on assumptions that the organ has a smooth inner surface uniformly covered by a respiratory epithelium. As shown previously, SEM studies (Hughes and Munshi, 1978) reveal this is not the case, particularly in the micro-ridge air sacs. Additional work should account for variables such as the prominence of the islets, the contribution of their microvilli to area (or possibly the effect of a mucus layer on diffusion distance), the area of the non-vascular lanes, and the presence of air sac ridges.

ABO ontogeny and development. Singh *et al.* (1981) report that *H. fossilis* begins to air breathe at a body length of 11–12 mm (about 10 mg mass), about 18–20 days post hatching. At this time the SBCs, but not the air sacs, are fully formed. The gill mass (an embryonic precursor formed by the rudiments of gill arch 5 and ectoderm that migrates from the dorsal epidermis) gives rise to the SBC, the cucullaris muscle, the air sacs and their respiratory epithelium, and to blood channel structures including the endothelial and pillar cells. The epithelium lining the chamber is derived from gill lamellae that form on a single hemibranch at the dorsal end of each gill arch and migrate into the SBC. The hemibranchs occurring opposite to those that become fused form the individual gill fans. The air sacs continue to grow as the fish increases in size and an "active" gill mass thus remains at the terminus of the air sac, even in adult fish (Singh *et al.*, 1981).

Anabantoidei

All fishes in the five families of this order have, as an ABO, paired SBCs containing an intricately laminated bony element, the labyrinth apparatus. The walls of the SBC and labyrinth are covered with a respiratory epithelium (Figure 3.31).

Anabantoids were given the name "labyrinth fishes" by Cuvier and Valenciennes (1831) who noted similarities between the labyrinth and the turbinated mammalian ethmoid. These authors illustrated (plate 205) the organs of six genera (*Spirobranchus*, *Macropodus*, *Polyacanthus*, *Colisa*, *Osphromenus* [old spelling], *Anabas*). It was thought, at the time, that the labyrinth functioned for water storage which, through seepage, kept the gills hydrated when fish became exposed to air, ". . . à peu près comme le réseau de la panse des chameaux" (Cuvier and Valenciennes, 1831, page 323).

Early anatomical descriptions of this ABO were done by Taylor (1831), who deduced its function in blood oxygenation; by Peters (1853), who detailed the bony labyrinth; and Hyrtl (1863), who studied blood-vessel patterning and also verified that the chambers did not contain water. Day's (1868) experiments also proved important in demonstrating the labyrinth's aerial respiratory function. Both Zograff (1886, 1888) and Henninger (1907) made comparative histological studies of structure and microcirculation.

Das (1927) was the first to describe the development of the anabantoid ABO, and Bader (1937) conducted comparative experimental studies of labyrinth development in *Betta*, *Macropodus*, *Colisa*, and *Luciocephalus*. Subsequent contributions have focused on comparative labyrinth–organ anatomy (Munshi, 1968; Hughes and Munshi, 1973b; Elsen, 1976), air ventilatory mechanics and functional morphology (H.M. Peters, 1978; Liem, 1963a, 1980, 1987), and details of microcirculation as revealed by vascular casting (Munshi *et al.*, 1986).

Structure. In addition to the genera listed thus far, observations of ABO structure and function have been

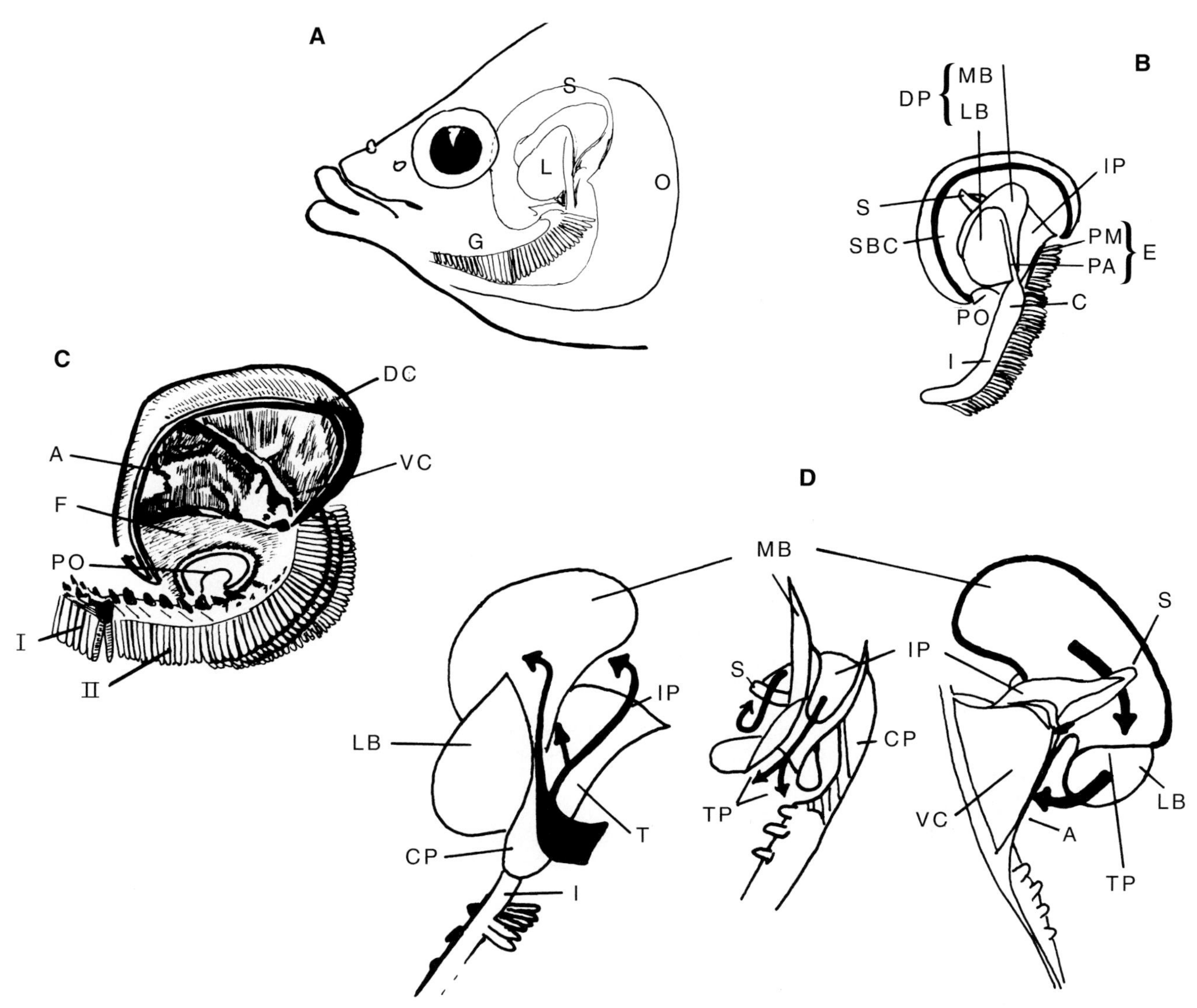

FIGURE 3.31 (**A–D** are redrawings of Peter's [1978] original illustrations of the anabantoid ABO. Used with permission of Springer-Verlag New York, Inc. from *Zoomorphologie.*) **A,** Lateral view of an anabantoid showing position of the suprabranchial chamber (S) labyrinth (L), gill (G), and operculum (O). **B,** Lateral view of the left labyrinth of *Osphronemus.* Abbreviations: suprabranchial chamber (SBC); gill arch I (I), ceratobranchial (C), processi muscularis (PM) and articularis (PA) of the epibranchial (E), inner plate (IP), lateral (LB) and medial (MB) blades of outer double plate (DP), stylet or axis (S), shuttered pharyngeal opening (PO). **C,** Lateral view of the opened left SBC of *Osphronemus* with labyrinth removed. Abbreviations: dorsal chamber (DC), atrium (A), ventro-caudal compartment (VC), floor (F), pharyngeal opening (PO), gill arches I and II. **D,** Three views of the left labyrinth of *Osphronemus* with arrows showing the probable route of water flow from the dorso-caudal compartment to the atrium and ventro-caudal compartment during the reversal phase of quadruphasic ventilation. *Left panel:* Lateral view showing water flow from the opercular chamber into dorso-caudal compartment. *Center:* Superior view showing flow over the stylet and through the trumpet and, as seen from the medial side *(right)* into the atrium and ventro-caudal compartment. Abbreviations: ceratobranchial of arch I (I), inner plate (IP), medial blade (MB), lateral blade (LB), closure plate (CP), stylet (S), axis (A), ventro-caudal compartment (VC), trough (T), trumpet (TP). **E,** Oblique section through a respiratory islet in the SBC wall of *Anabas* showing the parallel capillaries. RE, respiratory epithelium; B, basement membrane; CT, connective tissue pavement; EN, endothelial cell lining the capillary with valve-like projections (V); S, supporting epithelial cell; R, red cell. (Based on Hughes and Munshi, 1973b, used with permission of G.M. Hughes and J.S.D. Munshi.) **F,** SEM of the respiratory epithelium (RE) of *Anabas.* RE has been removed to reveal two parallel channels containing endothelial valves (EV) and red cells (*). (From Munshi *et al.*, 1986. Reprinted by permission of Wiley-Liss, Inc., a subsidiary of John Wiley & Sons, Inc., from *Developmental Dynamics* [formerly *American Journal of Anatomy*].) **G,** Contrasts of the basement membrane configuration in the fish gill *(lower),* where blood channels are formed by endothelial pillar cells (P), and in the respiratory islets of *Anabas (upper),* where capillaries are supported by epithelial support cells (S). Other symbols as in **E.**

(continues)

E

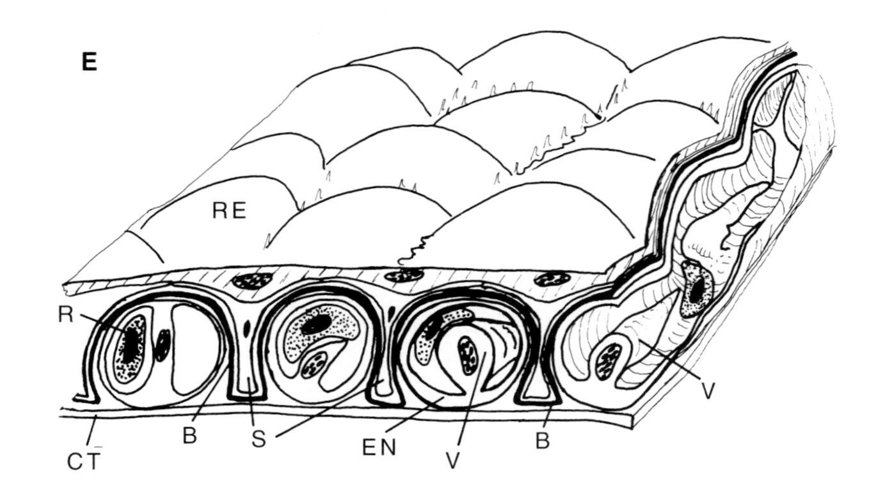

F

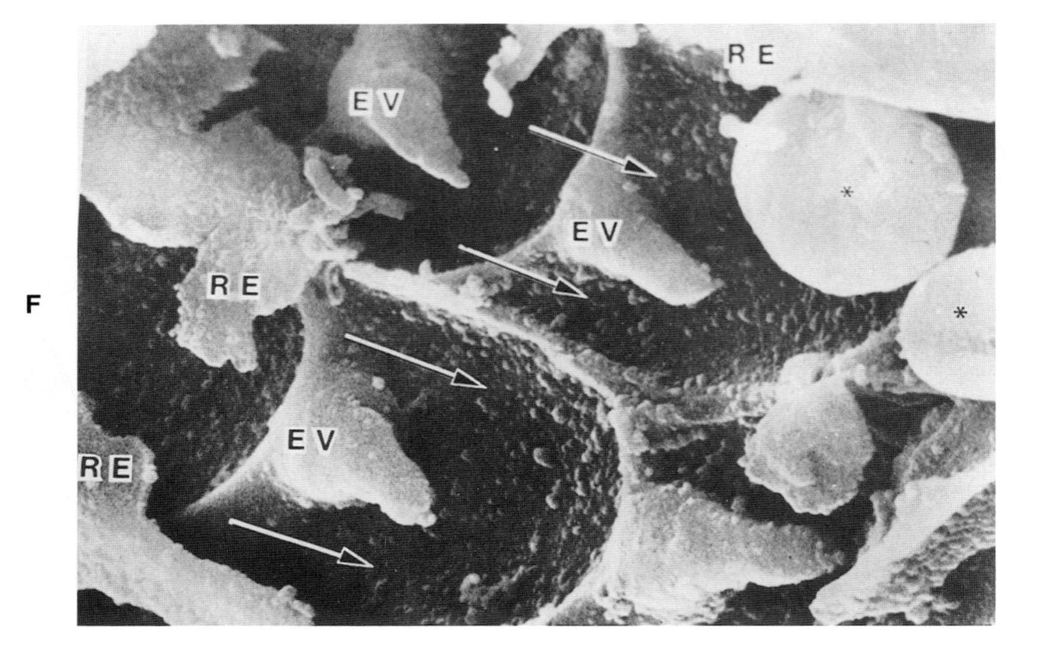

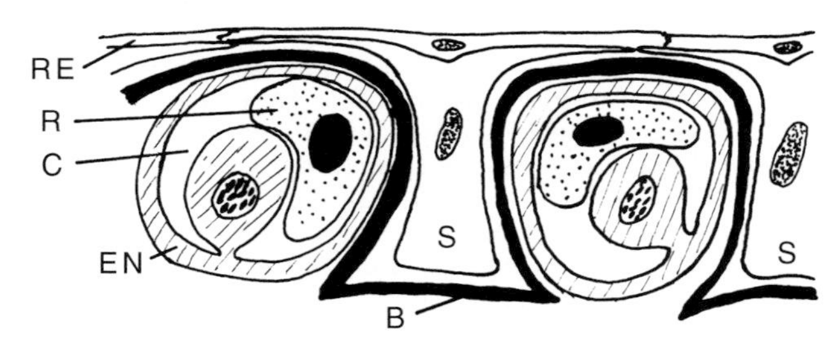

G

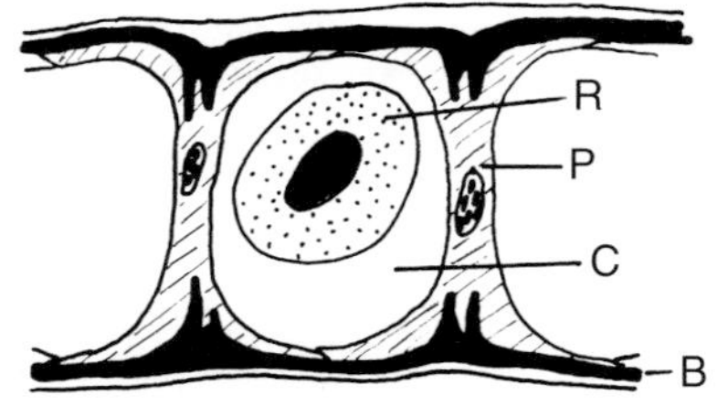

FIGURE 3.31 *(continued)*

made on *Helostoma*, *Trichogaster*, *Ctenopoma*, *Trichopsis*, and *Sandelia*. The most detail is known for *Anabas* (Munshi, 1968; Hughes and Munshi, 1973b; Peters, 1978). The paired SBCs are located adjacent to the skull above the gills (Figure 3.31A and B) and extend from behind the orbit to as far as the pectoral girdle where they contact both the coelom and head kidney. The upper and lateral margins of the SBC are bordered by bones of the skull and operculae. The chamber floor is formed by the roof of the buccopharynx and a thin membrane extending posteriorly from gill arch 2.

Each SBC consists of three compartments, the anterior atrium being larger than either the caudo-dorso or ventro-caudal compartments (Figure 3.31C). Air passage to and from the SBC is via three openings (one exception is *Luciocephalus* [Liem, 1987]) and, in nearly all anabantoids, each of these can be sealed to allow SBC isolation from other actions (feeding, aquatic ventilation) involving the mouth, pharynx, and gills (Peters, 1978). The pharyngeal opening, or shutter, is in the middle of the atrial floor and connects the atrium with the buccal cavity. A projection on the medial side of the first gill arch serves as the shutter's valve (Figure 3.31B). The second opening, between the atrium and branchial chamber, is via the modified cleft between arches 1 and 2. The third opening is the modified septal space between the operculum and gill arch 1. This connects the upper-posterior region of the SBC and the opercular chamber and is sealed by the opercular bulge, a fat-filled swelling on the medial opercular surface that nests into the SBC and labyrinth apparatus.

The labyrinth apparatus is a bony structure formed by the epibranchial (processus articularis) segment of gill arch 1. Although appearing highly convoluted, the labyrinth surface is essentially continuous (Bader, 1937; Elsen, 1976; Peters, 1978), consisting of an inner plate and an outer double plate (Figure 3.31D). The labyrinth largely fills the SBC and both its topography and spatial arrangement within the SBC are important for ventilation.

Respiratory surfaces. Characterized by parallel rows of vascular islets bordered by nonvascular lanes, the anabantoid respiratory membrane appears highly similar to that in both *Clarias* and *Heteropneustes* (Munshi, 1976). Anabantoid islets, however, lack the same degree of fine-scale lamellar structure found in these species. Specifically, the islet capillaries in both *Heteropneustes* and *Clarias* are supported by pillar cells (structures which form the sub-epithelial blood spaces in gill secondary lamellae); whereas, the anabantoid islet capillaries are formed by ring-shaped endothelial cells lying in parallel and connected in series (Hughes and Munshi, 1973b). These occur below the basement membrane of the overlying respiratory epithelium which, in a manner highly analogous to the pillar cells, sends extensions (pillars) of supporting epithelium in between each endothelial row (Figures 3.31E and G). The large endothelial nuclei project as tongue-like processes into the capillary lumen in the direction of flow (Figures 3.31E and F) and may act to regulate flow or generate microturbulence and enhance gas transfer (Hughes and Munshi, 1973b; Munshi, 1976). The respiratory membranes of the labyrinth and SBC have the same morphology. The nonvascular regions of the respiratory membrane are generally thicker and contain what Hughes and Munshi (1973b) describe as single chemoreceptor cells similar to those in the skin.

Morphometrics. A respiratory epithelium one or two cells thick occurs in both *Anabas* and *Macropodus* (Zograff, 1888; Henninger, 1907; Bader, 1937; Hughes and Munshi, 1973b). Estimates of the air–blood diffusion pathway for both the SBC wall and labyrinth of *Anabas* range from 0.1 to 0.3 μm with most of this space taken by the epithelium (0.18 μm) and the basement membrane (0.05 μm). In all cases the capillaries remain below the basement membrane (Hughes and Munshi, 1973b).

Table 3.3 contains scaling data for the ABO surfaces of *Anabas* (Hughes *et al.*, 1973b). A significant difference was found in the mass coefficient for SBC surface area for fish weighing 45 g or more ($M^{0.78}$) and smaller fish ($M^{0.33}$). The combined value for all fish ($M^{0.574}$) is much less than the labyrinth; in a 100 g fish the SBC surface area is only 7.8 cm^2, whereas labyrinth area is 29 cm^2. With its large area and small diffusion distance, the estimated *D* for the labyrinth (0.23) exceeds that of most other aerial-exchange surfaces (Table 3.3).

Observations show that anabantoids usually empty the ABO on each exhalation (Peters, 1978; Burggren, 1979; Schuster, 1989). It thus becomes possible to estimate both tidal and total ABO volume by correcting for net volume changes resulting from gas transfer during the time air is held. Burggren (1979) determined that *Trichogaster trichopterus* (8 g) inspired between 27–32 μl/g and lost (consumed) 11–15% of this volume over the average time (4.7 min) the gas was held. Similarly, for *Colisa lalia* ranging in length from 40 to 60 mm (about 1–3 g), Schuster (1989) estimated ABO volumes of between 20 and 100 μl at two minutes post inspiration. His power equation relating the 2 minute ABO volume (V) to body length (L) is $V = (2.65 \cdot 10^{-4})L^{3.1}$. There have been no studies to determine if nest building anabantoids, which can hold more air in their mouths (Peters, 1978), have larger inspired volumes.

Air-ventilatory mechanics. Studies by Peters (1978) vastly altered our understanding of anabantoid air ventilation. Previous workers had all concluded that compression of the ABO by overlying muscles caused exhalation and that ABO filling was by recoil. Although these muscles play a role in ventilation, it is clear from X-ray films taken by Peters that the SBC (which is largely surrounded by bone and also contains a bony labyrinth) does not change size during expiration and that the chamber is completely emptied of its gas on most exhalations. Based on these observations and high speed films, Peters suggested that forces external to the ABO had to cause ventilation and proposed a double pump mechanism derived from combination of the originally independent water-ventilation and coughing (reversed gill flow) reflex patterns. He surmised that ABO ventilation was driven by the alternating sucking and pumping actions of the buccal and opercular chambers.

For most of the anabantoids Peters (1978) described a double (biphasic) mechanism in which the expansion and then rapid contraction of the opercular chamber forces a reversed stream of water up into the SBC, displacing its gas. As this occurs, an abrupt buccal expansion aspirates the displaced gas into the mouth where it is exhaled as a bubble (if the mouth is below the surface) or directly into the air. In biphasic air ventilation, inhalation follows exhalation with new air being forced into the SBC by buccal compression aided by opercular-chamber expansion. Overflow of new air (if any) is ejected via the operculae. This description clearly implicates the valved SBC apertures in establishing the pressure gradients needed for bidirectional flow. Also, the configuration of the labyrinth is important in directing the water stream through each SBC compartment to assure the displacement of all gas (Figure 3.31D).

Peters (1978) also described monophasic air ventilation in a few species (adult *Anabas*, occasionally in both *Helostoma* and certain species of *Ctenopoma*). In this ventilation, the buccal pump pushes the newly aspired air into the ABO displacing the old gas to the opercular chamber from where it is expelled. Monophasic ventilation is thus simpler than biphasic in lacking the reversed water flow phase, and inspiration precedes expiration in monophasic ventilation. From the standpoint of natural selection, monophasic air ventilation has functional advantages over biphasic ventilation in requiring less time (0.2 versus 1 second), not "giving advanced notice" of fish surfacing by the release of expired air (used and overflow air bubbles are exhaled out the operculae after the new air has been taken) and, in the case of *Anabas*, enabling air ventilation while out of water (Peters, 1978; Liem, 1987). On the other hand, monophasy probably results in the dilution of new air through brief contact with gas about to be released.

Using cineradiographic and emg data, Liem (1987) added two additional phases, preparatory and compressive, to Peters' monophasic and biphasic patterns, thus elevating ventilatory complexity to triphasic and quadruphasic. The components of these two ventilatory modes are:

Triphasic (= monophasic)—Preparatory, Expansive, Compressive
Quadruphasic (= biphasic)—Preparatory, Reversal, Expansive, Compressive.

Prior to the displacement and the release of ABO-gas, anabantoids make what Liem refers to as preparatory movements (muscle tensing and buccal compression). Ventilation is then initiated with the reversal phase (in quadruphasic breathers) followed by the expansive phase (in triphasic fish this follows the preparatory phase), when new air is inspired. Ventilation concludes with the compressive phase, when the inspired air is forced into the SBC.

Comparative aspects of ABO structure. Beginning with Cuvier and Valenciennes (1831), many workers noted marked differences among the anabantoids in SBC size, shape, and position and in labyrinth complexity (Figure 3.32). Liem (1963a) described an evolutionary trend for ABO position. In the most primitive family Anabantidae (*Anabas*, *Ctenopoma*, *Sandelia*), the SBC extends dorsally and is enclosed by bony elements of the skull. In the other families by contrast, the SBC roof does not extend as high; among the belontiids (*Betta*, *Trichopsis*, *Macropodus*, and *Trichogaster*), the SBC rarely surpasses the upper margin of the eye.

The shape, size, and surface complexity of the labyrinth varies markedly among the anabantoids (Figure 3.32) and is also affected by body size (Hughes and Munshi, 1973b; Peters, 1978). While the labyrinth of *Anabas* is highly crenulated, that of *Sandelia* consists of only a simple inner plate. Extreme variation in labyrinth structure occurs within each family and genus, however, without a phyletic pattern (Liem, 1963a). In many cases, alterations in ABO morphology result from changes in body shape as well as in both skull and mouth structure. While the link between

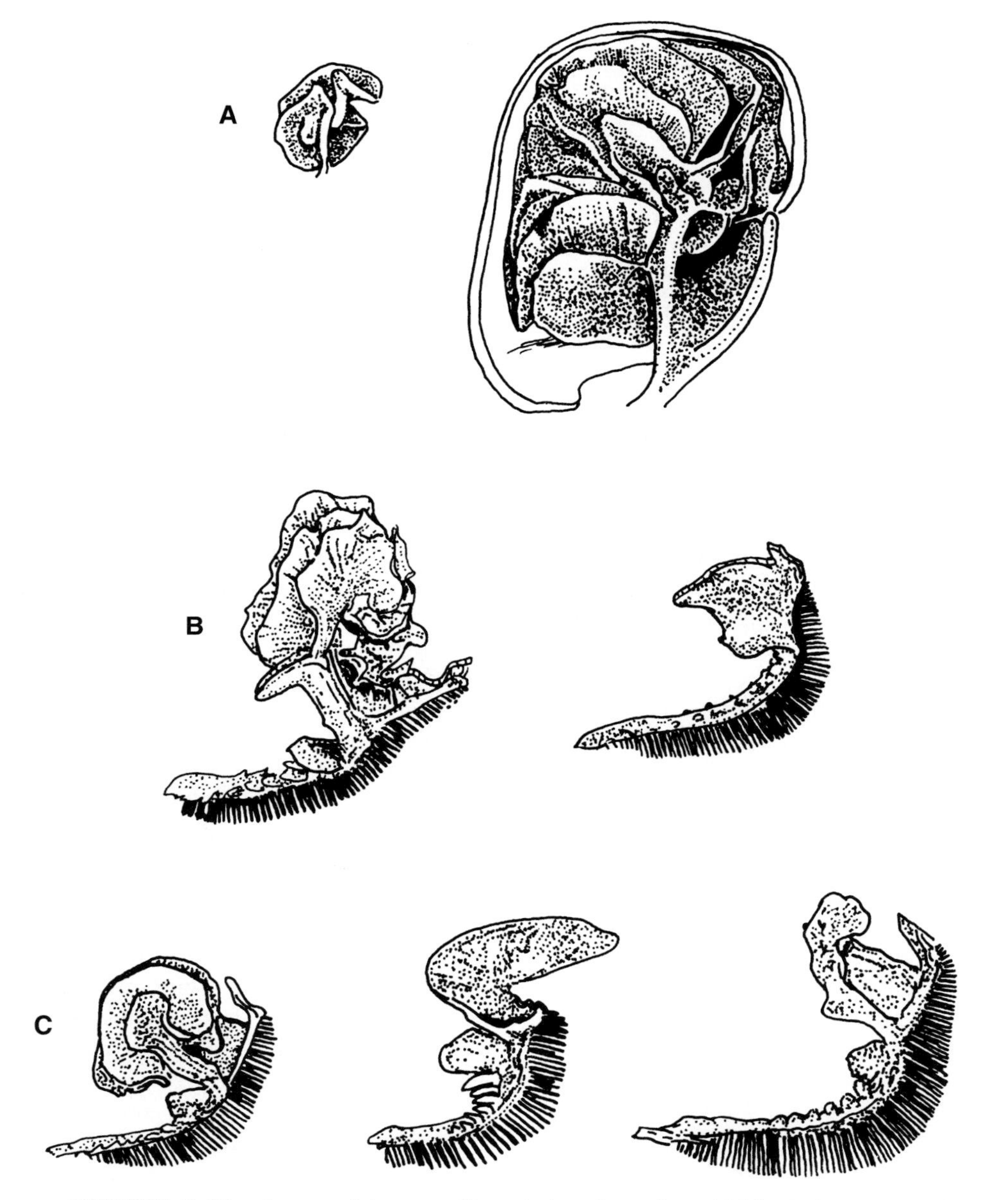

FIGURE 3.32 Intra- and interspecific variations in anabantoid labyrinth structure. **A,** Lateral view of the left labyrinth of *Osphronemus goramy:* left, 8 cm total length, right 20 cm. (Reproduced from Peters, 1978.) **B,** *Anabas* (left) and *Sandelia* (right). **C,** *Ctenopoma nigropannosum* (left), *Ctenopoma nanum* (center), *Ctenopoma kingsleyae* (right). (**B** and **C** redrawn from Elsen, 1976, by permission of the Academie Royal des Sciences, Belgique.)

ABO surface area and aerial respiration is clear, we have little understanding of how the overall structural variation seen in the anabantoid ABO may affect air breathing. In view of the labyrinth's role in air ventilation, these modifications must affect air-breathing efficacy as well as the complex interrelationships between air breathing and factors such as gill ventilation, feeding, buoyancy control, maneuverability, sound reception, and nest building (Liem, 1963a; Elsen, 1976; Peters, 1978).

ABO ontogeny and development. Anabantoids thus far examined all begin air gulping at about 10 mm length, which is long before both the labyrinth apparatus and the respiratory epithelium in the SBC are fully formed (Das, 1927; Peters, 1978; Prasad and Prasad, 1985). Initial observations of ABO development in *Anabas* (Das, 1927) indicated a progressive ontogeny toward general anatomical structures seen in adults. However, observations by Singh and Mishra (1980) indicate that the SBC forms separately and not as an extension from either the opercular or branchial chambers as was stated by Das (1927). They further assert that the labyrinth forms out of the the lateral wall of the SBC and grows across the chamber to connect with the gill arch. This claim is similar to that made for the respiratory trees of *Clarias* (page 104) but, as in the case of that earlier observation, it is not well documented in their paper. The accepted version is that the labyrinth originates from the epibranchial segment of the first branchial arch (Das, 1927).

Peters (1978) states that the ontogeny and evolution of monophasic and biphasic patterns is unclear. The youngest (10 mm) air breathing fish are incapable of biphasic air ventilation, and their emergency air gulping ("Notatmung") loosely resembles the monophasic pattern. While this suggests that ontogenetic development of air ventilation proceeds from monophasic to biphasic, there are notable exceptions (Peters, 1978). *Anabas*, for example, goes through an early phase of biphasic ventilation which gradually changes to exclusively monophasic by the time fish reach 80 mm (Peters, 1978). There is also some plasticity in the use of the two modes; both *Helostoma* and *Ctenopoma* can turn on monophasic ventilation in shallow water.

Regarding the differentiation of the vascular epithelium, the strong resemblance between anabantoid respiratory islets and those of the catfishes *Clarias* and *Heteropneustes* prompted the suggestion that the anabantoid ABO surface was also derived from the gills. This seemed certain when Munshi (1968) described what, at the time, were thought to be pillar cells in the ABO of *Anabas*. However, follow-up TEM studies (Hughes and Munshi, 1973b) showed these were pillar-like cells of epithelial origin (described previously), and each of the individual ABO capillaries had an endothelial cell lining (Figure 3.31F). According to Singh and Mishra (1980), the superficial resemblance of structures in the different organs stems from their origins in similar embryologic tissues (basophilic gill mass tissue and ectoderm), while their structural differences reflect the relative stage of development (pre- versus post yolk stage) when the ABO epithelium is formed in each group. In catfishes, normally developed gills differentiate and then migrate into the SBC where they form islets. In *Anabas* and *Colisa*, the vascular layer is formed when the ectoderm migrates into the SBC and combines with the gill mass of branchial arch 5 (Singh, 1988, 1993; Singh *et al.*, 1990).

Bader (1937) experimented with how a lack of air access affected ABO differentiation in *Macropodus opercularis*. He placed small fish (12–18 mm, i.e., too small for the ABO to have formed) in net covered tanks (to prevent air breathing) for periods of from one to seven months. During this period, specimens were removed to obtain a histological series. Bader found that lack of air access retarded ABO development. Relative to comparably sized control fish, fish without air access had greatly modified vascular linings in their SBCs. In young fish without air for as little as four weeks, the lining remained largely undifferentiated and much of the tissue that would have ordinarily gone into it was combined in large tufts covered by a sparse capillary bed. These tufts occupied much of the SBC lumen and apparently precluded labyrinth formation. Larger fish (i.e., growing without air breathing for a longer time) similarly developed tufts and lacked the regular islet lane pattern, and the number of capillaries in their SBC linings was much less than in controls.

At the end of the seven month experiment, Bader returned fish to tanks where air breathing was again possible and, after four weeks, sacrificed them to determine how air access affected ABO morphology. These fish responded to air by commencing air breathing and decreasing their rate of gill ventilation. Histological analysis revealed that air access stimulated the proliferation of capillaries over the surface of the tufts as well as along most surfaces of the ABO. There are no quantitative data on the extent that these fish "recovered" their potential air-breathing capacity, and Bader did not specifically address the extent to which the labyrinth itself became developed. Thus, additional work would be important in showing the extent that ABO development is contingent upon air contact. In this regard, Bader did conclude that air was a necessary environmental stimulus (Umweltreiz) for ABO differentiation.

Aerial Respiration and Gills

Gills as ABOs

Table 3.1 lists fish genera known to use gills for air breathing. Carter and Beadle (1931) observed that air held in the mouths of both *Hypopomus* and *Synbranchus* during air breathing also contacted the gills. They thus assumed that, even though other adjacent vascularized surfaces were present, some aerial O_2 uptake occurred at the gills. This conclusion has not been challenged (Lüling, 1958; Johansen, 1966; Liem, 1987) and is difficult to test because of the nearly impossible task of experimentally distinguishing between the aerial respiration occurring at the gills and in adjacent vascular tissues. Vargas and Concha (1957) smeared an algal paste on the gills of air-exposed *Sicyases* in order to gauge, by a reduced (12 versus 72 h) survival time in air, the organ's importance (relative to the skin) in aerial gas exchange. Their experiment was feasible because of the relatively large size of the specimens they studied.

In most cases, gill respiration has been inferred on the basis of negative evidence. In a situation where gas exchange can be measured in the cephalic end of a partitioned respirometer (Chapter 5), and the buccal and cutaneous epithelial surfaces surrounding the gills are neither heavily vascularized nor engorged with blood during air exposure, it can be concluded that gills are an important aerial gas exchanging surface. This rationale has been applied to *Mnierpes*, *Entomacrodus*, *Blennius pholis*, *Mastacembelus*, clingfishes, and several other species commonly exposed to air but lacking a well-developed ABO.

Influences of Air Breathing on Gill Structure

This section examines how gills are modified for aerial respiration and, in the case of species using other ABOs, how air breathing has affected gill structure and function. After first contrasting the functional requirements imposed on gills by different types of fish air-breathing, both gill structure and morphometrics will be reviewed.

Amphibious and aquatic air breathing have very different effects on gill structure. Species using gills as ABOs (Table 3.1) are either amphibious or hold air over them while in water. These fishes, and those periodically exposed to air in drying habitats, are best served by gills that can function in both water and air, and while in air resist desiccation and collapse.

The functional requirements placed on the gills of aquatic air-breathing species are more diverse mainly because in most species the ABO occurs in the systemic circulation (Chapter 4). The significance of this for the gills is that O_2-laden blood draining the ABO must pass through the heart and gills before entering the systemic circulation, and in the absence of gill modifications, the potential exists for aerially obtained O_2 to be lost to hypoxic water during gill transit (Randall *et al.*, 1981b, Chapter 4).

Additionally, the diverse environments and behaviors of aquatic air-breathers further complicate requirements for gill performance, as illustrated by comparing gill function in a facultative air breather and in a continuous air breather living in a chronically hypoxic habitat. The facultative air-breather (e.g., *Ancistrus* or *Hypostomus*) may spend the majority of its life never needing to breathe air and would thus be best served by well-developed gills fully adapted for aquatic respiration (Figures 3.33 and 3.34). However, when hypoxia necessitates air breathing, the facultative air breather's gills must remain functional (e.g., for CO_2 release, other gill-mediated exchanges, and possibly some O_2 capture [Graham, 1983]) but must also retard transbranchial O_2 loss. In contrast, sustaining the various branchial exchange processes while preventing transbranchial O_2 loss is a permanent requirement for the gills of air-breathing species residing in chronically hypoxic habitats and these fishes would likely benefit from a reduced gill area. Thus, the type and degree of gill modification can be expected to differ in amphibious and aquatic air breathers as well as among the latter, depending on environmental setting.

Gill Structure and Morphometrics

Branchial structure in most fishes consists of four gill arches, each bearing an array of primary and secondary gill lamellae (Figure 3.33). Primary lamellae (also termed filaments) are closely spaced rods of tissue extending perpendicularly from the arches and forming two posterior-directed rows, termed hemibranchs. Each primary lamella contains the afferent and efferent vasculature supplying the series of secondary lamellae lying across its dorsal and ventral surfaces. Gas exchange occurs through the secondary lamellae which are thin, blood-filled epithelial channels straddling and extending vertically from primary lamellae (Figures 3.33 and 3.34). Secondary lamellae are closely spaced and form narrow, parallel channels through which water passes during ventilation. Gas exchange across these surfaces is regulated by pillar cells (Figure 3.31G) which control lamellar-channel size and thus affect gill perfusion and diffusion distance.

Knowledge of air-breathing effects on gill morphology has benefitted from comparative morphometric and scaling analyses, initially by E. Schöttle (1931) and later by Hughes, Munshi, and others. Air-breathing fish gill morphometrics are summarized in several

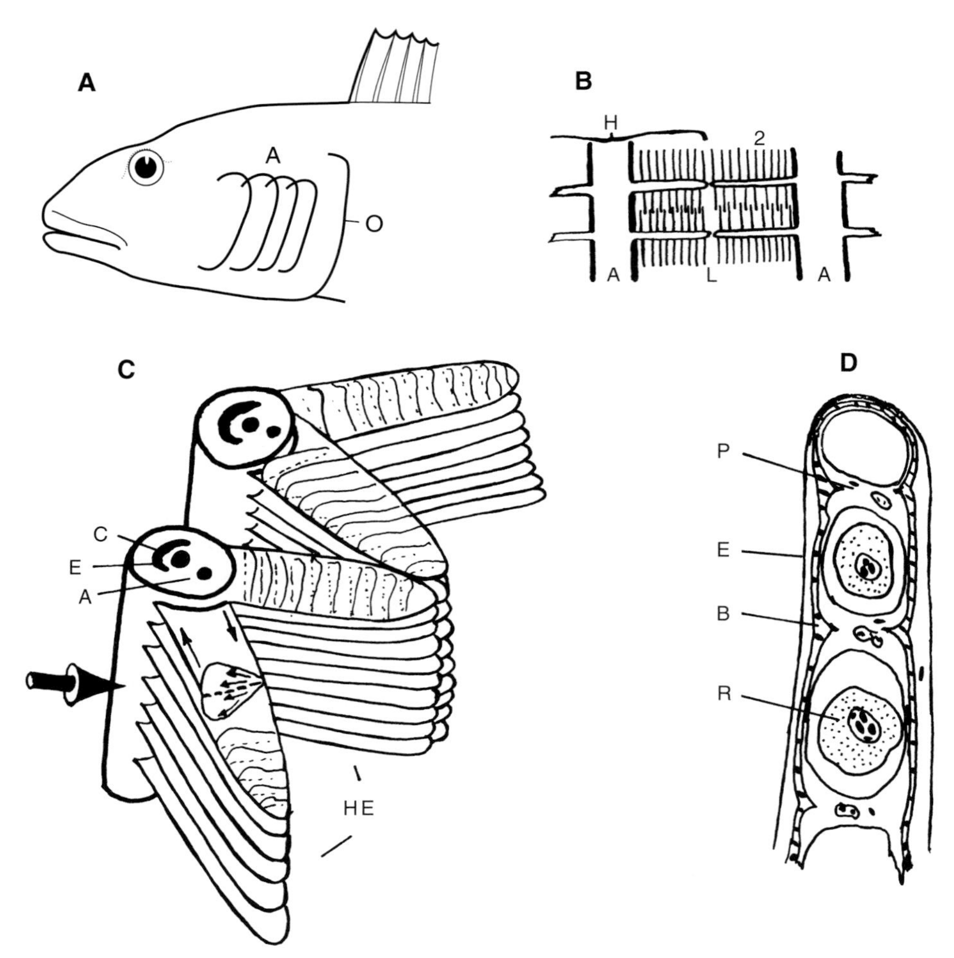

FIGURE 3.33 **A,** Relationship of the four teleost gill arches (A) and the operculum (O). **B,** Front-view diagram showing the sieve formed by the gill archs (A), their projecting lamellae (L) and secondary lamellae (2). H, holobranch. **C,** Oblique section through two arches showing the separation of the primary lamellae (filaments) into two left and right hemibranchs (HE). Arrows on the filament show paths of blood flow down the filament, through an upper and lower secondary lamella, and then back up the filament. Large arrow indicates direction of water flow. Arch detail shows the cartilage support (C), and the afferent (A) and efferent (E) arteries. **D,** Section through a secondary lamella showing parallel blood channels supported by pillar cells (P). The outer respiratory epithelial layer (E) surrounds the lamella and is supported by basement membrane (B). Note that pillar cell extensions line the blood channels. R, red blood cell.

works (Tamura and Moriyama, 1976; Hughes and Munshi 1979; Hughes, 1972, 1984; Roy and Munshi, 1986; Low *et al.*, 1990; Santos *et al.*, 1994). The major objective of most recent studies has been determination of total gill area in relation to body size.

Investigations to this point establish convincingly that, compared to non-air-breathing fishes, air breathers have a reduced gill area (Fullarton, 1931; Dubale, 1951; Tamura and Moriyama, 1976; Hughes and Munshi, 1979; Roy and Munshi, 1986). Fernandes *et al.* (1994), for example, determined that the non-air-breathing characins *Hoplias malabaricus* and *H. lacerdae* have twice the gill area of a comparably sized air-breathing characin *Hoplerythrinus unitaeniatus*, and that the mass exponent for total gill area (defined in the following section) was much less for the air breather (also see Cameron and Wood, 1978).

Taylor (1831) was probably first to deduce the relation between gill area and the need for aerial respiration and Schöttle (1931) first documented the trend for reduced gill area in air-breathing species. Her work established this pattern among gobies and in relation to the advent of mudskipper terrestriality. A significant finding by Schöttle was that in mudskippers the ratio

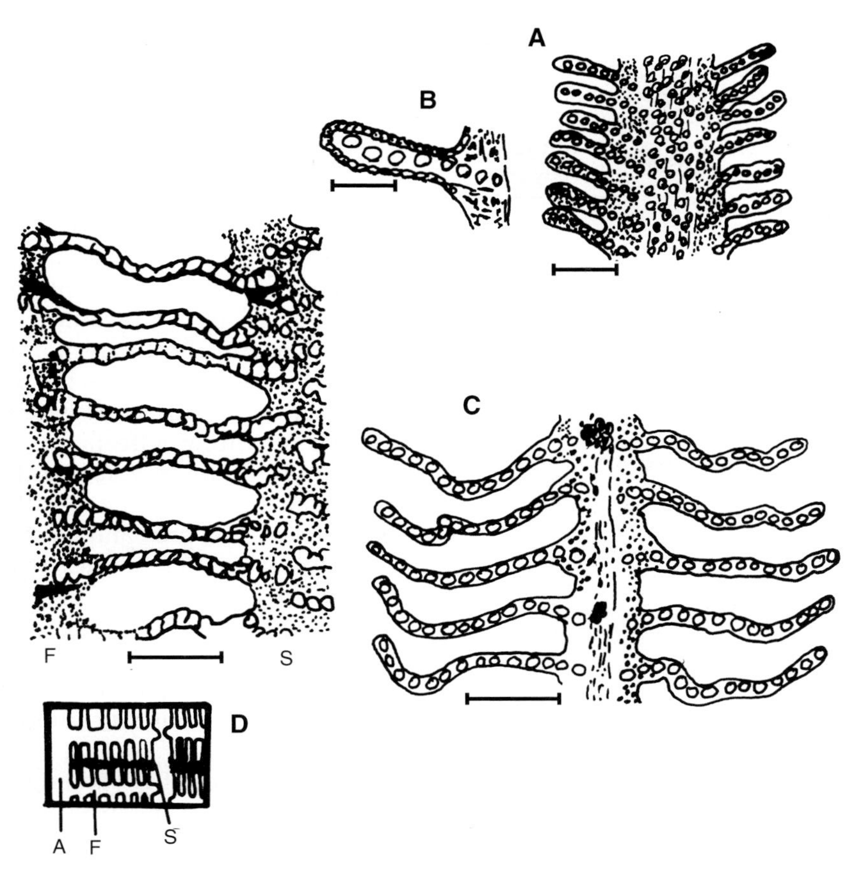

FIGURE 3.34 Modified version of Carter's (1935) comparison of a gill filament and branching secondary lamellae: **A,** *Electrophorus* (scale = 100 μm). **B,** Secondary lamellar detail for *Electrophorus* showing larger blood-channel diffusion-distance and the location of some channels within the filament (scale = 50 μm). **C,** *Hypostomus* (scale = 50 μm) has larger gills than *Electrophorus.* (Printed with permission of the Journal of the Linnean Society of London.) **D,** Longitudinal section through a filament (F) and support bar (S) of *Amia.* (Box shows both structures in relation to an arch, A). Note that blood channels at both ends are buried in the connective tissue and that the tips of the secondary lamellae on the adjacent filament occur within the support bar. Scale approximately 0.1 mm. (From Olson, 1981.)

of total gill area to total body surface area was at or below 1.0; whereas, in non-air-breathers this ratio was usually closer to 2 or higher (Table 3.4). Although varying with body size, the general utility of the gill:body surface-area index in distinguishing air-breathing and non-air-breathing fishes is apparent in Table 3.4.

Quantification of gill area. Estimation of total gill area requires determination of the total number and the bilateral surface area of the secondary lamellae. This can be done in several ways. It is usual to count the total number of primary lamellae on all arches on both sides of the body. Following this, estimates of average lamellar length and the average number of secondary lamellae/unit length of lamella are made by sampling lamellae at regular spatial intervals along all arches. Isolation of secondary lamellae and estimation of their surface area is then done and total gill area, *A,* is calculated as:

$$A = N \cdot L \cdot SLn \cdot SLa \qquad (3.7)$$

where N is the total number of gill lamellae, L is average lamellar length, SLn the mean number of secondary lamellae/unit length on both sides of a lamella, and SLa is mean bilateral secondary lamellar surface area (Hughes and Morgan, 1973). Interspecific comparisons of gill area and its constituent parameters (Figure 3.33) have most usually been made using allometric equations. To do this, mass-regressions are formulated for each variable in Equation 3.7 and then

TABLE 3.4 Ratios of Gill Area (cm^2) to Body Mass (g) and Total Body Surface Area (cm^2) of Some Air-Breathing and Non-Air-Breathing Fish Species

Species	Gill area/ body mass	Gill area/ body area
Amphibious Air Breathers		
Periophthalmus cantonensis[1]	1.24	0.38
P. koelreuteri[5]		0.46
P. dipus[5]		0.35
P. chrysospilos[5]		0.34–0.36
P. vulgaris[5]		0.27–0.32
P. schlosseri[9]		0.20–0.50
P. chrysospilos[9]		0.25–0.3
Boleophthalmus chinensis[1]	0.94	0.56
B. viridis[5]		0.72
B. boddarti[5]		0.68–0.83
B. boddarti[6]		0.52
B. boddarti[9]		0.65–0.75
Aquatic Air Breathers		
Heteropneustes fossilis[2]	0.32	0.34
Anabas testudineus[3]	0.39	0.40
Channa argus[1]	0.85	0.38
Clarias batrachus[7]	0.83	0.48
C. mossambicus[8]	0.17	
Non-Air Breathers		
Gobius jozo[5]		1.00
G. auratus[5]		1.17
G. caninus[5]		1.40
Lophius piscatorius[4]	1.96	2.99
Anguilla japonica[1]	3.32	1.45
Tautoga onitus[4]	3.92	4.35
Cyprinus carpio[1]	4.16	1.74
Carassius auratus[1]	4.49	2.91
Stenotomus chrysops[4]	5.06	4.78
Sarda sarda[4]	5.95	11.55
Mugil cephalus[4]	9.54	6.54
Scomber scombrus[4]	11.58	8.38
Brevoortia tyrannus[4]	17.73	18.28
Gymnosarda alleterata[4]	19.39	48.54

[1]Tamura and Moriyama, 1976; [2]Hughes *et al.*, 1974b; [3]Hughes *et al.*, 1973b; [4]Gray, 1954; [5]Schöttle, 1931; [6]Biswas *et al.*, 1981; [7] Munshi, 1985; [8]Maina and Maloiy, 1986; [9]Low *et al.*, 1990.

combined (by adding exponents) to describe gill area in relation to mass (Hughes and Al-Kadhomiy, 1986; Santos *et al.*, 1994).

Comparative gill area. The generally low mass-scaling coefficient of air-breathing fish gill area is apparent in Table 3.5 where intercept (*a*) and exponent (*b*) values of gill-area power equations are given for several species. Also shown in this table are regression-calculated gill areas for 100 g specimens (A_{100}) and secondary lamellar-diffusion distances where known. Except for some of the mudskippers (discussion follows), the mass exponents in Table 3.5 are generally below 0.85, and the mean for all 17 entries is 0.72 ± 0.02 (± SE). This is less than the mean slope values for non-air breathers (Prasad, 1988). The A_{100} values in Table 3.5 (22–274 cm^2, mean = 111 ± 17) also lie well below the range (400–2000 cm^2) for most non-air-breathing fishes (Roy and Munshi, 1986, figure 3).

Although Table 3.5 establishes that gill allometry is influenced by air breathing, it should be kept in mind that this effect may not always be evident. In young fishes, for example, the presence of a relatively large cutaneous surface area for respiration may obviate air-breathing effects on branchial allometry. Prasad (1988) found that even though *Colisa fasciatus* begins air breathing at about the same time it metamorphoses, the pre- and post-metamorphic scaling of gill area in this species was not different from non-air breathers of the same body size. In addition, intergeneric as well as interspecific differences in body and head shape and gill–arch anatomy doubtlessly affect parameters of gill area and may thus complicate interpretation of the air-breathing influences.

Future studies of gill scaling need to address some basic questions arising from existing data. Referring to Table 3.5, it is unclear why there are no apparent gill scaling differences among species of *Lepisosteus* but marked interspecific differences in the gill areas of *Clarias batrachus* compared to *C. mossambicus*, as well as differences among the three species of *Channa*. The data show that *Channa punctata* has a higher A_{100} but a lower mass coefficient than either *C. gaucha* or *C. striatus*. Is this difference correlated with obligatory air breathing or the partitioning of aerial and aquatic respiration? Do the aquatic respiratory capabilities of *Clarias mossambicus* and *C. batrachus* differ as markedly as the gill-mass coefficients and A_{100} values (data of Maina and Maloiy, 1986) suggest they should? Or, are the morphological differences due to artifacts (i.e., tissue shrinkage during preservation and fixation, counting errors, etc.)?

Three data sets for *Boleophthalmus boddarti* serve to illustrate the basic problem with comparative gill morphometrics (Table 3.5). First, there is the problem of mudskipper identification; Hughes and Al-Kadhomiy (1986), worked with fish from the northern Persian Gulf which, from Murdy's (1989) study, would be *B. dussumieri* and not *B. boddarti* (Table 2.4). This may be one reason why these workers reported a quite different gill area scaling relationship and A_{100} than was found for *B. boddarti* by Biswas *et al.* (1981), who studied a population from the Ganges Delta. In a

TABLE 3.5 Comparative Gill Morphometrics of Air-Breathing Fishes Showing Intercept (*a*) and Exponent (*b*) Values for Surface–Area Allometric Power Equations and Estimated Areas of a 100 g Fish (A_{100})[a]

	Total area		A_{100} (cm^2)	Secondary lamellar diffusion distance (μm)	Diffusing capacity D_t (ml O_2/min/torr/kg)
	a	*b*			
Lepisosteus oculatus[4,b]	3.94	0.738	118		
L. osseus					
L. platostomus					
Lepisosteus osseus[14]	3.148	0.71	83		
Amia calva[14]	3.986	0.84	191		
Lepidocephalus guntea[1]	4.936	0.745	152		
Hoplerythrinus unitaeniatus[19]	5.99	0.66	125		
Clarias batrachus[8,10]	2.278	0.781	83	7.67–12.4	0.014
C. mossambicus[11]	1.201	0.628[c]	22	1.97	
Heteropneustes fossilis[3,10]	1.86	0.746	58	3.58	0.024
Hypostomus plecostomus[18]	4.36	0.666	94		
Rhinelepis strigosa[20]	6.17	0.757	202		
Macrognathus aculeatus[5]	2.173	0.733	64	1.74	0.050
Boleophthalmus boddarti[7,h]	2.813	0.709	74	1.43, 4.66[d]	0.077
B. boddarti[15,f]	0.927	1.050	117	4–12[g]	
B. boddarti[17]	6.79	0.481	62		
Periophthalmus chrysospilos[17]	0.976	0.958	80		
P. schlosseri[17]	1	0.931	73		
Periophthalmodon schlosseri[16]	3.002	0.934	222		
Pseudapocryptes lanceolatus[12]	6.079	0.827	274		
Anabas testudineus[2,10]	2.783	0.615	47	10–15	0.007
Colisa fasciatus[13,e]	5.00	0.802	199	1.1	
Channa punctata[6,10]	4.704	0.592	72	2.03–2.44	0.053
C. gachua[9]	1.49	0.757	49		0.0382
C. striatus[9]	1.919	0.718	52		0.0115

[a]Secondary lamellar–diffusion distances and D_t are indicated where known.
[b]Values combined for three species.
[c]Allometric equation in the reference is wrong and has been recalculated.
[d]Thick in areas with lymphatic spaces.
[e]Very small fish, just above size where air breathing begins (12–25 mm, 30–290 mg).
[f]Probably *B. dussumieri.*
[g]Lowest at margin and highest near base.
[h]See Table 2.4 for updated mudskipper synonymies.
[1]Singh *et al.*, 1981; [2]Hughes *et al.*, 1973b; [3]Hughes *et al.*, 1974b; [4]Landolt and Hill, 1975; [5]Ojha and Munshi, 1974; [6]Hakim *et al.*, 1978; [7]Biswas *et al.*, 1981; [8]Sinha, 1977; [9]Munshi, 1985; [10]Hughes and Munshi, 1979; [11]Maina and Maloiy, 1986; [12]Yadav and Singh, 1989; [13]Prasad, 1988; [14]Crawford, 1971; [15]Hughes and Al-Kadhomiy, 1986); [16]Yadav *et al.*, 1990; [17]Low *et al.*, 1990; [18]Perna and Fernandes, 1996; [19]Fernandes *et al.*, 1994; [20]Santos *et al.*, 1994.

third study, Low *et al.* (1988, 1990), working with *B. boddarti* from Singapore, also obtained very different scaling and A_{100} values than either of the preceding works.

Figure 3.35 shows that the gill area regressions for *Boleophthalmus* determined in these studies have little in common. The mass exponents vary from 0.48 to 1.05. Hughes and Al-Kadhomiy (1986), who reported the highest slope value, related it to the presumed need for a higher $\dot{V}O_2$ in the case of *B. boddarti* compared to other mudskippers. Further analysis of these three data sets, moreover, indicates that there is also a considerable range in the estimated values for each of the separate components (listed in Equation 3.7) contributing to the scaling estimates. This suggests that variations in measurement techniques have also led to errors. For example, the Hughes and Al-Kadhomiy estimate for the scaling of secondary lamellar size ($M^{0.85}$) is nearly twice the value ($M^{0.43}$) obtained by Biswas *et al.* (1981). Not only is a large exponent for secondary lamellar scaling inconsistent with general findings for air-breathing fishes, a value this large is rare among non-air-breathers (data in Biswas *et al.*, 1981). In addition, the Hughes and Al-Kadhomiy estimate for the scaling of numbers of secondary lamellae/primary lamella is much less ($M^{-0.229}$) than the

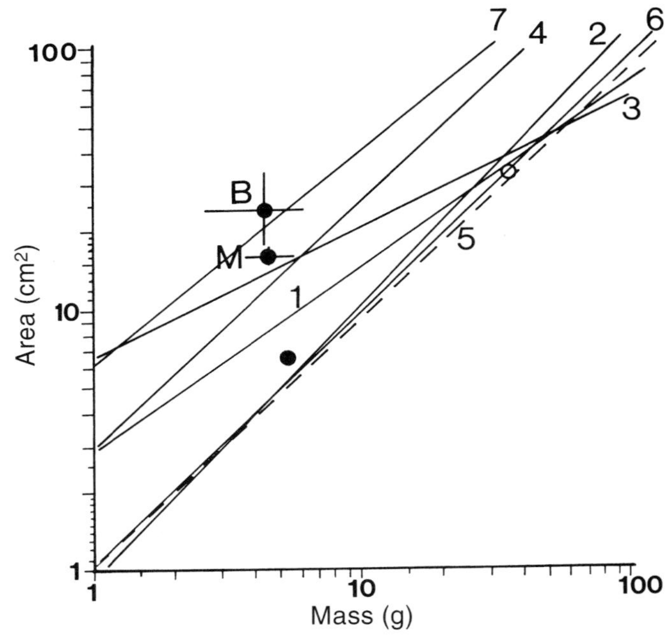

FIGURE 3.35 Gill area-body mass regressions for several mudskippers with comparative data for other species. Numbered regression lines and equations for each species are: 1) *Boleophthalmus boddarti*, Area = $2.813M^{0.709}$ (Biswas *et al.*, 1981). 2) *B. boddarti (dussumieri)*, A = $0.927M^{1.05}$ (Hughes and Al-Kadhomiy, 1986). 3) *B. boddarti*, A = $6.79M^{0.481}$ (Low *et al.*, 1990). 4) *Periophthalmodon schlosseri*, A = $3.002M^{0.9341}$ (Yadav *et al.*, 1990). 5) *Periophthalmus schlosseri*, A = $1M^{0.931}$ (Low *et al.*, 1990). 6) *P. chrysospilos*, A = $0.976M^{0.958}$ (Low *et al.*, 1990). 7) *Pseudapocryptes lanceolatus*, A = $6.079M^{0.827}$ (Yadav and Singh, 1989). Individual data points: B, *Blennius (Lipophrys) pholis;* M, *Mnierpes macrocephalus* (for both data sets horizontal and vertical lines indicate observed ranges of mass and area [Graham, 1973]). Circles: Values obtained by Tamura and Moriyama (1976) for *Periophthalmus cantonensis* (solid circle) and *B. chinensis* (open circle).

Biswas *et al.* value ($M^{-0.083}$). Thus, even though the allometric equations for total gill area obtained in these two studies (lines 1 and 2 in Figure 3.35) appear to lie within the statistical error limits of one another, we are left with uncertainty as to why the constituents of gill morphometrics differ so markedly. This picture is further complicated by the data of Low *et al.* (1990) (line 3, Figure 3.35). Even if Hughes and Ad Kadhomiy (1986) studied *B. dussumieri* and not *B. boddarti*, the three data sets provide no indication of a common generic-level similarity of gill morphometrics in *Boleophthalmus*. Thus, what, if any, functional significance do these observations have for *Boleophthalmus* and, just as important, for the body of theory surrounding gill and ABO allometry in air-breathing fishes?

Finally, the gill area regressions for the other mudskippers (*Periophthalmodon*, *Periophthalmus*) in Figure 3.35 are discouraging because they fail to support the generalization that these more terrestrial mudskipper genera have a reduced gill area relative to the more aquatic *Boleophthalmus*. As could be expected, the regression data for *Pseudapocryptes* and the data points for *Mnierpes* and *Blennius*, all of which are less amphibious than mudskippers, occur to the left and indicate a greater gill area. There is also fairly good agreement between individual gill area determinations for *P. cantonensis* and *B. chinensis* (Tamura and Moriyama, 1976) and the various mudskipper regressions. (Unfortunately, the location of the single-point gill area determination for *B. chinensis* made by these workers is such that it "fits" all three of the *Boleophthalmus* regressions.) However, the regression for *Periophthalmodon schlosseri* from the Andaman Islands (line 4, data of Yadav *et al.*, 1990) is to the left of all other mudskippers which is surprising in view of the consensus that *Boleophthalmus* is more aquatic than either *Periophthalmus* or *Periophthalmodon*, the most aquatic of the mudskipper genera. Yadav *et al.* did not discuss their findings in relation to this question or the data for other mudskippers.

Other Comparative Aspects of Gill Structure

Although not formalized by scaling relationships, many additional observations on gill structure support the hypothesis that gill area reduction is widespread among air-breathing fishes. These observations, many fragmentary and isolated, provide comparative insight into the structural bases for area reductions and the different effects of amphibious and aquatic air-breathing on gill structure. Table 3.6 summarizes this information for different gill variables (Equation 3.7) and compares their presence in (primarily) aquatic and amphibious (i.e., in the sense of having gills exposed to air) air-breathing groups.

General. Beginning at the level of the branchial arches, air-breathing related modifications are evident in both groups (Table 3.6). As will be detailed in Chapter 4, some branchial arches in the aquatic air breathers *Anabas* and *Channa*, the lungfishes, *Electrophorus*, and the frequently amphibious *Monopterus* function mainly as gill shunts; these have lost most or all gill lamellae and most of the cardiac output they receive heads directly into the systemic circulation. In *Clarias* (Maina and Maloiy, 1986), *Anabas*, *Channa*, and others, branchial arches have been shortened by the dorsal position of the ABO. In *Periophthalmus* and *Boleophthalmus*, gill arch 4 is reduced by a posterior ABO (Tamura and Moriyama, 1976; Biswas *et al.*, 1981). Finally, gill arch 4 is entirely absent in the clingfish (Gobiesocidae); however, whether or not total gill area is reduced in the more commonly air-exposed members of this family is unknown.

Reductions in the number of primary lamellae occur in both amphibious and aquatic air breathers (Table 3.6). Approximately 50% fewer secondary lamellae on the gills of the air-breathing *Hoplerythrinus* accounts for its having about 50% less gill area than the non-air-breathing *Hoplias* (Cameron and Wood, 1978; Fernandes *et al.*, 1994). Böker (1933) determined that in the trout (*Salmo*) gill total primary lamellar length amounted to 17% of head length but found that this percentage was highly reduced in the following air-breathing genera: *Arapaima* (7%), *Hoplerythrinus* (8%), *Synbranchus* (8.5%), *Electrophorus* (3%), *Gymnotus* (13%), and *Clarias* sp. (14%). Several other air-breathers (*Heteropneustes*, *Channa*, and *Anabas*) also have fewer and smaller primary lamellae and in *Pseudapocryptes* the lateral hemibranch of arch 1 is stunted under the ABO. Schöttle (1931) also concluded that support of the respiratory branchiostegal membrane reduced the respiratory function of the first arch in *Periophthalmus*.

Air-breathing fish generally have fewer and smaller secondary lamellae. In some species, these are short and stubby and even appear "vestigial" (Singh and Munshi, 1969; Laurent *et al.*, 1978). In *M. cuchia*, they are small, thick, few in number, and lack pillar cells (Hughes and Munshi, 1979; Munshi *et al.*, 1989). Carter (1935) described the secondary lamellae of *Electrophorus* and *Hoplosternum littorale* as narrow and poorly developed and thus in stark morphologic contrast to those of *Hypostomus* (Figure 3.34A, see Fernandes and Perna, 1995; Perna and Fernandes, 1996). Likewise, the secondary lamellae on arches 3 and 4 of *Anabas* are thick (Dube and Munshi, 1974) and have from 1 to 3 large-bore channels which reduce overall surface contact (Olson *et al.*, 1986). Olson *et al.* (1994) demonstrated a similar pattern for *Channa* and have also shown a greater arch and lamellar size and a greater number of channels in the obligatory air-breathing *C. marulius* relative to the non-obligatory *C. punctata*. Arterioles rather than capillaries comprise the secondary lamellar blood paths in *Protopterus* which reportedly lacks pillar cells (Laurent *et al.*, 1978).

Several secondary lamellar traits likely retard diffusive O_2 loss. Secondary lamellae much thicker than those of non-air-breathing fishes (e.g., 1–3 μm, Table 3.5) occur in several groups and in some species the base of the secondary lamellae is buried within the filament or a support (Figure 3.34B and D) or under mucus or cell organelles. The presence of numerous large acidophilic mucous cells also increases the secondary lamellar diffusion distance in *P. vulgaris* to between 8 and 32 μm (Singh and Munshi, 1969). Related to this, the exposure of some species to ranges of salinity and water quality can affect diffusion distances by triggering changes in epithelial cell layers or the appearance of specialized cells (mucus, granular, ionophores, see also Skobe *et al.*, 1970) and lymphatic channels in the epithelium, or affect lamellar mucus layer thickness (Hughes and Munshi, 1979). In contrast, Laurent *et al.* (1978) found that estivation decreased gill epithelial thickness in *Protopterus*. Also, species like *Lepidocephalus*, *Arapaima*, *Heteropneustes*, *Amia*, and *Lepisosteus* may depend upon the combination of a limited amount of anatomical gill occlusion and physiological control of perfusion and ventilation to prevent transbranchial O_2 loss (Randall *et al.*, 1981b; Olson, 1981; Olson *et al.*, 1990).

Gill structure and amphibious air breathing. Modifications reducing the potential for gill-water loss include reductions in the size and number of both primary and secondary lamellae and the thickening and mucus–sequestering of secondary lamellae.

Mechanisms that prevent gill collapse include secondary lamellar supports in *Amia* (Figure 3.33D), the relatively thick cartilagenous rods in the lamellae of *Mnierpes* (Graham, 1973) and *Alticus* (Brown *et al.*, 1992), and the presence of a "cytoplasmic stiffening material" in the pillar cells of *Boleophthalmus* (Munshi, 1985). The comparative study of mudskipper gill structure by Low *et al.* (1988) revealed the dramatic degree of secondary lamellar fusion in *Periophthalmodon schlosseri* (Figure 3.36). These elaborate fusions may aid the gill in retaining water during emersion but may also retard ventilatory water flow.

For *Synbranchus*, Liem (1987, figure 9) observed that the lamellae of each hemibranch arise from the arch in alternating series, which may stiffen them and prevent collapse. I observed this in the galaxiid *Neochanna*. Schmidt (1942) reported that the hemibranchs of the stichaeid *Chirolophis polyactocephalus* originate alternatively. He further noted that the secondary lamellae of this fish were large and densely packed, making its gill volume two to three times larger than expected for a fish in this size range. While the gill modifications seen in *Chirolophis* have been frequently interpreted as adaptive for air exposure and respiration, there are no supporting natural history observations for this species.

The majority of gill air breathers do have short, stubby lamellae and small, thick secondary lamellae

TABLE 3.6 Comparative Structural Variability in the Gills of Aquatic and Amphibious Air-Breathing Fishes

GROUP/GENUS	AQUATIC										AMPHIBIOUS													
Structural Component	*Protopterus*	*Amia*	*Arapaima*	*Clarias*	*Heteropneustes*	*Electrophorus*	*Hoplerythrinus*	*Hoplosternum*	*Gymnotus*	*Channa*	*Anabas*	*Monopterus*	*Synbranchus*	*Neochanna*	*Gobiesocidae*	*Chirolophis*	*Periophthalmodon*	*Periophthalmus*	*Boleophthalmus*	*Pseudapocryptes*	*Gillichthys*	*Alticus*	*Mnierpes*	*Blennius*
I. Branchial Arches A. No. <4															●									
B. Shortened by ABO				●	●					●	●													
C. Significant gill reduction, a.4	●		●	●						●		●					●	●	●					
D. Gills absent: 1 arch											●	●												
2+ arches	●					●							●											
E. Shunts	●						●				●	●												
II. Primary Lamellae A. Short	●				●	●	●	●	●	●	●		●				●	●	●	●	●			
B. Few	●			●	●	●						●												
C. Small diameter				●	●						●							●						
D. Reduced by ABO											●													
E. Rigidity		●[a]										●[c]		●[c]		●[c]						●[b]	●[b]	
III. Secondary Lamellae A. Few	●											●					●	●	●				●	●
B. Small (short/flat)	●			●	●	●		●		●	●	●	●				●	●	●	●			●	●
C. + Diffusion distance: Thick	●			●	●						●	●												
Blood channel reduction Few and large bore	●			●	●						●	●												
Buried in p. lamellae or mucus		●	●	●	●																			
D. No pillar cells	●										●	●												
E. Rigidity																	●[d]	●[e]				●[e]		
F. Physiological shunts likely	●	●	●		●		●	●	●	●	●													

[a]Interlamellar fusion in *Amia.*
[b]Cartilage rods.
[c]Staggered.
[d]Intralamellar fusion (see Figure 3.34).
[e]Pillar cells contain stiffening material.

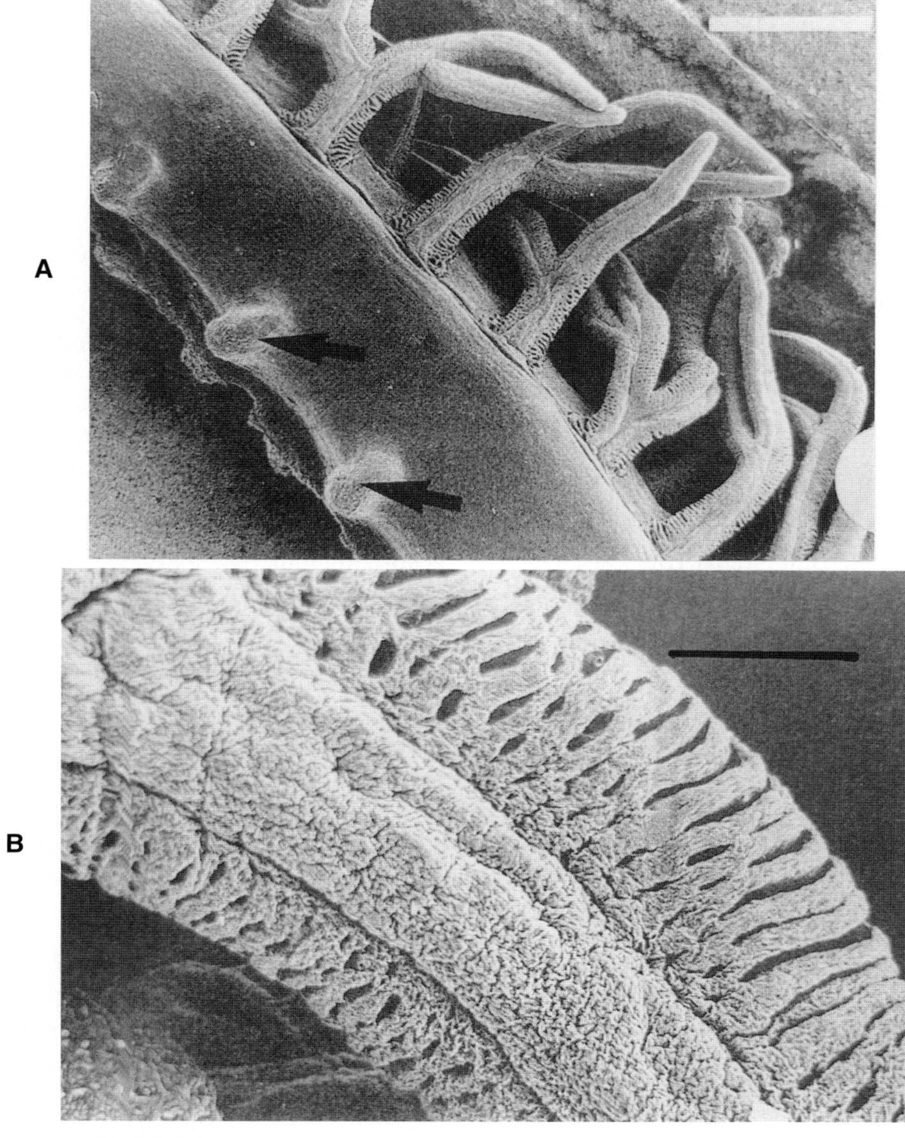

FIGURE 3.36 SEM photographs of mudskipper gill filaments. **A,** Branched gill filaments of *Periophthalmodon schlosseri* (scale = 1000 μm). Arrows show relative size of gill rakers and filaments. **B,** Interlamellar fusions in *Periophthalmus schlosseri* (scale = 100 μm). (Reprinted from Low *et al.*, 1988, with permission of Marine Biological Laboratory, from the *Biological Bulletin*.)

(Table 3.6). However, there is little to support the assertion by Carter and Beadle (1931) that, as an adaptation for air breathing, the gills of both *Synbranchus* and *Hypostomus* were modified in a fashion that resulted in a large ratio of secondary lamellae breadth (i.e., height) to thickness. Johansen (1966) remarked that the Carter and Beadle ratio did not make sense in view of the likely tendency for high, thin lamellae to fold or coalesce under their own weight in air. Moreover, later studies of *Hypostomus* gills show they have a large surface area relative to other air breathers (Table 3.5) as well as a microstructure that reflects the normally aquatic respiratory mode of this facultative air breather (Fernandes and Perna, 1995; Perna and Fernandes, 1996).

Gill structure and aquatic air breathing. Gill reduction in aquatic air-breathing fishes is manifest at several levels of organization, from the arches to secondary lamellae (Fullarton, 1931; Dubale, 1951). Two factors, space occlusion by an ABO and the need to reduce transbranchial O_2 loss and thereby increase air-breathing efficiency, have played a role in this. Modifications reducing the potential for O_2 loss include gill shunts, decreased gill area, and secondary lamellar thickening. The specific differences in these parameters reflects air-breathing dependence. For example, other than having deep proximal blood channels, *Hypostomus* has virtually no other branchial modifications to indicate it is a facultative air breather (Fernandes and Perna, 1995). The sequestration of secondary lamellar channels (Figure 3.34) occurs in many species, but modifications that increase gill rigidity are less developed in this group (although seen in *Amia* and *Protopterus* [Laurent *et al.*, 1978]) than in the amphibious air breathers.

ABOs in or Adjacent to the Digestive Tract

Esophagus

Dallia

Dallia pectoralis (Umbridae) uses its esophagus as an ABO (Crawford, 1971, 1974). The respiratory section of this tube occurs between the branch point of the pneumatic duct and the sphincter separating the esophagus and stomach (about 80% of esophagus length). The region is heavily invested with vessels, both externally and internally, giving it a distinctive red color. The luminal margin of the ABO is densely populated with capillaries and mucous cells. The respiratory epithelium is mostly stratified, and thin extensions of endothelial and epithelial cells increase total surface area. Figures in Crawford's (1974) paper indicate blood-air distances to be less than 1 μm.

Muscular layers in the wall of the respiratory esophagus are well developed. Also, the esophagus and the gas bladder of *Dallia* are both ensheathed in a layer of striated muscle which originates in the externa of the esophagus and inserts dorsally on the vertebrae, ribs, and trunk musculature. Crawford (1974) suggests this muscle layer functions for compression of the ABO and gas bladder during expiration.

Dallia, unlike *Umbra* (the other air-breathing genus in this family), lacks any trace of respiratory epithelium in its gas bladder and thus depends exclusively on its esophagus for air breathing. The gas bladder of *Dallia* is, however, similar to that of *Umbra* in having a *rete* and gas gland.

Blennius pholis

X-rays taken by Laming *et al.* (1982) revealed air bubbles within the distended esophagus of air-exposed *B. pholis*. Dissections revealed that the longitudinal folds of this organ contained an extensive capillary network positioned within a few micrometers of the lumen.

Pneumatic Duct

Anguilla

The lung-like structure of the pneumatic duct of *A. anguilla* has been implicated as a possible site of air breathing (Rauther, 1937; Mott, 1951; Fange, 1976). The lumen of this organ is lined by a heavily vascularized, smooth epithelium and its venous return is directly into the sinus venosus (Chapter 4). The major uncertainty is if, or how, *Anguilla* uses this structure for air breathing. This fish is known to use both its skin and gills for air breathing while emergent (Berg and Steen, 1965); however, the opening from the gut to the pneumatic duct is small and controlled by a sphincter, suggesting that air ventilation would not be easy. The gas bladder of this fish contains both a *rete* and gas gland, and Mott (1951) suggests that *Anguilla* might be able to absorb O_2 from gas bladder contents expelled into the pneumatic duct.

Stomach

Air-breathing loricariids and trichomycterids use their stomachs as an ABO. There has been only limited morphological and histological study of these organs in relation to air breathing.

Loricariidae

Structure. Carter and Beadle (1931) observed that the stomachs of *Hypostomus* (= *Plecostomus*) and *Ancistrus* were thin, papery, transparent, and prominent with blood vessels but lacked digestive gland tissue. In *A. anisitsi*, respiratory vessels are concentrated in a transparent dome region of the stomach, and in *Hypostomus* the respiratory plexus is more dense in the posterior part of the organ. Cross-sectional drawings of the stomach walls of both genera (Carter and Beadle, 1931; Carter, 1935) show dense capillary layers within 0.5 μm of the luminal boundary. Carter (1935) noted several differences in the two genera; while the stomach of *Ancistrus* was set off in a separate visceral compartment on the right side of the body (Figure 3.37), that of *Hypostomus* was in the center of the body and had more glandular cells. He also observed that *Hypostomus* seems to continually gulp air and always has air in its stomach. When it is air filled, the side position of the stomach in *Ancistrus* affects the center of gravity of the fish and raises its right side (Gee, 1976 and pers. obs.).

Morphometrics. Gradwell (1971) estimated that gas volumes of about 2.0 ml were released from 61 g

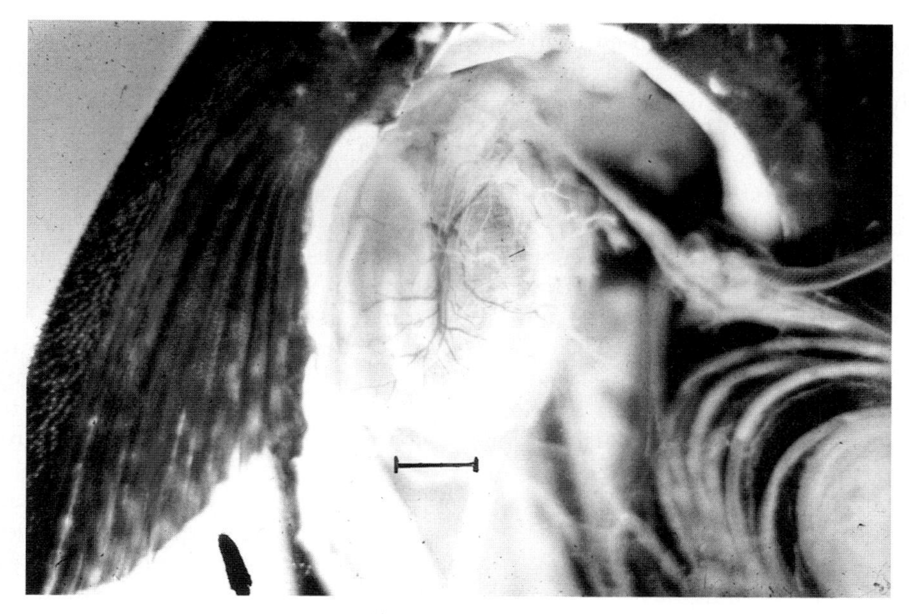

FIGURE 3.37 Ventral view of the air-inflated stomach of *Ancistrus chagresi (spinosus)* located on the right side of the abdomen. Structure to the left of the stomach is the right pectoral fin; on the opposite side is the spiral intestine. Scale approximately 0.5 cm. Photo by J.H. Gee.

(18 cm) *Hypostomus*. His values are in general agreement with those of Gee (1976) who surveyed ABO volumes of five loricariid genera. Gee found a positive relationship between body length and ABO volume, with the largest fishes he examined (30 cm *Hypostomus*) having an ABO volume of over 10 ml. He also found that *Ancistrus* and *Hypostomus* had much greater size-specific ABO volumes than did *Chaetostoma*, *Loricaria*, or *Sturisoma*. Studies with *Ancistrus chagresi* (*spinosus*) (Graham, 1983) permitted estimation of the stomach–ABO volume in relation to mass (Figure 3.38) and additionally demonstrated that fish acclimated to hypoxia (i.e., engaged in facultative air breathing for at least 10 days) had stretched their ABO volume by 25% compared to controls (just induced to breathe air). This work also showed that the expired ABO volumes of individual fish were nearly constant and closely matched the ABO volumes of comparably sized fish determined by Gee (1976), thus supporting the conclusion that *Ancistrus* completely empties its ABO during expiration (Graham, 1983).

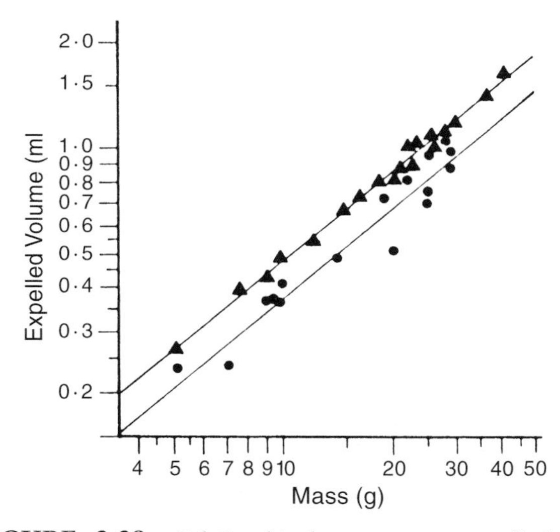

FIGURE 3.38 Relationship between mean expelled ABO volume and body mass (M) in control (dots) and hypoxia-acclimated (triangles) *Ancistrus chagresi (spinosus)* (25 C). Regression line equations: Control volume = $0.054M^{0.85}$, r = 0.99, N = 16. Hypoxia acclimated volume = $0.069M^{0.84}$, r = 0.99, N = 19. (From Graham, 1983, by permission of the Company of Biologists, Ltd., from the *Journal of Experimental Biology*.)

Air ventilation. Gradwell (1971) described four behavioral stages in the process of "gastroventilation:" Resting on the bottom, restless stage prior to gas release, gas release, and rapid ascent for air. Loricariids often release air while resting on the bottom, which reduces their requisite surface time. Numerous independent observations now verify that gas release is via the operculae and mouth and not via the anus as reported by various workers. Although several exhalation mechanisms have been suggested (Gradwell, 1971), the most probable is the contraction of the muscles lining the stomach. The ascent for air is rapid, and less than one second is needed at the surface for inspiration (Gradwell, 1971). After air is taken, the fish flips around, often flicking its tail against the surface, and quickly swims back to depth.

Trichomycteridae

Cala (1987b) found that the stomach of *Eremophilus mutisii* contained gas, and that the ventro–central area of this organ is perfused by two vessels. The esophagus opens into the stomach which is larger in diameter, but has a thin wall without much muscle, and is separated by a sphincter from the intestine. *Pygidium striatum* holds air in the central part of its stomach while air-breathing, and Gee (1976) found a linear relationship between ABO volume and fork length for this species, with specimens between 40 and 80 mm (1–3 g) holding between 0.02 and 0.07 ml.

Intestine

Reports of intestinal air breathing have been made for several fishes. Marusic *et al.* (1981) suggested this for *Sicyases* because they found that air exposed fish had air in their intestines and intestinal capillaries were engorged with blood. Aerial respiration through a specialized segment of the intestine has been well documented for several cobitids (*Misgurnus, Cobitis, Acanthophthalmus,* and *Lepidocephalichthys*) and callichthyids (*Hoplosternum, Corydoras, Callichthys, Brochis, Dianema*). In cases where it has been studied, ventilation of this ABO is one way, with gas being released from the vent at the same instant new air is swallowed (Gee and Graham, 1978).

Cobitididae

Gut function in air breathing has been known for loaches since the early 1800s (Chapter 2), and there have been several descriptions of intestinal structure related to air breathing for *Cobitis* and *Misgurnus* (Leydig, 1853; Calugareanu, 1907; Lupu, 1908; Wu and Chang, 1945; Jeuken, 1957). As reviewed by Dorn (1983), aspects of the structure and physiology of the ABO's of *Brochis* and *Dianema* were reported in dissertations by Kopp (1978) and Blaese (1980). Recent descriptions for *Lepidocephalus guntea* (Yadav and Singh, 1980; Moitra *et al.*, 1989) and *M. anguillicaudatus* (McMahon and Burggren, 1987) are summarized here.

Structure and morphology. According to Yadav and Singh (1980) the intestine of *Lepidocephalus guntea* makes up 85% of the total digestive tube and is divided into two parts. The anterior part, which functions for digestion, is thick-walled and has a folded mucosa. The posterior part, accounting for two-thirds of intestinal length (i.e., about 60% of the digestive tube), has a respiratory function. This section has a thin wall consisting of an outer collagenous layer, a single layer of non-ciliated epithelial cells (18–21 μm thickness), mucous cells (6–12 μm), and a luminal mucus film (1 μm). The absence of muscular tunic layers and a thin collagen layer appear important in increasing the organ's air capacity. The respiratory segment is also densely vascularized and generally filled with air. Blood vessels from the dorsal intestinal artery penetrate all along the dorsal intestinal wall and form transverse feeder vessels (about 150 capillaries/mm^2) positioned within 0.9 to 1.1 μm of luminal air. For this same species, Moitra *et al.* (1989) report a larger mean diffusion distance (2.6 μm), but do not discuss possible reasons for this disparity with the results of Yadav and Singh (1980).

The respiratory region in *M. anguillicaudatus* also occurs behind the digestive segment of the intestine and, similar to *L. guntea,* constitutes about 60% of alimentary canal length (McMahon and Burggren, 1987). The respiratory section has smooth inner walls, and lumen diameter remains fairly constant along its length (Figure 3.39). Unlike the condition reported for *L. guntea*, the respiratory section in *M. anguillicaudatus* has a thin muscular layer. The volume of the respiratory intestine is small (0.026 ml/g, for a 2 g fish), and from 50–80% of this volume is turned over on each breath. Blood supply to this ABO is generally similar to that in *L. guntea* (Chapter 4). Capillaries in the respiratory intestine of *M. anguillicaudatus* have a uniform diameter (3–4 μm) in all regions, but reach their greatest number (about 60/linear mm) in the rear of the organ where diffusion distance is also lowest (overall range within organ 1–2 μm). McMahon and Burggren (1987) note that the occurrence of capillaries among the intestinal epithelial cells is similar to the condition described for *M. fossilis*. They emphasize the important function played by the spiral region of the gut, the section immediately anterior to the respiratory intestine, in compressing the fecal material and encapsulating it in mucus which in turn enhances air contact in the respiratory segment.

Callichthyidae

Some structural details for these fishes are contained in Jobert (1878). Studies of the respiratory intestine in *Hoplosternum thoracatum* by Huebner and Chee (1978) expand earlier observations for *H. littorale* (Carter and Beadle, 1931). As in the loaches, the respiratory section in these fishes occurs posterior to the digestive intestine and continues to the rectum, and amounts to about 50% of total alimentary tract length. Also, most species have a prominent sphincter just anterior to the respiratory sector which appears important in both compressing fecal material and seal-

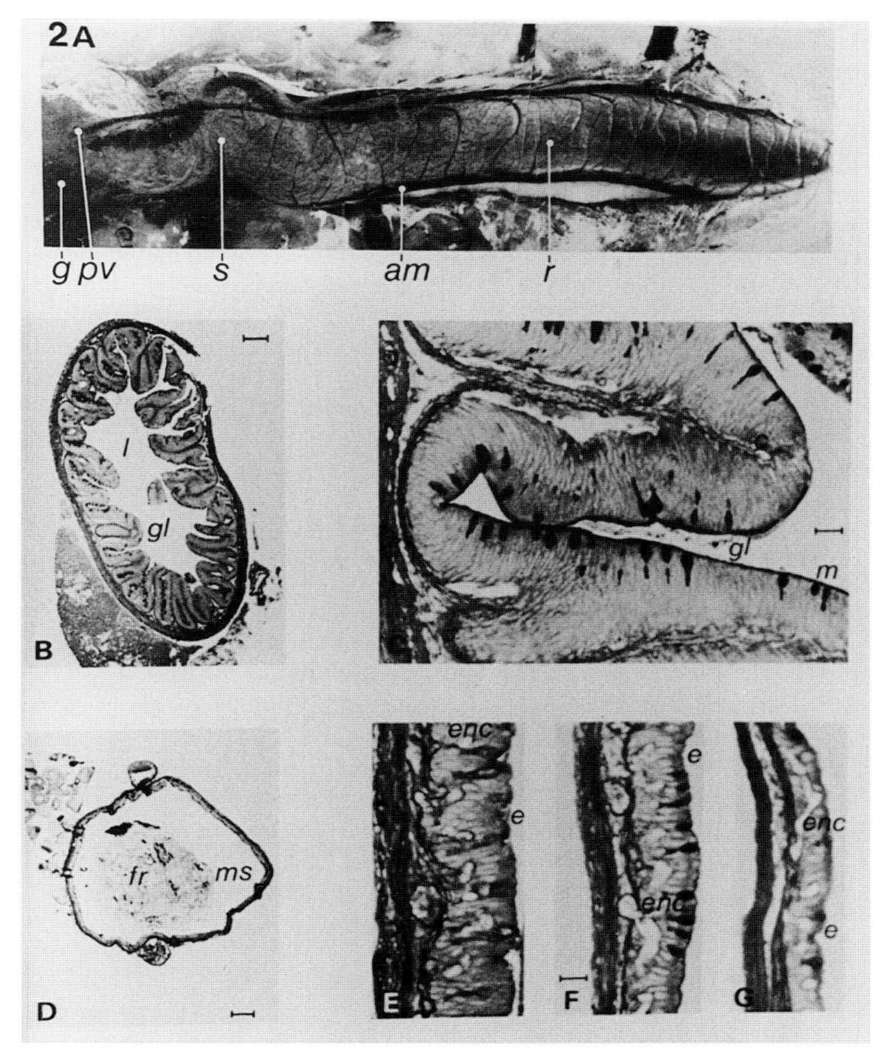

FIGURE 3.39 General and microanatomy of the alimentary canal of *Misgurnus anguillicaudatus.* (Reprinted from McMahon and Burggren, 1987, by permission of the Company of Biologists, Ltd., from the *Journal of Experimental Biology.*) **2A,** Anterior glandular stomach region (g), spiral (s), and respiratory (r) regions and their blood supply; hepatic portal vein entering the liver (pv), and the anterior mesenteric artery (am). **B,** Cross section of the anterior stomach region, scale = 100 μm. **C,** Higher magnification of B showing cell detail, scale = 10 μm. **D,** Cross section of respiratory region, scale = 100 μm. **E-G,** Higher magnifications of the respiratory epithelium at progressively more posterior positions showing wall and capillary detail. Abbreviations: e, intestinal epithelium; enc, intestinal capillary network; l, lumen; ms, deeply stained mucus sleeve; gl, glandular region of the stomach; fr, food residue; scale = 10 μm.

ing the ABO (Kramer and McClure, 1980). Gee (1976) found a linear relationship between body length and ABO volume in *H. thoracatum* with fish ranging from 50 to 90 mm fork length (about 1–8 g) holding between 0.3 and 1.0 ml.

Huebner and Chee (1978) found that the respiratory section of *H. thoracatum* is specialized for gas exchange, for limited nutrient uptake, as well as for the facilitation of fecal transport. They showed that delimiting digestive function to a relatively short region of the alimentary tract—an obvious necessity for the elaboration of an aerial-exchange section—has been compensated by the rich amplification of digestive surface area within that region. They further observed that the respiratory section has thin walls; however, both circular and longitudinal muscle layers are present, and the organ is typically air-filled. This section has a smooth surface of chiefly cuboidal epithelial cells, numerous mucous-producing goblet cells, and cells rich in the organelles (cytosomes, endoplasmic reticulum, and mitochondria) involved in nutrient uptake. The thin respiratory epithelium is penetrated by branching capillaries, which in many sectors form labyrinth sheets that are separated from the luminal space by only a single cell layer (< 1–2 μm). Most of the diffusion distances illustrated for *H. littorale* by Carter and Beadle (1931) are also less than 1 μm.

The Skin of Air-Breathing Fishes

Respiration

The ability of fishes to exchange respiratory gases through their skin was demonstrated by Krogh (1904a,b). There have been recent experiments showing cutaneous respiration in a variety of species (Kirsch and Nonnotte, 1977; Nonnotte, 1984) and this subject was reviewed in 1985 by Feder and Burggren. Compared to the other gas exchanging surfaces, fish skin is relatively thick and neither particularly well ventilated nor perfused; its role in gas exchange is thus always auxiliary. Skin is a metabolically active tissue (Kitzan and Sweeny, 1968; Banerjee and Mittal, 1975), and the cutaneous respiration of some species may only serve the metabolic needs of the tissue itself and not otherwise contribute to the organism's respiratory gas exchange (Nonnotte, 1984). Also, in view of studies pointing to red cell skimming from blood perfusing the cutaneous (secondary) circulatory network (Steffensen *et al.*, 1986), the potential for vascular transport of cutaneously obtained O_2 to other body regions is reduced.

Cutaneous air breathing. Table 3.1 lists the genera for which cutaneous air breathing has either been

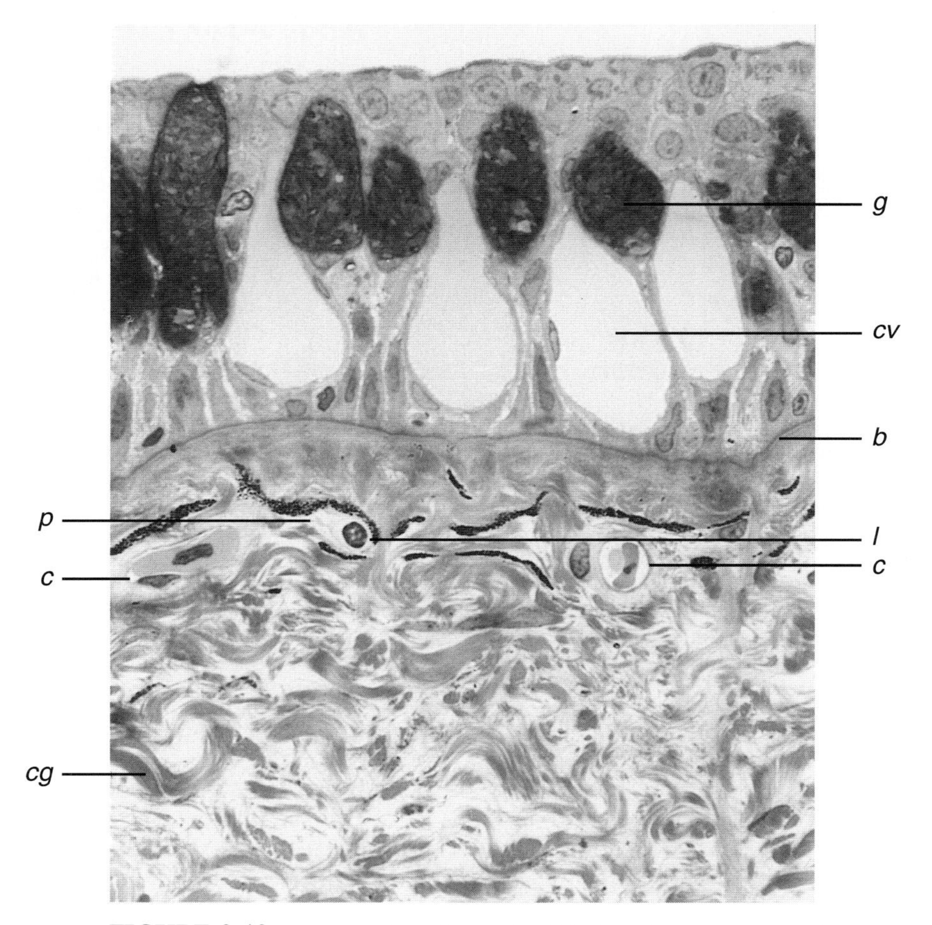

FIGURE 3.40 Cross-section of epaxial mid-body skin from a 260 g (57 cm) *Synbranchus marmoratus*, *c*, capillary; *p*, pigment; *l*, lymphocyte; *b*, basal lamina; *cg*, collagen bundle; *cv*, central vacuole of a serous gland; *g*, goblet cell. Scale = 100 μm. (From Graham *et al.*, 1987, by permission of the Company of Biologists, Ltd., from the *Journal of Experimental Biology*.)

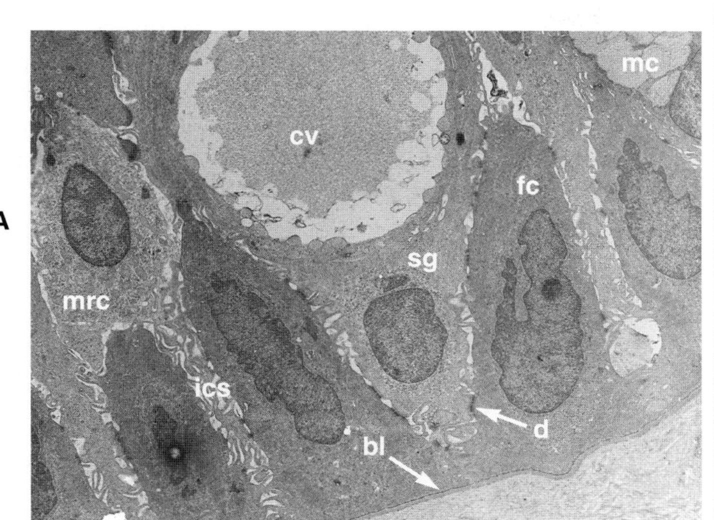

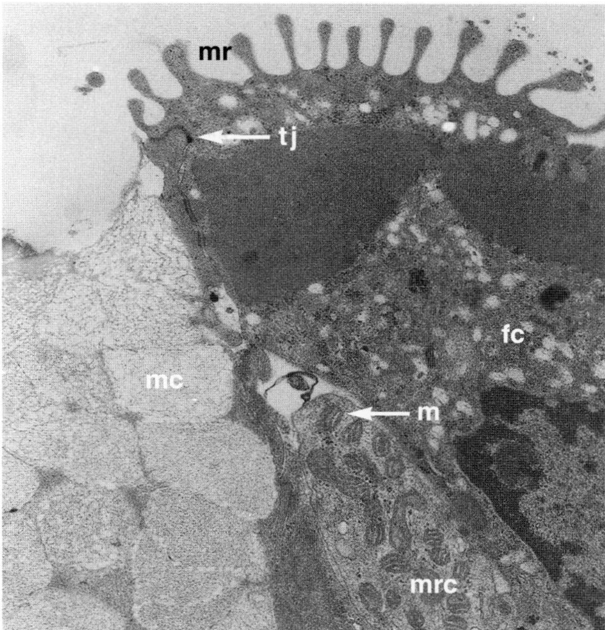

FIGURE 3.41 *Synbranchus marmoratus*. **A,** EM of skin from the basal region of epidermis (x 5,069). bl, basal lamina; fc, filament-containing cell; mrc, mitochondria-rich cell; mc, mucous cell; sg, serous gland cell with central vacuole (cv); d, desmosomes; ics, intercellular space . **B,** EM of surface skin (x 17,322). mc, mucous cell; fc, filament-containing cell; mrc, mitochondria-rich cell; tj, tight junction; mr, microridges; m, mitochondrion. (From Stiffler *et al.*, 1986, reprinted with permission of the University of Chicago Press.)

confirmed by respirometry or is strongly suspected on the basis of skin structure. This respiratory mode has been quantified for at least 13 air-breathing fishes but likely occurs in all amphibious species. Some air-breathing species can make use of cutaneous O_2 uptake while in water (Graham *et al.*, 1987), but this capability has limited utility in stagnant, hypoxic habitats and is most useful during amphibious air breathing. Cutaneous release of CO_2, however, is probably a widespread occurrence among aquatic air breathers, particularly in species with reduced gill areas.

Structure and morphometrics. Fish skin is composed of epidermal and dermal layers (Figures 3.40 and 3.41). The epidermis consists of a living epithelial cell layer, sensory cells, various types of glandular secretory cells, and ionocytes. Scales, when present, are embedded within the dermis which is also the normal location of cutaneous blood vessels. Histological examinations in relation to respiration have emphasized the skin's potential for gas exchange, as shown by diffusion distance, capillary density, and total area. Table 3.7 summarizes data available on these points for various species and shows the power equations relating body mass and skin surface area. The ranges of the intercept and exponent values in the equations for air breathers show some intraspecific variation but are generally similar to values for non-air breathers.

Skin capillaries. Both diffusion distance and skin-capillary density affect cutaneous respiration. Zander (1972a,b, 1983), found cutaneous vessel density correlated with the degree of amphibious life among the blennies *Lipophrys trigloides*, *Coryphoblennius galerita*, and *Alticus kirki*. A similar trend was established for mudskippers in relation to other gobies (Schöttle, 1931; Tamura *et al.*, 1976). Cutaneous capillaries are most abundant in areas of the body (e.g., the dorsal body surfaces of gobies and blennies, the throat of *Sicyases*) where respiratory gas transfer would be favored when fish are exposed to air or are at the water surface.

Given that fish cutaneous capillaries generally lie in the dermis, what structural mechanisms lessen diffusion distance? Table 3.7 shows that diffusion distances range from as little as 1 μm in *Rivulus* to over 370 μm in *Anguilla*. In most species, epithelial thickness is large and data for both *Rivulus* and *Misgurnus* show that thickness is not regionally modified to favor cutaneous respiration (Jakubowski, 1958; Grizzle and Thiyagarajah, 1987). Rather, the penetration of vessels

TABLE 3.7 Respiratory Properties of Air-Breathing Fish Skin[a]

Species	Total area (cm^2)	Diffusion distance (μm)	Capillaries /mm
Protopterus annectens[12]		55	45
Notopterus notopterus[15]		120[d]	14
Heteropneustes fossilis[1]	$8.51M^{0.684}$	98	16
Clarias batrachus[11]	$5.65M^{0.743}$	137[b]	12
Anguilla australis[4]		373[b,c]	16
A. anguilla[3]	$9.40M^{0.683}$	263[b,c]	13
Misgurnus fossilis[5]	$5.42M^{0.759}$	339[c]	25
Rivulus marmoratus[2]		1[c]	16
Coryphoblennius galerita[10]		60	11
Boleophthalmus boddarti[8]	$6.01M^{0.687}$	22.5[b]	
B. boddarti[16]	$8.41M^{0.523}$		
B. boddarti[17]		5–230[e]	
B. chinensis[9]		17[b]	
Periophthalmus cantonensis[9]		9[b]	
P. schlosseri[16]	$6.33M^{0.689}$		
P. chrysospilos[16]	$4.39M^{0.808}$		
Pseudapocryptes lanceolatus[14]	$7.99M^{0.601}$		
Mugilogobius paludis[18]		102[f]	8,16[h]
		26[g]	14,19[i]
Arenigobius bifrenatus[18]		75	8[i]
Chlamydogobius sp.[18]		61	3,6[i]
Mastacembelus pancalus[1]		7–39[b]	18
Macrognathus aculeatus[13]	$7.05M^{0.705}$		
Synbranchus marmoratus[6]	$1.31M^{0.618}$	90	16
Monopterus albus[7]		82	25
Monopterus cuchia[1]	$8.77M^{0.707}$	119	12

[a]Data presented include allometric power equations for total skin surface area in relation to mass (M), skin diffusion distances, and estimated capillary density.

[b]Epithelial capillaries present.

[c]Capillary loops within dermal papilla penetrate epidermis.

[d]Outer epidermis keratinized.

[e]Thickness varies regionally.

[f]Mean value, between eyes on snout, normoxic water.

[g]Mean value, same area, hypoxic water.

[h]Skin on snout and between eyes, normoxia.

[i]Skin on snout and between eyes, hypoxia.

[1]Mittal and Munshi, 1971; [2]Grizzle and Thiyagarajah, 1987; [3]Jakubowski, 1960; [4]Smith *et al.*, 1983; [5]Jakubowski, 1958; [6]Graham *et al.*, 1987; [7]Liem, 1967c; [8]Biswas *et al.*, 1981; [9]Tamura *et al.*, 1976; [10]Zander, 1983; [11]Banerjee and Mittal, 1975; [12]Kitzan and Sweeny, 1968; [13]Ojha and Munshi, 1974; [14]Munshi, 1985; [15]Mittal and Banerjee, 1974; [16]Low *et al.*, 1990; [17]Ad-Kadhomiy and Hughes, 1988; [18]Gee and Gee, 1995.

into the epidermis, and thus closer to the skin surface, is the mechanism used to enhance gas exchange.

Cutaneous vessels in amphibious fishes may occupy space between the dermis and epidermis or even invade the epidermis itself. *Rivulus marmoratus*, for example, has an average epidermal thickness of 10 μm over its body; however, epidermal capillaries on its dorsal head, nape, and anterio-dorso trunk lie within 1 μm of the skin surface (Grizzle and Thiyagarajah, 1987). Epidermal vessels are known to occur in mudskippers (Salih and Al-Jaffery, 1980; Al-Kadhomiy and Hughes, 1988). Some areas of the epidermis of *Mastacembelus pancalus* are so thin that diffusion distance is composed of only a few layers of epithelial cells, the basement membrane, and the endothelial wall of the capillary.

In species with a thick epidermis (e.g., *Heteropneustes*, *A. anguilla*, and *Misgurnus*), papillary loops of blood vessels frequently penetrate the epidermis and in some species closely approach the skin surface (Smith *et al.*, 1983). The pattern of vascular perfusion and capillarity in the dorsal head and snout skin of *Dormitator* (this area is exposed to air during facultative air breathing) has not been examined. Gee and Gee (1995) found dermal capillaries in *Mugilogobius paludis*, *Arenigobius bifrenatus*, and *Chlamydogobius* sp. and observed that exposure to aquatic hypoxia stimulated ASR in all three species and opened capillary tracts in the upper reaches of the epidermis in *Mugilogobius* and *Chlamydogobius* (Table 3.7).

Even with epidermal blood vessels, the transcutaneous diffusion distances in most air-breathing fishes vastly exceed those of gills and other vascular surfaces (Table 3.7). Fish skin diffusion distances also exceed those of most amphibians (23–54 μm, Jakubowski, 1958). Nevertheless, cutaneous air-breathing and cutaneous respiration in aerated water are important gas exchange modes for these fishes (Chapter 5), supplying in many cases over 50% of the total O_2 uptake. The answer to this enigma apparently lies in the makeup of the skin itself; although fish epidermis is thick, it consists primarily of glandular cells and has a watery consistency. Although there are no data, it has been suggested that this quality increases the respiratory gas diffusion constant of fish skin relative to that of the frog (the latter contains leathery layers that limit desiccation, Krogh, 1904a,b, 1941; Mittal and Munshi, 1971).

Table 3.7 also presents data for cutaneous vessel density, expressed in terms of the number of vessels per linear mm. In most cases, vessel density was not reported in the references cited in this table, and I estimated the numbers from counts and measures of histological sections figured therein. This analysis shows cutaneous vascular densities ranging from 11 to 45/mm. For calibration purposes, it is noted that Jakubowski (1958) found the cutaneous vascular density of *Misgurnus fossilis* was practically the same as that of the amphibians *Rana terrestris*, *Bufo bufo*, and

TABLE 3.8 Cutaneous Vessel Dimensions Expressed on a per Gram Body Mass Basis for Two Air-Breathing Fish Species[a]

Species	Total length of capillaries (mm)	Vessel surface area (mm^2)	Volume (mm^3)	Total skin area covered by vessels (%)
Misgurnus fossilis	5972	145.8	0.28	74.5
Anguilla anguilla	3922	99.1	0.20	44.6

[a]Data from Jakubowski (1958, 1960).

Xenopus laevis. The two dimensional vascular maps of Jakubowski (1958) enabled me to estimate a vessel density of 25/linear mm for *Misgurnus* (Table 3.7). Extrapolation from this leads to the conclusion that the skin vascularity of most air-breathing fishes is less than frogs. A notable exception to this conclusion, however, is suggested by my calculation, based on a lateral body wall section in Kitzan and Sweeny (1968), that *Protopterus annectens* contains up to 45 supradermal capillaries/mm, more than all other species in Table 3.7. Additional lungfish data are needed to verify this; nevertheless, this finding corroborates the subjective impression about lungfish cutaneous vessel density, which I gained from inspection of microvascular casts and is in line with the conclusions of numerous investigators (Chapters 2 and 5) that lungfishes carry on high rates of cutaneous gas transfer.

Exceptionally detailed studies of fish skin vascularity were done by Michael Jacobowski (1958, 1960) whose reports pall most others in their wealth of detail. Jacobowski determined total skin surface areas and the percentage of skin area occupied by blood vessels. He also measured regional skin thicknesses and vascular densities as well as total cutaneous vessel lengths, surface areas, and volumes. Table 3.8 summarizes his data for two air breathers, *M. fossilis* and *A. anguilla*. Counts made from two dimensional vascular maps in his papers were used to estimate single plane cutaneous vessel density for these species (Table 3.7).

Amphibious Life and Protective Functions of the Skin

Unicellular mucous glands, extremely abundant in the epidermis of most fishes, exude a mucus coat which is important for protection from abrasion, desiccation, and infection. Several types of mucus and the presence in some species of epidermal pouches, or scale cups, that hold water during emersion, are the principal external features retarding water loss from fish exposed to air. In some amphibious, intertidal species (*Blennius*, *Periophthalmus*), relatively thick pads or caps, formed by mucus and cornified epidermal cells, cover ventral body surfaces and fins for protection from abrasion (Harms, 1929; Whitear and Mittal, 1984; Al-Kadhomiy and Hughes, 1988).

Light and electron microscopic studies by Kitzan and Sweeny (1968) showed that the skin of *Protopterus annectens* has three types of goblet cells, each with a characteristic mucus. These workers suggested that the mucus products of the various goblet cells and other epidermal mucous cells might be combined in different proportions to produce the different mucoidal properties required by this lungfish, from the mucus coat typical of a fish in water to the formation of the dry season cocoon. Kitzan and Sweeny also observed that the outer epidermal cells, although apparently fully functional, were held together by desmosomes and hemidesmosomes, features commonly found in the dead, keratinized epidermis of higher vertebrates.

Mittal and Munshi (1971) correlated the presence of the sulphated acid mucopolysaccharides (substantia amorpha) in the epidermis and dermis of *Monopterus cuchia* with the ability of this fish to resist desiccation during amphibious excursions and seasonal drying. They found this substance had many of the same histochemical properties of amphibian ground substance, which Elkan (1968) had previously shown to play a role in desiccation resistance. Mittal and Banerjee (1974) also found sulphated acid mucopolysaccharide secreting cells in the epidermis of *N. notopterus* and further reported that as these polygonal cells mature and migrate towards the outer epidermal layer, their sulphated acid mucopolysaccharide content increases, and their shape and orientation change to the extent that they fuse and form a keratinized layer. These workers did not do the TEM studies required to detect desmosomes and other ultrastructural details. Their histochemical analyses, however, did show that cells in the outer layer are senescent and contain keratin. Mittal and Banerjee (1974) concluded that the skin cornification process in *Notopterus* is similar to that of amphibians and aquatic reptiles and that keratinization, along with sulphated acid mucopolysaccharides prevent desiccation. The mean epidermal thickness reported for *N. notopterus* (120 μm) is comparable to other air-breathers (Table 3.7); however, no documented accounts of terrestrial exposure exist for this species. Illustrations in Mittal and Banerjee (1974) show significant concentrations of blood vessels throughout the dermis of *N. notopterus*, and my estimates of vessel density compare well with other fishes (Table 3.7); however, there are no data for cutaneous respiration in this fish, and we do not know the effect of keratinization on gas transfer. Finally, practically all cutaneous gas transfer measurements and diffusion distance estimates have ignored the resistance imposed by the mucus layer which may change, depending upon habitat and fish stress.

SUMMARY AND OVERVIEW

ABO Structure and Design

Phylogeny and descent with change have knit an elaborate tapestry of fish ABO diversity in which the early origin of the vertebrate lung and its evolutionary transition to first a respiratory, and then a non-respiratory, gas bladder is the central thread. The lung and respiratory gas bladder are the functional ABOs of the primitive fishes, including the more primitive teleosts. However, the gradual loss, through evolutionary canalization, of the gas bladder's capacity to serve a respiratory function led to the acquisition of novel ABOs in higher teleosts. The ostariophysan grade of fish evolution marks the approximate "point of no return" for use of the gas bladder in respiratory function. At this level, there is diversity in the organ's degree of specialization for respiration (i.e., from *Umbra* to *Phractolaemus*), suggesting that both retention and secondary reacquisition of air breathing have occurred.

Morphology

Further complexity in the fabric of ABO structure has been woven by the independent evolution of air breathing and novel ABO acquisition among several ostariophysans and higher teleosts. However, in spite of this diversity there has been a remarkable convergence in ABO structure owing to evolutionary constraints imposed on devices needed for air capture, for air storage space, and for adequate exchange surfaces. Thus, and with few exceptions, ABOs depend upon the mouth and jaws for acquisition and manipulation of air, while air storage usually occurs along the digestive tube or in spaces adjacent to the head and gills where the perfusion potential is high.

Ventilation

The mouth and jaws have a central role in ventilation. Although the "recoil aspiration" of the Polypteridae is an exception, inhalation by means of a buccal force pump and either passive or active exhalation is the common method of lung and respiratory gas bladder ventilation. Also, in most forms, exhalation precedes inspiration; the exceptions being the characins, *Gymnotus*, *Pantodon*, *Umbra* and perhaps others where, for reasons that are not entirely clear, new air is inspired prior to the release of old air.

Among species possessing novel ABOs, buccal force inhalation usually follows a compressive exhalation in fishes using esophageal, stomach, and intestinal ABOs. In others (e.g., in clariids, *Heteropneustes*, and *Channa*), the ABOs are partially wrapped by layers of muscle providing for compressive deflation, but in these fishes as well, inhalation and exhalation are the primary result of jaw and branchial motions similar to those used during feeding and water ventilation. In cases of quadruphasic air ventilation, exhalation occurs first, driven by the reversed flow coughing reflex. In triphasic ventilation inspiration precedes exhalation, and there is no flow reversal. Water is thus not required, and this may be preadaptive for amphibious air breathing.

Morphometrics

Considerable information has been obtained on the structure and morphometrics of the novel ABOs; however, more data are needed. Interspecific ABO comparisons within both the anabantoids and clariids are particularly important. Comparative studies of synbranchids with, and without, respiratory sacs would also be useful. Reexamination of *Heteropneustes* is warranted because earlier studies failed to clarify the impact of microridged surfaces on total ABO surface area. Also, few data exist for the stomach-breathing Loricariidae and Trichomyteridae. Studies to date show significant ontogenetic changes in the scaling of the SBC in *Anabas* and the air sacs of *Monopterus*, however, not *Heteropneustes* (Hughes *et al.*, 1973b, 1974a,b). Finally, with the exception of *Clarias batrachus* and *Anabas testudineus*, most morphometric data sets do not describe the entire respiratory system (i.e., gills, skin, and all air-breathing surfaces).

A surprising revelation of this survey is the sparsity of data on the fine structure and morphometrics of fish lungs and respiratory gas bladders where, apart from the lungfishes and *Lepisosteus*, only cursory observations have been made, and no data exist for most genera. Especially important for future study are the African teleosts *Phractolaemus* and *Gymnarchus* and the Asian *Pangasius* which have ABO respiratory surface complexities rivaling those of the Sarcopterygii.

Surfactants and Mucus

Cell types I (the main exchange surface) and II (ciliated and mucus producing) are found in all fish lungs and respiratory gas bladders examined thus far. There is also evidence for the presence of linings with surfactant-like properties in these organs, and in certain groups, affinity with the tetrapod lung is certified by the presence of lamellated osmophilic bodies.

However, because fish lungs and respiratory gas bladders lack both the degree of respiratory surface complexity and alveolar dimunition common in higher tetrapod lungs, important questions arise about what additional roles surfactants may play and what selection pressures might have accounted for their origin. Moreover, the presence of surfactant-like properties in and on the epithelial linings of the physoclistous gas bladders of the higher teleosts (Liem, 1988) further complicates the idea that surfactants evolved and serve primarily for surface-tension reduction. Surfactants may have evolved for other functions such as the prevention of desiccation, anti-edema agents, or as a mucociliary adjuvant.

We lack comparative information about the surfactant properties of other air-breathing species which is important to know in view of the range of surface complexity seen in their lungs and respiratory gas bladders. In most of the novel ABOs of higher teleosts, surfactants are not present suggesting that their presumed function would have to be taken over either by mucous cells or the regular flushing of the ABO with water. In the respiratory sac of *Heteropneustes*, which is seldom flooded with water, no surfactant has been found on either the islet or lane surfaces; however, their microvillous and ciliated cell surfaces suggest that a mucus layer may be normally present. No studies have compared ABO mucus and lung-gas bladder surfactant for their chemical and physical properties in relation to ABO function.

ABO Surface Area

Morphological analyses of fish lungs and respiratory gas bladders reveal their fundamental similarities with the lungs of tetrapods. The diffusion distances in most of these organs range from 0.5 to 1.0 μm and are thus comparable to the mammalian lung. However, fish lungs and respiratory gas bladders uniformly lack the high degree of alveolar surface complexity found in the mammals. It is thus not surprising that both estimated surface areas and diffusing capacities of these ABOs fall short of those for mammal lungs.

What is surprising is that these values are also uniformly less than those found for the lungs of most amphibians and reptiles. With the exception of the doubtful value of 14 cm^2/g for *Protopterus*, most estimates place fish ABO surface areas at less than 1 cm^2/g (Tables 3.2 and 3.3), well below values common for amphibians (about 2.5 cm^2/g) and mammals (> 5 cm^2/g). It may be that the continued presence of auxiliary respiratory surfaces (i.e., gills and skin) has obviated ABO elaboration to the extent seen in amphibians and higher tetrapods.

Gills and Air Breathing

Surface Area

The generally reduced gill area of most air-breathing fishes has been well documented. An interesting paper by Dubale (1951) suggested that gill regression was an independent mutation that in turn drove the evolution of air breathing and ABO development in various groups. A more parsimonious interpretation is that gill regression has been an integral part of aerial and aquatic respiratory partitioning that, for optimal respiratory function, underwent natural selection concurrently with ABO development. Another view, that reduced gill surface area is a consequence of the incorporation of large quantities of rudimentary gill tissue into ABO structure (Singh, 1988) may be possible, but does not account for the observed reduction in gill areas among species possessing ABOs derived from other than embryonic gill tissue.

Studies Relating ABO and Gill Morphometrics

It is noteworthy that while *Anabas* has among the highest values for ABO surface area and diffusing capacity (Table 3.3), its gill values for these parameters are among the lowest of all air-breathing fishes (Table 3.5). Also, the lung surface area (0.85 cm^2/g) of *Lepidosiren*, a species generally regarded as having a relatively small gill area, exceeds all values for teleost ABOs (Table 3.3). The gill area of *Lepidosiren* (and all lungfishes) is reduced (Fullarton, 1931), however, this has not been quantified and needs to be in order to examine the relationship between functional gill and ABO surface areas. Further insight would also come from studies of *Phractolaemus*, *Pangasius*, *Gymnarchus* and other genera with developed ABOs but for which nothing is known about either ABO or gill dimensions.

Assuming that gill area reduction is a given for air-breathing fishes, where do we go from here? There are many more groups, for example, most of the lung and gas bladder air breathers, the Loricariidae, Trichomycteridae, and Callichthyidae, for which few gill data are available. However, little insight will come from merely compiling more area data. Future studies need to compare, more fully and critically, morphometric interrelationships between the ABO, the gills, and skin. Although early success came from morphometric analyses of individual respiratory surfaces, this approach is no longer sufficient; mass coefficient and diffusing capacity data are not ends in themselves, they are comparative entry points. New and more penetrating questions need to be asked about tissues that are prepared and analyzed in similar fashion.

The dilemma facing this field is illustrated by data for *Boleophthalmus* and *Periophthalmodon* (Figure 3.35) and the marked interspecific differences in both ABO and gill surface area scaling for three species of *Channa* and two species of *Clarias* (Tables 3.3 and 3.5). Data show that *Channa punctata* has both the largest gill and ABO areas as well as the highest and lowest (respectively) mass coefficients for these variables. Similarly, the ABO and gill areas of *Clarias batrachus* are much greater than for *C. mossambicus*. To sort out these confusing trends, workers acquiring such information need to critically concern themselves with their data, methods, and statistics. They also must integrate previous data for the group under study and thoughtfully consider the functional implications, for both gills and ABO, of generic and interspecific differences in area and mass coefficients. New questions need to be asked. What relationships should be expected for the scaling of gill and ABO, and what factors might cause this to vary among different groups? Can hypotheses be erected and tested about the scaling of surface areas and diffusing capacities of gills, the ABO, and skin? Do methods now in use provide sufficient precision to answer comparative inquiries about the functional and anatomic relationships between aerial and aquatic respiration?

My opinion is that a fundamental shortcoming of many of the scaling studies to date is the absence of a critical view of the statistical and biological errors associated with, for example, the additive coupling of separate regression functions to obtain area, A_{100}, and other morphometric estimates. There are statistical errors associated with extrapolation to values beyond the range of the actual data and biological errors caused by extensions beyond the maximum size of a species. Connected to this, too much faith is often placed in the accurate predictive capability of a regression slope value; that is, to the extent that the statistical variability of the original data (i.e., the highest and lowest values possible for an extrapolation, in view of the 95% confidence interval around the slope value) is often ignored.

Amphibious versus Aquatic Air Breathing

While illustrating the unevenness of our knowledge of air-breathing fish gill morphology, Table 3.6 draws attention to changes that have taken place in these structures and permits qualitative comparison of the different effects of amphibious (i.e., gill surfaces exposed to air) and aquatic air breathing.

At the level of the branchial arches, modifications are more common among aquatic air breathers. These extend from various classes of reductions (shortening of arches or hemibranchs to accommodate the ABO) to the near loss of gills on certain arches, thus rendering them branchial shunts. Reductions in primary and secondary lamellar dimensions are also more apparent among aquatic air breathers. Whereas amphibious air breathers do have diminutive lamellae, they also have a variety of gill stiffening mechanisms not found among aquatic air breathers. In both groups, reductions in secondary lamellar size and number and increases in diffusion distance appear to be convergent structural changes driven by different requirements (conservation of water versus O_2). Data in Table 3.5 show no differences in gill diffusion distance for the two groups, and ratios of gill area to body area (Table 3.4) imply that both groups decrease gill area to about the same relative amount.

An enigma among aquatic air breathers such as *Amia*, *Lepisosteus*, *Hypostomus*, *Heteropneustes*, the characins, and probably others, is their apparently lesser degree of secondary lamellar modifications compared to species such as *Anabas*, *Channa*, and *Clarias* (Table 3.6). *Amia* does lose aerial O_2 from its gills. In *Amia* and other species, this may be reduced through physiological mechanisms (i.e., neural and hormonal controllers) that regulate gill ventilation and perfusion in relation to air breathing (Chapter 6).

Conversely, one consequence of the strong selective pressure to reduce transbranchial O_2 loss is an extremely small gill diffusing capacity for some genera (*Anabas*, *Arapaima*, *Electrophorus*, *Monopterus*, *Protopterus*, and *Lepidosiren*), as well as some species of *Clarias* and *Channa* and many of the anabantoids, the majority of which are obligatory air breathers (Chapter 2). Nevertheless, in spite of such selective pressures and regardless of the environmental extremes, no air-breathing fish has evolved to the point of completely abandoning a dependence on gills. This is because the gills carry out indispensible functions in aquatic CO_2 release, ion and acid-base balance, and NH_3 excretion, and are the site of many metabolic and biosynthetic functions, even in the air-breathing fishes (Oduleye, 1977; Stiffler *et al.*, 1986; Stiffler, 1989; Olson *et al.*, 1994). In this regard it is noteworthy that *Protopterus*, *Channa*, *Anabas*, *Trichogaster*, and *Monopterus* have been reported to have few, or even no, pillar cells in their gills. In view of the role of these cells in deactivating catecholamines and bradykinin and in the activation of angiotensin II, it may have been necessary to recruit other tissues for this function (Olson *et al.*, 1987).

ABO Development and Differentiation

Comparative data for genera (*Clarias*, *Heteropneustes*, *Channa*, *Monopterus*, *Anabas*, and several gobiids) accounting for no less than six Indian air-breathing fish families has given Singh (1988, 1993) and his colleagues (Singh *et al.*, 1988, 1990) an exceptional evolutionary perspective to ABO ontogeny. All of these fishes begin air breathing before their ABO and its vascular tissues are fully developed and, in certain cases, the ABO continues its development, in terms of both size and respiratory surface, for a period extending well into adult life. Both the timing and the course of development undertaken by ABO epithelia vary among these fishes as well as regionally within the organ. Because the ABO respiratory surface of *Clarias* takes the form of abbreviated gill lamellae and has pillar cells, there is no doubt that this tissue originated from differentiated lamellae that fused and migrated into the SBC. While this is also the case for the SBC epithelium of *Heteropneustes*, the epithelium formed in its respiratory sacs is derived from the embryonic gill mass (mesenchyme) but still takes the form of abbreviated gills, and pillar cells are present, although the intralamellar circulation is absent (Olson *et al.*, 1990).

In *Channa*, *Monopterus*, *Anabas*, and *Trichogaster* the embryonic gill mass forms an ABO epithelium that does not closely resemble gill lamellae, and pillar cells are absent (Figure 3.31G). Rather, the capillaries are bounded by endothelial cells that are both supported and covered by pillar-shaped epithelial cells. According to Singh (1988) this difference can be traced to the later occurrence, during ABO ontogeny, of contact between the embryonic ectoderm and the endo- and mesodermal elements of the gill mass. In both *Clarias* and *Heteropneustes*, this occurs sufficiently early (6–8 mm length) to result in normal lamellar development. In other genera mixing occurs too late (post embryonic) for the ectoderm to dictate tissue morphology. It is uncertain what developmental forces shape the different papillar capillary patterns found in *Monopterus* and *Channa* (long and spiraling or wave-shaped, Figures 3.23 and 3.25) as opposed to the pavement-like parallel tubes (Figures 3.31E and F) of the anabantoids.

CHAPTER

4

Circulatory Adaptations

In order to avoid efficiency loss, the cardiovascular system should secure that the gas exchanger be perfused with the most deoxygenated blood while the systematic vascular beds should receive the most oxygen rich blood. Less than complete separation of the two qualities of blood not only results in less than optimal O_2 transport to the tissues, but the consequent recirculation in the systemic and/or gas exchange circuits is wasteful on the energy cost of blood propulsion.

K. Johansen, *Respiration Physiology XIV*, 1972

INTRODUCTION

Chapter three has detailed the microcirculatory modifications of different air breathing organs (ABO). The present chapter considers how air breathing has affected circulatory specializations of fishes. It catalogs the afferent and efferent blood conduits for different ABOs and examines the influences that ABO position in the circulatory system may have on blood pressure, gill function, respiratory gas transport efficiency, and other physiological functions. Evidence for functional anatomic shunts in different species will be discussed. As in previous chapters, the comparative phylogenetic approach is emphasized, and the primitive lunged air-breathers are considered first.

Whereas gas exchange within the ABO depends upon the matching of air ventilation and blood perfusion, the contribution of air breathing to total respiration depends upon integration of aerial gas exchange with other and usually concurrently operating respiratory exchanges at the gills, skin, or both. This integration is determined by the pattern of blood flow between the ABO and other respiratory surfaces and the systemic circulation. Also, many species have the capacity to temporally regulate the perfusion of different surfaces in conjunction with phasic changes in the predominating respiratory mode.

The importance of blood circulation pattern to ABO function has been recognized since the earliest investigations. Mention of ABO tributaries and drainages was provided in most organ descriptions and included in reviews by Rauther (1910) and Das (1927). Carter and Beadle (1931) tabulated the major afferent and efferent conduit vessels to air-breathing surfaces in 27 genera. This was later updated (Carter, 1957) and Satchell (1976) presented a substantive review of ABO circulation patterns and analyzed the problems of pressure regulation and phasic gill and ABO shunting. Randall (1994) has briefly reviewed the effects of the aquatic-terrestrial transition on circulation while Olson (1994) compared ABO circulation in a number of air-breathing fishes.

Early observations by Johannes Müller (1842, 1844) had an important effect on thinking about ABO circulation. Clearly influenced by the situation in higher vertebrates, Müller held that to function as an ABO, an organ had to receive "dark" blood and return "light" blood to the heart. He allowed that while the lung of *Lepidosiren* could function for aerial respiration, organs in *Polypterus*, *Erythrinus*, and other species could not, because, as indicated by their afferent and efferent circulations, "oxygenated" blood entered the ABO and, on the venous side, the oxygenated and deoxygenated blood streams would become mixed during their return to the heart. Müller influenced thinking about air breathing for several years. Even Hyrtl (1852a), while remarking on the lung-like qualities of the respiratory gas bladder of *Lepisosteus*, concluded that, in view of its arterial input and venous effluent, the organ could not function for aerial respiration. Absent from these early deliberations was knowledge of the environmental conditions under which fish ABOs would need to function. It is clear that gills ventilated in hypoxic water would not supply O_2-rich blood to an ABO. Similarly, venous outflow from an organ regularly ventilated with air would likely be more oxygenated than its arterial supply.

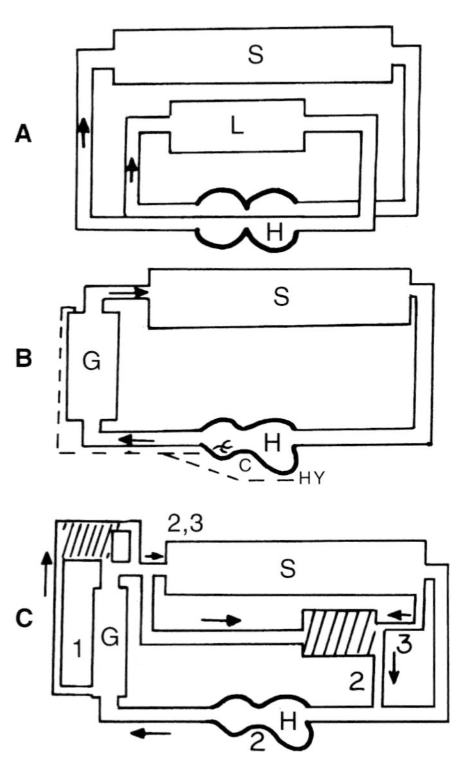

FIGURE 4.1 Circulatory flow diagrams. **A,** Double loop circulation of mammals and birds. **B,** In-series flow of fishes. Dashed line shows the efferent branchial origin of the hypobranchial (HY) and coronary (C) arteries (see text). **C,** Illustration of three potential consequences of various in-series placements of the ABO (diagonal lines): **1,** gill O_2 loss, **2,** forced unsaturation of ABO blood, **3,** pressure loss. Arrows indicate flow direction. S, systemic circulation; G, gills; L, lung; H, heart.

DEFINING THE PROBLEMS

Figure 4.1 contrasts the circulation patterns of a mammal and fish in order to illustrate problems imposed on fish circulation by the presence of an ABO. Mammal (and bird) circulation consists of two separate, parallel loops (Figure 4.1A); the pulmonary loop serves for respiratory gas exchange between blood and the atmosphere, while the systemic loop provides for blood–tissue exchange. Although separate, the pulmonary and systemic loops are serially linked at the heart, which is axially subdivided into parallel left and right pumps. Thus, a double passage of blood through the heart is needed for a complete circuit, and the full separation of the O_2-rich and poor blood channels maximizes the gas diffusion gradients on both surfaces. In contrast, fish circulation is in a single loop and is termed "in-series flow" because blood leaving the heart goes to the gills and systemic tissue beds before returning (Figure 4.1B; Johansen, 1972).

As seen in Figure 4.1C, insertion of an ABO into this loop creates at least three basic problems not encountered with a separate pulmonary loop. These are now treated.

Enforced Partial Unsaturation

Placement of an ABO at any point along the systemic circulation would lead to the partial deoxygenation of efferent blood as it joins less saturated tributaries. Satchell (1976) termed this "enforced partial unsaturation of ABO blood." Venous admixture of O_2 saturated and unsaturated blood is common and may be especially critical in the case of stomach and intestinal ABOs (Figure 4.1C), which normally drain via the hepatic portal system and thus infiltrate the metabolically active liver before reaching the heart.

The circulation pattern of most air-breathing fishes subjects them to some level of enforced unsaturation, and adaptations lessening this effect include blood streaming (i.e., the unmixed flow of two streams within the same lumen), more direct venous return from ABO to the heart, and the enhancement of blood O_2 transport capacity (Satchell, 1976; Munshi *et al.*, 1990).

Transbranchial Oxygen Loss

Another problem posed by in-series ABO placement is the potential for the transbranchial loss of aerially obtained O_2, as described in Chapter 3. The only fishes not encountering this problem are the amphibious air breathers and those that hold air over their gills (and thus not water-ventilating) while air breathing.

Possession of—and dependence upon—functional aquatic gill surfaces during all life stages distinguishes air-breathing fishes from all tetrapods including most amphibians (viz., only a few aquatic amphibians have concurrently functional gills and lungs, and in most adult amphibians using both lungs and skin, the latter is more often exposed to air than water). The need, moreover, to regulate circulatory flow between the ABO and gills, and very likely the skin, further complicates the circulation of air-breathing fishes relative to amphibians. While elimination of gills solves this problem, this organ's vital importance in metabolic, respiratory, and osmotic functions assures its retention (although usually with a lower diffusing capacity, Chapter 3).

Pressure Regulation

Allowing that gills are essential, another option would be placement of the ABO immediately posterior to the gills and in series with the systemic circulation (Figure 4.1C). This would compensate for transbranchial losses, but this circulatory pattern is rare probably because the in-series flow resistance imposed by three successive capillary networks (gills, ABO, systemic capillaries) affects perfusion pressure (Satchell, 1976; Olson *et al.*, 1990; Olson, 1994). Raising pressure sufficiently to counter the losses has the potential of affecting fluid balance and damaging intravascular tissues. An intermediate option, establishing a parallel gill and ABO circuit (e.g., as in the clariids, *Heteropneustes*, *Channa*, and *Anabas*, Figure 4.2), lessens requisite pressure but results in enforced blood unsaturation.

COMPARATIVE CIRCULATORY SPECIALIZATION

Figure 4.2 shows the general circulation plans for fish ABOs and Table 4.1 lists the major afferent and efferent conduit vessels for the ABOs and respiratory surfaces in 37 genera. Different systems have been used in naming and numbering the branchial arches (Goodrich, 1930; Foxon, 1950, 1955) and the protocol adopted here is now clarified. Gill arches are derived from six pairs of embryonic pharyngeal arches or pouches. Each pouch contains an aortic arch (AA) linking the ventral and dorsal aortae as well as connective tissue elements that differentiate into various components of the visceral skeleton. AA I, the mandibular arch, forms the jaw and related structures including the circulatory elements. AA II, the hyoid, forms the jaw support. In most sharks and the primitive actinopterygians, functional gills develop on the branchial arches derived from AA II–VI while in teleosts, gills form on arches stemming from AA III–VI. By convention, the aortic arches are designated by Roman numerals and the branchial arches by Arabic numerals. For purposes of comparing different groups, the gill-bearing arches are numbered, beginning after the hyoid, as 1–4. Thus, lungfish and polypterids have five branchial arches (hyoid and 1–4) and teleosts four (1–4). For comparative purposes, branchial arch 4 is homologous (derived from AA VI) in both groups.

PULMONARY CIRCULATION

Among lungfishes (*Protopterus*, *Lepidosiren*, and *Neoceratodus*) and three primitive actinopterygians (*Polypterus*, *Erpetoichthys*, and *Amia*), a pulmonary circulation permits close approach to the double loop (Figures 4.2A–C). In these fishes, paired pulmonary arteries originate from the efferent arteries of branchial arches 3 and 4 while pulmonary veins conduct blood either directly into or to the general vicinity of the heart, thereby lessening contact with the systemic venous drainage. The lungfishes are the most advanced, in the latter respect, because their pulmonary venous return is into the heart which is almost completely divided into right and left halves (Johansen *et al.*, 1968a).

Lungfish

The circulatory anatomy, pattern of blood flow, and mechanisms of cardiorespiratory regulation in lungfishes are treated in several reviews (Johansen *et al.*, 1968a; Szidon *et al.*, 1969; Burggren and Johansen, 1987; Fishman *et al.*, 1985, 1989) and need not be fully detailed here. More is known about these phenomena in lungfishes than any other air-breathing fishes, and it has become apparent that, due to the need to sustain the simultaneous function of both gill- and lung-exchange surfaces, *Protopterus* and *Lepidosiren* achieve levels of pulmonary systemic separation surpassing that in most amphibians (Goodrich, 1930; Foxon, 1955; Burggren and Johansen, 1987).

The Double Circulatory Loop

Figure 4.3 diagrams the circulation of *Lepidosiren* and *Protopterus*. Beginning with different ports of entry into the heart, the pulmonary and systemic streams are kept largely separated. In both genera, the common opening of the left and right pulmonary veins flows into the "left side" of the atrium. Blood ejected from the "left side" of the ventricle flows through the bulbus cordis mainly to the anterior three

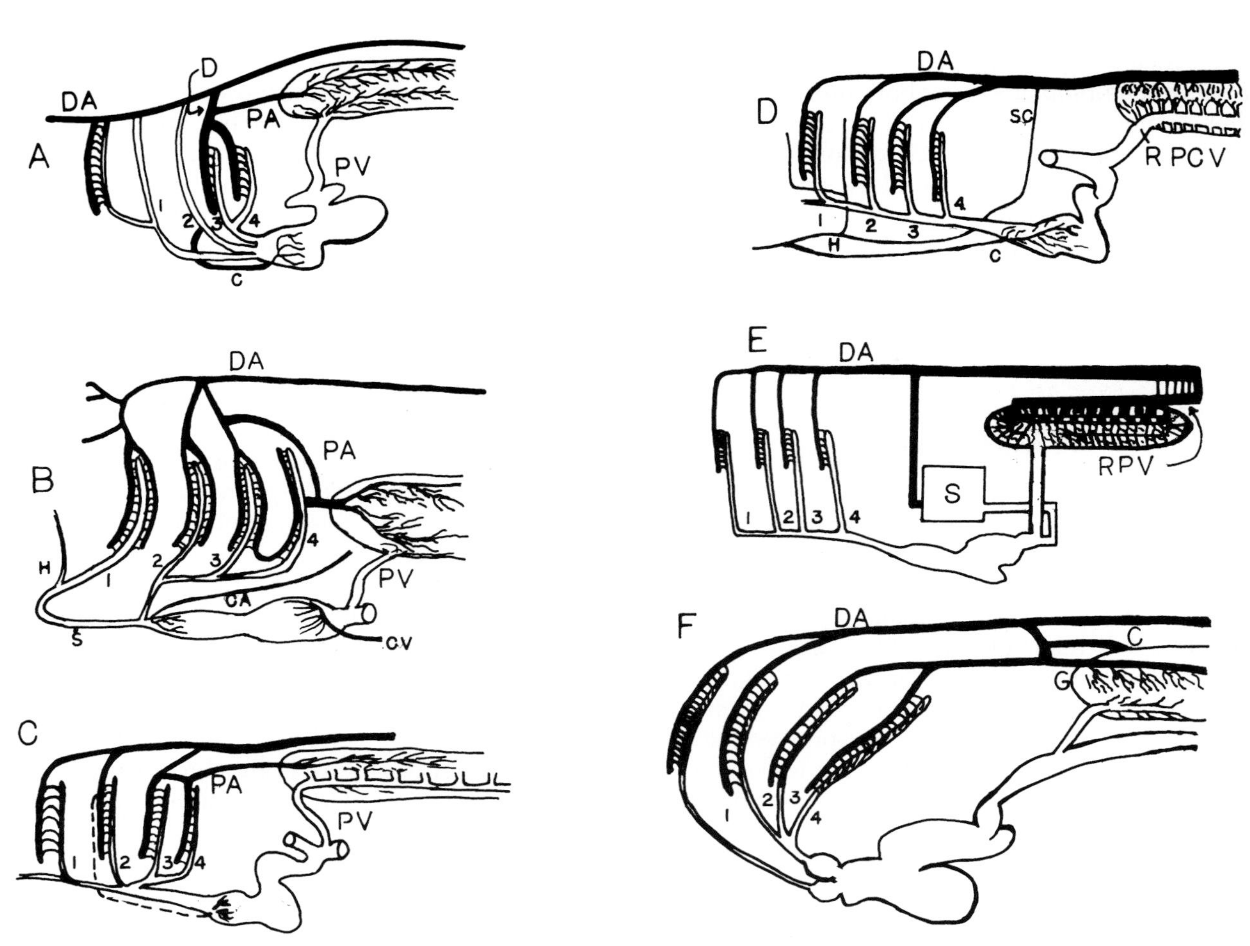

FIGURE 4.2 Representative blood circulation patterns of the air-breathing fishes. Each diagram views fish from the left side and shows tracks of the major vessels leading to and from the ABO, and their relationships to the branchial arches (1–4) and the heart (H). **A,** *Protopterus.* Note presence of pulmonary circulation and absence of gill filaments on arches 1 and 2. Coronary artery (C) is a branch of the hypobranchial which originates on arch 2. DA, dorsal aorta; D, ductus arteriosus; PA, pulmonary artery and PV, pulmonary vein. **B,** *Polypterus.* Arches 3 and 4 contribute to PA formation. S indicates the approximate position of the ventral aortic strap (see text and Figure 4.6). Tracks of the paired coronary arteries (from subclavian artery) and veins (CA, CV) are also shown. **C,** *Amia.* PA originates from arches 3 and 4. The PV enters the left side of duct of Cuvier. Dashed line shows coronary artery supply via the hypobranchial artery. **D,** *Lepisosteus.* ABO drainage is to the right post cardinal vein (RPCV). Note that the coronary artery is formed with combined branches of the subclavian (SC) and hypobranchial (H) arteries. **E,** *Arapaima.* Afferent supply via the DA to the renal portal vein (RPV). S, systemic circulation. **F,** *Gymnarchus.* Note anastomoses between DA and the coeliac (C), and gas bladder (G) arteries (also see Figure 4.7). **G,** Generalized ostariophysan interrenal vein tracks (dashed lines) showing connections from the gas bladder, stomach, and intestinal (G,S,I) ABOs to the PCV. The shunt between branchial efferents 3 and 4 and the DA described for *Hoplerythrinus* is shown (see text). C, coeliac artery. **H,** *Clarias.* Dashed line marks the periphery of the ABO. Note presence of an ABO element on each branchial arch and that ABO efferent flow is into the DA. **I,** *Heteropneustes.* Dashed line marks the periphery of the ABO. Gill fans occur on each afferent branchial. Afferent 4 goes into the air sac (AS) and efferent drainage (arrow) is into the DA. **J,** *Electrophorus.* A. Branchial arch arteries give rise to small superior and inferior vascular epithelial branches but are essentially ventral–dorsal conduits. V. Venous drainage is into the systemic circulation via the jugular (J) and anterior cardinal (ACV). **K,** *Boleophthalmus.* Branchial circulation similar to most non-air-breathing fishes, with systemic capillaries being the site of oxygen uptake and venous return (dashed lines) occurring via the ACV and J. **L,** *Anabas* and *Channa.* Efferent branchials 1 and 2 go into the ABO and ABO drainage is systemic (ACV and J, dashed line). **M,** *Monopterus.* Arrows show flow direction. *Top part of diagram.* Arterial outflow from heart through elongated ventral aorta (VA) to branchial arches. Arch 1 serves air sac (AS), arch 2 traverses a small gill and forms the carotid artery (CA). Arches 3 and 4 branch into the respiratory epithelium (note that arch 4 forms as a side branch of the ventral–dorsal aortic conduit). Blood supply to the head is also via paired lateral dorsal aortae (LDA). The coronary artery (CO, curved arrow) emerges from the ventral aorta just beyond the bulbus. C, coeliac. *Lower diagram.* Venous return to the heart is via the superior jugular (SJ), and median (MV) veins into the ACV. The above drawings have been compiled based on data from various sources including diagrams in the following references: Assheton (1907) (**F**); Johansen (1970) (**A–D, G–M**); Satchell (1976) (**S,B,H,I,J,L**); Burggren and Johansen (1987) (**A**); Munshi *et al.* (1990) (**M**); Olson (1994) (**A–M**).

(continues)

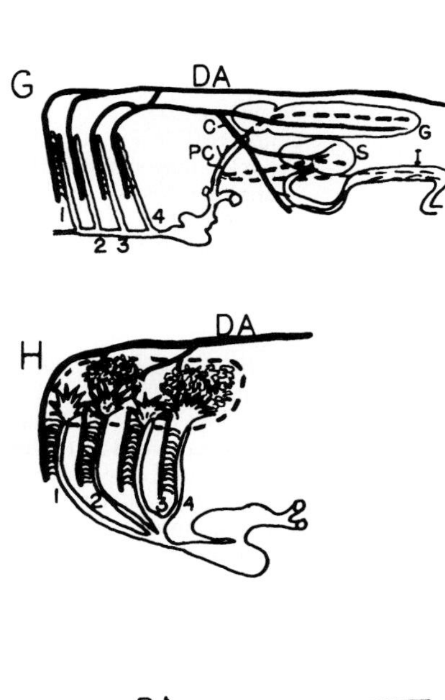

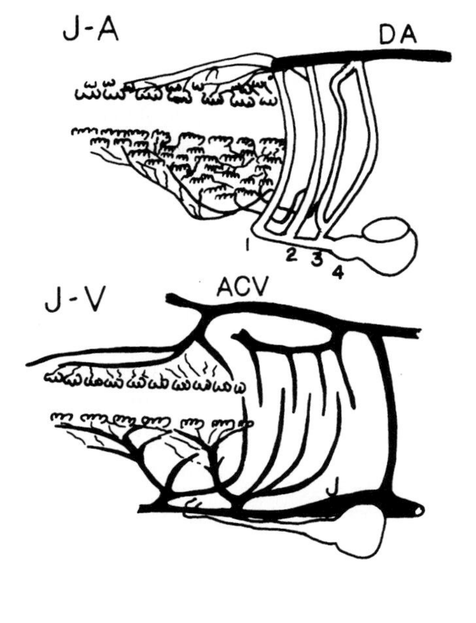

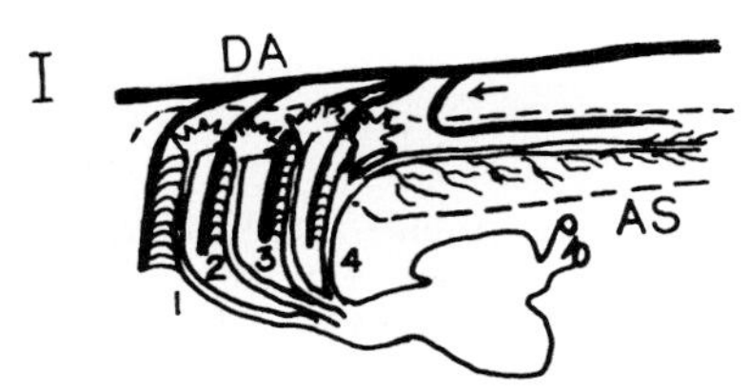

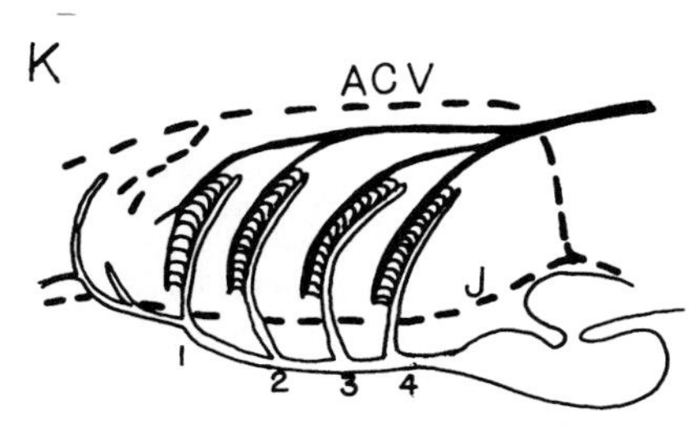

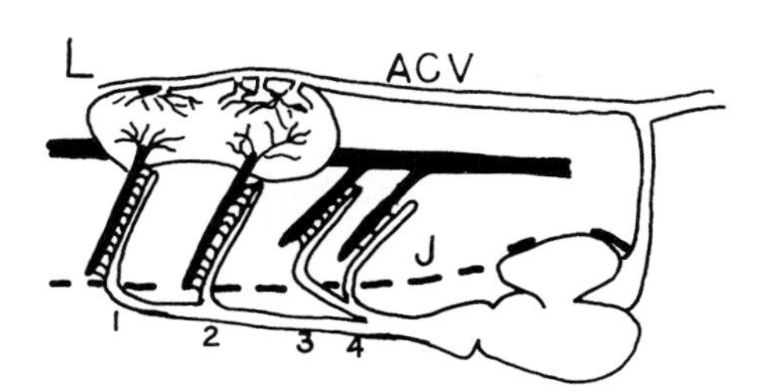

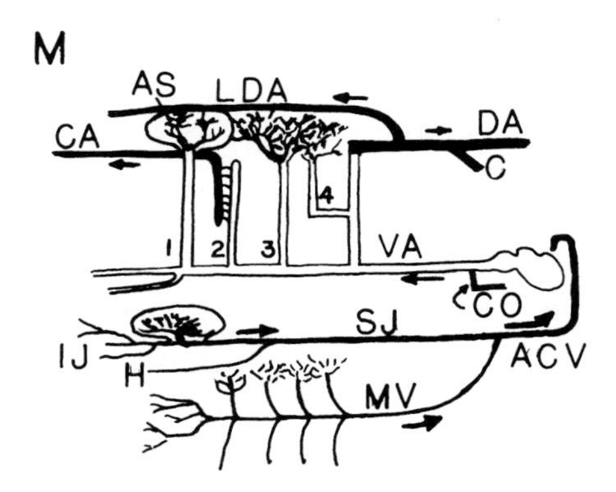

FIGURE 4.2 *(continued)*

TABLE 4.1 Afferent and Efferent Blood Supplies to Fish ABOs and Aerial-Respiratory Surfaces

Genus	Afferent	Efferent	Notes and references
Lung			
Lepidosiren	Eff BA 3,4 to PA	PV to left atrium	Robertson, 1913
Protopterus	Eff BA 3,4 to PA	PV to left atrium	Bugge, 1961. In the four species, LPA supplies lung ventral surfaces, RPA supplies the dorsal surfaces.
Neoceratodus	Eff BA 3,4 to PA	PV to sinus venosus	Spencer, 1892
Polypterus	Eff BA 3,4 to PA	PV to hepatopulmonary	Magid, 1967; Poll and Deswattines, 1967
Gas Bladder			
Lepisosteus	DA branches	RPCV to DC	Potter, 1927
Amia	Eff BA 3,4 to PA	PV to DC	Goodrich, 1930; Olson, 1981
Gymnarchus	Eff BA 3,4 to "PA"	PV to LPCV to L atrium	Assheton, 1907
Arapaima, Heterotis, Pantodon	Caudal art. to RPV	R and LPCV	Poll and Nysten, 1962; Schaller and Dorn, 1973; Greenwood and Liem, 1984
Notopterus	DA to intercostal arts.	ABO V to PCV	Dehadrai, 1962a
Hoplerythrinus	Eff BA 3,4 to COA	IRV to RPCV	Carter and Beadle, 1931; Smith and Gannon, 1978
Umbra krameri	Vesicular branches from DA	RPCV	Jasinski, 1965
Gymnotus	COA	HPV	Liem *et al.*, 1984
Phractolaemus	"PA"	"PV" to RPCV	Thys Van Den Audenaerde, 1961
Buccopharyngeal Epithelium			
Gillichthys	Branches of four arts. (carotid, hyomandibular, hypobranchial, and segmental)	Opercular and orbital Vs to the ACV	Todd and Ebeling, 1966
Pseudapocryptes	Eff BA 1,2 to buccal and opercular arts.	Opercular V to ACV	Das, 1934
Periophthalmus koelreuteri[a]	Orbitonasal branch of Int. CA	JV	Rauther, 1910; Schöttle, 1931
Boleophthalmus boddarti[a]	Aff BA 1-4	ACV, JV	Biswas *et al.*, 1981
Monopterus cuchia	Aff BA 1-4 and the hyomandibular branch of VA	JV	Munshi *et al.*, 1990
Buccopharyngeal and Branchial Air Chambers and Associated Structures			
Monopterus cuchia (air sac)	Aff BA 1	JV	Singh *et al.*, 1984; Munshi *et al.*, 1990
Electrophorus	Aff BA 1-4	JV, ACV	Carter, 1935
Clarias	Aff BA 1-4	Eff BA to DA	Munshi, 1961
Heteropneustes	Aff BA 4	Eff ABO art. to DA	Olson *et al.*, 1990
Channa	Eff BA 1,2 to suprabranchial arteries	JV	Ishimatsu *et al.*, 1979; Ishimatsu and Itazawa, 1993; Munshi *et al.*, 1994; Olson *et al.*, 1994
Anabantoids (*Anabas, Betta, Macropodus* and *Osphronemus*)	Eff BA 1,2	JV	Olson *et al.*, 1986
Gills			
Synbranchus	Aff BA and head systemic vessels	Eff BA and systemic venous return	Carter and Beadle, 1931; Spiropulos Piccolo and Sawaya, 1981
Hypopomus	Aff BA	Eff BA	Carter and Beadle, 1931
Esophagus			
Dallia	COA to gastrointestinal art. —gastrosplenic and gastric branches	PCV	Crawford, 1974
Stomach			
Ancistrus	COA	IRV to PCV	Carter and Beadle, 1931
Intestine			
Misgurnus	DA to anterior mesenteric art.	HPV	McMahon and Burggren, 1987
Cobitis	COA	HPV	Lupu, 1914
Lepidocephalus	COA to mesenteric to dorsal intestinal art.	Ventral intestinal vein into HPV	Yadav and Singh, 1980
Callichthys	DA	IRV to PCV	Carter and Beadle, 1931
Hoplosternum	DA	IRV to PCV	Carter and Beadle, 1931

[a]See Table 2.4 for updated species names.

Abbreviations: ABO, air-breathing organ; Aff, afferent; ACV, anterior cardinal vein; Art, artery; BA, branchial arch; CA, carotid artery; COA, coeliac artery; DA, dorsal aorta; DC, ductus Cuvier; Eff, efferent; HPV, hepatic portal vein; Int, internal; IRV, interrenal vein; JV, jugular vein; L, left; PA, pulmonary artery; PV, pulmonary vein; PCV, post-cardinal vein; RPV, renal portal vein; R, right; V, vein.

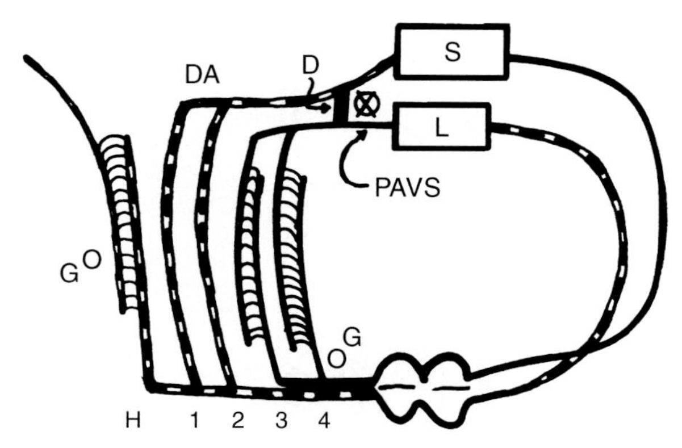

FIGURE 4.3 Lung (L) and systemic (S) vascular loops of *Lepidosiren* and *Protopterus*. Oxygen-rich blood (alternating black and white lines) leaves L, enters the left side of the heart, and goes into the branchial arches (hyoid, 1, and 2) and then to either the head or the DA. Deoxygenated blood from the systemic (S) loop (solid black line) goes through the right side of the heart to arches 3 and 4 and to the lung. Diagram shows location of the shunts (circles) and their configuration during air breathing: Gill and PAVS shunts are open and the ductus arteriosus (D) is closed. Based on information and diagrams in Szidon *et al.* (1969), Johansen (1970), Fishman *et al.* (1985), and Burggren and Johansen (1987).

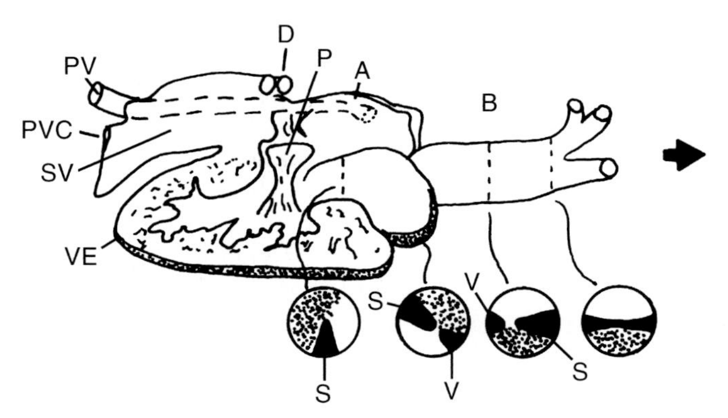

FIGURE 4.4 View of the heart of *Lepidosiren* from which the right side has been removed. PV, pulmonary vein: Dashed lines track PV advance to the anterior left atrium (A) which is obscured by the septum dividing this chamber. PVC, posterior vena cava: Heart entry is via the sinus venosus (SV). Other abbreviations: D, ductus Cuvier; P, atrioventricular plug; VE, ventricle; B, bulbus cordis (extending from VE to the arterial branches). Circles are cross sections through the bulbus at different positions showing how the rotation of the spiral (S) and ventrolateral (V) folds affects the distribution of oxygenated blood (stippled area). Note that the two folds become fused at the end of the bulbus. (Arrow = rostral direction.) Based on data and illustrations in Bugge (1961), Fishman *et al.* (1985), and Burggren and Johansen (1987).

gill arches (hyoid, 1, 2) and then on to the systemic circulation. Systemic venous return enters the right atrium, passes through the "right side" of the ventricle, and then into branchial arches 3 and 4 and on to the pulmonary arteries.

Heart Structure

The atrium, ventricle, and bulbus cordis are all important in maintaining flow separation in *Lepidosiren* and *Protopterus* (Figure 4.4). The atrium has a partial septum (better developed in *Lepidosiren*). Rather than an A–V valve, an atrioventricular plug sits between the atrium and ventricle and likely functions to reduce turbulence. The ventricle is almost fully divided along its length by a ventral ridge. Spiral and ventrolateral folds within the bulbus direct flow into the arches leading to either the pulmonary or systemic loops. There is no ventral aorta; rather, three branchial arch pairs emerge from the bulbus, with the first and third of these dividing to form the five arches (Figure 4.2A).

Although the pulmonary and systemic circulations are not completely separate, measurements of blood O_2 levels in various locales, as well as the flow paths taken by injections of India ink and contrast agent, demonstrate that there can be little mixing between the right and left streams (Johansen *et al.*, 1968a; Szidon *et al.*, 1969). Contrast agent injected into the pulmonary vein was seen to mainly opacify the ventral channel of the bulbus and the first three arches but was only faintly apparent in the pulmonary artery. Johansen *et al.* (1968a) estimated that, immediately after an air breath is taken, 90% of the blood leaving in the pulmonary vein of *Protopterus* remains cohesive during its journey through the heart and anterior arches to the dorsal aorta. After air has been held several minutes, this fraction reduces to 50–60%.

Temporal Shifts in Loop Flow: Shunts

As shown in Figures 4.2A and 4.3, anatomical junctions lie between the systemic and pulmonary loops. These are facultatively controlled to achieve optimal flow separation in phase with the air breath and inter-air breath cycles (Fishman *et al.*, 1987, 1989, and Chapter 6). When air is inspired, the pulmonary artery (specifically the pulmonary artery vasomotor segment, PAVS) is fully opened as are the gill shunts on arches 3 and 4 (i.e., these shunts permit the bypass of gill lamellae), while the ductus arteriosus, the main pulmonary-systemic conduit, is closed. With reduction in lung O_2 or exhalation, the PAVS expands to restrict pulmonary flow, and both the ductus arteriosus and the branchial shunts open. This directs blood into the systemic circulation, and the incomplete ventricular septum facilitates a temporary pulmonary shunting by relieving any right and left ventricle systolic pressure differentials resulting from changes in pulmonary flow resistance (Fishman *et al.*, 1989).

Comparative Aspects

The degree of separation of pulmonary and systemic flow in *Lepidosiren* and *Protopterus* reflects their obligatory dependence upon aquatic breathing and on the branchial exchange of CO_2, NH_3, and ions. In contrast to *Lepidosiren* and *Protopterus*, *Neoceratodus* has gills on all arches, a pulmonary venous return only into the sinus venosus, and less developed atrial- and ventricular septa and a smaller spiral fold. The degree of pulmonary and systemic separation achieved in *Protopterus* and *Lepidosiren* ensures pulmonary O_2 conservation because most of the O_2-laden blood destined for the systemic circulation transits arches lacking lamellae. Lastly, in *Neoceratodus* the coronary artery branches off the right hypobranchial artery which originates from the efferent artery of branchial arch 2. This contrasts with *Lepidosiren* in which coronary origin is not post-branchial; the hypobranchial artery branches from the lower part of arch 2 (which lacks gills, Figure 4.2A) and gives rise to a coronary branch (Robertson, 1913; Foxon, 1950, 1955).

The functional importance of systemic and pulmonary separation to *Lepidosiren* and *Protopterus* can be illustrated by comparison with the amphibian heart. In contrast to the nearly complete left–right separation seen in these two lungfish genera, most lunged amphibians have a single ventricle where pulmonary and systemic bloods mix freely (Foxon, 1955). This may not pose a limitation for most amphibians because the systemic loop of their circulation also perfuses an effective gas exchange surface, the skin, that more often than not contacts air. Amphibians regulate pulmonary (West and Burggren, 1984) and cutaneous (Malvin, 1988) perfusion in order to alter the percentage of gas exchange taking place through these two respiratory systems, and most amphibians, including all frogs, lack gills as adults, and thus do not have to contend with the same potential for transbranchial O_2 loss that plagues most aquatic air-breathing fishes. This suggests that the increment of cutaneously derived O_2 present in the systemic venous blood of the amphibian may offset the chronic undersaturation of pulmonary blood brought about by intraventricular mixing.

The Pulmonary Loop of Primitive Actinopterygians

In *Polypterus*, *Erpetoichthys*, and *Amia*, the pulmonary artery comes mainly from the efferent of branchial arch 4 (Figure 4.2B and C). Branchial arch 3 also connects the pulmonary artery in most specimens; however, notable variation was seen for this in *Amia* (Olson, 1981) and both among and within species of *Polypterus* (Poll and Deswattines, 1967).

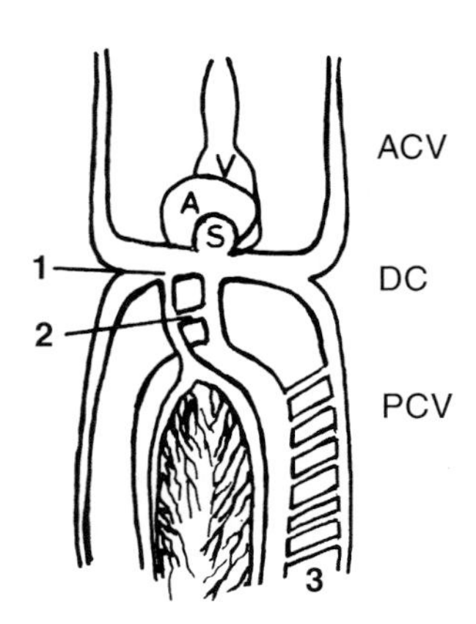

FIGURE 4.5 Generalized diagram of central venous return to the heart showing the patterns of pulmonary vein flow for three primitive actinopterygians. **1,** *Amia*; PV empties into left duct of Cuvier (DC). **2,** Polypterids, into hepatic vein. **3,** Gar; into right post cardinal vein (PCV). ACV, anterior cardinal vein; S, sinus venosus; A, atrium; V, ventricle.

The pulmonary veins of these fishes empty at sites upstream from the heart (Figure 4.5); in *Polypterus* and *Erpetoichthys*, the pulmonary and hepatic veins combine to form the hepatopulmonary which goes into the sinus venosus (Kerr, 1910; Magid, 1967; Poll and Deswattines, 1967). In *Amia*, the pulmonary vein flows into the left ductus Cuvier (Olson, 1981). Another difference is seen in the coronary circulation. In *Polypterus* (Figure 4.2B), the coronary artery branches from the subclavian artery, passes forward and through the pericardium to the anterior conus, and then back to the ventricle. This condition is similar to, but not exactly the same, as that in *Polyodon*, a non-air breathing polypteriform (Danforth, 1916). The coronary vein of *Polypterus* goes from the ventricle to the right side of the duct of Cuvier (Poll and Deswattines, 1967). In *Amia* (Figure 4.2C), the coronary artery branches from the hypobranchial which usually originates from efferent branchial arch 2, but may be connected to more than one efferent by a commissural artery (Danforth, 1916; Olson, 1981).

Hearts of these primitive actinopterygian air breathers lack medial septation and thus depend upon streamlined blood flow and vascular shunting to augment aerial respiratory efficiency. Even so, it can be expected that these fishes have some capacity to regulate gill and ABO perfusion in conjunction with air breathing. Figures 4.2B and C show that the major ABO conduits in polypterids and *Amia* are similar to lungfish. These likely work to phasically shunt the ABO. While this has been investigated to some extent for *Amia* (Olson, 1981; Randall *et al.*, 1981b), little is known about the actual mechanisms for either the polypterids or *Amia*, although the latter is subject to significant transbranchial O_2 loss (Randall *et al.*, 1981b).

My study of *Polypterus* in the British Museum did reveal one possible gill shunt mechanism (Figure 4.6), a muscle actuated tendonous strap that underlies the ventral aorta proximal to the hyoid and first branchial arches. My observations suggest tension on this strap could constrict the ventral aorta, thereby increasing flow through its posterior branches (Figure 4.2B). Depending upon other anatomical and physiological mechanisms in place, this could favor lung perfusion and even permit retrograde flow via the dorsal aorta to compensate flow reductions caused by clamping the anterior ventral aorta.

Gars and Primitive Teleosts

With the exception of *Gymnarchus* (see p. 144), in all remaining species using a gas bladder as an ABO, the afferent ABO blood comes via the dorsal aorta or one of its first visceral branches, while the ABO-efferent enters the systemic venous return at a site somewhat distant from the heart (Table 4.1).

Gars

Illustrations showing gars (*Lepisosteus* and *Atractosteus*) with a pulmonary circulation (Johansen, 1970; Wood and Lenfant, 1976; Burggren and Johansen, 1987) are not correct. As shown by Potter (1927), the gar's respiratory gas bladder receives blood via multiple branches from the dorsal aorta (Figures 4.2D and 3.8). Venous return to the heart is via the right posterior cardinal vein (Figure 4.5) which, though lacking the directness of a pulmonary vein, does bypass the hepatic portal system. In *Lepisosteus*, the coronary artery arises from the hypobranchial artery which is derived from more than one branchial arch and in turn connects to the subclavian artery (Figure 4.2D; see Danforth [1916]).

Osteoglossomorpha

Osteoglossomorph fishes have diverse afferent and efferent ABO vessels, in some cases featuring circulation patterns not found in any other species. Continuing the anatomical trend first seen in the gars, afferent flow to the gas-bladder ABO in these fishes is, in almost all cases, derived from the dorsal aorta or more distally. Also, a single artery and vein rather than paired pulmonary vessels connect the ABO.

The ABO completely encloses the kidneys of *Arapaima*, *Heterotis*, and *Pantodon* (Chapter 3), and the renal portal system is the major afferent ABO supply (Figure 4.2E). There has been some controversy over this anatomy (Greenwood and Liem, 1984). Poll and Nysten (1962) found close communication between the renal portal and ABO circulations of *Arapaima*, *Heterotis*, and *Pantodon*, as did Schaller and Dorn (1973) for *Arapaima*. Nevertheless, Randall *et al.* (1978a, figure 4) diagrammed a dorsal–aortic supply for the ABO of *Arapaima*. A reexamination of the question by Greenwood and Liem (1984) again confirmed that

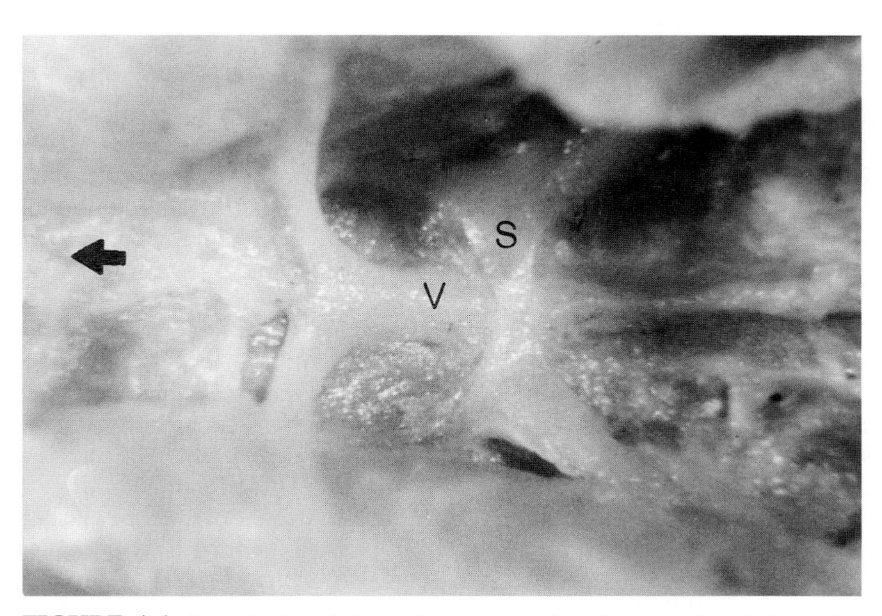

FIGURE 4.6 Ventral view of the tendonous ventral–aortic strap (S) underlying the ventral aorta (VA) of *Polypterus bicher*. The anterior-most branches of the ventral aorta (see Fig. 4.2B) can be seen to left of the strap. (Arrow = anterior.)

afferent ABO supply in *Arapaima* is mainly via the renal portal system.

Efferent ABO flow in *Arapaima*, *Heterotis*, and *Pantodon* is partly into the right but mainly via the left posterior cardinal vein to the ductus Cuvier. Venous mixing does occur because the hepatic vein and ductus enter the sinus venosus through the same opening (Greenwood and Liem, 1984). Schaller and Dorn (1973) observed that a gigantic balloon valve guarded the entry point of the hepatic vein into the sinus venosus and speculated that selective control of this valve could minimize unsaturation. In any case, Greenwood and Liem (1984) point out that, even in the absence of this mechanism, the afferent ABO circulation of *Arapaima* offers a significant functional advantage in that virtually all blood returning from the posterior systemic circulation does so via the renal portal system. Venous blood would therefore pick up O_2 in the ABO and thereby reduce the effects of venous mixing on blood unsaturation. Support for this idea is contained in observations that the ventral–aortic PO_2 of *Arapaima* is relatively high (45–50 torr, about 80–90% saturation, Randall *et al.*, 1978a; Johansen *et al.*, 1978a). Lastly, it is likely that either the volume of blood flowing through this ABO, or the remote positioning of this organ in the renal portal circulation (where pressure would be low), may partially account for the apparent absence of air-breath-induced changes in dorsal aortic pressures, pulse pressure, or heart rate noted by Farrell (1978). In all other fishes that have been tested (where in all the ABO conduit is either separate or derived more proximally from the dorsal aorta), taking an air breath usually causes increased ABO perfusion and elevated heart rate (Chapter 6).

Hulbert *et al.* (1978a) reported several circulatory modifications thought to affect flow and diffusion in the gills of *Arapaima*. These include the presence of thick smooth muscle layers in both the afferent and efferent branchial arteries, the presence of discrete secondary lamellar blood channels (i.e., in contrast to the more lacunar secondary lamellar flow in most gills), complete with smooth-muscle valves at their origins, and as in *Amia*, the recession of blood channels to deep within a secondary lamella.

Blood flow to the respiratory gas bladder of *Notopterus* is through several intercostal arteries branching from the dorsal aorta. Drainage is via the left post cardinal vein (Dehadrai, 1962a). The blood flow patterns of the other air-breathing notopterid genera *Xenomystus* and *Papyrocranus* are not described.

"Pulmonary" Circulation in *Gymnarchus niloticus*

Early works by Hyrtl (1856) and Assheton (1907) showed that the circulatory system of *Gymnarchus* (Figures 4.2F and 4.7) has undergone several modifications. Surprisingly, there are no studies of the cardiorespiratory physiology of this fish. In *Gymnarchus* the ductus Cuvier is divided into left and right parts. The atrium (the sinus venous is highly reduced) is also partially divided by a small protrusion of tissue from its inner wall (Figure 4.7). Efferent flow from the respiratory gas bladder is via the left post cardinal vein to the left ductus Cuvier and directly into the left atrium. The right ductus, carrying blood from the posterior body, hepatic veins, and kidney, enters the right part of the atrium. It is presumed that the two flow streams are kept somewhat separated by the atrial septum.

As seen in Assheton's (1907) illustrations (Figure 4.7), afferent ABO blood comes from efferent branchials 3 and 4. The efferent branches of arches 1 and 2 distribute blood to the systemic circulation via the dorsal aorta which has small anastomoses with both the gas bladder and coeliac arteries. *Gymnarchus* lacks a ventral aorta. Rather, and similar to lungfish (Figure 4.4), ventricular ejection is into four separate arteries, each of which appears to have its own bulbus-like swelling or capacitance chamber (Figure 4.7). A short distance from the heart the dorsal pair of these vessels branch to form branchial arch 4; then, a little further along this divides into arches 2 and 3. Branchial arch 1 is formed by the ventral vessel pair (Figure 4.7).

This anatomy and the well-developed ABO of *Gymnarchus* suggest that it has the capacity to partially separate its systemic and ABO circulations. This needs testing. My study of *Gymnarchus* in the British Museum suggests that its unique aortic diverticulae might function in the selective distribution of blood flow to different aortic tracks. There are several muscular attachments on the anterior pericardium, and these might, by applying tension in different directions, enable selective occlusion of one or more of the root vessels and thus regulate blood flow to different sets of the branchial arches. These muscle attachments are found only on the anterior pericardium (Assheton, 1907), and the need to differentially occlude the four root arteries may explain the presence of separate diverticula.

There are, nevertheless, several anatomical impediments to the efficacy of flow separation in *Gymnarchus*. Mixture of efferent ABO and systemic (anterior left quadrant) blood in the left ductus Cuvier should cause unsaturation. Also, from Assheton's drawings, there appears to be no gill reduction on any of the arches. It is also unclear how flow separation to different afferent arches would be controlled because, as seen in Figure 4.7, flow to arches 2, 3, and 4 originates from the same aortic root.

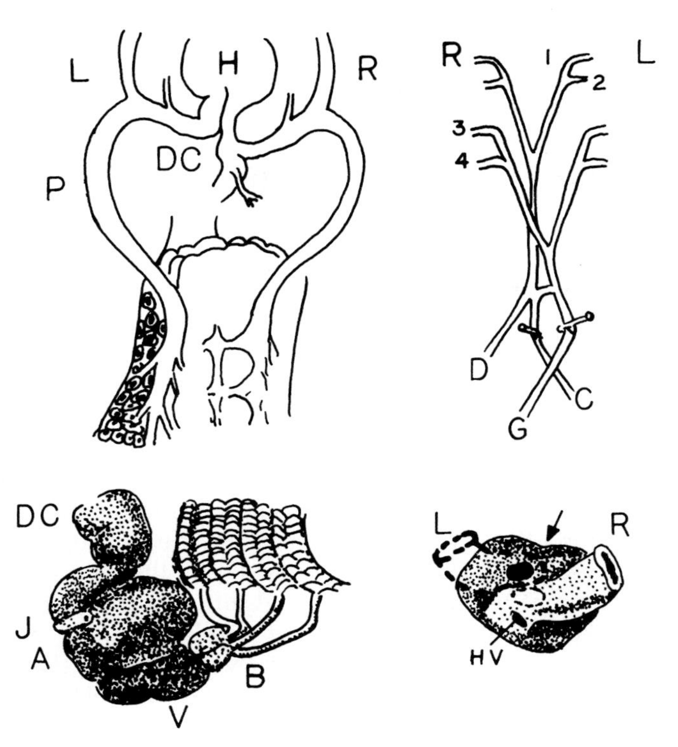

FIGURE 4.7 Circulatory detail for *Gymnarchus niloticus*. *Top left.* Dorsal view of the central venous return showing the separated duct of Cuvier. H, heart (note the partial atrial septum); P, post cardinal vein; L, left side; R, right side. ABO drainage is via the left P and into the left atrium. *Top right.* Ventral view of the efferent branchial arteries (1–4) showing their connections to the dorsal aorta (D), the gas bladder (G) and coeliac (C) arteries. Note anastomoses between three vessels. *Lower left.* Right side view of the heart showing the right ductus Cuvier (DC), jugular vein (J), the atrium (A), ventricle (V), and two of the four bulbus chambers (B) that lie at the base of each root artery. *Lower right.* Posterior view of the atrium showing the left (L) and right (R) Cuvierian ducts. The left duct is indicated by the dashed line and its entry port is shown as a black hole. The right duct is drawn and its entry port is shown as a white area outlined by a dashed line. Note that the hepatic vein enters the right duct. Arrow shows position of the atrial septum dividing the two entry ports. Based on Assheton's (1907) descriptions and drawings.

Other Primitive Teleosts

No details are known about the gas-bladder circulation of *Megalops*. For *A. anguilla*, Mott (1951) found that one branch of the gas bladder artery enters the expanded pneumatic duct and gives off branches resembling the vascular supply to a simple lung. She also found that venous return from this area is by means of the pneumatic duct vein which goes directly into the sinus venosus. This suggests that *A. anguilla* could use the pneumatic duct region of its gas bladder for aerial gas exchange and Mott (1951), using angiography, was able to detect intracardiac streaming of blood entering from the pneumatic duct vein. Berg and Steen (1965) found that when it was transferred to air, *A. vulgaris* did in fact initially consume O_2 from its gas bladder. *A. anguilla* likely has a similar capability, and direct circulation between the ABO and heart may play an important role.

The respiratory gas bladder of *Umbra krameri* (Jasinski, 1965) receives six feeder arteries from the median dorsal aorta. Venous return is by visceral veins that bypass the hepatic circulation and enter the post cardinal. There are no reports of a specialized ABO circulation in *Neochanna*.

Ostariophysans

The diversity of ABO structure and position among ostariophysans is matched by a range of ABO circulatory designs (Figure 4.2 G–J).

Intestinal and stomach ABOs. Among all stomach and intestinal air breathers, afferent ABO supply is via the coeliac or other arterial branch from the dorsal aorta (Figure 4.2G and H; Table 4.1). As reported by Carter and Beadle (1931), venous return in *Callichthys*, *Hoplosternum*, *Ancistrus*, and *Hypostomus* is via the interrenal to the post cardinal vein, thus bypassing the hepatic portal system. There have been no recent anatomical or hemodynamic studies of the interrenal circulation.

In *Misgurnus* and *Cobitis*, venous outflow from the ABO enters the hepatic portal system (McMahon and Burggren, 1987). *Lepidocephalus* also has a hepatic portal return; however, Yadav and Singh (1980) reported the presence of a gill–shunt vessel connecting the prebranchial ventral aorta and the intestine. More details are needed about this vessel's function.

Gas bladder. For characins air breathing with a gas bladder, early reports by Jobert (1878) and Carter and Beadle (1931) (probably for *Hoplerythrinus*) indicated the coeliac artery was the afferent supply. Corrosion casts (Smith and Gannon, 1978) verified this for *Hoplerythrinus* and also revealed that the coeliac forms a junction with the efferent branches of branchial arches 3 and 4 (Figure 4.2G). Venous return in *Hoplerythrinus* is to the post cardinal via the interrenal vein (Carter, 1957; Smith and Gannon, 1978). No comparable details are known for the other air-breathing characins which probably have similar circulations.

The efferent branchial–coeliac and dorsal–aortic junction seen in *Hoplerythrinus* is reminiscent of the lungfish ductus arteriosus and likely serves a shunt function. Farrell (1978) reported that air inhalation by this fish increased both heart rate and cardiac output to the ABO and caused a transient drop in dorsal aortic pressure. Impedance changes in both the ABO and the branchial circulation may also occur during air breathing. Smith and Gannon (1978) concluded that most of the flow through arches 3 and 4 goes to the ABO. Their experiments showed various vasoactive agents modulated flow through the coeliac and that acetylcholine administered to the gills tended to divert more blood through arches 3 and 4 while reducing flow through arches 1 and 2.

In *Gymnotus* the afferent ABO supply is from a branch of the coeliac, and venous drainage is into the hepatic portal system (Liem *et al.*, 1984). *Phractolaemus* has what Thys van Audenaerde (1961) termed a "pulmonary artery" (Figure 3.15) which branches off the dorsal aorta just anterior to the coeliac. Venous drainage is via a "pulmonary vein" to the post cardinal, with some minor branching into the kidney.

***Clarias* and *Heteropneustes*.** Vascular arrangements among the clariids and *Heteropneustes* are unique in featuring parallel gill and ABO circulation, as well as efferent ABO drainage into the dorsal aorta (Figure 4.2H and I). The gill fans and branchial trees of *Clarias* are supplied completely from within the afferent and efferent gill circulation, and the efferent air sac circulation of *Heteropneustes* empties into the dorsal aorta. The clariids and *Heteropneustes* have gills on all branchial arches, and as a consequence, efferent gill and ABO bloods become mixed enroute to the systemic circulation. Hyrtl (1854b) had initially shown (Figure 3.30A) that afferent flow to the air sacs of *Heteropneustes* was from branchial arch 4 on the left side and from arch 1 on the right side. Later work (Burne, 1896) showed the afferent supply to both air sacs was from arch 4, the pattern confirmed by Olson *et al.* (1990).

There are no data on shunting or selective perfusion of the gills or ABO by air-breathing clariids. For this to occur, lamellar shunts would be needed because elements of the ABO (fans, trees, chamber walls) are perfused from the distal end of each branchial arch.

A different situation occurs in *Heteropneustes* where the highly modified bulbus arteriosus gives rise to three ventral aortae (Figure 4.8), and these are thought to be the structural basis for branchial shunting (Olson *et al.*, 1990; Olson, 1994). The smallest and most anterio–ventral of these vessels originates from the left side of the bulbus and forms gill arch 1. The middle branch gives rise to arch 2, and the third and largest branch divides into arches 3 and 4. The side view (Figure 4.8) suggests that, with expansion, the bulbar area at the base of arches 3 and 4 can compress arches 1 and 2 and send more blood into the ABO (Olson *et al.*, 1990). While partial occlusion of branchial arch 1 would appear to affect brain perfusion, it may be that this could be compensated by retrograde flow from the ABO along the dorsal aorta. The pouched sac

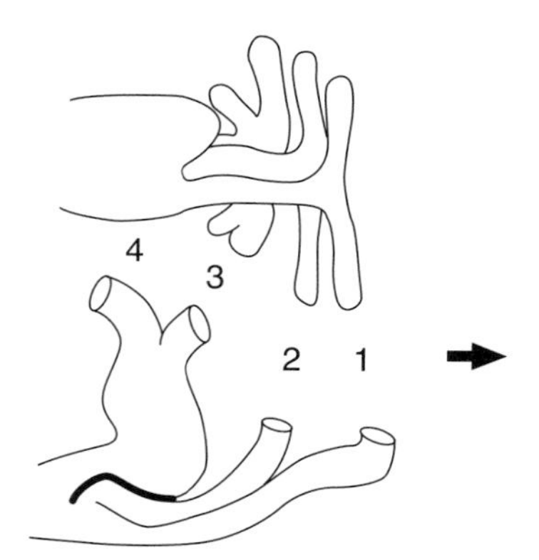

FIGURE 4.8 Specialized bulbus arteriosus of *Heteropneustes*. *Top*. Ventral view showing the tiered emergence of three ventral aortae. *Bottom*. Right side view showing the three ventral aortae and indicating how the lower bulbus section (thick black line), giving rise to branchial arches 3 and 4 may compress aortae 1 and 2 to cause partial branchial shunting. (Arrow = rostral.) (Olson *et al.*, 1990.)

extension of the bulbus described by Thomas (1967) may serve a capacitance function.

***Hypopomus* and *Electrophorus*.** In *Hypopomus*, the gills, along with the buccopharyngeal and opercular epithelia become aerial gas exchangers, and aerially derived O_2 enters capillaries formed in both the branchial and systemic circulations. Oxygenated blood thus enters both the arterial and venous sides of the systemic circulation.

An analogous but much more derived condition occurs in *Electrophorus* (Figure 4.2J-A and J-V) where, with highly reduced gills, branchial arches function principally as blood conduits to dorsally located ABO surfaces and to the systemic circulation. With hardly any branchial circulation, O_2 uptake from air held in the mouth is mainly by extensive systemic capillaries in the buccopharyngeal epithelium that feed venous return to the heart (Carter and Beadle, 1931; Carter, 1935). As in other species, cutaneous blood flow is highly important for gas exchange.

Experiments by Johansen *et al.* (1968b) confirm that *Electrophorus* does regulate blood flow to its ABO vasculature, primarily in response to ABO–O_2 level. Immediately after a breath, 75% of cardiac output flows to the ABO, but over a period of about two minutes, this falls to about 50% as ABO–O_2 declines. It is notable that neither breaths of pure O_2 nor N_2 markedly affected ABO perfusion (see Chapter 6).

Higher Teleosts

Many of the air-breathing higher teleosts are littoral inhabitants that are either exposed to air during low tide or are active amphibious species. Amphibious air breathing enables O_2 pickup across any one or all of several perfused surfaces in close proximity to air (Carter, 1957; Gans, 1970). The risk of transepithelial O_2 loss is not present and circulatory specializations for air breathing are thus not numerous (Figure 4.2K; Table 4.1).

The asymmetrical heart of clingfish (Gobiesocidae) needs to be mentioned. In these fishes, the ventricle is displaced to the right side and the atrium to the left. The sinus venosus is greatly reduced; however, a pair of accessory common cardinal chambers extend laterally from the atrium and may function as volume reservoirs for atrial diastole (Lauder and Liem, 1983a). Clingfish also differ from most other teleosts in having only three branchial arches. It is unknown if any of these modifications relate specifically to amphibious air exposure. The dorsal–ventral body flattening of most clingfish has probably had an effect on heart symmetry and cardinal chamber formation.

An exception to amphibious air breathing among higher teleosts is *Dormitator latifrons* which uses a vascularized patch of skin on its head for aerial gas exchange while floating at the surface of hypoxic water (Todd, 1973; Figure 2.9). Blood flow to this organ is thought to be from the carotid artery with a systemic venous return.

Anabantoidei and Channidae. Except for *Anabas*, few details are known about circulation patterns in most of the anabantoids (Olson *et al.*, 1986; Olson, 1994). Studies with *Channa* have been much more extensive (reviewed by Ishimatsu and Itazawa, 1993) and now include microvasculature details for the gills and ABO (Munshi *et al.*, 1994; Olson *et al.*, 1994). The circulation in *Channa* is very similar to *Anabas* and thus exceedingly specialized for air breathing with close approach to the double loop ideal. The basis for separation is the presence of two ventral aortae (anterior and posterior) that route cardiac output into either the anterior (1,2) or posterior (3,4) branchial arches (Figure 4.2L).

Das and Saxena (1956) first showed that the ventral aorta of *Channa* was subdivided. Vascular castings have now documented this condition for *Channa* (Ishimatsu *et al.*, 1979; Munshi *et al.*, 1994) and *Anabas* (Olson *et al.*, 1986). In both genera (Figure 4.2L), blood entering the anterior ventral aorta flows to arches 1 and 2, passes through the gill circulation on these arches, and then enters the suprabranchial chamber via the efferent branchial arteries. Venous return from the ABO is to the heart via the anterior cardinal vein and ductus Cuvier. Blood entering the posterior ventral aorta flows into branchial arches 3 and 4 (arch 4 has few or no lamellae) which feed into the dorsal aorta and systemic circulation. There are specific differences in some features of this general flow pattern; in *C. punctatus* and *C. striatus*, the afferent as well as efferent arteries continue into the ABO (Lele, 1932; Das and Saxena, 1956), whereas only the efferent arteries do this in *C. argus* and *C. maculatus* (Ishimatsu *et al.*, 1979). Also, because the entire buccopharyngeal surface is covered with respiratory islets, it is undoubtedly the case that vessels other than in branchial arches 1 and 2 perfuse the ABO surface, and for this reason O_2-rich blood also enters the systemic venous circulation (Munshi *et al.*, 1994).

The question posed earlier for *Gymnarchus* can now be asked for *Channa*, *Anabas*, and the other anabantoids. Does efferent ABO blood entering the anterior cardinal vein somehow remain separate through the ductus Cuvier and heart to then be preferentially channeled into the posterior dorsal aorta for systemic dispersal? Even though this blood is subject to mixing with several sources of deoxygenated venous return, including the post cardinal vein, experiments with *C. argus* indicate the answer to this question is yes! Ishimatsu and Itazawa (1983a) found both the O_2 content and PO_2 of blood in the anterior ventral aorta to be lower than that in the posterior ventral aorta. The differences in oxygenation were not large (PO_2 ranges: Anterior 10–20 torr; posterior 25–42) but did confirm that partial separation occurs. The anterior and ventral aortic pH values also differed as expected (anterior more acidic); however, PCO_2 values in both streams were high and not significantly different (Ishimatsu and Itazawa, 1983a, figure 3; 1993).

Although there is no direct evidence, a number of anatomical features within the cardiac chambers of *Channa* have been suggested to promote axial flow separation during the cardiac cycle. These include the absence of a sino–atrial valve, extensive ventricular trabeculation, and longitudinal muscular ridges on the lumen of the bulbus arteriosus (Ishimatsu and Itazawa, 1983a, 1993; Olson *et al.*, 1994). Thomas (1976) described vertical and oblique trabeculae subdividing the bulbus arteriosus of *Anabas* and *Channa* into a central and two lateral channels. This modification likely relates to the presence of two ventral aortae in these genera; however, it is presently unknown how these channels feed into the paired aortae or how they might influence blood flow.

Andresen *et al.* (1987) used angiography to study blood flow in *C. argus*. They were unable to determine if efferent ABO blood flowed preferentially into the posterior ventral aorta, nor could they discern any changes over the course of an air-breath cycle in relative flows through the anterior and posterior ventral aortae. The heart of *Anabas* similarly lacks dramatic structural modifications for separating ABO and systemic flows (Olson *et al.*, 1986; Olson, 1994).

Synbranchiformes. With aerial-respiratory surfaces in the mouth and branchial chamber, the synbranchid circulation pattern (Figure 4.2M) has been shaped by the same factors affecting *Electrophorus* and *Hypopomus*. Transbranchial O_2 loss is not a problem for the synbranchids because they do not ventilate water while holding air in the buccal-chamber ABO.

Depending upon the contribution made by gills to aerial gas exchange, O_2 can enter through blood in both the branchial and the systemic capillaries. As illustrated in Rosen and Greenwood (1976, figures 51–56), there is considerable inter-specific variation in the synbranchid branchial arch and ABO circulation patterns. In *Monopterus* (Figure 4.2M), the large caliber conduit between the ventral and dorsal aortae, which gives off a side branch to form arch 4, clearly favors distribution of O_2-rich blood to the body. Other notable circulatory modifications include paired lateral dorsal aortae, a large median vein, and a highly derived ventral–aortic position of the coronary artery (Munshi *et al.*, 1990).

The synbranchid heart is elongated and has a large sinus venous and atrium (Liem, 1961; Munshi and Mishra, 1974; Rosen and Greenwood, 1976; Spiropulos Piccolo and Sawaya, 1981). This organ is also displaced posteriorly (i.e., to about 20% of body length behind the snout) relative to other fishes (including many eel-shaped species), and this is generally regarded as a specialization to protect the organ from the compressive forces resulting from burrowing and holding air in an expanded branchial chamber (Munshi and Mishra, 1974). Synbranchids also spend considerable periods with their heads held vertically, and the more posterior heart position may be an adaptation, ensuring adequate venous return (Liem, 1961). Other features of the synbranchid (*Monopterus*) heart that may affect air-breathing hemodynamics include the dorso–lateral encirclement of the bulbus by the atrium and the exceptional occurrence of a positive heart–mass allometry ($M^{1.25}$, Munshi and Mishra, 1974). Although there are no hemodynamic data, the latter correlates with the increased utilization of air breathing by larger *Monopterus* (Chapter 5), and the scaling of air-sac surface area ($M^{1.19}$, Chapter 3).

SUMMARY AND OVERVIEW

The independent origin of air breathing has resulted in a range of circulatory modifications. In total, about eight basic patterns of afferent and efferent ABO circulation are known (Figure 4.2). Some of these approach, to varying extents, the ideal condition of separation as represented by the double circulatory loop of mammals and birds. Nevertheless, no air-breathing fish has a completely separated left and right heart, nor can any species function totally without gills. Thus, the problems of enforced blood desaturation, gill–O_2 loss, and vascular-pressure regulation present significant functional limitations for the circulation of most aquatic air-breathing species.

Blood Shunting

Because of these limits, the need to phasically regulate ABO, gill, and systemic perfusion is a major functional requirement for aquatic air-breathing fishes. This chapter has described various structures and circulation patterns favoring shunting, and Table 4.2 summarizes these for the different taxa. As documented for *Lepidosiren* and *Protopterus*, branchial and ABO shunting are phasically integrated with air breathing and the control of both cardiac performance and vascular resistance. Although less detail is known about shunting and its control in the actinopterygian air-breathers, Table 4.2 establishes the presence of mechanisms that favor this action. Future research will expand this information and our understanding of shunt control.

Comparative Circulation Patterns

The lungfishes (*Lepidosiren*, *Protopterus* and to a lesser extent *Neoceratodus*) come closest to the double loop system in having almost completely separate pulmonary and systemic circulations. In fact, *Lepidosiren* and *Protopterus* employ the combinations of pulmonary venous return of blood to the atrium, nearly complete (functionally) heart septation, and both pulmonary and branchial-arch shunts to attain a level of pulmonary and systemic isolation exceeding that in most amphibians (Table 4.2).

The presence of a pulmonary circulation also enables some primitive actinopterygians (*Polypterus*, *Erpetoichthys*, and *Amia*) to partially separate ABO and systemic flows. However, unlike lungfishes, pul-

TABLE 4.2 Structural Modifications in the Circulatory Anatomy of Air-Breathing Fishes Related to the Shunting or Selective Perfusion of the ABO, Systemic, or Branchial Vascular Beds during Bimodal Respiration[a]

Genus	Structural feature	Action
Lungfishes		
Protopterus, Lepidosiren	Ductus arteriosus	Lung shunt to systemic circulation
	PAVS	Regulate pulmonary artery blood flow
	Aff branchial arch shunts	Regulate branchial flow and pulmonary perfusion
	Incomplete ventricular septum	Balance pulmonary and systemic loop pressures
	Spiral and ventrolateral folds in bulbus	Guide flow to "systemic" and "pulmonary" branchial arches
(Less derived versions of all these mechanisms are likely present in *Neoceratodus*)		
Polypterus	"Analog" ductus arteriosus connections	Lung shunt?
	Ventral-aortic strap	Anterior branchial shunt?
Amia	Commissural vessels	Allows ABO shunt?
	Recessed lamellar channels	Partial gill shunt
Gymnarchus	Dorsal aortic–efferent branchial arch junctions	ABO shunt?
	Four separate arteries with individual capacitance valves	Selective branchial arch perfusion?
	Muscle insertions onto pericardium	Heart torsion may facilitate blood shunting?
Arapaima	Valves on systemic veins near heart entrance	Control venous admixture?
	Thick layers of smooth muscle in afferent and efferent arteries	Regulate branchial circulation?
	Valved and discrete lamellar blood channels that are also recessed	Regulate branchial circulation?
Clariidae	Highly separated ventral aortic branches	Lamellar shunts needed to perfuse ABO?
Heteropneustes	Manifold bulbus with separate ventral aortic openings and dead-end sac extension	Selective branchial perfusion? Pressure compensation?
Hoplerythrinus	"Ductus arteriosus" type junction of dorsal aorta and coeliac artery	ABO shunt
	Vasoactive branchial impedance	Regulation of branchial shunting and ABO perfusion
Lepidocephalus	Vessel connecting pre-branchial ventral aorta to intestine	Branchial shunt?
Channa, *Anabas*, and other anabantoids	Anterior and posterior ventral aortae	Optimal routing of ABO and systemic venous returns
	Reduced gill lamellae on arch 4	Branchial shunt
Channa	Vessel connecting the arch 2 suprabranchial artery to the lateral aorta	ABO shunt
Electrophorus and *Monopterus*	Ventral–dorsal aortic conduits	Shunts

[a]Question marks signify experiments not done to verify presumed action. See text for details.

monary return in these forms is into the great veins at sites anterior to where they join the heart. Also, none of these primitive hearts have medial septation.

In what appears to be a secondary specialization for air breathing, the teleost *Gymnarchus*, has a "quasi-pulmonary" circulation to its respiratory gas bladder and can likely return partially (minimally?) unsaturated blood into its left atrium. The unique heart structure (divided atria, four ventral–aortic branches) of *Gymnarchus* likely sustains the flow separation needed to enhance aerial respiration; however, no data exist (Table 4.2). Among other species possessing a respiratory gas bladder (gars, *Arapaima*, *Pantodon*, *Notopterus*, *Phractolaemus*, *Hoplerythrinus*), ABO circulation is generally via the dorsal aorta, coeliac artery or an affluent (including the renal portal), and venous return is into the post cardinal vein.

While no air-breathing fish can function without a branchial circulation, some species have lessened the separate problems of vascular pressure drop and transbranchial O_2 loss by setting the ABO and gills in parallel circulation. In clariids and *Heteropneustes* (Figures 4.2H and I), the ABO receives afferent branchial blood which it passes into the arterial side of the systemic circulation via the dorsal aorta. Passage of ABO-enriched blood into the systemic circulation also occurs in genera with branchial ABO surfaces (*Electrophorus*, *Hypopomus*, synbranchids), but is a rare pattern. *Channa* and the anabantoids (Figure 4.2L) have circulation patterns approaching a double loop; one fraction of the cardiac output passes through the two anterior gill arches and then into the ABO, while the other circulates directly to the body via branchial arches having minimal respiratory surfaces. This sequence enhances ABO function by preparing blood (i.e., by gill removal of CO_2) for maximum O_2 absorption; however, efferent ABO blood enters the venous circulation and, in the absence of a medially septated heart, maintenance of separate high and low O_2 blood streams remains problematic. Optimal function of both the clariid and *Channa*-type circulations is thus dependent upon streamlined flow in the great vessels and heart as well as on the capacity to selectively regulate branchial and ABO perfusion (Table 4.2).

In species with ABOs located in visceral organs, modified routes of venous return appear to lessen effects of enforced unsaturation and circumvent the hepatic portal circulation. ABO drainage via the interrenal to the post cardinal forms a functional hepatic–portal shunt in *Hoplosternum*, *Callichthys*, and *Ancistrus*. On the other hand, some species (*Gymnotus*, *Misgurnus*, *Cobitis*) breathe-air effectively, even though having an hepatic–portal ABO drainage.

While the interrenal route and other vascular shunts probably heighten the efficiency of ABO–O_2 transfer, these modifications do not eliminate the problems of venous mixing nor do they lessen the potential effects of gills on O_2 loss. It is thus apparent that anatomical modification of the circulation to permit shunting and flow separation provides a stage for the physiological regulation of ventilation and organ perfusion and for adaptations enhancing blood-respiratory gas transport to act in concert for effective integration of aerial and aquatic respiration.

Comparative Heart Structure and Coronary Circulation

It is unknown if the evolution of actinopterygian heart structure was at all influenced by air breathing. In primitive actinopterygians, a large and contractile conus arteriosus forms the terminal cardiac chamber linking the ventricle and ventral aorta. Comparisons reveal that relative to non-air-breathers, the conus in both *Lepisosteus* and *Polypterus* is longer and has extra sets of pocket valves (Goodrich, 1930). In higher forms, the conus is replaced by the non-contractile bulbus arteriosus; however, both *Amia* and *Megalops*, which have a bulbus, also retain sets of conal valves (Parsons, 1930). Future work can decide whether or not the conal valves play a role in the streaming and distribution of blood flow in relation to air breathing.

Medial heart septation is a key element preventing the mixture of pulmonary and systemic flows in the hearts of lungfishes. This is probably also true for *Gymnarchus* which has a partially septated atrium. Beyond this, we have little insight as to how most of the remaining air-breathing fishes, which uniformly lack medial heart septation, are able to prevent intracardiac mixing of ABO and systemic blood streams.

How does the coronary circulation relate to air breathing? In most fishes, the coronary perfusion enters the outer, compact myocardial layer, whereas the inner, spongy endocardium is supplied by diffusion from venous blood within the coronary chambers. The extent of coronary perfusion into the compact layer correlates with a species' activity. Little is known about how the epi- and endocardial layers of air-breathing fish hearts may differ from other species. There are, however, suggestions that air breathing has affected coronary circulation pattern.

In most non-air-breathing fishes, the coronary artery commonly branches from the hypobranchial artery which originates on the efferent (oxygenated) side of an anterior branchial arch (Figure 4.1B). This pattern is found in *Neoceratodus* where the hypobranchial originates from the efferent branches of arch 2. Both *Lepidosiren* and *Protopterus*, however, lack gills on arch 2 but, because of the separate pulmonary loop, blood in this arch is oxygenated. The origin of the hypobranchial is thus direct and more proximal to the heart (Figure 4.2A), which assures an oxygenated source and should also heighten perfusion pressure. In *Polypterus* (Figure 4.2B) coronary supply is from the dorsal aorta via the subclavian artery and not from the branchial arches. This would ensure cardiac perfusion in the event of branchial shunting (caused by the ventral–aortic strap). However, a somewhat similar coronary flow also occurs in some of the non-air breathing polypteriforms and comparative studies are needed to determine the degree of specialization related to air breathing in *Polypterus*. In *Lepisosteus* (Figure 4.2D), connections between the hypobranchial and subclavian arteries ensure coronary flow from either route, suggesting a facultative connection for reduced hypobranchial flow during air breathing. Coronary flow in *Amia* (Figure 4.2C) is similar to most modern fishes.

In most of the higher teleosts coronary circulation is via the hypobranchial artery which originates on the efferent branchial artery of arch 2. Whereas details are lacking for most air-breathing species, this coronary circulation pattern is present in *Channa* (Munshi *et al.*, 1994) which is noteworthy because in this genus the efferent branchial arch blood has yet to enter the ABO and is therefore not O_2 enriched (Figure 4.2L). This implies that there is sufficient O_2 within the venous circulation to sustain the heart and seems reasonable in view of the potential for axial flow mixing and the presence of a significant ABO drainage into the systemic venous return (Munshi *et al.*, 1994).

A radical departure for the origin of the coronary blood supply is seen in *Monopterus* where the artery originates on the ventral aorta (i.e., "hypoxic," venous site) just distal to the heart. This origin is not surprising in view of the diffuse branchial circulation of *Monopterus* but is nevertheless significant in its implication that the levels of O_2 in venous blood are sufficient to nourish the heart. This correlates with the large blood O_2 storage capacity of *Monopterus* (Chapter 7). Beyond these facts, we lack knowledge of how the coronary circulations of most other air-breathing species have been modified to augment O_2 supply during air breathing. Comparative data for other synbranchids (which have functional gills) and for *Arapaima* (as well as other osteoglossids with a large, low pressure, venous ABO flow) would be particularly important for the insight they provide about how circulatory modifications for air breathing have affected the roles of blood pressure and O_2 content in regulating coronary perfusion.

CHAPTER

5

Aerial and Aquatic Gas Exchange

Few events in the life history of vertebrates stand out in importance like the transition from life in water to life on land. Such a transition is inseparable from the development of accessory organs of respiration making possible the direct utilization of O_2 in atmospheric air. The fishes occupy a crucial phylogenetic position in this respect, and they demonstrate numerous and extremely varied examples of how the fundamental process of breathing with gills in water is modified or assisted by accessory organs for aerial respiration. Aerial respiration in fishes therefore merits considerable attention, not only because of its profound effect on the direction of vertebrate evolution, but as much for light thrown on the physiological mechanisms of gas exchange in water and air.

K. Johansen, *Comparative Biochemistry and Physiology XVIII*, 1966

INTRODUCTION

This chapter reviews the diverse information found under the general subject of "air-breathing fish respiration." Previous treatments of this topic include surveys by Carter (1957) and Johansen (1970) and reviews of bimodal respiration (Lenfant *et al.*, 1970; Johansen, 1972; Lenfant and Johansen, 1972; Rahn and Howell, 1976). Among the most complete comparative gas-exchange summaries is one dealing primarily with aquatic air breathing (Singh, 1976) and another emphasizing amphibious air breathing (Graham, 1976a). Ar and Zacks (1989) reviewed respiratory partitioning in air-breathing fishes and provided an especially detailed analyses of this in *Clarias lazera*. Works by Bridges (1993a,b) and Martin (1993, 1995) explore the relationships between the rates of oxygen consumption ($\dot{V}O_2$) and carbon dioxide release ($\dot{V}CO_2$) during amphibious air breathing.

The explosion of air-breathing fish respiratory data over the past thirty-five years invites a resynthesis to determine if the underlying principles guiding this field remain intact and to discover areas needing additional study. This chapter begins with a historical overview which is followed by a comparison of air-breathing cycles (i.e., breath frequency and duration) and a review of whole-organism respiratory gas exchange. The final section examines specialized aspects of respiratory organ function. Unlike the primarily phylogenetic approach adopted in previous chapters, this review emphasizes the comparative physiological aspects of air breathing. A similar approach is followed in subsequent chapters on respiratory control and integration (Chapter 6), on blood–respiratory properties (Chapter 7) and biochemical adaptations (Chapter 8).

HISTORICAL

The need for studies of gas exchange was recognized by the early naturalists who wondered if air breathing was sufficient to completely replace or only supplement the normal aquatic mode. Taylor (1831), in noting atrophied gills of *Monopterus*, drew attention to the necessity for bimodal respiration, a process described by Day (1868) as "compound breathing."

Physiological investigation of air breathing began with the pond loach (*Misgurnus fossilis*) for which Erman (1808) and Bischof (1818) made the earliest measurements of ABO gas levels. Baumert (1853), also working with *Misgurnus*, did the first air-breathing fish respirometry, estimated the $\dot{V}O_2$ and $\dot{V}CO_2$ of forcibly submerged fish, and obtained data suggesting an unequal partitioning of aerial and aquatic gas exchange (i.e., aerial $\dot{V}O_2 >> $ aerial $\dot{V}CO_2$). Based on this and their own studies, Jolyet and Regnard (1877a,b) concluded that *Misgurnus* used its ABO primarily for O_2 absorption while remaining dependent upon gills for CO_2 release into water. This was the first enunciation of what is now regarded as a basic precept of bimodal respiratory partitioning and preceded by 27 years the studies of bimodal respiration by August Krogh (1904b) who is generally credited with the original insight.

Although determination of ABO–gas content had been done on various species for well over a century (Erman, 1808; Bischof, 1818; Jobert, 1877; Mark, 1890) this work came of age with the experiments of Potter (1927) on *Lepisosteus*, by Carter and Beadle (1931) on the South American genera (*Ancistrus*, *Erythrinus*, and *Hoplosternum*), and with development of techniques for ABO cannulation (Johansen, 1966; Johansen *et al.*, 1968a). Interest in the evolutionary aspects of air-breathing physiology was sparked by Homer Smith's (1930) studies of estivating lungfish metabolism (Chapter 8) and, at about the same time, Schöttle (1931) made the first amphibious fish $\dot{V}O_2$ measures on mudskippers as well as the first estimates of respiratory exchange ratio ($R = \dot{V}CO_2/\dot{V}O_2$) for fish in air.

The 1960s, 1970s, and 1980s can be considered the golden age of comparative air-breathing fish respiratory physiology. Progress was stimulated by the conviction that these investigations would trace the physiological evolution of vertebrate air breathing and lead to basic discoveries of biomedical significance. Aided by technical advances leading to smaller and more portable instruments and ones requiring smaller sample sizes, techniques used in mammalian physiology became applicable to air-breathing fishes. The advent of modern jet transportation also made it possible for specimens to be routinely transferred from remote locations to modern laboratories. In the United States of America, major initiatives in comparative biology, by both the National Institutes of Health and the National Science Foundation (NSF), gave impetus to air-breathing fish research. The NSF developed the floating laboratory vessel RV *Alpha Helix* and supported its numerous expeditions to habitats (Great Barrier Reef, Amazon Basin) containing air-breathing fishes and species with allied physiological capacities.

During this period, comparative investigations examined almost every aspect of aerial and aquatic gas exchange including rates, aerial and aquatic R, air convection requirement, and the partitioning within each mode of gill, skin, and ABO respiration (Johansen *et al.*, 1968a, 1970a; Johansen, 1972; Lenfant and Johansen, 1968; Farber and Rahn, 1970; Hughes and Singh, 1970a,b, 1971; Rahn *et al.*, 1971; Burggren and Haswell, 1979). Krogh's (1904a) early work had demonstrated the respiratory capacity of *Anguilla* skin, and studies by Berg and Steen (1965) showed that in air-exposed eels, cutaneous O_2 uptake may account for as much as 40% of total $\dot{V}O_2$. Subsequent experiments with partitioned respirometers have confirmed the generally large contribution made by the skin to amphibious O_2 consumption (Teal and Carey, 1967). In fact, recent data from Pelster *et al.* (1988) show that the gill–skin partitioning of aerial CO_2 release by *Blennius* (*Lipophrys*) *pholis* correlates closely with the partitioning of aerial $\dot{V}O_2$ between these two surfaces (Bridges, 1993a,b). Important papers on amphibious respiration (Stebbins and Kalk, 1961; Bandurski *et al.*, 1968; Gordon *et al.*, 1968) fueled interest in the physiology of the transition to air breathing as did studies showing how air breathing related to circulation, cardiovascular function, and blood–respiratory properties (Chapters 4, 6, and 7).

The major contribution of these whole-animal air-breathing studies was to expand our comparative perspective and broaden the basis for approaching certain aspects of mammalian physiology; the control of respiration and the regulation of acid-base physiology are prominent examples (Rahn, 1966a,b; Rahn and Howell, 1976; Heisler, 1982; Ar and Zacks, 1989). However, the primarily biomedical and "anthro-evolutionary" (i.e., "from fish to man") perspectives of these works left a number of important physiological aspects of fish air breathing unstudied. For example, a tendency of many studies was to categorize a species regarding variables such as air-breathing rate and percentage of respiratory partitioning. This chapter will show that in most cases these variables are unlikely to be constant.

Another indication of the "anthropocentric view" of fish air breathing was the approach taken to the question of the energetic costs of aerial versus aquatic respiration: The classical physiological analysis of this issue mainly considers the capacitances and densities of the two media, ignoring biologically important cost factors such as gravitational effects on terrestrial locomotion, the vertical distance a fish might need to swim for air, the effect of air breathing on buoyancy, and the increased risk of predation associated with surfacing. Only in the past two decades have biologists begun to note the importance to air-breathing physiology of both biological and behavioral influences on this activity (Kramer, 1983a,b). Research in this area has shown that air-breathing physiology is affected by crowding and predation (Kramer and Graham, 1976; Liem, 1987) as well as environmental influences (e.g., water temperature and respiratory gas levels, Johansen *et al.*, 1968b, 1970a; Rahn *et al.*, 1971; Ar and Zacks, 1989). Acclimatory adjustments to hypoxia also affect respiration physiology (Johnston *et al.*, 1983; Graham, 1983, 1985). The challenge posed by this chapter is to weave the various findings of investigations on a broad but interrelated set of topics into a synthesis of air-breathing fish respiration that emphasizes major principles and sets goals for future research.

THE AIR-BREATHING CYCLE

Generalized air-breathing cycles are shown in Figure 5.1. Their central feature is the air gulp by the aquatic air breather or emergence from water by the amphibious air breather. In anabantoids, *Channa*, the callichthyids, *Polypterus*, and most other aquatic air breathers, the mechanical coupling of exhalation and inhalation (Chapter 3) has essentially reduced the air-breath cycle to a single component, an air-holding phase. In other species, for example, the loricariids and synbranchids, ABO evacuation is decoupled from inspiration and the two processes may often be temporally separated by as little as a few seconds to an hour or more (Johansen, 1966; Gee, 1976; Kramer and Graham, 1976; Graham and Baird, 1982, 1984). Also, under some circumstances an aquatic air breather may hyperventilate (a series of rapid inhalations and exha-

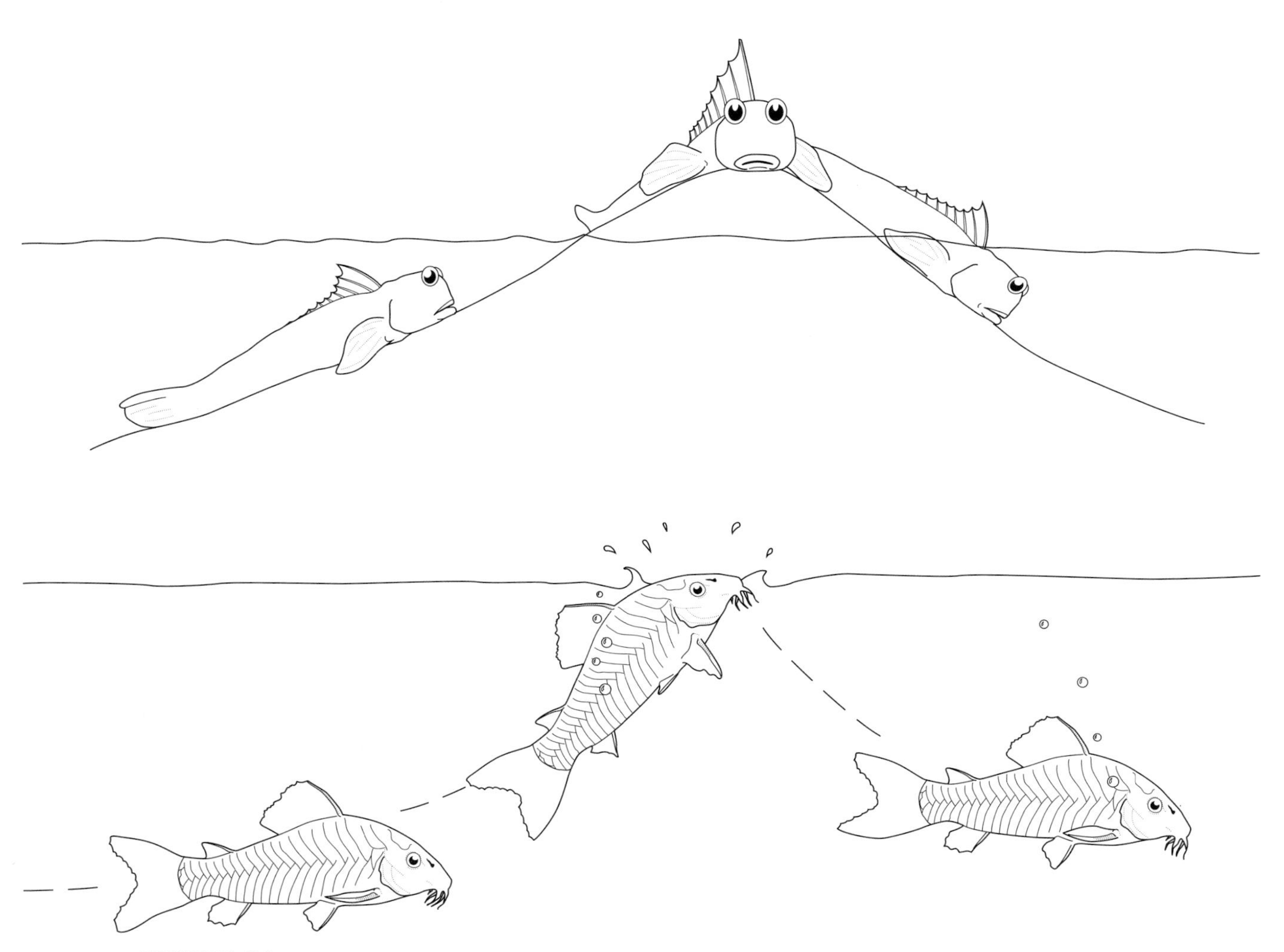

FIGURE 5.1 Air-breathing cycles. *Top*, An amphibious blenny; note that fish may, at times, be only partially submerged. *Lower*, Aquatic air breather (*Hoplosternum*). Note anal expulsion of used respiratory gas during inspiration and post-inspiration gas release from branchial region.

lations) before submerging, or small amounts of gas may be released from the ABO at various times during the breath-hold phase (Lomholt and Johansen, 1974; Hedrick and Jones, 1993, and preceding references).

Amphibious cycles are not always as simple as depicted in Figure 5.1. While emergent, some fish periodically ventilate their gills with air (*Mnierpes*), hold their mouths open for long intervals (*Synbranchus*), take periodic air gulps (*Entomacrodus*), adopt a tidal ventilatory mode by sealing their operculae (*Pseudapocryptes, Blennius*), or seal their branchial chambers over a small volume of water which is then periodically sloshed around (some mudskippers). Also, some of the active intertidal amphibious forms have been frequently observed to move to the water's edge, submerge only their heads and ventilate their gills for a short period. Lastly, when it comes to remoistening body and respiratory surfaces—or perhaps shedding heat—these fishes will usually avail themselves of any pocket of water encountered along their amphibious course (Graham, 1976a; Sayer and Davenport, 1991; Clayton, 1993).

Aquatic Air Breathing

The amount of time a species spends air breathing reveals information about its respiratory physiology. Garden's (1775) report that *Electrophorus* held air in its buccal chamber for between four and five minutes is probably the earliest measurement of air–breath duration. Many observations of this type have now been made. Depending upon the experimental setting, air-breathing times have been reported as either breath–hold duration or as the number of air breaths taken over a period of time (rate or frequency). Some workers report both data forms, and in cases where there is zero time between exhalation and inhalation, the two measures are inversely related (1/duration = frequency). Air–breath durations have been most often recorded during controlled respirometry or during related air-breathing experiments on individual fish. Air-breathing frequencies are usually obtained in experiments with individuals or groups held in less confining circumstances but subject to ambient controls.

Tables 5.1 and 5.2 show that there is wide variation in air–breath interval, both among and within species. These tables also establish that frequency and duration are affected by both experimental setting and ambient conditions. In general, shorter air–breath durations occur among the smaller species, those that are obligatory and continuous air breathers, and species that often swim near the water surface.

The tables indicate that stress associated with handling, surgery (even post-surgical effects and the constraints imposed by the attachment of various monitoring devices), the conditions used to induce air breathing, and the design of the experimental or observation chamber can all affect cycle duration. For example, *Electrophorus* fitted with both vascular and ABO cannulae had shorter breath–hold durations (data of Johansen *et al.*, 1968b; Table 5.1) than did an aquarium specimen (Farber and Rahn, 1970). Furthermore, I observed much longer breath–holds for *Electrophorus* in display tanks at Sea World, San Diego and the Vancouver (British Columbia) Public Aquarium. Similarly, the air–breath durations reported for *Arapaima* in a respirometer or with vascular and ABO cannulae attached (1–4 minutes, Table 5.1) contrast markedly with my observations of 13–29 minute durations for fish in the Vancouver Aquarium. A range of values is also seen for *Synbranchus*; however, both the greatest range in duration and the longest mean durations were obtained in a study by Graham and Baird (1984) in which observations were made long after fish were handled and acclimated to an experimental chamber that was fairly isolated from laboratory disturbances. Similarly, eight-hour video records of type I air breathing in *Amia* yielded data (Table 5.1) that, when analyzed by time series, are strongly periodic and suggest a basis for respiratory control (Hedrick and Katz, 1991; Hedrick *et al.*, 1994).

On the other hand, much of the variability seen for fish air–breath duration appears to be real. Even in controlled laboratory conditions, rhythmic or periodic aerial respiration patterns and uniform inter-breath intervals are seldom seen. Thus, the numerous attempts to discern quantitative links between sequences of air–breath duration, or between durations of breath hold and the intervening periods of aerial apnea, have not been successful (Johansen 1966; Schwartz, 1969; Johansen *et al.*, 1970a; Rahn *et al.*, 1971; Kramer and Graham, 1976; Lomholt and Johansen, 1976; Itazawa and Ishimatsu, 1981; Graham and Baird, 1982, 1984; Hedrick and Jones, 1993). The lesson of the work on *Amia* by Hedrick and co-workers is that long-term studies on undisturbed fish in controlled ambient conditions, combined with sophisticated computational methods, can reveal relationships that are not otherwise discernible.

It is evident in Tables 5.1 and 5.2 that there is not a species-specific rate of air-breathing. Rather, this process is affected by numerous physical and biological factors. The direct effects of aquatic O_2 are well documented: In nearly all cases, the time that an air breath is held decreases with declining aquatic PO_2 and is longer in hyperoxia. Studies with *Amia* (Johansen *et al.*, 1970a) and *Lepisosteus* (Rahn *et al.*,

TABLE 5.1 Air-Breath Durations Reported for Aquatic Air-Breathing Fishes

Species	Air breath duration (min)	Reference
Lepidosiren paradoxa	4–5	Sawaya, 1946
	3–10	Johansen and Lenfant, 1967
Protopterus aethiopicus	2–9	Lenfant and Johansen, 1968
Lepisosteus osseus	3–165[a]	Rahn *et al.*, 1971
L. oculatus	3–60[a,p]	Smatresk and Cameron, 1982a,b,c,
Amia calva	8.5–240[a]	Johansen *et al.*, 1970a
	10–12[o]	Randall *et al.*, 1981b
	31 ± 9[r]	Hedrick *et al.*, 1994
	16 ± 5[s]	
Arapaima gigas	1–2[b]	Stevens and Holeton, 1978a
	4[b]	Randall *et al.*, 1978a
	8–29[c], 15[m]	Graham, pers. obs.
	8–9[t]	Lüling, 1964a
	4–7[u]	Lüling, 1964a
Notopterus chitala	6.5–8.5	Ghosh *et al., 1986*
Pantodon buchholzi	1–5[d,n]	Schwartz, 1969
Heteropneustes fossilis	2–30[d]	Hughes and Singh, 1971
Clarias batrachus	3–20[d]	Singh and Hughes, 1971
C. mossambicus	1–31[h], 9.5[m]	Johnston *et al.*, 1983
	2–18[g], 7.4[m]	
C. lazera	1.5–16.5[d]	Ar and Zacks, 1989
Electrophorus electricus	4–5	Garden, 1775
	1–2.5[b,q]	Johansen *et al.*, 1968b
	2.5[e]	Farber and Rahn, 1970
	5–12[f]	Graham, pers. obs.
Hoplerythrinus unitaeniatus	1–2[b]	Stevens and Holeton, 1978b
	2–3[b]	Randall *et al.*, 1978b
	3–23[d,j]	Lüling, 1964b
Ancistrus chagresi (spinosus)	7.4[c,g]	Graham, 1983
	8.5[c,h]	
Gillichthys mirabilis	1–8	Todd and Ebeling, 1966
Channa argus	11.2[k] (normoxia)	Itazawa and Ishimatsu, 1981
	7.6[k] (hypoxia)	
	2–5[d,l] (15 C)	Glass *et al.*, 1986
	7–25[d,l] (25 C)	
Anabas testudineus	< 1–15[d]	Hughes and Singh, 1970a
Trichogaster trichopterus	4–6, 4.7[m]	Burggren, 1979
Osphronemus olfax	4.5–30	Natarajan, 1980
Monopterus cuchia	1–8	Singh and Thakur, 1979
	7–9[a]	Lomholt and Johansen, 1974, 1976
Synbranchus marmoratus	8–35	Johansen, 1966
	5–10	Bicudo and Johansen, 1979
	2–55[d], 15.5[m]	Graham and Baird, 1984

[a] Temperature and O_2 effects.
[b] Constrained or cannulated.
[c] Three or more fish in aquarium.
[d] O_2 effect.
[e] One aquarium fish.
[f] Two aquarium fish.
[g] Hypoxia acclimated.
[h] Normoxia acclimated.
[i] Affected by light level and temperature.
[j] Feeding effect.
[k] 410–800 g, 25 C.
[l] 1–2 kg.
[m] Mean value.
[n] Activity effect.
[o] 30 C.
[p] Salinity effect.
[q] Aerial (ABO) O_2 and CO_2 effects.
[r] mean ± SE, normoxia.
[s] mean ± SE, hypoxia.
[t] small fish (2.5–5 cm).
[u] larvae (< 2.5 cm).

TABLE 5.2 Air-Breathing Frequencies of Aquatic Air-Breathing Fishes

Species	Air breaths/h	Reference
Protopterus annectens	6–20[b,c]	Babiker, 1979
P. aethiopicus	16–17[r]	Jesse *et al.*, 1967
	75,12,5[u]	Lahiri *et al.*, 1970
P. amphibius	8.6	Lomholt, 1993
Polypterus senegalus	3–20[c]	Babiker, 1984a
Amia calva	1 vs 32[b,t]	Horn and Riggs, 1973
Megalops atlanticus	4–12[d]	Shlaifer and Breder, 1940
Lepisosteus platyrhincus	8 vs 18[e]	Smith and Kramer, 1986
Clarias lazera	10–20[b,c]	Babiker, 1979
C. batrachus	5–57[b,f,s]	Jordan, 1976
C. macrocephalus	1–2 vs 12[c,f]	Bevan and Kramer, 1987a,b
	10–12[c,p]	
	7–9[c,i]	
	3–4 vs 20[c]	Kulakkattolickal and Kramer, 1988
	40[g]	
Umbra limi	< 1 vs 28[n]	Gee, 1980, 1981
	< 1 vs 12[o]	
Pangasius sutchi	1 vs 30[c]	Browman and Kramer, 1985
Piabucina festae	12–36[d]	Kramer and Graham, 1976
	15–150[c,q]	Graham *et al.*, 1977
Hoplosternum thoracatum	2–20	Gee, 1976
	7–12[c,d]	Kramer and Graham, 1976
	0 vs 8[c]	Gee and Graham, 1978
Corydoras aeneus	2 vs 58[c,k]	Kramer and McClure, 1980
Misgurnus anguillicaudatus	5–26	McMahon and Burggren, 1987
Ancistrus chagresi	8	Gee, 1976
	6–11[c,d]	Kramer and Graham, 1976
	2–20[c,d]	Graham and Baird, 1982
Hypostomus plecostomus	3	Gee, 1976
	1.5–3[c,d,m]	Kramer and Graham, 1976
	2–22[c,d]	Graham and Baird, 1982
Loricaria uracantha	3	Gee, 1976
	3–6[c,d,m]	Kramer and Graham, 1976
Sturisoma citurense	22	Gee, 1976
Chaetostoma fishcheri	24	Gee, 1976
Pterygoplichthys multiradiatus	6.5 (25–27 C)	Val *et al.*, 1990
Pygidium striatum	4–15[c,d,m]	Kramer and Graham, 1976
Gymnotus carapo	< 1–90[s]	Liem *et al.*, 1984
Channa micropeltes	93 vs 133[c,j]	Wolf and Kramer, 1987
C. maculata	4 vs 9[c] (15 C)	Yu and Woo, 1985
	9 vs 35[c] (25 C)	
Colisa lalia	< 1 vs 40 [c,e]	Wolf and Kramer, 1987
C. chuna	17 vs 30[k]	Bevan and Kramer, 1986
C. fasciatus	10 vs 140[e]	Mustafa and Mubarak, 1980
Trichogaster leeri	14–23[c,d]	Kramer and Graham, 1976

[a]A dash between numbers signifies a range of values whereas "vs" contrasts control values with the condition specified in the footnote.
[b]Diurnal effect.
[c]Hypoxia effect.
[d]Grouped behavior/synchrony effect.
[e]Also affected by predator (fish/bird) presence.
[f]Negative size effect.
[g]Water-borne toxin effect; same in both normoxia and hypoxia.
[h]1–7 day starvation effect.
[i]8–14 day starvation effect.
[j]Feeding on *Colisa lalia*.
[k]Negative depth effect.
[l]Slight CO_2 effect.
[m]Synchronous air breathing not observed.
[n]Undisturbed fish, normoxia versus hypoxia.
[o]Predator simulation, normoxia versus hypoxia.
[p]Normal diet.
[q]Slight CO_2 effect.
[r]Size range 45–120 g; possibly *P. dolloi*.
[s]Light, O_2, and thermal effects.
[t]Temperature effect.
[u]Respective frequency differences breathing pure N_2, air, and pure O_2.

1971) convincingly detail the combined and separate effects of O_2 and temperature on breath duration. Fish in cool water have lower air-breathing frequencies; some may hold a breath for as long as 3–4 hours. Other air–breath duration factors are light level, time of day, and the depth (surface proximity) of a fish (Tables 5.1 and 5.2). The presence of water-borne toxins also markedly increased the air-breathing frequency of *Clarias* (Kulakkattolickal and Kramer, 1988). Although implicated, the effects of aquatic CO_2 on air breathing rate are not well documented (Dejours, 1981; Graham and Baird, 1982; Chapter 6).

D. L. Kramer and his colleagues at McGill University have made fundamental contributions to our understanding of the biological factors influencing fish air–breath frequency and duration. Working with *Clarias*, *Lepisosteus*, *Colisa*, *Corydoras*, *Channa*, and other genera, Kramer's group has chronicled the influences of distance to the water surface (in most cases frequency decreases with depth), competition for food, diet, and satiation (Tables 5.1 and 5.2). The behavioral factors that Kramer and co-workers found to affect air-breathing frequency include activity level, group-induced synchronous (i.e., socially facilitated) air breathing, and the presence of a potential predator or predator-mimicking disturbances. In addition, experiments with *Ancistrus* and *Hypostomus* have shown that brief acclimation to hypoxia and hypercapnia will reduce air-breathing frequency (Graham and Baird, 1982; Graham, 1983).

Regarding body size, studies with *Clarias* (Jordan, 1976; Bevan and Kramer, 1987a) and other genera show that smaller fish tend to air breathe more frequently than do larger ones. This is opposite to the body size effect on respiratory partitioning (see following, larger fish usually obtain more O_2 aerially) and implies that smaller fish have different air convection requirements or respiratory efficiencies (Jordan, 1976; Burggren, 1979; Liem *et al.*, 1984).

Amphibious Air Breathing

The categories of amphibious air breathing are detailed in Table 1.2. The natural air exposure times endured by these species range from a few seconds to a large part of the dry season. Unfortunately, there are few data describing air-breathing time budgets. Even among the active amphibious gobies, blennies, and clinids, we have little information about the duration of aerial exposures or the percentage of time spent out of water. Maximum daylight exposure times of 10 seconds and 5 minutes were observed for *Entomacrodus* and *Mnierpes* (Graham, 1973; Graham *et al.*, 1985), with lengths increasing at night. Reports (Chapter 2) for *Coryphoblennius*, *Alticus*, and *Andamia*, imply that air exposures seldom exceed a few minutes, although *Coryphoblennius* sleeps above the water line at night (Louisy, 1987). Exposure times for some mudskippers seem generally longer; few activity data exist, and in recent experiments measuring their bimodal $\dot{V}O_2$ (following), the percentage of time fish spent in either air or water was not recorded. In laboratory settings mimicking natural field situations, Todd (1968) observed emersions lasting 100 minutes for *Gillichthys*. Shuttle–box records indicated that small *Erpetoichthys* would crawl from an aquarium to a terrarium as frequently as 17 times/day. These excursions lasted from less than 1 to up to 74 minutes (mean 2.3 minutes, Sacca and Burggren, 1982).

RESPIRATION

Although amphibious and aquatic air-breathing cycles have similarities, different factors affect them. Amphibious air breathers make sequential use of one and then the other respiratory mode. In air, there is unlimited access to O_2; however, time in that media is limited by either or both the capacity to release CO_2 aerially and the risk of desiccation. By contrast, aquatic air breathers must simultaneously partition O_2 uptake and CO_2 release between air and water, and while dehydration is not a factor, ABO–O_2 supply does limit breath duration.

In both types of air breathing, more than one respiratory surface (e.g., the ABO, gills, and skin) actively participates in the two modes, and the rates of O_2 and CO_2 exchange at these sites may differ. Factors affecting the latter include the relative phase (i.e., early or late) of the respiratory cycle, ambient conditions, and body size.

Respirometry

Basic respiratory methods developed for fishes and small air-breathing animals have been applied to air-breathing fishes. Cech (1990) has reviewed the methods and protocols of fish respirometry and illustrates apparati employed in air-breathing. Most respirometry studies use estimates of $\dot{V}O_2$ as an index of fish metabolism.

Monitoring aquatic O_2 is done by Winkler titration or by a polarographic electrode mounted in either open (flow-through) or closed respirometers. Aquatic CO_2 is measured by volumetric titration, by conductometric analysis, or by measurement of PCO_2 and pH (Cech, 1990; Steeger and Bridges, 1995). Monitoring gas-phase O_2 and CO_2 changes may be done in volumetric or manometric chambers. O_2 and CO_2 electrodes have also been employed and the Scholander 0.5 cc gas analyzer has been used to measure respiratory gas levels in samples extracted from a sealed chamber. Electrochemical and paramagnetic methods have been used to sample the flow stream through an open respirometer (Ar and Zacks, 1989; Martin and Lighton, 1989; Steeger and Bridges, 1995).

Experiments have been done with both unimodal and bimodal respirometers (Cech, 1990; Edwards and Cech, 1990). Unimodal systems measure gas exchange in only one phase. Bimodal systems can monitor simultaneous respiration in both phases. The different O_2 capacitances and densities of water and air (Chapters 1 and 6) pose limitations on unimodal chamber design. Because of water's low O_2 content, chambers used in aquatic tests need to have a larger volume (i.e., so that O_2 is not depleted too fast) and must be gently mixed. Conversely, aerial chambers need to be smaller so that a relatively small rate of O_2 decline can be detected. As a result, differences in handling and chamber size make it hard to compare the separate aerial and aquatic determinations on the same fish. One alternative is to use the same chamber and decrease its volume (by use of a plunger or valved shunts) during aerial measurements. In this case, fish handling is reduced and changes from water to air can be done by flooding and draining. The latter, however, remains this system's major limit in that altering water level usually disturbs the fish, and it is hard to remove all water. Moreover, even with this technique, it is impossible to quantify rapid respiratory responses to immersion and emersion. Flow-through systems are potentially applicable to the latter question; however, these are complex and need precision flow and respiratory gas-measuring equipment, as well as computer-regulated flow and data acquisition systems (Martin and Lighton, 1989).

Bimodal respirometry is even more challenging because simultaneous changes occur in two media, and system design must take into account respiratory gas diffusion between them. This variable is intrinsic to bimodal systems and needs to be calibrated (e.g., by altering gas–phase contents, using a non-air-breathing fish, etc.) over the range of air and water gas tensions at which the system operates. In addition, and depending upon how aerial determinations are made, the high humidity in the air phase (owing to the presence of a water phase below) can affect measurement accuracy.

Air-Breathing Fish Metabolic Rate

Using bimodal respirometry, it is possible to determine the standard $\dot{V}O_2$ of a fish (i.e., its $\dot{V}O_2$ under typical habitat conditions of temperature, water O_2, and with air access) and then compare this rate to its $\dot{V}O_2$ when access to either air or water is denied. This "respiratory limitation approach" enables estimation of the maximum capacity of a particular system, and indicates how bimodal respiration has affected the efficacy of its two component parts (Graham *et al.*, 1978b; Graham, 1983).

Effects of Air-Exposure on $\dot{V}O_2$

Air-exposure effects on $\dot{V}O_2$ are shown in Table 5.3. An earlier review (Graham, 1976a) indicated that naturally emergent species, particularly the active amphibious forms, can generally sustain aerial $\dot{V}O_2$ at or near the same level (i.e., within 10–15%) or higher than in water. However, interspecific variability in the $\dot{V}O_2$ estimates is large and, while much of the newer work supports this generality (Bridges, 1988; Chan, 1990; Clayton, 1993), some studies do not, and new facets of this question have arisen.

Among the usually aquatic air breathers, *M. cuchia*, which is frequently amphibious, has an aerial $\dot{V}O_2$ nearly double its aquatic rate. *Channa argus*, which is rarely amphibious, also has an aerial $\dot{V}O_2$ much higher than its aquatic rate. In addition, *Trichogaster*, a non-amphibious anabantoid, has the same $\dot{V}O_2$ in air as in water, which is surprising in view of its quadruphasic respiratory mode (presumed to be less efficient in air, Chapter 3) and because *Anabas*, a triphasic breather with a demonstrated amphibious behavior, has a reduced $\dot{V}O_2$ in air (Table 5.3).

Among the amphibious air breathers, most of which occur in the intertidal, the pattern I described in 1976 has been generally found (Bridges, 1988; Chan, 1990). Much of the newer data, however, are for species that would be expected to remain quiescent during air exposure (*Clinocottus*, *Helcogramma*, *Lipophrys*, *Xiphister*, *Cebidichthys*, *Anoplarchus*, Table 5.3), and therefore might normally have a reduced metabolism in air. Some results are highly variable, and a near equality of aerial and aquatic $\dot{V}O_2$ was not always found. For example, the finding of Daxboeck and Heming (1982) that *Xiphister* had an aerial $\dot{V}O_2$ nearly twice its $\dot{V}O_2$ in water is probably an artifact of having made aerial measures for only one hour. For *Alticus kirki*, Brown *et al.* (1992) measured an aerial $\dot{V}O_2$ that was about 50% of the aquatic rate. The aerial rate estimates obtained by these workers agree with other data for this species (Martin and Lighton, 1989), however, neither the body size differences among the fish tested in air and water nor the small differences in test temperature are sufficient to account for the more than doubled aquatic $\dot{V}O_2$ observed by Brown *et al.* (1992).

Mudskipper Aerial $\dot{V}O_2$

Of the 14 air–water comparative data sets now available for mudskippers, seven indicate a higher $\dot{V}O_2$ for fish in water, six for fish in air, and one shows the same rate in both media (Table 5.3). However, there are also many discrepancies among the data amassed for the different species as well as for the same species by different workers. Whereas Gordon *et al.* (1978) measured a 25% higher $\dot{V}O_2$ for small *Periophthalmus cantonensis* in air, Tamura *et al.* (1976), working with larger specimens of the same species, measured a lower rate in air than in water (Table 5.3). Lee *et al.* (1987) similarly measured a lower rate for *P. chrysospilos* in air than in water. Both Tamura *et al.* (1976) and Natarajan and Rajulu (1983) measured the $\dot{V}O_2$ of mudskippers that were forcibly emerged, forcibly submerged, and respiring bimodally. As seen in Table 5.3, the aerial $\dot{V}O_2$ of *Boleophthalmus chinensis*, *Periophthalmus koelreuteri*, and *P. cantonensis* (see Table 2.4 for the revised mudskipper nomenclature) is much less than their rates in water. This surprising result conflicts with the earlier reports for mudskippers. Unfortunately, neither team of investigators discussed their findings in a way that bridges the discrepancy with previous data for mudskippers, and their methods do not suggest any obvious factors contributing to different results. A study by Chan (1990) demonstrated higher rates for *Scartelaos viridis*, *Boleophthalmus pectinirostris*, and *Periophthalmus cantonensis* in air, however, Steeger and Bridges (1995) reported the opposite result for *P. barbarus*. Results are also mixed for the scaling of mudskipper metabolism in air and water. Among the five studies reporting a power equation for $\dot{V}O_2$ and body mass in air and water (reviewed by Chan, 1990), three show a lower mass exponent in air (*Periophthalmus cantonensis*, *B. pectinirostris*, *Scartelaos viridis*), one indicates the exponent in air is greater than water (*Periophthalmodon schlosseri*), and one result finds no difference for the exponent in either water or air (*P. vulgaris*).

Another surprising element in these studies is that

TABLE 5.3 Oxygen Consumption Rates of Aquatic and Amphibious Air Breathers during Air Exposure Compared to Their Control $\dot{V}O_2$[a]

Species	Mass (g)	°C	Total $\dot{V}O_2$ (water control)	(ml/kg h) (in air)	Reference
Aquatic Air Breathers					
Lepidosiren paradoxa[c]	100–200	20	54	12–39[m]	Johansen and Lenfant, 1967
Protopterus aethiopicus[d]	500+	20	13.5	13.8	Lenfant and Johansen, 1968
Erpetoichthys calabaricus	20–29	27	117	286	Sacca and Burggren, 1982
Heteropneustes fossilis	46–69	25	84.5	54.5	Hughes and Singh, 1971
Clarias batrachus[e]	76–157	25	93.4	71.7	Singh and Hughes, 1971
	10–300	25	88	90	Jordan, 1976
Misgurnus fossilis	10–31	10	18.8	14.2	Jeuken, 1957
	—	20	75.5	60.9	
Gillichthys mirabilis	8–10	20–21	104	104	Todd and Ebeling, 1966
Channa argus	400–800	25	45.7	60.0[i]	Itazawa and Ishimatsu, 1981
Anabas testudineus[e]	29–40	25	119.7	105	Hughes and Singh, 1970b
	10–25	29	215	164	Natarajan, 1978
Trichogaster trichopterus	8.1	27.5	177	320	Burggren and Haswell, 1979
Monopterus cuchia[e]	240–450	30	33.3	44.6	Lomholt and Johansen, 1976
Synbranchus marmoratus	70–200	25	39.8	37.4	Bicudo and Johansen, 1979
Amphibious Air Breathers					
Anguilla vulgaris	200–400	7	12.4	7.34	Berg and Steen, 1965
	—	15	26.6	11.54	
Neochanna burrowsius	4–10[l]	17	36.6	57.4	Meredith *et al.*, 1982
Sicyases sanguineus	1–5	14–15	60	85(140)[j]	Gordon *et al.*, 1970
	30–80	14–15	40	32(18)[k]	
Tomicodon humeralis	1–5	20	107	104	Eger, 1971
Clinocottus recalvus	6–14	15.5	100	93	Wright and Raymond, 1978
C. analis	11–12	15	85	65	Martin, 1991
C globiceps	8	13–14	57.7	56.9	Yoshiyama and Cech, 1994
Oligocottus snyderi	8	13–14	44.9	30.0	Yoshiyama and Cech, 1994
Oligocottus maculosus	7	13–14	50.3	32.0	Yoshiyama and Cech, 1994
Ascelichthys rhodorus	7	13–14	51.1	28.8	Yoshiyama and Cech, 1994
Xiphister atropurpureus	5–19	11-13	30.2	53.8	Daxboeck and Heming, 1982
Cebidichthys violaceus	0.5–7	15	117	89	Edwards and Cech, 1990
Anoplarchus purpurescens	6	13–14	30.9	40.2	Yoshiyama and Cech, 1994
Helcogramma medium	1–15	15	106	70	Innes and Wells, 1985
Mnierpes macrocephalus	1–15	30	310	310	Graham, 1973
Alticus kirki	< 1	24	360	—	Brown *et al.*, 1992
	1–2	26	—	160	
Blennius (Lipophrys) pholis	13–35	12.5	38	39	Pelster *et al.*, 1988
	32(10)[n]	15	62	45	Steeger and Bridges, 1995
Periophthalmus cantonensis[b]	4–8	20	196[g]	119	Tamura *et al.*, 1976
	0.5–1.5	20	85	106	Gordon *et al.*, 1978
	0.7–4.0	30	126(5)	156(6)[n]	Chan, 1990
P. sobrinus	15	23–30	84	94	Gordon *et al.*, 1969
P. koelreuteri	20–35	29	88[h]	48	Natarajan and Rajulu, 1983
P. chrysospilos	6–12	25	306	378	Lee *et al.*, 1987
P. barbarus	17(1)[n]	28	90	76	Steeger and Bridges, 1995
P. vulgaris	2–9	20	65	63	Milward, 1974
P. expeditionium	5–7	25	103	90	Milward, 1974
Periophthalmodon schlosseri	12–51	20	53	48	Milward, 1974
Boleophthalmus chinensis	37–47	20	72[f]	47	Tamura *et al.*, 1976
	6–10	20	57	50	Milward, 1974
	1–9	30	86(4)	118(4)[n]	Chan, 1990
Scartelaos histophorus	1–5	30	84(2)	115(4)[n]	Chan, 1990

[a]Either exclusively aquatic or bimodal $\dot{V}O_2$.
[b]See Tables 2.4 and 2.5 for the updated mudskipper classification.
[c]Can also occur in humid burrows.
[d]Estivates seasonally.
[e]Also amphibious at certain times.
[f]Bimodal rate even higher than aquatic rate (110).
[g]Bimodal rate even higher than aquatic rate (236).
[h]Bimodal rate even higher than aquatic rate (127).
[i]Two values not significantly different.
[j]1–3 h in air (11–13 h).
[k]0–12 h in air (15–23 h).
[l]Also gulps air.
[m]Rate drops with time in air.
[n]Parenthetical values are SE.

mudskipper bimodal $\dot{V}O_2$ (see footnotes in Table 5.3) exceeds separate rates in either air or water. Biswas *et al.* (1979) similarly measured a much higher bimodal–$\dot{V}O_2$ than aquatic-only $\dot{V}O_2$ for *B. boddarti*. This suggests that bimodal access, the normal condition for mudskippers and many amphibious fishes, may trigger an elevated metabolic activity and that there may be measurable costs associated with climbing in and out of water and remaining poised on the land. Such a response and the metabolic costs of bimodal behavior would not be detected by unimodal respirometry. Regrettably, the bimodal respirometers used by these workers did not permit full aerial emersion of the fish and the $\dot{V}O_2$ data that were presented were not accompanied by data indicating how long fish spent in air and water. Observations of time spent in air and water and the degree of exposure in each medium, together with bimodal respirometry, would enable progress toward understanding these presently anomalous, but potentially important, findings.

Effects of Forced Submersion on $\dot{V}O_2$

As reviewed in Chapters 1 and 2, many of the early practitioners of forced submergence experiments (e.g., Ghosh, 1934; Das, 1934; Hora, 1935; and others) used this method to determine if a fish would drown without air access, thus indicating whether or not a particular species was an obligatory air breather. However, many of these experiments were not adequately controlled and the forcibly submerged fish often suffocated as a result of localized aquatic deoxygenation rather than a lack of air access.

Later experiments with forcibly submerged, aquatic air-breathing fishes ask the question: To what extent can aquatic respiratory surfaces be used to sustain $\dot{V}O_2$ at levels near that of bimodal respiration? This emphasis on the scope for increasing aquatic $\dot{V}O_2$ focuses attention on the extent that air breathing has relaxed selection pressures for a robust branchial respiratory capacity.

Table 5.4 shows that in all cases but one (*Channa gachua*), the $\dot{V}O_2$ of forcibly submerged fish is less than their total (bimodal) $\dot{V}O_2$. The table shows good agreement between independently obtained values for both *Clarias* and *Monopterus*, as well as some variability for *Anabas*. The overall mean relative $\dot{V}O_2$ is 52%; the lowest value is for *Protopterus aethiopicus* (17%), and the highest value (discounting *C. gaucha*) is above 80%.

The scope of a species for elevating aquatic $\dot{V}O_2$ can be further gauged by comparing its $\dot{V}O_2$ during forced

TABLE 5.4 The $\dot{V}O_2$ of Air-Breathing Fishes during Forced Submergence Compared to Their "Standard" Total $\dot{V}O_2$[a] and the Normal Aquatic Bimodal $\dot{V}O_2$[b]

Species	°C	Forced submergence $\dot{V}O_2$ as percentage of total $\dot{V}O_2$[a]	Ratio of forced submergence $\dot{V}O_2$: aquatic $\dot{V}O_2$ with air access	Reference
Protopterus aethiopicus	24	17	2.04	McMahon, 1970
Heteropneustes fossilis	25	79	1.33	Hughes and Singh, 1971
Clarias batrachus	25	80	1.94	Singh and Hughes, 1971
	25	84	—	Jordan, 1976
	25	—	1.55	Munshi *et al.*, 1976
Notopterus chitala	29	43	1.47	Ghosh *et al.*, 1986
Gymnotus carapo	29–33	69	2.75	Liem *et al.*, 1984
Umbra lacustris	16	15	1.62	Geyer and Mann, 1939b
Channa gachua	29	170	2.04	Ojha *et al.*, 1978
	29	32	1.23	Natarajan, 1979
C. punctata	21	56	1.23	Ghosh, 1984
	32	66	1.36	
C. marulius	30	50	2.25	Ojha *et al.*, 1979
Anabas testudineus	25	81	1.73	Hughes and Singh, 1970a
	29	18	1.00	Natarajan, 1978
	27.5	—	1.97	Munshi and Dube, 1973
Osphronemus olfax	28	34	1.10	Natarajan, 1980
Monopterus cuchia	30	49	1.46	Lomholt and Johansen, 1976
	25	63	2.44	Singh and Thakur, 1979
	Mean	52.3	1.73	
	SE	6.0	0.11	

[a]With air access.
[b]Fish body masses reported in other tables.

submergence with its aquatic $\dot{V}O_2$ during normal bimodal breathing. The mean ratio for increased aquatic O_2 uptake in these fishes (Table 5.4) is 1.73, and values range from a low of 1.10 (*O. olfax*, i.e., 10% above its bimodal aquatic rate) to 2.75 (*G. carapo*).

As reviewed by Ojha *et al.* (1978), several investigators have sought to establish a relationship between the allometric scaling of gill and ABO surface areas and the retardation of total $\dot{V}O_2$ during forced submergence. Munshi and Dube (1973) found that forced submergence significantly reduces the $\dot{V}O_2$ mass exponent (M^b, see Equation 3.1) of *Anabas* (from 0.67 with air access to 0.53 during submergence). This reflects the relatively small gill–area mass exponent of larger fish, as well as an increased dependence on aerial $\dot{V}O_2$ in larger specimens (Chapter 3).

Even though forced submergence is not a natural circumstance for these fishes, the two $\dot{V}O_2$ indices revealed by this experimental technique are a window on the effect that bimodal air-breathing evolution has exerted on aquatic respiratory capacity. If the lack of air access merely cancelled the metabolic costs associated with surfacing for air and bimodal behavior, then forcibly submerged fishes should have a reduced $\dot{V}O_2$. However, all of the fish tested elevated their aquatic $\dot{V}O_2$ during forced submergence, although with one exception, the increase did not match the bimodal rate. Thus, the ability of a bimodal fish to compensate for the lack of air access is dependent upon its capacity to utilize aquatic respiratory surfaces which are often diminished as a consequence of air-breathing reliance. For the obligatory air breathers (*Anabas*, *Osphronemus*, *Protopterus*) the consequence of prolonged forced submergence would be suffocation because aquatic ventilation cannot sustain aerobic maintenance metabolism. For other species, an increased $\dot{V}O_2$ and gill respiration factor during submergence may appear sufficient to sustain maintenance metabolism, although in such cases a disproportionate fraction of the $\dot{V}O_2$ increase may go into powering ventilation.

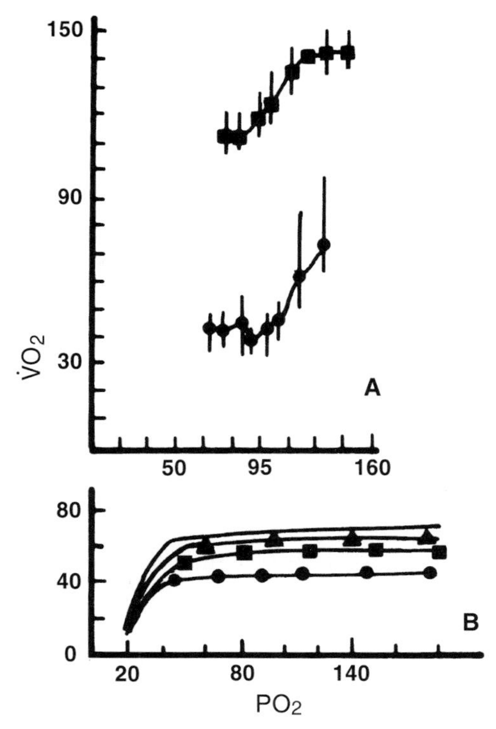

FIGURE 5.2 **A,** Respiratory dependence curves for *Anabas* (28–40 g, 25 C), in water with access to an N_2 gas phase (squares) and without access to air (dots). (From Hughes and Singh, 1970a, by permission of the Company of Biologists, Ltd., from the *Journal of Experimental Biology,* volume 53, pp. 265–280.) **B,** Respiratory dependence functions for *Clarias batrachus* (25 C), top line, 77–83 g; triangles, 172 g; squares, 195 g; dots, 206 g. PO_2 is in torr, $\dot{V}O_2$ is ml/kg h. (By permission from the publisher. The influence of body weight on gas exchange in the air-breathing fish, *Clarias batrachus.* Jordan. *Comparative Biochemistry and Physiology,* copyright 1976, by Elsevier Science, Inc.)

Responses to Hypoxia during Forced Submersion

In hypoxia, most fishes can sufficiently increase gill ventilation and O_2 extraction to regulate $\dot{V}O_2$ at near normal levels down to a critical O_2 tension, below which $\dot{V}O_2$ is reduced (Hughes, 1973b; Ultsch *et al.*, 1981). Fishes able to do this are termed O_2 regulators, while those that passively allow $\dot{V}O_2$ to drop with PO_2 are termed O_2 conformers.

Exposure of forcibly submerged, air-breathing fish to progressive hypoxia assesses their ventilatory scope and indicates how air breathing itself has altered behavioral and physiological responses to hypoxia. Figure 5.2 shows the $\dot{V}O_2$ of *Anabas* and *Clarias batrachus* in progressive hypoxia. Experiments with *Anabas* contrasted aquatic $\dot{V}O_2$ when air was present in the respirometer with the $\dot{V}O_2$ when N_2 gas had been substituted for air. Figure 5.2A shows that the aquatic $\dot{V}O_2$ of *Anabas* under N_2 was higher than under air. It is apparent that in progressive hypoxia, fish with air access let their aquatic $\dot{V}O_2$ drop fairly abruptly; whereas fish under N_2 sustained a higher $\dot{V}O_2$ until about 120 torr and then reduced it more gradually. In progressive hypoxia, both the depth and frequency of ventilation increased, although the scope of ventilatory frequency was small (50–70/min).

Experiments by Moussa (1957) confirm that *Clarias lazera* is an obligatory air breather and that larger fish are much more dependent upon air than are smaller ones. Jordan (1976) determined that *C. batrachus* is an O_2 regulator in hypoxia and not an obligatory air breather. $\dot{V}O_2$ dependence curves for a range of body sizes (Figure 5.2B) show that critical O_2 tension for this species is about 40 torr. (As discussed in Chapter 2, Jordan's findings for *C. batrachus* differ markedly from

Singh and Hughes [1971] who described this species as an O_2 conformer and obligatory air breather.)

Heteropneustes was also found to be an O_2 regulator (Hughes and Singh, 1971); its $\dot{V}O_2$, although variable, was largely unaffected by a decline in PO_2 from saturation to 40 torr. Both the ventilatory amplitude and frequency of this species increased steadily until about 30 torr and then declined. As ambient PO_2 decreased below 100 torr, some fish made repeated attempts to breathe air. The drop in $\dot{V}O_2$ below 40 torr is dramatic; however, fish were capable of surviving in this condition for several hours.

Neoceratodus forsteri was reported to be an O_2 conformer by Johansen *et al.* (1967). This has also been shown for *Synbranchus marmoratus* (Figure 5.3), also a non-obligatory air breather. Because of its modified branchial chamber, *S. marmoratus* has relatively large ventilatory costs and thus ventilates intermittently. This fish also does not elevate gill-ventilation frequency in hypoxia, although the percentage of the time gills were used did rise (Graham *et al.*, 1987). *Synbranchus* has a large blood hemoglobin (Hb) concentration which functions for O_2 storage (Chapter 7). In view of these features, O_2 conformity in *S. marmoratus* has been interpreted as an energy saving mechanism normally complemented by air breathing. (Findings of O_2 conformity and stable ventilation for *Synbranchus* are at variance with observations by Bicudo and Johansen [1979], who described this species as an O_2 regulator with a critical PO_2 between 30 and 50 torr.)

Ancistrus chagresi, a facultative air breather, is an O_2 regulator in hypoxia (Figure 5.4); its capacity for this is enhanced (i.e., a higher $\dot{V}O_2$ at a lower O_2 tension) after at least 10 days of exposure to hypoxia and acclimatory air breathing. (The underlying mechanism for this enhancement is a left-shift of the hemoglobin–O_2 dissociation curve [Graham, 1983; and Chapter 7]).

Two interspecific comparisons serve for final emphasis of the link between air breathing and a limited aquatic respiration capacity. First, comparison of progressive hypoxia effects on the non-air-breathing *Piabucina panamensis* and its air-breathing congener *P. festae* (Figure 5.5) shows that the air breather has a reduced ventilatory scope and cannot sustain its aquatic $\dot{V}O_2$ at lower O_2 tensions (Graham *et al.*, 1978b). Exactly the same pattern has been recently shown for other characins, the non-air-breathing *Hoplias malabaricus* and the air-breathing *Hoplerythrinus unitaeniatus* (Mattias *et al.*, 1996). In addition, a comparative study of *Lepisosteus osseus* and *Amia calva* (Crawford, 1971) shows specific differences in critical PO_2 and its relationship to the onset of air breathing (Figure 5.6). In normoxic water, these species have about the same aquatic $\dot{V}O_2$. In hypoxia, the $\dot{V}O_2$ of *Lepisosteus* begins to drop at a higher PO_2 and at a faster rate than does that of *Amia*. To compensate for this loss, *Lepisosteus* normally switches on air breathing at about 120 torr, and it becomes increasingly dependent upon air in progressive hypoxia. By contrast, the air-breathing threshold for *Amia* is about 60 torr (Crawford, 1971).

Respiratory Partitioning

O_2 Uptake

The partitioning of $\dot{V}O_2$ between air and water differs among species and is affected by biological and

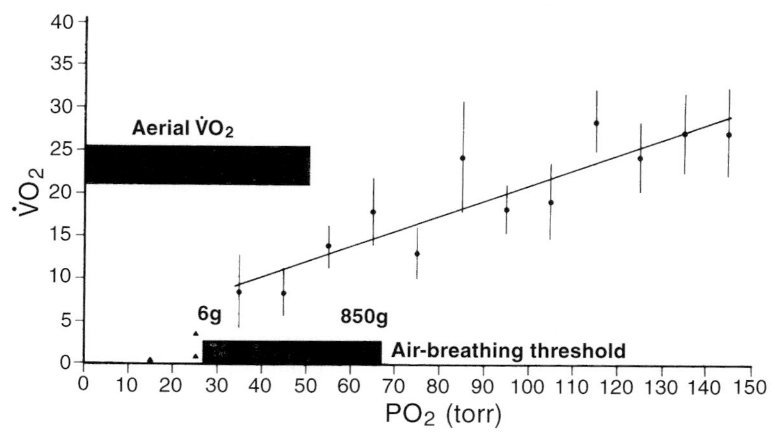

FIGURE 5.3 Aquatic respiratory dependence of *Synbranchus marmoratus* without air access (25 C). Regression equation: $\dot{V}O_2 = 3.022 + 0.179\ (PO_2)$. Triangles are data for one fish at $PO_2 < 30$ torr. Shaded areas show mean aerial $\dot{V}O_2$ and the air-breathing threshold (torr) for fish between 6 and 850 g. $\dot{V}O_2$ units are ml/kg h. (From Graham and Baird, 1984, by permission of the Company of Biologists, Ltd., from the *Journal of Experimental Biology*, volume 108, pp. 357–375.)

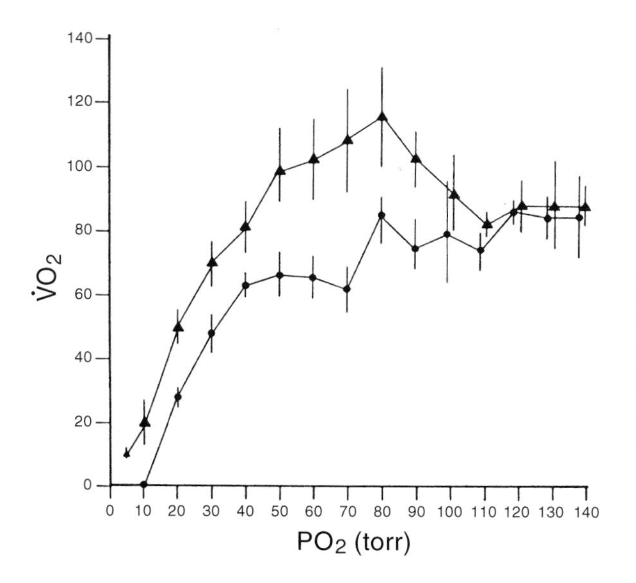

FIGURE 5.4 PO_2 effects on the mean relative $\dot{V}O_2$ of control (dots) and hypoxia acclimated (triangles) groups of *Ancistrus chagresi* (*spinosus*) without air access. Vertical lines are ± SE (25 C). (From Graham, 1983, by permission of the Company of Biologists, Ltd., from the *Journal of Experimental Biology*, volume 102, pp. 157–173.)

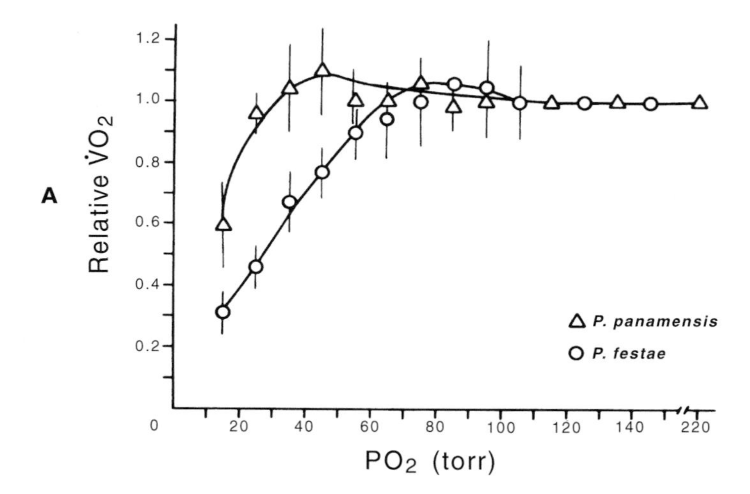

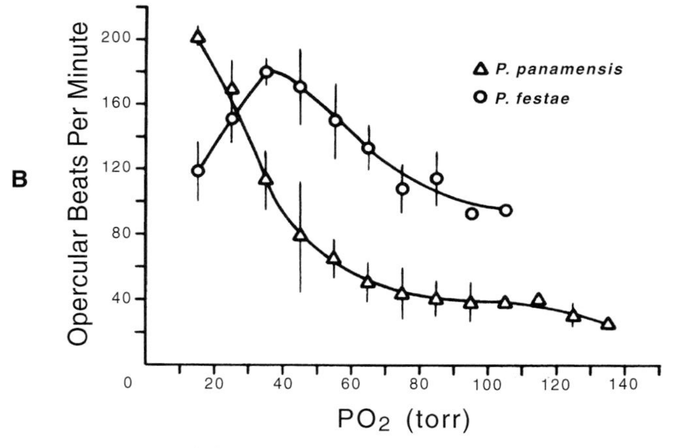

FIGURE 5.5 **A,** Relative respiratory dependencies of *P. panamensis*, a non-air breather, and *P. festae* an air breather. **B,** PO_2 effects on the opercular ventilation rate of *P. festae* and *P. panamensis* (25 C). (From Graham *et al.*, 1977, by permission of the University of Chicago Press.)

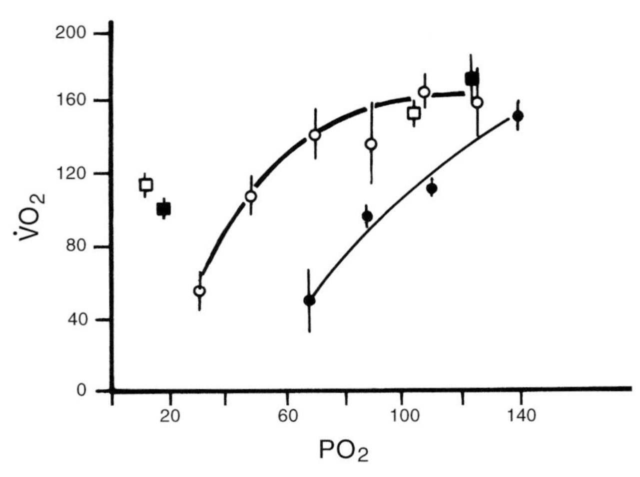

FIGURE 5.6 Comparative respiration rates of *Amia* (open symbols) and *Lepisosteus osseus* (closed symbols) determined by Crawford (1971) at 30 C. Circles show effects of PO_2 on aquatic respiration without air access. Squares show bimodal respiration rates at low and high PO_2s. PO_2 units are torr; $\dot{V}O_2$ units are ml/kg h.

physical factors (Table 5.5). The highest aerial use rates are among the obligatory lungfishes. Partitioning is not notably different in the obligatory air breathers *Channa argus* and *Clarias mossambicus* and their non-obligatory congeners *Channa maculata* and *Clarias lazera*. Also, in both *Clarias mossambicus* and *Ancistrus*, aerial O_2 use is reduced by hypoxia acclimation.

In many cases, a single value has been used to describe respiratory partitioning; however, a trend in several of the later studies was to specify a range of partitioning values arising as a result of variation in one or more factors (Table 5.5). Therefore, much of the later data suggest, as was shown for air–breath cycles (Tables 5.1 and 5.2), that the respiratory partitioning of a species is not accurately described by a single value. An objective for future work should be to determine if some of the species presently listed with a single partitioning percentage can in fact modulate this parameter.

Table 5.5 notes several instances where aquatic O_2

TABLE 5.5 Respiratory Partitioning of O_2 Uptake in Aquatic and Amphibious Air-Breathing Fishes

Species	Mass[a]	°C	% Total $\dot{V}O_2$ via air	Reference
Aquatic Species				
Lepidosiren paradoxa	450	21–29	98	Sawaya, 1946
	104–112	20	85	Johansen and Lenfant, 1967
	—	22	95	Lenfant *et al.*, 1970
Protopterus aethiopicus	500–6*	20	89	Lenfant and Johansen, 1968
	150–600	24	92	McMahon, 1970
P. annectens	80–1*	28–32	18–85[b,c]	Babiker, 1979
Polypterus senegalus	12–460	28	6–100[b,c]	Babiker, 1984a
Lepisosteus osseus	300–900	22	75	Rahn *et al.*, 1971
L. oculatus	600–1.4*	20–30	42–100[b,d]	Smatresk and Cameron, 1982a,b
Amia calva	800–1.7*	10–30	0–100[d]	Johansen *et al.*, 1970a
	700	30	40[j]	Randall *et al.*, 1981b
Arapaima gigas	(2–3)*	27–29	50–100[b]	Stevens and Holeton, 1978a
Notopterus chitala	35–105	29	70	Ghosh *et al.*, 1986
Heteropneustes fossilis	60	25	41	Hughes and Singh, 1971
Clarias batrachus	99	25	58	Singh and Hughes, 1971
	140–210	25	2–99[b,c]	Jordan, 1976
C. mossambicus	50	20	25–78[b]	Johnston *et al.*, 1983
	—	—	59[b,f]	
C. lazera	1.7*	28.5	2–64[b]	Ar and Zacks, 1989
	40–6*	28–32	5–40[b,c]	Babiker, 1979, 1984a
Misgurnus fossilis	30–40	5–20	0–95[b,d]	Jeuken, 1957
M. anguillicaudatus	2	10–30	20[d]	McMahon and Burggren, 1987
Electrophorus electricus	2.8*[e]	26	78	Farber and Rahn, 1970
Hoplerythrinus unitaeniatus	99–500	27–30	20–100[b]	Stevens and Holeton, 1978b
Erythrinus erythrinus	243[e]	27–30	44[b]	Stevens and Holeton, 1978b
Piabucina festae	17–80	25	10–75[b]	Graham *et al.*, 1977
Hoplosternum thoracatum	5–10	24	42–72[b]	Gee and Graham, 1978
Ancistrus chagresi (*spinosus*)	2–200	25	37–100[b,f]	Graham, 1983
Gymnotus carapo	12–29	29–33	36–74[b]	Liem *et al.*, 1984
Umbra lacustris	0.1–10	14–22	25–92[b,c,g]	Geyer and Mann, 1939a,b
Channa punctata	—	31.5	52	Ghosh, 1984
	—	21	55	

(continues)

TABLE 5.5 *(continued)*

Species	Mass[a]	°C	% Total $\dot{V}O_2$ via air	Reference
C. maculata	150–200	15	31–90[b]	Yu and Woo, 1985
	—	25	60–90[b]	
C. marulius	13–18	30	83	Ojha *et al.*, 1979
	93	30	78[g]	
C. gachua	3–74	29	55	Ojha *et al.*, 1978
	12–29	28	71	Rama Samy and Reddy, 1978
	12–15	29	79	Natarajan, 1979
C. argus	400–800	25	61–84[b]	Itazawa and Ishimatsu, 1981
Anabas testudineus	28–51	25	55	Hughes and Singh, 1970a
	10–25	29	82	Natarajan, 1978
Trichogaster trichopterus	8	27	40	Burggren, 1979
T. pectoralis	10–15	29	64	Natarajan and Rajulu, 1982
Osphronemus olfax	10–15	28	69	Natarajan, 1980
Monopterus cuchia	2–200	25	40–80[c]	Singh and Thakur, 1979
	250–450	20	65	Lomholt and Johansen, 1976
	250–450	30	66	
M. albus	34–50	23	75	Liem, 1967c
Synbranchus marmoratus	—	22	89	Lenfant and Johansen, 1972
	6–850	25	35–80[b,c]	Graham and Baird, 1984
Amphibious Species[h,i]				
Periophthalmus koelreuteri	20–25	29	41	Natarajan and Rajulu, 1983
P. cantonensis	4–8	20	34	Tamura *et al.*, 1976
Boleophthalmus boddarti	3–13	26	42	Biswas *et al.*, 1979
B. chinensis	33–47	20	30	Tamura *et al.*, 1976

[a]Mean or range, units are g or kg(*).
[b]PwO_2 affects partitioning.
[c]Body size affects partitioning.
[d]Temperature affects partitioning.
[e]$n = 1$.
[f]Air fraction reduced by hypoxia acclimation.
[g]Circadian effects on partitioning.
[h]Tables 2.3 and 2.4 contain the updated mudskipper names.
[i]Respirometer design permitted fish to move freely between land and water.
[j]Values in normoxic water are much less than measured by Johansen *et al.* (1970a).

is a determining factor in respiratory partitioning, and Figure 5.7 documents this more fully for eight species. Even the obligatory air-breathing *Protopterus annectens* varies its aerial O_2 uptake from a low of 85% of total in normoxic water to 100% in hypoxia. A more dramatic change occurs in *Polypterus senegalus* which uses air breathing for only 5% of total $\dot{V}O_2$ in normoxic water, but for a reported 100% at O_2 tensions below 60 torr. *Clarias lazera* also uses very little aerial O_2 in normoxic water. An extreme case is seen for *Ancistrus* which does not commence facultative air breathing until PO_2 drops to about 35 torr. Below this tension, aerial O_2 utilization increases steeply, although air-breathing dependence is reduced by hypoxia acclimation (Figure 5.7).

Temperature affects respiratory partitioning. Johansen *et al.* (1970a) describe the additive effects of hypoxia and temperature on partitioning in *Amia* (Figure 5.8). Depending upon ambient conditions, *Amia* may not need air at all, or be totally dependent on it.

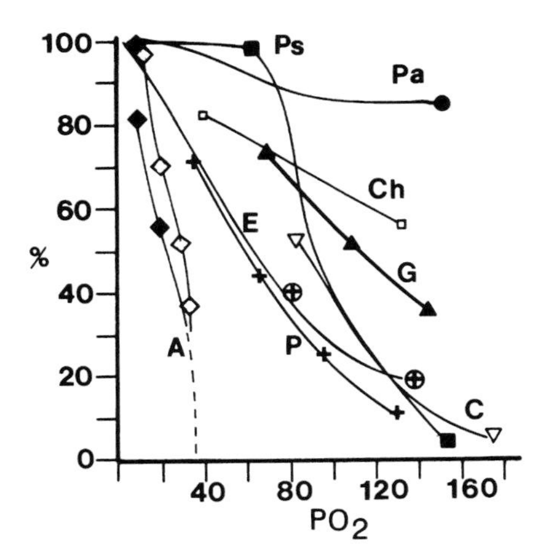

FIGURE 5.7 Aquatic PO_2 effects on the percentage of total $\dot{V}O_2$ taken via aerial respiration in nine genera. Abbreviations: Ps, *Polypterus senegalus*; Pa, *Protopterus annectens*; Ch, *Channa*; G, *Gymnotus*; E, *Erythrinus* and *Hoplerythrinus*; C, *Clarias lazera*; P, *Piabucina*; A, *Ancistrus*; hypoxia acclimated (solid symbols); control (open symbols). See text for references. PO_2 units are torr.

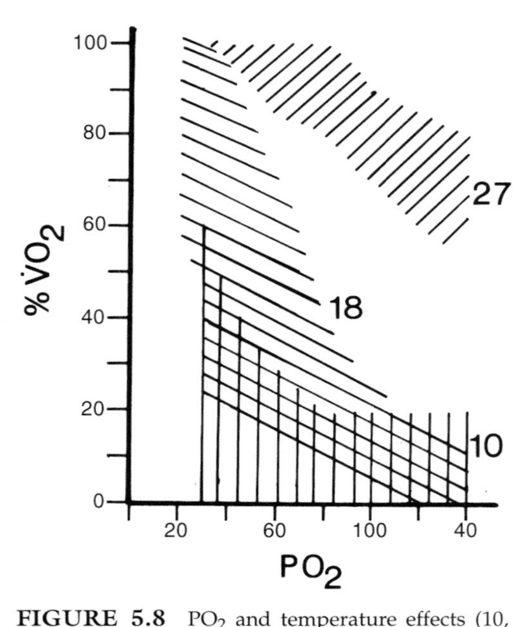

FIGURE 5.8 PO_2 and temperature effects (10, 18, 27 C) on the percentage of total $\dot{V}O_2$ obtained via air breathing in *Amia*. PO_2 units are torr. (Data modified from Johansen *et al.*, 1970a).

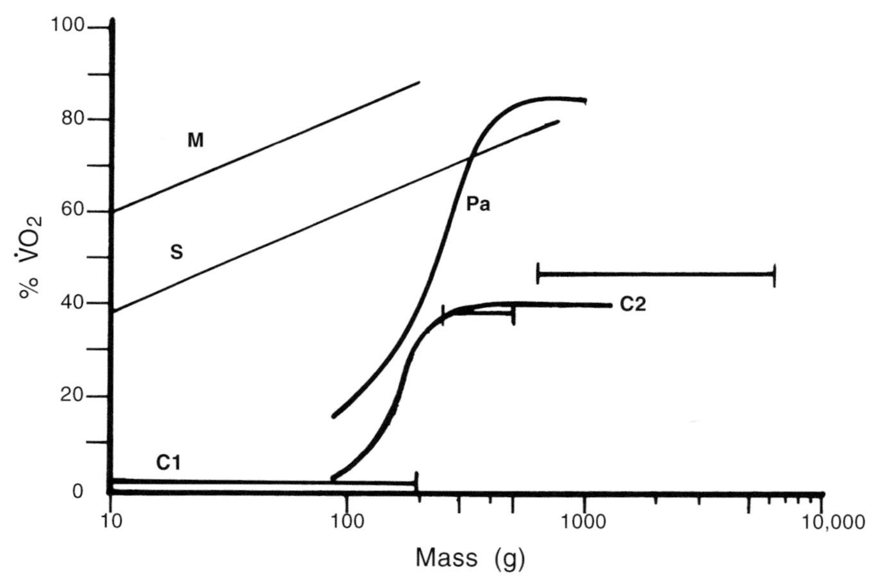

FIGURE 5.9 Body mass effects on the percentage of total $\dot{V}O_2$ via air breathing in four genera. Abbreviations: M, *Monopterus* (Singh and Thakur, 1979); S, *Synbranchus* (Graham and Baird, 1984); Pa, *Protopterus annectens* (Babiker, 1979); C1 and C2, *Clarias lazera* (Babiker, 1979).

Estimates of a species' respiratory partitioning vary among different workers. Johansen *et al.* (1970a) reported that at 30 C *Amia* obtained 80% of its $\dot{V}O_2$ aerially and had air-breathing frequencies near 60. Crawford (1971) and Randall *et al.* (1981b) measured a much lower level of aerial respiration for this species at 30 C and much longer breath durations (Table 5.1). Such differences may be due to experimental techniques and variables that affect behavior. They might also reflect seasonal and population differences, or variables such as body size. Johansen's group worked with 800–1700 g *Amia*, while smaller fish were studied by both Crawford (60–180 g) and Randall *et al.* (700 g).

Body mass effects on respiratory partitioning are indicated for several species in Table 5.5 and illustrated for four species in Figure 5.9. In both *Monopterus* and *Synbranchus*, dependence increases linearly with mass; the greater percentage utilization of aerial O_2 by all sizes of *Monopterus* presumably reflects its obligatory air-breathing requirement. *Clarias lazera*, an obligatory air breather, undergoes a dramatic increase in air-breathing dependence above 200 g (Moussa, 1957; Magid and Babiker, 1975; Babiker, 1979). This also occurs in *Protopterus annectens* and closely parallels the stages of increased aerial dependence with size described for *P. aethiopicus* (Table 5.6) by McMahon (1970). There is a strong reduction in aquatic $\dot{V}O_2$ in larger *Polypterus senegalus* (Magid and Babiker, 1975).

CO_2 Release

Measurements of aerial-CO_2 release during air breathing are presented (Tables 5.7 and 5.8) as respiratory exchange ratios ($R = \dot{V}CO_2/\dot{V}O_2$). This R value, termed total or metabolic R, is the functional equivalent of the respiratory quotient (RQ) and is not to be confused with R estimates for a single respiratory

TABLE 5.6 Ontogenetic Changes in the Bimodal Respiratory Capacity of *Protopterus aethiopicus*[a]

Stage	Mass (g)	Maximum viable submergence time	Respiratory organ development
Larval	0.1	At least 15 days	External gill, mouth open, lungs?
Young Juvenile	1.7	At least 6 days	External gills reduced, internal gills functional
Juvenile	50	One to 4 days	Gills and lung functional
Adults	500+	Up to 1 day	Gills and lung functional

[a]Adapted from McMahon, 1970.

TABLE 5.7 Aerial Respiratory Exchange Ratios (R) Determined for Aquatic Air-Breathing Fishes[a,b]

Species	R	Reference
Protopterus aethiopicus	0.27	McMahon, 1970
P. amphibius	0.87[p]	Lomholt, 1993
Lepidosiren paradoxa	0.4	Lenfant *et al.*, 1970
Neoceratodus forsteri	0,0.2[k]	Lenfant *et al.*, 1966
Polypterus senegalus	0.08[c]	Magid *et al.*, 1970
	0.12[d]	
Lepisosteus osseus	0.09	Rahn *et al.*, 1971
L. oculatus	0.10	Smatresk and Cameron, 1982a,b
Amia calva	0.4[e]	Johansen *et al.*, 1970a
	0.21	Randall *et al.*, 1981b
Arapaima gigas	0.38	Randall *et al.*, 1978a
Misgurnus anguillicaudatus	0.08[e]	McMahon and Burggren, 1987
Electrophorus electricus	0.25	Farber and Rahn, 1970
	0.2	Johansen *et al.*, 1968b
Hoplerythrinus unitaeniatus	0.2	Randall *et al.*, 1978b
	0.4[f]	
Ancistrus chagresi	0.01[g]	Graham, 1983
Clarias lazera	0.05	Ar and Zacks, 1989
C. batrachus	0.11	Singh and Hughes, 1971
	0.51[i]	
	0.52[f]	
Heteropneustes fossilis	0.17	Hughes and Singh, 1971
	0.38[i]	
	0.58[f]	
Gillichthys mirabilis	0.28	Todd and Ebeling, 1966
Neochanna burrowsius	0.32[o]	Meredith *et al.*, 1982
Channa argus	0.17–0.18[l]	Itazawa and Ishimatsu, 1981
	0.75[f]	
	0.16[n]	Glass *et al.*, 1986
	0.29[g]	
C. maculata	0.20[j]	Yu and Woo, 1985
Anabas testudineus	0.2	Hughes and Singh, 1970a
	0.15[l]	
	0.71[f]	
Trichogaster trichopterus	0.25	Burggren, 1979
	0.17[l]	
	0.80[f]	Burggren and Haswell, 1979
Monopterus cuchia	0.2	Lomholt and Johansen, 1976
Synbranchus marmoratus	0.1	Bicudo and Johansen, 1979
	0.02–0.05[m]	Graham and Baird, 1984

[a]Unless otherwise noted, indicated R values were obtained in near normoxic water.
[b]Other experimental details as in Table 5.1 or footnoted.
[c]Left branch of lung.
[d]Right branch.
[e]Slight thermal effect.
[f]Value for fish in air.
[g]In hypoxic water, 25 C.
[h]Some values in hypoxic water were this high.
[i]In deoxygenated water with a hypercarbic, hypoxic air phase.
[j]No difference at 15 and 25 C, but significantly more aerial CO_2 release in hypoxia at 25 C.
[k]No air breathing in normoxic water, 0.2 after 25 min in air.
[l]Slight rise in hypoxia.
[m]Expired CO_2 values affected by aquatic PCO_2.
[n]In hypoxic water, 15 C.
[o]Head in air, body in hypoxic water.
[p]Estivating, held in laboratory for six years.

TABLE 5.8 Respiratory Exchange Ratios for Some Amphibious Air-Breathing Fishes during Air Exposure

Species	Mass (g)	°C	R	Reference
Neochanna burrowsius	4–10	17	0.53	Meredith *et al.*, 1982
Lepidogalaxias salamandroides	< 1	22	0.82	Martin *et al.*, 1993
Gobiesox meandricus	12.6	15	0.73	Martin, 1993
Clinocottus analis	9–13	15	0.82 (0)[a]	Martin, 1991
	—	—	0.73 (6)	
	—	—	1.10 (24)	
C. recalvus	10.3	23	0.89	Martin, 1991
	6–14	15.5	0.46	Wright and Raymond, 1978
C. globiceps	5.4	15	0.83	Martin, 1993
Oligocottus snyderi	4.5	23	0.98	Martin, 1993
	—	3.6	0.95	Martin, 1993
Mnierpes macrocephalus	1–17	30	0.73	Graham, 1973
Blennius (Lipophrys) pholis	13–81	12.5	0.80	Pelster *et al.*, 1988
	32	15	0.80	Steeger and Bridges, 1995
Alticus kirki	2.5	23	0.73	Martin and Lighton, 1989
Periophthalmus[b] *vulgaris*	7.2	24.5	0.78	Schöttle, 1931
P. dipus	16–17	24.5	0.77	Schöttle, 1931
P. schlosseri	110	24.5	0.73	Schöttle, 1931
P. barbarus	17	28	0.77	Steeger and Bridges, 1995
Xererpes fucorum	1.8	15	0.86	Martin, 1993
Xiphister atropurpureus	5–19	11–13	0.34	Daxboeck and Heming, 1982
	31	15	0.85	Martin, 1993
X. mucosus	13	15	1.05	Martin, 1993
Cebidichthys violaceus	175	23	0.93	Martin, 1993
Anoplarcus purpurescens	2	15	0.77	Martin, 1993

[a]Data for different exposure durations (initial 24 hours).
[b]Tables 2.3 and 2.4 have updated mudskipper species names.

mode or organ (see following). Assuming an organism is in respiratory equilibrium (i.e., not hypoxic or acidotic), its total R should lie between 0.7 and 1.0, depending upon diet. However, as was shown by early workers, the partitioning of bimodal CO_2 release is often the inverse of O_2 uptake; whereas O_2 is removed from the air, CO_2 is released predominantly to the water.

That water is the primary exchange media for CO_2 in aquatic air-breathing fishes is shown by the aerial R values of 18 species (Table 5.7) which range from 0.01 to 0.3 and only seldom reach 0.5. Although this table reveals discrepancies among R estimates obtained by different workers (e.g., *Amia*), it is nevertheless clear that aerial R is less affected by factors known to influence O_2 partitioning. Aquatic hypoxia does slightly elevate R in a few species (*Channa maculata, Clarias lazera*) and manipulations such as raising water PCO_2 can affect the apparent R (Table 5.7), but short of exposing a fish to air, aerial R generally remains low.

The low R of aquatic air breathers implies that, even though gill ventilation is often reduced (to prevent O_2 loss), aquatic ventilation remains sufficient for CO_2 release by these forms. This is because CO_2 is about 30 times more soluble in water than is O_2 (Chapter 1). However, a point exists in theory, where, with progressive gill disuse, branchial CO_2 release is retarded and chronic hypercapnia results. The list of chronically hypercapnic species is not presently well defined, but probably includes *Protopterus, Lepidosiren, Arapaima, Electrophorus*, and others (Chapter 7). Both evolutionary and physiological compensations for this condition have revolved around an elevated blood bicarbonate (Rahn and Garey, 1973) and the transfer of branchial functions to the kidney (Chapter 8).

Exposure of an aquatic air breather to air generally raises R. However, except for a few species (*Trichogaster, Anabas, Channa*, and estivating *P. amphibius*, Table 5.7), aerial R does not approach levels needed to void all respiratory CO_2, and the fish would probably become hypercarbic and acidotic (Pelster *et al.*, 1988). Burggren and Haswell (1979) and others have hypothesized that the ability to raise R during air exposure may be limited to species having an ABO derived from the gills because such an organ is more likely to be richly supplied with carbonic anhydrase (CA, the enzyme catalyzing the dehydration of bicarbonate). The role of CA is discussed subsequently; however, as seen in Table 5.7, some species with ABOs originating from embryonic gill tissue (*Clarias*, *Heteropneustes*, and others; see Chapter 3) fail to exhibit high R values during air exposure. On the other hand, Ishimatsu and Itazawa (1983b) found that artificial ventilation of the ABO elevated the aerial R of *Channa* and thus relieved hypercapnic acidosis.

Amphibious air breathers must have the capacity to remove CO_2 aerially, and my (Graham, 1976a) review suggested that this capability would be especially developed in the active intertidal forms. Table 5.8 shows aerial R values measured for fishes that are frequently amphibious. While data are sparse, most values exceed 0.7. Many of the newer data are from a survey done by Martin (1993, 1995). Her findings of a high R for both *Xiphister* and *Clinocottus* at least equivocates earlier low R estimates for these genera (Table 5.8). However, it is somewhat surprising that studies to date do not show differences in R between the highly active amphibious forms (*Alticus*, *Mnierpes*, and mudskippers) and those that are facultative amphibious air breathers during low tide (*Cebidichthys*, *Xererpes*, and *Anoplarcus*) or fishes (*Helcogramma*) that only emerge in response to aquatic hypoxia. Additional studies, comparing respiration over short-term and tidal-phase length air exposures may provide this information.

TABLE 5.9 Cutaneous $\dot{V}O_2$ As a Percentage of the Total $\dot{V}O_2$ of Air-Breathing Fishes Exposed to Air (a) or Water (w)

Species		% Total $\dot{V}O_2$	Reference
Erpetoichthys calabaricus	w	32	Sacca and Burggren, 1982
	a	41	
Anguilla vulgaris	w	10–15[a]	Berg and Steen, 1965
	a	27–41[a]	
Misgurnus fossilis	w	17[a]	Jeuken, 1957
	a	33–67[a,b]	
Heteropneustes fossilis	w	17	Hughes and Singh, 1971
Clarias batrachus	w	17	Singh and Hughes, 1971
	a	50	Jordan, 1976
Neochanna burrowsius	a,w[e]	43	Meredith *et al.*, 1982
Clinocottus analis	a,w[e]	22–29	Martin, 1991
Periophthalmus sobrinus	w	55	Teal and Carey, 1967
	a	63	
P. cantonensis	w	48	Tamura *et al.*, 1976
	a	76	
Boleophthalmus chinensis	w	36	Tamura *et al.*, 1976
	a	43	
Blennius pholis	a	50[c]	Pelster *et al.*, 1988
Mnierpes macrocephalus	a	40–48	Graham, 1973
Trichogaster trichopterus	a	10–12	Burggren and Haswell,1979
Monopterus albus	w	14	Liem, 1967c
	a	24	
Synbranchus marmoratus	w	31–94[d]	Graham *et al.*, 1987

[a]7–15 C.
[b]Also a temperature effect.
[c]50% of total $\dot{V}CO_2$ also via skin.
[d]Cutaneous $\dot{V}O_2$ scales with $Mass^{0.651}$ (1–1000 g).
[e]Same estimate for air and water.

Cutaneous Respiration

Cutaneous gas exchange has been examined with unimodal respirometers divided in such a way as to separate gas exchange measurements over the front (gills and head skin with an appropriate skin–area correction factor) and back (skin) halves of the body. A variation on this approach isolates the body and branchial surfaces by wrapping part of the fish body in plastic sheeting or even a condom (Berg and Steen, 1965; Jordan, 1976; Laming *et al.*, 1982; Sacca and Burggren, 1982) or coating the gills or skin with diffusion retardants (Vargas and Concha, 1957; Tamura *et al.*, 1976).

Cutaneous $\dot{V}O_2$ estimates (Table 5.9) range from 10 to 90% of total $\dot{V}O_2$, depending upon species. Most variability is seen among the freshwater and aquatic air breathers. Values for *Clarias* vary considerably, and the estimated 32–41% rate of cutaneous $\dot{V}O_2$ for *Erpetoichthys* is surprising in view of its thick layer of scales and functional lung. Body size has an effect on cutaneous respiration. In *Synbranchus marmoratus*, cutaneous $\dot{V}O_2$ scales with body surface area, and larger fish become increasingly dependent upon branchial and aerial $\dot{V}O_2$ (Graham *et al.*, 1987).

Cutaneous rates of the marine intertidal genera are grouped between 40 and 60% of total $\dot{V}O_2$. Studies on mudskippers (Teal and Carey, 1967; Tamura *et al.*, 1976) indicate that cutaneous $\dot{V}O_2$ increases during air exposure. In a comparative study of *Boleophthalmus* and *Periophthalmus*, Tamura and co-workers (1976) found that the more amphibious *Periophthalmus* uses its skin for a greater percentage of aerial respiration than does *Boleophthalmus*. With *Blennius pholis*, Pelster *et al.* (1988) discovered that proportional amounts of O_2 and CO_2 were exchanged in the head and body sections of the partitioned respirometer; thus, about 50% of the $\dot{V}O_2$ and $\dot{V}CO_2$ occurs through the skin of this fish.

THE METABOLIC SCOPE OF AIR-BREATHING FISHES

Activity

The relation between fish activity and air-breathing frequency has been noted for a number of species. During activity, *Amia* increases its air-breathing frequency and depletes its ABO–O_2 more rapidly (Johansen *et al.*, 1970a). *Amia* is more active and breathes air more at night while in search of prey (Horn and Riggs, 1973). Observations relating air-breathing frequency and activity also have been made for *Neoceratodus* (Grigg, 1965), *Lepisosteus* (Rahn *et al.*, 1971), *Clarias* (Jordan, 1976), and several other species (Tables 5.1 and 5.2).

Activity clearly affects air breathing. Beyond this, however, is there a relationship between air breathing and the capacity to be active? Recognizing that air-breathing evolved in most cases as a mechanism to augment O_2 supply in hypoxic water (Carter, 1957; Graham *et al.*, 1978a), the question arises: Can air breathing also function to augment metabolic demand above levels sustainable by aquatic respiration? This answer is relevant in view of the active nature of many amphibious forms and because the generally reduced gill areas and ventilatory capacities of some air-breathing fishes lessens their aquatic respiratory scope, even in normoxic water. It may be that under certain circumstances natural selection has favored the contribution of air breathing to aerobic scope. Johansen (1970) suggested increased scope as a reason for air breathing in *Amia* and *Megalops*; however, this has never been tested. The higher $\dot{V}O_2$ of the air-breathing *Piabucina festae* compared to the non-air-breathing *P. panamensis* also suggests a difference in scope (Graham *et al.*, 1978b; and see Mattias *et al.*, 1996).

Only one experiment (Grigg, 1965) was designed to test if metabolic rate could be elevated through air breathing. Grigg measured the $\dot{V}O_2$ of small *Neoceratodus* placed in a respirometer with a magnetic stirrer at the bottom. Stirring at a constant rate forced the fish to fight the turbulence and Grigg compared the $\dot{V}O_2$ of fish with and without air access. Tests were done at both 18 and 25 C. Fish without air access had a mean $\dot{V}O_2$ of 51 ml/kg h at both temperatures. With air, the rate at 18 C was 71, and at 25 C it was 66. There was a marked difference in gill ventilation frequency at 25 C when air access was denied (80/min) and when allowed (67/min). Air-breathing thus increased the mean $\dot{V}O_2$ by 30–40% and lowered gill frequency at 25 C; Grigg concluded that affording an increased metabolic scope rather than emergency respiration is the principal function for aerial respiration in *Neoceratodus*.

The metabolic scope question has also been approached by determining the effects on air breathing of increased metabolic demand brought on by elevated ambient temperature. Results for *Amia* (Johansen *et al.*, 1970a), *Lepisosteus* (Rahn *et al.*, 1971; Smatresk and Cameron, 1982b), and *Channa* (Yu and Woo, 1985; Glass *et al.*, 1986) show that increased metabolic needs at higher temperature could not be sustained without air breathing.

Findings for *Channa* additionally verify its use of air breathing to enhance metabolic scope as well as for emergency respiration. When subjected to an increase in water temperature from 15 to 25 C, both *C. argus* and *C. maculata* underwent minimal changes in aquatic ventilation but substantially raised their aerial ventilation frequency, indicating that elevations in metabolic demand are met by aerial respiration. Yu and Woo (1985) determined that 25 C did not affect the aquatic $\dot{V}O_2$ of *C. maculata* in either normoxia or hypoxia; its 2.3-fold increase in total $\dot{V}O_2$ at 25 C was sustained entirely by elevated aerial $\dot{V}O_2$. Also, and related to the increased dependence on aerial respiration, these workers found that the aerial $\dot{V}CO_2$ of *C. maculata* increased significantly in hypoxic water (Table 5.7). Glass *et al.* (1986) also observed a large increase in aerial R for *Channa argus* in hypoxia at 25 C.

The relationship between aerial respiration and metabolic scope may differ for aquatic and amphibious air breathing. The metabolic rates of *Clarias batrachus* and *C. mossambicus* air breathing in hypoxic water are reduced (Jordan, 1976; Johnston *et al.*, 1983). On the other hand, *C. batrachus*, which is known to migrate on land, increased its aerial $\dot{V}O_2$ by threefold in air (Jordan, 1976). *Monopterus* also increased its $\dot{V}O_2$ during air-exposure (Liem, 1967c; Lomholt and Johansen, 1976). Air-exposed *Periophthalmus* stimulated to do two minutes of exercise in a gently rotated respirometer chamber increased $\dot{V}O_2$ by a factor of 3.1; whereas, no change in $\dot{V}O_2$ was measured for goldfish given this treatment (Hillman and Withers, 1987). A similar treatment nearly doubled the aerial $\dot{V}O_2$ of *Alticus* (Martin and Lighton, 1989). Finally, the demonstration of an increased bimodal $\dot{V}O_2$ in *Boleophthalmus* and *Periophthalmus* (Tamura *et al.*, 1976; Natarajan and Rajulu, 1983) has implications for metabolic scope, but corresponding aerial–aquatic time budgets are needed.

Estivation and Fasting Effects

Studies of the effects of prolonged emersion have been done on *Protopterus* by Smith (1930, 1935a,b) who found the $\dot{V}O_2$ of cocoon-dwelling fish to be about 20% of normal values in water. This result has been confirmed in more recent studies of estivating lungfish (Lahiri *et al.*, 1970; Fishman *et al.*, 1987) and Lomholt (1993), using specimens collected in the estivating state and held in the laboratory for up to six years, measured metabolic reductions to less than 5% of the rate of awake fish. (Swan *et al.* [1968] extracted an antimetabolic factor from the brains of estivating *Protopterus*, however, its nature has not been further investigated [Chapter 8]).

Metabolic reductions have also been measured for starved fish (Greenwood, 1987). For *Synbranchus marmoratus* kept in mud burrows without access to water and starved for three months, Bicudo and Johansen (1979) measured a $\dot{V}O_2$ that was about 20–25% that of fish in water. Notably, and unlike the lungfish, the synbranchids were not estivating, but were actively moving about in their burrow systems. All of these observations suggest that starvation, and not strictly water deprivation, may play a key role in the metabolic reductions measured during long-term air exposures. However, the importance of both factors is suggested by the finding of Liem (1967c) that starved *Monopterus* did not have a reduced metabolism. Clearly more data are required.

RESPIRATORY ORGAN PHYSIOLOGY

Respirometry and ABO–gas analyses enable close examination of respiratory organ function during air breathing.

The ABO

Temporal Changes in Respiratory Gases

Figure 5.10 illustrates the time course of O_2 and CO_2 changes in the ABO of *Channa argus* (Ishimatsu and Itazawa, 1981). While similar observations are available for several air-breathing species, results for *C. argus* were selected for this general description of ABO gas flux because the data are well documented statistically, and the overall pattern is similar to other species. Another advantage offered by the *C. argus* data is that ABO function is compared for fish in water and in air.

Figure 5.10 shows that the inspired O_2 and CO_2 levels (149/1 torr) of *Channa argus* closely approximate those of air, indicating that when respiring air aquatically, tidal volume nearly equals ABO volume. (This finding is expected based on the ventilatory mechanism of this fish [Chapter 3].) In the course of a 10 minute breath hold, ABO–O_2 declines steadily (about 9 torr/min) to 63 torr, meaning that 57% of the available O_2 is extracted. In contrast to the gradual decline seen for O_2, ABO–CO_2 rises abruptly to 21 torr in the first two to three minutes after inspiration and then remains fairly steady for the duration. The ABO R value (not total R) during the first three minutes is about 0.6, but then falls to zero over the remainder of the breath and organ R for the entire cycle is 0.2.

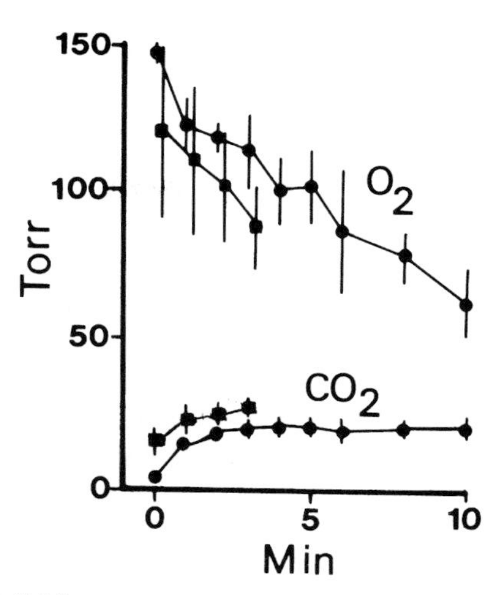

FIGURE 5.10 Time course of the changes in the ABO gas of *Channa argus* in air (squares) and water (dots). (From Ishimatsu and Itazawa, 1981, by permission of the *Japanese Journal of Ichthyology*.)

A steady decline in O_2 over the entire breath hold indicates that O_2 absorption continues unimpeded by changes within the organ. If a breath were to be held longer, the rate of O_2 removal would gradually taper off as the O_2 diffusion gradient was further reduced; the lower limit for O_2 extraction is determined by the O_2 affinity of blood. In contrast, the post-inspiration rise in organ CO_2 followed by a leveling-off suggests rapid equilibration between blood PCO_2 and ABO gas. Once this has occurred, there is little net CO_2 flux into the ABO.

Air exposure reduces the tidal volume of *C. argus*. This is demonstrated by a lowered inhalation O_2 level (120 torr) and elevated CO_2 (16 torr, Figure 5.10). Although the inspired O_2 is less, the rate of O_2 removal from the ABO (10 torr/min) approximates that for breaths taken in water and, about 25% extraction takes place in the brief time the breath is held. In a pattern similar to that for the aquatic air breaths, post-inhalation CO_2 level rises abruptly (16–27 torr) in three minutes, R is 0.7 during the early phase but falls to 0.2 over the entire breath period; there is little net CO_2 release into the ABO. (The higher organ CO_2 levels in the air-exposed fish are attributable to reduced ventilation and probably hypercapnia [Ishimatsu and Itazawa, 1981].)

With *Channa argus* as background, major comparative features of ABO gas-exchange physiology can now be examined for 13 species (Table 5.10) based on gas extracted from cannulae or expelled from the ABO. It is clear that fishes capable of a 100% tidal volume exchange (viz., the lower part of the table from *Electrophorus* to *Gillichthys*) have higher inspiration–O_2 levels and thus a larger relative O_2 supply. ABO–O_2 extraction ranges from a low of 15% (*Lepisosteus* in hypoxia) to 97% (forcibly submerged *Protopterus*). An extensive analysis of O_2 extraction in *Clarias lazera* yielded a mean value of 82.9% (range 57–96) and showed no effect of either body size (1.6–3 kg) or aquatic O_2 tension (Ar and Zacks, 1989). Rates of change of ABO gas, although modified with time, are inversely related to breath duration. The maximum observed rate of O_2 decline is 20 torr/min (*Synbranchus*), the maximum CO_2 deposition is 3.9 torr/min (*Electrophorus*).

R Values

As described above, ABO R is an instantaneous index of ABO gas exchange. This R is not the same as total or metabolic R, although some workers have mistakenly equated the two. R values for the ABO, while not large, generally exceed total R (Tables 5.7 and 5.10).

Changes in ABO R over the course of the air-breath cycle are shown by the O_2/CO_2 diagram and Figure 5.11 compares these for five species including *Channa argus*. The thick diagonal lines in the figure describe the theoretical R functions 0.7 and 0.1: It is evident that R's for these species are less than 0.7, and some are even below 0.1. In most cases, R values are initially high and then taper off. As previously described for *Channa* in water (see line 7 in Figure 5.11), ABO–CO_2 rises abruptly to 21 torr and levels off; the flattening of the R function indicates no more CO_2 is effusing in the ABO, even though O_2 is still being removed.

The R line for *Channa* in air shows how ambient

TABLE 5.10 Time Course of Respiratory Gas Changes in Fish Air-Breathing Organs

Species		Duration (min)	Gas	Partial pressure		% O_2 extraction	torr/min	R	Reference
				Initial	Final				
Protopterus aethiopicus[a]	C[b]	160	O_2	118	4	97	0.71		McMahon, 1970
			CO_2	3	7	—	0.03	0.04	
	C	50	O_2	83	15	82	1.36		Lenfant and Johansen, 1968
			CO_2	3.3	4.1	—	0.02	0.002	
Lepidosiren paradoxa	E	16	O_2	150	25	83	7.81		Johansen and Lenfant, 1967
Polypterus senegalus[c]	C	70	O_2	125	10	92	1.64		Magid *et al.*, 1970
Lepisosteus oculatus[d]	N,C	30	O_2	85	70	18	0.5		Smatresk and Cameron, 1982a
			CO_2	7.5	7.5	—	0	0	
	H,C	5	O_2	72	61	15	2.2		Smatresk and Cameron, 1982a
Amia calva	C	30	O_2	135	75	40	2.0		Johansen *et al.*, 1970a
Arapaima gigas	C	5	O_2	140	50	64	18		Randall *et al.*, 1978a
			CO_2	8	38	—	6	0.3	
Electrophorus electricus	E	9	O_2	145	10	93	15.0		Farber and Rahn, 1970
			CO_2	0	35	—	3.9	0.25	
Clarias lazera	C&E	10	O_2	150	27	82	12.3	0.02–0.18	Ar and Zacks, 1989
Ancistrus chagresi[f]	N,E	12	O_2	150	32	79	9.8		Graham, 1983
	HA,E	30	O_2	150	15	90	4.5		
Synbranchus marmoratus	C	5	O_2	150	50	67	20		Bicudo and Johansen, 1979
			CO_2	0	1	—	0.2	0.01	
	E	20	O_2	150	25	83	6.25		Graham and Baird, 1984
			CO_2	0	11	—	0.55	0.09	
Monopterus cuchia	C	15	O_2	150	35	77	9		Lomholt and Johansen, 1976
			CO_2	0	2.9	—	0.19	0.03	
Gillichthys mirabilis	E	10	O_2	148	96	35	5.2		Todd and Ebeling, 1966
Anabas testudineus	C	10	O_2	120	35	71	8.5		Singh and Hughes, 1973
			CO_2	15	30	—	1.5	0.18	

[a]Forced submergence.
[b]Gas sampled by catheter (C) or was an end-expired sample (E).
[c]Right lung, anterior.
[d]N, in normoxia; H in hypoxia.
[e]Some values in hypoxia reached 0.5.
[f]N, normoxia fish in hypoxic water; HA, hypoxia-acclimated fish.

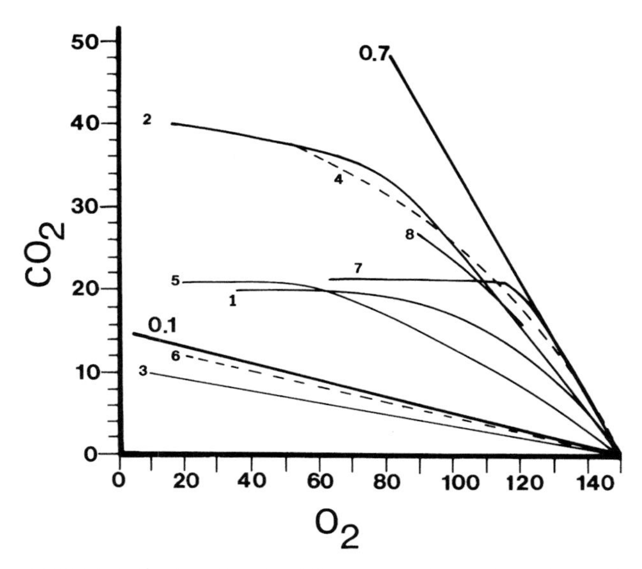

FIGURE 5.11 ABO–O_2/CO_2 diagram for various air-breathing fishes. Thick diagonal lines are functions for R = 0.1 and 0.7. Line identifications: 1, *Monopterus* (Lomholt and Johansen, 1976); 2, *Clarias lazera* in hypoxia and 3, in normoxia (Ar and Zacks, 1989); 4, *Arapaima* (Randall *et al.*, 1978a); 5, *Synbranchus* in PCO_2 19–20 torr and 6, in 6–7 torr (Graham and Baird, 1984); 7, *Channa argus* in water and 8, in air (Ishimatsu and Itazawa, 1981).

conditions alter instantaneous R. Ar and Zacks (1989) also reported that hypoxic water markedly elevated the instantaneous R of *Clarias lazera* (compare lines 2 and 3); however, their data revealed that this had little effect on its total aerial R. Experimentation reveals that ambient CO_2 can also affect apparent R. Dehadrai (1962a, figure 7) found a direct relationship between ambient water and gas-bladder CO_2 contents. In addition, the large instantaneous CO_2 diffusion rates seen for *Channa*, *Clarias*, *Arapaima*, and other species are a measured function of the post-inhalation PCO_2 diffusion gradient and not a specialized property of the ABO. This can be illustrated by comparing the R functions for *Synbranchus* in water with a relatively low PCO_2 (e.g., 5–6 torr, line 6) and in water with a PCO_2 of 19–20 torr (line 5). The higher instantaneous R of line 5 is solely attributable to the increased CO_2 diffusion gradient, owing to a higher PCO_2 in the blood of the hypercarbic fish (Graham and Baird, 1984). This comparison emphasizes the need to distinguish instantaneous R values from those determined by respirometry and underlines the importance of accounting for ambient CO_2 levels whenever variables such as organ R or blood PCO_2 are under investigation.

Integrated ABO Function

Data from two bimodal respirometry studies are used to illustrate how numerous ABO-variables, including ABO and tidal volume, breath duration, ABO–O_2 extraction, and air convection requirement [ACR], govern the organ's potential contribution to total $\dot{V}O_2$.

Table 5.11 compares these ABO respiratory parameters for *Clarias lazera* (Ar and Zacks, 1989) at three different ambient O_2 tensions. The first three terms, bimodal $\dot{V}O_2$, breath duration, and initial (post-breath) O_2, were given by Ar and Zacks (1989, tables 14.1–14.3 and figure 14.3). The rate of change in ABO–O_2 (torr/min) was calculated as in Table 5.10. With a known O_2 extraction percentage and assuming (conservatively) a mean aerial R of 0.1, the inspired ABO volume can be estimated from the expired ABO values given by Ar and Zacks. Recall that in clariids, air ventilation usually replaces the entire ABO volume (Chapter 3; and see Ar and Zacks (1989), figure 14.3). The clariid suprabranchial chamber has a rigid construction, and ABO volume does not vary greatly; although, how the PO_2 decline during the breath hold affects ABO pressure and volume remains unknown.

Knowing ABO volume and that organ contents are changed completely by each breath, the amount of O_2 inhaled per breath is calculated as 21% of inspired ABO volume. The product of this term and the O_2 utilization coefficient is the total (ml) per-breath O_2 consumption, which can also be expressed on a per minute or per hour basis. Aerial O_2 uptake per hour, divided by total $\dot{V}O_2$ gives the percentage of aerial

TABLE 5.11 Air-Breath Variables with an Analysis of Respiratory Partitioning: In Relation to PO_2 for *Clarias lazera*[a] and following Hypoxia Acclimation in *Ancistrus chagresi*[b]

	Water PO_2	Total $\dot{V}O_2$ (ml/h)	Air breath variables: Duration (min)	PO_2 (torr): Begin[c]	PO_2 (torr): End	O_2 Extraction (%)	O_2 (torr/min)	Expired tidal volume (ml)	Inspired ABO volume (ml)	Breath O_2 volume[f] (ml)	Aerial O_2 utilization: O_2/breath[g]	O_2/min[h]	O_2/h[i]	% Total $\dot{V}O_2$	ACR[j] (ml/ml)
Clarias lazera	117	96	150.0	150	23	85[d]	0.85	84[e]	102[e]	21.4	18.21	0.121	7.28	7.6	5.60
(1.8 kg, 28.5 C)	94	110	20.0	150	32	80[d]	6.00	79[e]	95[e]	19.9	15.95	0.80	47.86	43.5	5.96
	65	114	12.5	150	21	86[d]	10.32	76[e]	93[e]	19.5	16.80	1.34	80.62	70.7	5.53
Ancistrus chagresi N[k]	10	84	4.6	154	46	71	23.48	39[l]	39	8.2	5.79	1.26	75.52	89.9	6.74
(25 g, 25 C) HA[k]	10	84	5.5	154	31	80	22.36	49[l,m]	49	10.3	8.23	1.50	89.80	107.9	5.94

[a] Ar and Zacks, 1989.
[b] Graham, 1983.
[c] Assumes total turnover of ABO gas (see text).
[d] Specific extractions for the stated PO_2 are used in calculations. Average O_2 extraction for this fish was 84.9 ± 3% (SD); for all fish was 82.9 ± 5.5%.
[e] For *Clarias* specific volumes are used in calculations. End-expired volumes are shown; the average expired gulp volume for this fish is 80 ± 3 ml (SD). Inspired ABO volume calculated by correcting end-expired volume for O_2 extraction volume and assuming a mean R of 0.1.
[f] Inspired ABO volume · 0.21.
[g] ml O_2 · % extraction.
[h] O_2/breath duration.
[i] Term in $f \cdot 60$ min/h.
[j] Air convection requirement = Tidal volume/ml O_2 extracted from breath.
[k] N, normoxic fish in hypoxia; HA, hypoxia-acclimated fish.
[l] Complete expiration of each breath; Tidal volume = ABO volume. Volumes corrected to pre-O_2 depletion volume.
[m] ABO volume increased by air breathing.

$\dot{V}O_2$ (e.g., in Table 5.11, for 117 torr: $5.98/52 \cdot 100 = 11.5\%$) and ACR is the ratio of tidal volume to the per-breath O_2 utilization.

Table 5.11 confirms the assertion by Ar and Zacks that air–breath duration is the determining factor in the aerial $\dot{V}O_2$ of *C. lazera*. Although Ar and Zacks (1989) noted variation in both the air-gulp and O_2 extraction terms, these did not correlate with aquatic O_2 and the variation fell uniformly around the mean values reported for all fish (Table 5.11, footnotes). Given that both breath volume and O_2 extraction are nearly constant, the rise in the percentage of aerial $\dot{V}O_2$ at lower aquatic O_2 tensions is directly linked to breath frequency. Conservation of ACR at the three O_2 levels also occurs for these reasons.

Both the percentage of aerial $\dot{V}O_2$ obtained from an air breath and the ACR calculated for *C. lazera* in Table 5.11 differ markedly from values given by Ar and Zacks. This table, for example, shows that at 65 torr, the fish could extract its total O_2 requirement from an air breath. By contrast, Ar and Zacks (1989, tables 14.1 and 14.2) estimated that, at 65 torr, aerial respiration amounted to about 52% of total. These differences reside in the method of aerial $\dot{V}O_2$ calculation: Ar and Zacks used respirometry and integrated hourly $\dot{V}O_2$ and air-ventilation volumes to estimate both parameters; calculations in Table 5.11 are based on temporal changes in ABO O_2 content and a known ABO volume. Also, my Table 5.11 calculations show that 16–18 ml O_2 is available to a 1.8 kg fish from each air breath (i.e., 9.44 ml/kg), which is greater than the 6.74 ml/kg given by Ar and Zacks.

Ar and Zacks (1989) interpret the aerial $\dot{V}O_2$ of *Clarias lazera* as the sole function of breath frequency. Their data show that each air breath has the same volume and the same amount of O_2 extraction (6.74 ml/kg). The only variable, the time over which this occurs, is a direct function of aquatic respiratory proficiency or, in other words, aquatic PO_2. Ar and Zacks, however, did not explore the effects of hypoxia more extreme than 65 torr. My suspicion is that the respiratory physiology of *Clarias lazera* was not sufficiently challenged to reveal subtler respiratory controls acting on the ABO and partitioning. At 65 torr, the hemoglobin (Hb) of *C. lazera* would be fully saturated (data in Chapter 7) and its aquatic ventilation not severely affected (recall that *Clarias batrachus* regulates $\dot{V}O_2$ down to 40 torr). Moussa (1957) determined that although an obligatory air breather, *C. lazera* can survive in water with an O_2 content of 0.3 ml/l (i.e., about 7 to 8 torr at 20–25 C; Chapter 1).

It is likely that *Clarias lazera* frequently encounters much more extreme hypoxia, extreme enough to probably eliminate gill respiration entirely and cause transbranchial O_2 loss. Assuming a 1.8 kg fish was in 10 torr water and could obtain no O_2 aquatically, then, applying the 6.74 ml/kg breath constant of Ar and Zacks, this fish would need a new air breath every 7 minutes (8.6/h) to meet its standard metabolic demand of 1.7 ml/min (Table 5.11). However, as seen in Table 5.1, *C. lazera* may breathe air as frequently as every 1.5 minutes which, to satisfy the fixed O_2 extraction parameter, would necessitate an O_2 absorption rate of 8.1 ml/min. This rate would require increased organ perfusion and would supply the fish with nearly five times the amount of O_2 needed for its standard $\dot{V}O_2$. Thus, the time air is held can work against efficient O_2 utilization if it falls below the minimum time required for its complete extraction. Additional studies are needed to determine whether or not the relationship between air-breath duration and ABO–O_2 extraction can be altered in hypoxic water. One possibility is that in aquatic O_2 tensions of 5–10 torr, *C. lazera* may lose excess O_2 transbranchially. It may also be the case, as the following discussion for *Ancistrus* reveals, that a period of hypoxia acclimation is needed to enhance both the aerial and aquatic respiratory capacities of *C. lazera* in extreme hypoxia.

Table 5.11 also compares bimodal respiration parameters in two groups of *Ancistrus chagresi*, one that was hypoxia acclimated for 10 days minimum, and another kept in normoxic water and acutely exposed to hypoxia (Graham, 1983). This comparison shows that hypoxia acclimation brings about a number of morphological and physiological changes that increase the aerial-respiratory capacity of *Ancistrus* in hypoxia. The main differences are that hypoxia acclimated fish take a larger volume air breath, hold it for a longer time, extract a greater percentage of O_2 from it, and have a lower ACR. The significance of acclimation is that *Ancistrus* can potentially use aerial respiration to supply its entire $\dot{V}O_2$ in 10 torr water (Table 5.11). As seen in Figure 5.4, hypoxia-acclimated *Ancistrus* have a significantly upgraded capacity to respire aquatically (obtaining about 24% of their total $\dot{V}O_2$ at 10 torr). Hypoxia acclimation can thus elevate the potential for metabolic scope far above that of normoxia-adapted fish which can only supply about 90% of total $\dot{V}O_2$ via air and have a zero aquatic $\dot{V}O_2$ at this tension (Graham, 1983).

Gill Physiology in Relation To Air Breathing

Ventilation and O_2 Transfer

Relationships between rates of gill ventilation, hypoxia, and O_2 transfer during air breathing have been examined in several species.

Both *Amia* and *Lepisosteus* have limited gill modifications for air breathing and lack branchial shunts, but do reduce ventilation in hypoxia. Nevertheless, these fishes lose O_2 from their gills in severely hypoxic water. While the rate of loss estimated for *Lepisosteus* is a low 0.04 ml/kg min (Smatresk and Cameron, 1982a), it is a staggeringly large 4.3 ml/kg min in *Amia* (Randall *et al.*, 1981b).

As reported by Ar and Zacks (1989), *Clarias lazera* undergoes cyclic changes in gill-ventilation frequency that are in phase with air breathing. Just after air is taken, ventilation frequency is a low 24/min and gradually rises to 44/min near the end of the breath hold. Ar and Zacks measured instantaneous rates of aerial and aquatic $\dot{V}O_2$ (in normoxic water) and determined these were reciprocally related over the air-breath cycle. Thus, as aerial $\dot{V}O_2$ declines (owing to reduced ABO–PO_2), aquatic $\dot{V}O_2$ rises in direct proportion, resulting in the conservation of a stable total bimodal $\dot{V}O_2$ throughout the breath cycle. In their example, post-inhalation aerial $\dot{V}O_2$ was about 20 ml/kg h and steadily declined to zero over 40 minutes. Aquatic $\dot{V}O_2$ was about 30 ml/kg h at inhalation and gradually increased to about 50 in the span of 40 minutes.

Concerning gill O_2 utilization, Yu and Woo (1985) used the Fick method to determine that *Channa maculata* has a relatively large ventilation volume (705 ml/min, at 25 C) but an extremely low O_2 utilization of 21%. This utilization is low compared to non-air breathers and is surprising in view of the non-obligatory air-breathing status of this fish; nevertheless it correlates well with the findings (see preceding) that *C. maculata* has a limited scope for aquatic $\dot{V}O_2$.

The effects of hypoxia acclimation on gill–O_2 transfer have significance for earlier discussions of hypoxia responses and integrated ABO function. First, and contrary to findings for *Clarias lazera* (Ar and Zacks, 1989), Johnston *et al.* (1983) found that acute exposure to hypoxia lowered the bimodal $\dot{V}O_2$ of *Clarias mossambicus* (Jordan [1976] also found this for *C. batrachus*) relative to normoxic fish. Johnston and colleagues further determined that 27 days of hypoxia acclimation markedly improved the respiration capacity of *C. mossambicus* in hypoxia. Specifically, total $\dot{V}O_2$ in hypoxia increased (relative to acutely exposed fish) by 148%; the largest increase occurred in aquatic $\dot{V}O_2$ (+164%), while aerial $\dot{V}O_2$ increased by only 112%. All of this suggests that, at the very least, the air-breathing physiology of *Clarias mossambicus* differs substantially from that of *C. lazera*. Alternatively, hypoxia acclimation may significantly alter the air breathing and bimodal partitioning of *C. lazera*. The improved aquatic respiratory capacity of *Clarias mossambicus* following hypoxia acclimation is remarkably similar to findings for *Ancistrus* (Figure 5.4).

Gill Function during Air-Breathing Ontogeny

A final topic related to gill physiology is the apparent metabolic cost savings associated with air-breathing ontogeny determined for some Indian species. Singh and colleagues (references in Table 5.12) showed that the ontogenetic progression to air breathing led to reductions in aquatic $\dot{V}O_2$ and ventilation and that this significantly reduced the mass exponent of the $\dot{V}O_2$–scaling equation. As seen in Table 5.12, all five species commenced air breathing at a body length of 10–11 mm (10–40 mg) at which point aquatic $\dot{V}O_2$ dropped by an average of 41%. The transition to air breathing also reduced the mean mass exponent for aquatic $\dot{V}O_2$ from 0.899 to 0.641.

CARBONIC ANHYDRASE AND AIR BREATHING

Attention has been given to the possible role of carbonic anhydrase (CA) in aerial CO_2 excretion. CA catalyzes the dehydration of bicarbonate ion to water and CO_2, and in fish gills this enzyme is thought to play a

TABLE 5.12 Effects of the Ontogenetic Transition to Air Breathing on the Mass Exponent *(b)* of the Equation ($\dot{V}O_2 = aM^b$) Relating $\dot{V}O_2$ and Body Mass *(M)* in Five Indian Air-Breathing Fishes

Species	°C	Body size at air-breathing transition: Mass	Length	$\dot{V}O_2$-mass exponent: Before air-breathing	After	% Reduction in aquatic $\dot{V}O_2$	Reference
Heteropneustes fossilis	28	25	11–12	0.875	0.710	40	Sheel and Singh, 1981
Channa striatus	28	10	11–12	0.922	0.672	42	Singh *et al.*, 1986
Channa punctatus	28	10	11–12	0.912	0.590	45	Singh *et al.*, 1982b
Anabas testudineus	28	40	11–12	0.950	0.587	40	Mishra and Singh, 1979
Colisa fasciatus	28	30	11–12	0.836	0.645	36	Prasad and Singh, 1984
			Mean	0.899	0.641	41	

key role in the rapid conversion of plasma HCO_3^- to molecular CO_2, which then diffuses into water. Experiments have attempted to determine what limits aerial CO_2 release by testing whether the injection of CA, or the chemicals that block its action, will alter CO_2 flux and thus R.

Results to this point do not show a clear pattern. Randall *et al.* (1978b) found that CA injection into air-exposed *Hoplerythrinus* raised the ABO–R value from 0.25 to 0.51, suggesting that the enzyme's activity in this organ was quite low. Different results were obtained for *Arapaima* (Randall *et al.*, 1978a), where, following CA injection, ABO–R actually dropped from 0.45 to 0.39 even though blood PCO_2 increased slightly from 27 to 30 torr. For *Xiphister atropurpureus*, neither CA nor its inhibitor acetazolamide significantly altered $\dot{V}CO_2$ relative to the control rate (Daxboeck and Heming, 1982). Air-exposed *Trichogaster* (Table 5.7) has a large R which could not be further elevated by CA injection (Burggren and Haswell, 1979). Nor did CA affect the R of this fish in water. Acetazolamide administration did, however, reduce the R and $\dot{V}CO_2$ of both groups, suggesting that the labyrinth itself contains a large amount of CA. Smatresk and Cameron (1982c) also found no effect of acetazolamide injection on the aerial R of *Lepisosteus*, although its blood PCO_2 was raised and blood pH lowered. Similar results were found for *Blennius pholis* (Pelster *et al.*, 1988).

Comparative surveys of CA activity in the tissues of air-breathing and non-air-breathing fishes do not reveal any major differences. Singh *et al.* (1990) have suggested that a proliferation of CA containing cells has occurred in tissues lining the inner opercular walls of several Indian air-breathing genera (mudskippers, *Pseudapocryptes*), however, this has not been quantified nor is there evidence for the exclusivity or a high rate of CO_2 release from these sites. Daxboeck and Heming (1982) found no marked differences in the CA activity in the red cells, gills, or skin of *X. atropurpureus* and the rainbow trout (*Salmo gairdneri*). Pelster *et al.* (1988) similarly concluded that the tissue CA profiles gave no evidence of regional specializations for CO_2 excretion in *B. pholis*.

Table 5.13 summarizes Burggren and Haswell's (1979) comparison of CA levels in the tissues of five species (two non-air breathers and three air breathers). These data are highly variable, and few trends or significant differences are seen among values for whole blood or gills. Gas-bladder values are high for the two non-air breathers; whereas no enzyme activity was detected in the physoclistous organ of *Trichogaster*. Notably, the ABOs of *Trichogaster* and *Clarias* had relatively high CA activities, but this enzyme was not detected in the lung of *Erpetoichthys*.

TABLE 5.13 Carbonic Anhydrase Activities in the Tissues of Five Fishes[a]

Species	Tissue enzyme activity[b] ($\bar{x}$ ± SD)
	Whole Blood
Salmo gairdneri	319 ± 47
Cyprinus carpio	1148 ± 570
Trichogaster trichopterus	1564 ± 258
Clarias batrachus	1660 ± 258
Erpetoichthys calabaricus	289 ± 131
	Gill
S. gairdneri	972 ± 186
C. carpio	817 ± 269
T. trichopterus	826 ± 125
C. batrachus	1385 ± 327
E. calabaricus	1584 ± 1140
	Gas Bladder
S. gairdneri	25 ± 18
C. carpio	184 ± 59
T. trichopterus	n.d.[c]
	ABO
T. trichopterus	396 ± 128
C. batrachus	315 ± 111
E. calabaricus	n.d.[c]

[a]Burggren and Haswell, 1979.
[b]Activity units are amount of enzyme/g protein needed to double the uncatalyzed rate of CO_2 evolution from bicarbonate.
[c]n.d. = not detected.

Finding CA in the ABO of *Trichogaster* confirmed what Burggren and Haswell's (1979) inhibition experiments had indicated. These workers further offered that relatively high levels of CA activity might occur in all ABOs derived from gill tissue, and that these organs should be capable of sustaining a relatively high R when fish are in air. Although enzyme activity data are lacking, it is clear in Table 5.7 that for a number of genera (*Clarias, Anabas, Heteropneustes, Channa*) with ABOs derived entirely or partly from gills, R values are elevated, although to a lesser extent than in *Trichogaster*.

Nevertheless, when all the facts are considered, a general correlation between respiratory surface CA activity and aerial R cannot be supported. The findings for *Arapaima* are the opposite of expected. Daxboeck and Heming (1982) found no effect of CA or acetazolamide on the $\dot{V}CO_2$ of *Xiphister*. It is paradoxical that the skin of *B. pholis* has only a small fraction of the CA activity found in gills, but that partitioned respirometry shows the R values of the two surfaces are equal (Pelster *et al.*, 1988). Finally, if a high ABO–CA activity correlates with a high rate of aerial CO_2 release; then it could be expected that fish holding air in their mouths and thus in contact with the gills, vestigial gills, or gill-derived ABO surfaces would also have high aerial R values; however, this is not the case for *Synbranchus*, *Monopterus*, or *Electrophorus* (Table 5.7).

SUMMARY AND OVERVIEW

Studies of air-breathing fish respiration have identified factors regulating the duration of the air–breath cycle, determined respiratory capacity, and elucidated respiratory organ function during bimodal respiration. The emerging picture is that for most aquatic air breathers, aerial respiration is utilized to the extent needed to supplement aquatic respiration and sustain normal $\dot{V}O_2$. For amphibious air breathers, the complete transition from aquatic to aerial gas exchange ensures a period of total respiratory independence from water.

Measurements of air-breathing cycle duration and frequency date from the late eighteenth century. Air–breath durations differ among species, are highly variable, and are affected by several physical factors including aquatic O_2 level and temperature. In situations of chronic hypoxia, fish may ascend for air as frequently as 1200 times in 24 hours. Light level also affects air–breath duration as can the presence of water-borne toxins. An effect of aquatic CO_2 has also been suggested for some species, but comparative data lack a definitive pattern (Chapter 6).

Biological factors influencing air-breathing frequency include body size, activity, and physiological variables such as obligatory versus non-obligatory state, as well as degree of hypoxia acclimation. Behaviorally mediated effects on aerial respiration have been shown for predatory stimuli, food supply, crowding, and group induction (social facilitation). Distance to the water surface affects air-breathing frequency. In general, smaller species—those living in the water column and thus closer to the surface—and the obligatory and continuous air breathers tend to gulp air more frequently. There are hardly any data comparing the "aquatic versus terrestrial" time budgets of amphibious air breathers.

The metabolic data presented in this chapter are based on respirometric determinations of O_2 uptake and CO_2 release. Although there is considerable variability, the total metabolic rates measured for air-breathing fishes do not differ from those of other ectotherms over a similar temperature range. Not all respirometry data are of superior quality; however, and in many studies basic respirometry protocols (summarized in Cech, 1990; and Edwards and Cech, 1990) were not followed. For example, some workers did not allow sufficient time between fish handling and data acquisition. Reports also exist in which essential details such as fish size, experimental temperature, and ambient O_2 and CO_2 levels in the chambers were not given. Other frequently omitted details included the time fish had been in captivity, their diet and the time they were fasted prior to study, the length of time they were acclimated to the chamber prior to measurements, and the duration of the measurement periods. In some bimodal studies, the methods used to correct for air–water phase diffusion were not stated. Also, to be useful, bimodal data must be accompanied by simultaneous records of air-breathing frequency, which was not always done. Lastly, some results were reported without information on sample size and variability.

Unimodal respirometry has enabled definition of the limitations of gas exchange in air and water. Studies with forcibly exposed fish reveal aerial $\dot{V}O_2$ capacity, and thus provide an index of scope for terrestrial activity; these can also be used to measure cutaneous gas exchange. General results are that the naturally amphibious species (i.e., most of the active intertidal forms, *Monopterus*, *Synbranchus*, *Erpetoichthys*, and *Clarias*) have a high $\dot{V}O_2$ in air. There is large variation, however, in the responses of some aquatic air breathers; the often amphibious *Anabas* displays a reduced $\dot{V}O_2$ in air, as do *Misgurnus* and *Anguilla*, while species of *Channa*, which seldom leave water, have an elevated $\dot{V}O_2$. The findings that *Periophthalmus* and *Boleophthalmus* have a higher bimodal $\dot{V}O_2$ than when exclusively in air may reflect the added energetic costs of amphibious activity. Unimodal respirometry indicates that the skin of air-breathing fish has an important gas-exchange function in both water and air.

Respirometry during forced submergence verifies that most aquatic air-breathers having a diminished gill ventilatory capacity are not able to sustain their $\dot{V}O_2$ at normal bimodal levels. This is particularly evident in *Protopterus* and *Osphronemus*, both obligatory air breathers; however, there is wide variation in the aquatic respiratory capacities of different *Channa* species. The reduced $\dot{V}O_2$–mass exponent of forcibly submerged *Anabas* demonstrates how respiratory morphometrics can affect partitioning. An important task for future studies will be to quantify the relationship between an air-breathing fishes' aquatic respiratory capacity and morphometrics (including scaling effects), as well as to factor in the added effects of variables such as obligatory air breathing.

Further evidence that air-breathers have a generally reduced aquatic respiratory scope has come from forcibly exposing them to progressive hypoxia without air access. Under such conditions most fishes (*Heteropneustes*, *Clarias*, *Ancistrus* and others) respond similarly to non-air breathers and regulate $\dot{V}O_2$ down to a critical ambient O_2 tension. Data for *Electrophorus* (Farber and Rahn, 1970) suggest it is also an O_2 regulator, but this needs further verification. The $\dot{V}O_2$ of *Synbranchus*, by contrast, passively conforms and thus falls with progressive hypoxia. Forced submergence respirometry has also revealed species-specific differences (e.g., in *Amia* and *Lepisosteus*) in the threshold level of hypoxia that triggers air breathing, and in the case of *Piabucina festae* and its non-air-breathing congener *P. panamensis*, fully documents that "the air-breathing option" has selected against maintenance of gill-ventilatory and aquatic-respiratory capacities in hypoxia.

Little is known about the mechanisms by which metabolism is depressed in prolonged aerial exposure (*Protopterus*, *Synbranchus*) or in severe hypoxia without air access (*Heteropnuestes*, *Gymnotus*). There are no studies of the aquatic respiratory responses of amphibious marine air breathers, most of which would not naturally experience hypoxia in their habitat or have the capacity to avoid this condition by micro-habitat selection.

Considerable range exists in the bimodal partitioning of $\dot{V}O_2$ and $\dot{V}CO_2$, and most species appear capable of altering partitioning in correlation with ambient conditions. Major effects on partitioning have been shown for aquatic PO_2, temperature, and body size. The obligatory air-breathing *Protopterus* has a higher dependence on air than most species; however, among other species, correlations are not found between obligatory versus non-obligatory status and partitioning level (e.g., *Channa argus* versus *C. maculata*). In some cases, but not all, the size effect on partitioning closely correlates with respiratory surface allometry.

Aerial and aquatic R values compare differences in bimodal rates of O_2 and CO_2 transfer. For aquatic air breathers, aerial R is generally below 0.3 while aquatic R exceeds unity, indicating that the bulk of CO_2 is released aquatically. Aerial R can be elevated by hypoxia, warm temperature, and air exposure, but generally remains lower than needed to prevent hypercapnic acidosis in emergent fish. Many amphibious air breathers, on the other hand, appear capable of sustaining high rates of aerial CO_2 release and thus remain in respiratory equilibrium; however, studies have yet to correlate aerial $\dot{V}CO_2$ with the different normal emergence times and activity levels of the amphibious species. No clear relationship exists between the observed rates of CO_2 release from either gills, skin, or ABO surfaces, and the activity of carbonic anhydrase.

Monitoring temporal changes in ABO–respiratory gas levels demonstrates the importance of breath volume, organ perfusion, breath duration, and Hb–O_2 affinity on the extent of O_2 extraction from the ABO. Among these, the time that a fish holds air is the major determinant; however, because air-breathing frequency increases with hypoxia, organ perfusion rate, and very likely transbranchial O_2 loss, must play a role. In certain cases (*Lepisosteus* in hypoxia, *Channa maculata* in 25 C and hypoxia), the rate of O_2 extraction may be limiting or, as suggested for *Clarias lazera*, some O_2 taken up in the ABO may be lost via gills. At the other extreme, the O_2 level in breaths held for a long period may decline to the point that O_2 extraction becomes limited by blood Hb–O_2 affinity.

Fishes that exhale prior to inhaling and thus change more of their ABO volume on each breath (listed in Chapter 3) are able to gain the largest post-breath ABO–O_2 volume. While the ABO volume of most species is limited by the organ's noncompliant surfaces, the ABO–stomach of *Ancistrus* becomes stretched following air-breathing induction. Also, owing to increases in its Hb–O_2 affinity during hypoxia acclimation, *Ancistrus* extracts more O_2 from each breath and thus reduces both air-breath frequency and air-convection requirement.

In species such as *Arapaima*, *Electrophorus*, *Channa*, and others, or in ambient conditions where there is a large CO_2 diffusion gradient between blood and newly inspired air, the instantaneous rate of CO_2 release into the ABO can be rapid. However, in most cases, and when integrated over the entire breath duration, the amount of CO_2 released into the ABO is small.

Several facets of gill function during air breathing have been elucidated. Demonstrations of transbranchial O_2 loss (*Lepisosteus*, *Amia*) and low O_2 utilization (*Channa*) confirm a limited gill function for some air breathers. On the other hand, entirely different aspects of integrated aerial and aquatic respiratory function have been brought to light by findings that aerial and aquatic $\dot{V}O_2$ are reciprocally modulated during the air–breath cycle (*Clarias lazera*) and that both *Clarias mossambicus* and *Ancistrus* can improve aquatic respiratory capacity in hypoxia as a result of hypoxia acclimation and air breathing "experience." Unfortunately, we are presently unable to define the point of balance for the integrated function of aerial and aquatic respiration in CO_2 release. In all species studied thus far, aquatic R is much larger than aerial R, implying that morphological and physiological barriers to transbranchial O_2 loss do not retard the bulk of CO_2 release. On the other hand, and as will be seen in Chapters 6 and 7, some species appear to be chronically hypercapnic, an apparent consequence of limited CO_2 diffusion across the gills.

CHAPTER

6

Cardiorespiratory Control

In retrospect, it is easy to see how water-breathing animals became easily addicted to breathing air since its O_2 was easily accessible, always constant, and at a very high concentration compared with that dissolved in water. As long as they could eliminate the CO_2 through their gill-skin system, it would not build up in the air organ or primitive lung, and thus they could extract most of the inhaled O_2 by holding their breath for long periods and only occasionally replenishing their air supply.

Hermann Rahn and Barbara Howell,
Respiration of Amphibious Vertebrates, 1976

INTRODUCTION

This chapter describes the sensory interactions regulating fish air breathing and the environmental, physiological, and evolutionary factors affecting the integration of air breathing with aquatic respiration and with the cardiovascular system. For many species, information about respiratory or ventilatory control began accumulating with the first documentation of air-breathing behavior. Although many early experiments were done (Chapters 2 and 5), Willmer's (1934) frequently reproduced graph (Figure 6.1) depicting the combined and separate effects of ambient CO_2 and O_2 on the air breathing of the yarrow (*Hoplerythrinus*) probably constitutes the first ventilatory control data for a bimodal species. Willmer integrated diverse topics such as the ecology and physiology of a species into speculation about how CO_2 and pH might affect Hb–O_2 binding as well as the "brain" respiratory center. Reviews by Carter (1957) and Johansen (1970) focused attention on the mechanisms of fish air-breathing control. Rahn's (1966a) quantitative exposition of how physical and chemical differences between air and water affect the physiology of respiration is a benchmark for this field (Dejours, 1988, 1994). Several recent reviews of bimodal control in fishes and lower vertebrates (Shelton *et al.*, 1986; Shelton and Croghan, 1988; Milsom, 1989a, 1990, 1991) reflect current interest in this subject, and a series of papers by Smatresk (1988, 1990, 1994) demonstrate the linkage between basic studies with fishes and an understanding of the mechanisms and evolution of tetrapod ventilatory control.

This essay begins by considering how physical differences between air and water have influenced the physiology of ventilatory control. Basic features of neural respiratory control are then reviewed through a contrast of the unimodal breathing patterns in fishes and mammals. This is followed by a comparative synthesis of bimodal respiratory control in fishes.

PHYSICAL AND CHEMICAL ASPECTS

The low O_2 capacitance of water requires that fishes have high aquatic ventilation and O_2 extraction rates (Rahn, 1966a; Dejours, 1988, 1994; Perry and Wood, 1989). Conversely, the relatively high solubility of CO_2 in water (CO_2 is about twenty-eight

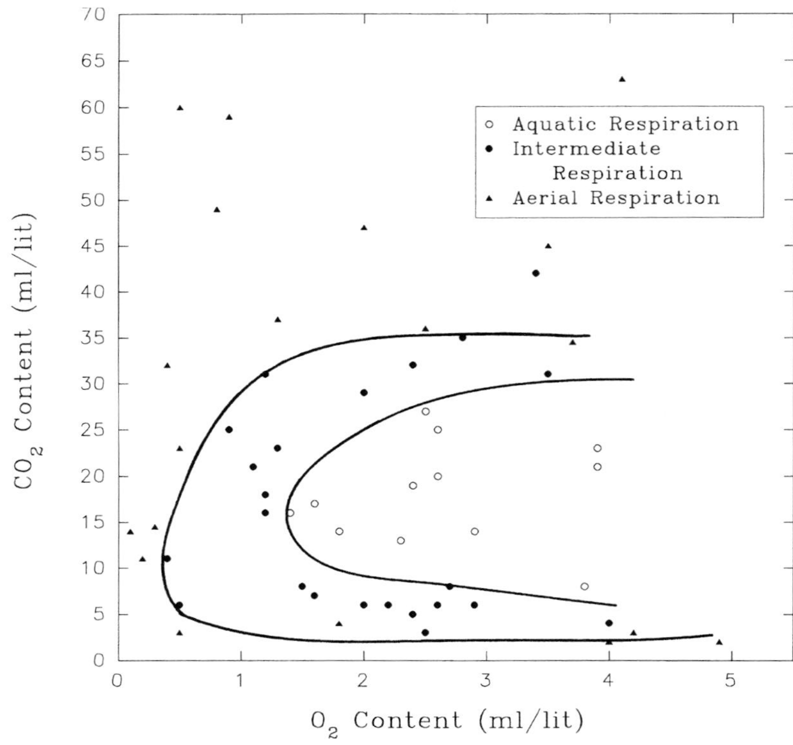

FIGURE 6.1 Reproduction of Willmer's (1934) graphical depiction of the effects of aquatic O_2 and CO_2 contents on the respiratory behavior of a single *Hoplerythrinus*. (Reprinted by permission of the Company of Biologists, Ltd., from the *Journal of Experimental Biology*, volume 11, pp. 283–306.)

times more soluble than O_2, Chapter 1) virtually assures that, at the ventilation volumes needed for O_2 uptake, PCO_2 levels in tissues, blood, and the exhalant water flow will usually be quite low, even at normally high aquatic R ($\dot{V}CO_2/\dot{V}O_2$, Chapter 5) values. For example, if respiratory O_2 extraction lowers the PO_2 in the branchial ventilatory stream of a fish by 10 torr, this would be matched by a PCO_2 rise of 10/28 torr (assuming R = 1.0). Therefore, a first principle of aquatic ventilatory control, one that has its basis in the physical properties of water, is that the relative difficulty in obtaining O_2 from water is of much greater significance than the relative ease with which the release of CO_2 is assured by its high solubility in water (Rahn, 1966a; Rahn and Howell, 1976).

Another facet of the control differences between the aquatic and terrestrial vertebrates arises from the nature of CO_2 in plasma and tissues. Because CO_2 readily combines with water to form carbonic acid, PCO_2 affects plasma and tissue pH, as shown by the Henderson-Hasselbalch equation:

$$pH = pK_1 + \log ([HCO_3^-] / \alpha CO_2 \cdot PCO_2) \quad (6.1)$$

where pK_1 is the negative logarithm of the carbonic acid dissociation constant, and α is the aquatic solubility coefficient for CO_2.

Because of the greater solubility of CO_2 relative to O_2, and because aquatically respiring fishes ventilate their gills more or less continuously, fish tissue and blood PCO_2 and HCO_3^- levels are generally very low (Garey, 1970; Holeton, 1980; Randall and Daxboeck, 1984). Fishes thus have little potential for modulating their rate of CO_2 release in order to regulate plasma pH. Respiratory control, therefore, has little to do with the chemistry of CO_2 unless, as is the case for some air-breathing species, aquatic ventilation becomes dramatically curtailed through a reduction in gill surface area, branchial apnea, or emersion (Garey and Rahn, 1970; Cameron, 1989). In contrast, the terrestrial vertebrate or tetrapod, generally has a higher level of plasma CO_2, as well as the potential to regulate blood pH by altering ventilation (pulmonary CO_2 release), by the excretion of bicarbonate through the kidney, or both of these. It seems clear that the evolutionary origin of the pH-PCO_2 ventilatory regulating mechanism of tetrapods was forged by a suite of factors that included a limited contact with environmental water, a dependence on lung ventilation for the release of CO_2, and the chemistry dictated by Equation 6.1 (Rahn, 1966a; Howell, 1970; Jones and Milsom, 1982; Shelton and Croghan, 1988).

COMPARATIVE ASPECTS OF VENTILATORY CONTROL: FISHES AND MAMMALS

The ventilatory control mechanisms of all vertebrates consist of three basic elements: Sensory receptors, a central (central nervous system) integrator or controller, and motor-neuron effectors. Unimodal fishes and the mammals are similar in that (with some notable exceptions, see Milsom, 1990) most species have a continuous ventilatory pattern. Much more is known about the central control of respiration in mammals than in fishes. The mammal ventilatory signal originates endogenously within a region of the medulla, termed the central rhythm generator (CRG). An adjacent medullary region, the central pattern generator (CPG) processes the CRG signal, integrates it with inputs from peripheral and central sensory receptors and higher brain centers, and sends the resulting rhythmic impulses down the motor column of the medulla to stimulate nerves contacting the ventilatory apparatus (Smatresk, 1990, 1994). Although many fewer details are known for fishes (for example, an endogenous rhythm generator has not been isolated), their respiratory control mechanism seems somewhat similar to that of mammals.

Ventilation is influenced by sensory feedback from both central and peripheral sensory receptors. A diversity of sensory cell types including chemoreceptors, mechanoreceptors, taste and olfactory receptors, and nociceptors (cells responsive to injury or toxic substances) can serve in this capacity. Peripheral-receptor feedback occurs via afferent (sensory) cranial nerve branches (Shelton and Croghan, 1988; Feldman *et al.*, 1990; Milsom, 1990; Smatresk, 1988, 1994). In most vertebrates sensory afferents of the vagus nerve extend throughout the body. In fishes the vagal network connects the gill arches, heart, and the organs within the coelomic cavity (Smatresk, 1990; Davies *et al.*, 1993) and, in view of the diversity of organs functioning as ABOs in the actinopterygians, this connection has played a critical role in coordinating aerial and aquatic respiration and cardiorespiratory interactions. A subsequent section will discuss the various receptor types affecting fish bimodal ventilation. The following account of fish and mammalian oxygen receptors underlines basic similarities in the sensory mechanisms of the two groups.

In mammals the third embryonic aortic arch forms the internal carotid arteries. Two types of peripheral O_2 chemoreceptors are located on the carotid arteries, the carotid and aortic bodies. The carotid body is a highly perfused tissue that senses deviations in arterial PO_2. Aortic bodies occur in the nutritive circulation (i.e., within the vessel wall) and are most sensitive to changes in blood-flow rate and O_2 content. These receptors are innervated by afferent branches of cranial nerves IX and X (Smatresk, 1990).

Fish O_2 chemosensors occur throughout the gills and in most species are concentrated in the more anterior branchial arches (Jones and Milsom, 1982). The homology of fish and mammalian O_2 chemosensors is suggested, although not confirmed, by their morphological similarity as well as a common site of embryonic origin (i.e., as with the carotid artery, branchial arch 1 is derived from aortic arch III, Chapter 4). Fish O_2 chemoreceptors are also innervated by afferent branches of cranial nerves IX and X, and there are both externally and internally orientated O_2 receptors. Internal sensors monitor blood O_2 content and, when stimulated, affect both ventilatory frequency and amplitude. External receptors monitor ambient O_2 level and affect both cardiac and respiratory activity (Smatresk, 1988). Smatresk (1994) suggested a functional homology in the chemoreflexive action of the external O_2 receptors, which have an action similar to the mammalian carotid body, and the internal receptors, whose actions resemble the aortic bodies.

The major difference in the ventilatory control of fishes and mammals is that fish ventilation is driven by peripheral oxygen receptors. Mammals also have peripheral O_2 receptors; however, these have a smaller effect on ventilation than central chemoreceptors sensitive to slight changes in medullary tissue CO_2, pH, or both of these variables. All tetrapod classes, including amphibians appear to have central CO_2 or pH receptors (Shelton *et al.*, 1986; Dejours, 1988, 1994; Smatresk, 1994) and in some mammals these appear to have direct input into the CRG itself (Feldman *et al.*, 1990; Smatresk, 1990). By contrast, no fishes, including the air breathers, are known to have central CO_2/pH receptors.

STUDIES OF BIMODAL CONTROL

Similarities, as well as contrasts, in the ventilatory control mechanisms of mammals and fishes have stimulated research with the bimodally breathing fishes and amphibians. One objective has been to find the point of evolutionary divergence for the aerial and aquatic respiratory control mechanisms (Smatresk, 1990; Pack *et al.*, 1992). With some exceptions, most of the unimodal fishes and the mammals ventilate continuously (Milsom, 1990). The bimodal breathers, however, have an intrinsically variable ventilation pattern driven largely by peripheral sensory feedback (Shelton and Croghan, 1988; Hedrick *et al.*, 1994). This difference has raised questions about the evolutionary sequence leading to tetrapod central control and has even suggested a shifting role for central and peripheral receptors in the evolutionary transition from bimodal to exclusively aerial vertebrate respiration (Shelton *et al.*, 1986; Smatresk, 1988, 1990, 1994; Feldman *et al.*, 1990; Pack *et al.*, 1990). The comparative study of extant air-breathing fishes allows some insight into these issues.

Experimental Approaches

Investigations of bimodal control have been as varied as the species investigated. Because fish air breathing is easy to observe, much information has come from studies of behavior in relation to physical conditions. Other research has included *in vivo* monitoring of physiological parameters (e.g., ABO pressure, volume, respiratory gas levels, blood respiratory properties) and their correlation with changes in either, or both, aerial and aquatic ventilatory patterns. There have also been more direct approaches such as cranial nerve sectioning or ablation, brain transection, or recordings from isolated efferent branches of the visceral vagus and intracranial perfusion with mock extradural fluid. Techniques such as infusing NaCN, which interferes with electron transfer in the mitochondrial respiratory chain, into different regions of the circulatory system have enabled localization and characterization of peripheral chemoreceptors affecting bimodal breathing. Thus, the challenge for this chapter is to develop an overview of bimodal control that encompasses the broad spectrum of data that have been obtained by employing a variety of techniques on a diversity of air-breathing species.

The Scope of Bimodal Control

Figure 6.2 illustrates the hierarchy of factors potentially influencing ventilatory control. Although it is rare when more than a few details are known for any

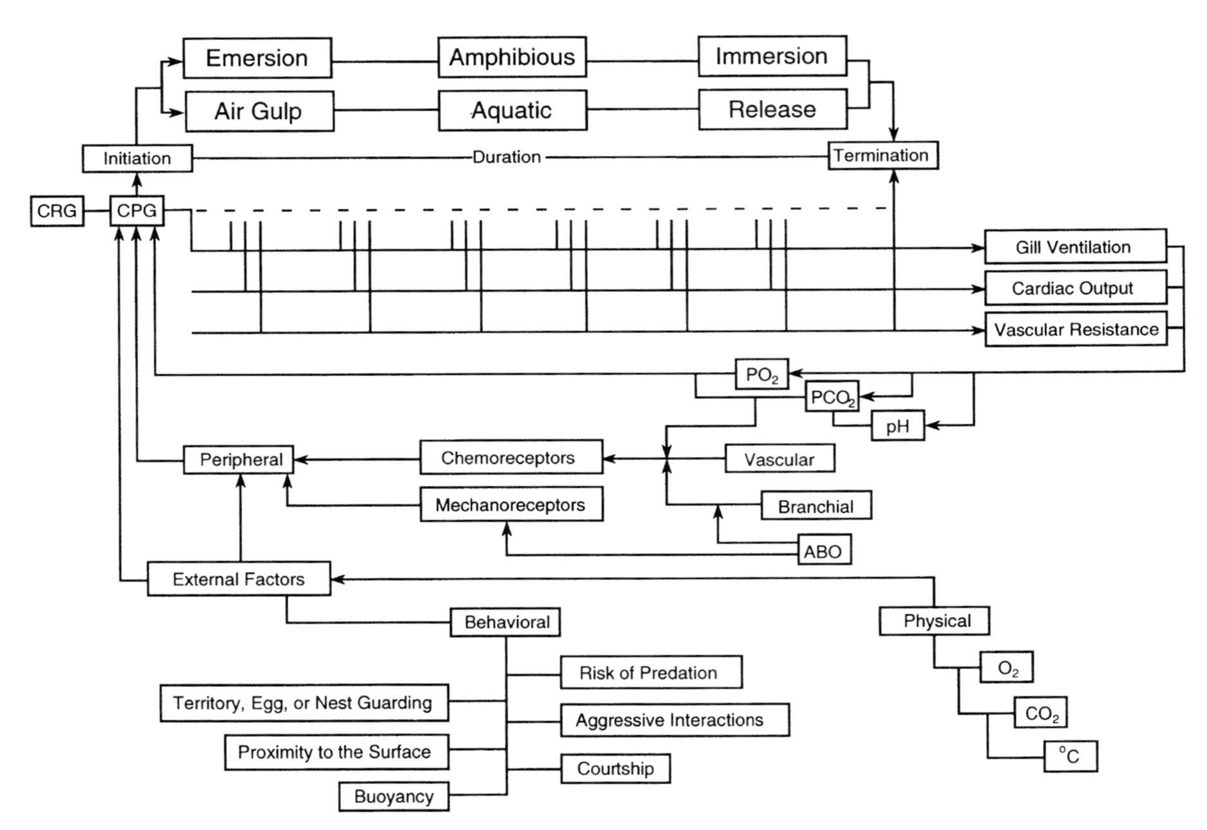

FIGURE 6.2 Integrated flow scheme showing the scope and diversity of factors likely to affect the onset, duration, and termination of an aquatic air breath or an amphibious excursion.

species, this figure shows that the onset, duration, and termination of the air-breathing interval could all be under the combined influence of many variables (Shelton *et al.*, 1986; Shelton and Croghan, 1988). Smatresk (1988), for example, lists no less than eleven separate physical factors or experimental manipulations (see following) affecting aspects of bimodal respiration in at least 15 species. Shelton and Croghan (1988) developed a respiratory control model for *Electrophorus* that integrated diverse features including the volume and rate of decline of ABO-O_2, blood O_2 affinity and storage capacity, respiratory frequency and partitioning, and metabolic rate. Katz (1996) proposed a signal conditioning model for air breathing in *Amia* and morphologically similar species in which, owing to phasic changes in ABO-venous return (Chapter 4), and a sigmoidally shaped O_2 dissociation curve, the gills amplify fluctuations in venous PO_2 and strengthen the air-breath eliciting response of arterial chemoreceptors.

This discussion of ventilatory control proceeds by examining the details of Figure 6.2 in order to show general organizational principles of the control mechanism and to document the breadth and diversity of variables potentially affecting this process in aquatic and amphibious air-breathing fishes. This is followed by an analysis and synthesis of experiments on fish bimodal respiratory control.

Figure 6.2 raises the possibility that bimodal control may reside within the CPG, be affected by peripheral receptors, and also be influenced by visual and other inputs. It is easy to imagine how the inputs from a variety of sensors could change in the short span of an air breath or an amphibious excursion. However, which factors have the greatest influence on breathing behavior or terrestriality is unknown. In addition, inputs from higher sensory modalities such as vision descend onto the CPG and are integrated with central and peripheral stimuli to behaviorally optimize air ventilation or emergence. Emergence may be prompted by stimuli including topography or wave action, the sighting of a conspecific, a predator, or potential prey, climate, time of day, crowding, and a diminished food supply (Chapter 2; Graham, 1970; Liem, 1987; Sayer and Davenport, 1991).

For aquatic air breathers, the threat of predation during surfacing for air means that auditory and visual stimuli may override the normal ventilatory signal (Kramer and Graham, 1976; Smith and Kramer, 1986). Air-breathing behavior thus becomes subject to either, or both, conspecific behavioral patterns and the activity of potential predators (Kramer and Graham, 1976). In total, a broad diversity of variables including physical factors (Chapter 5), buoyancy control, heterospecific aggression, conspecific synchronous air-breathing behavior, territorial defense, nest and larval protection, and even behavioral-ecological factors are elements of "higher" respiratory control (Spurway and Haldane, 1953; Halliday and Sweatman, 1976; Kramer and Graham, 1976; Liem, 1987; Kramer, 1988; Smatresk, 1988; and Chapter 2).

STUDIES OF CENTRAL AND PERIPHERAL CONTROL

Investigations of central control have sought evidence for central O_2, CO_2, and pH receptors, and to discriminate the presence of separate branchial and aerial CPGs (Smatresk, 1994). There has been no substantiation of central O_2, CO_2, or pH reception in any bimodal fish.

Evidence suggesting an aerial CPG has been reported for lungfishes. DeLaney and Fishman (1977) found that *Protopterus* has rhythmic bouts of air breathing that occur independent of branchial activity, thus implying the presence of two separate (branchial and aerial) CPGs. Pack *et al.* (1990) confirmed this by showing that, in a *Protopterus* with its lungs cannulated in such a way that *in vivo* pulmonary O_2 tension and pressure (volume) could be regulated, volitional air breaths by the fish were followed by a slowing of gill ventilation. Because these breaths caused only slight and transient changes in lung volume and pressure, Pack and co-workers concluded that, rather than being an inflation reflex, the post-breath reduction in gill ventilation displayed by *Protopterus* was a centrally patterned response. Further, the possibility that either the passage of air through the buccal pump or transient pressure changes associated with spontaneous air ventilation might have triggered peripheral receptors was ruled out by the finding that manual lung inflation did not cause a reduction in aquatic ventilation. ABO mechanoreceptors thus have little influence on gill ventilation in *Protopterus*, suggesting the presence of a central interaction between separate branchial and pulmonary CPGs (Pack *et al.*, 1990).

Among actinopterygians, most work has focused on *Amia* and *Lepisosteus*. Hedrick *et al.* (1991) found no evidence for central receptors in *Amia*. These workers applied a mock extradural fluid (EDF) through the meningeal spaces onto the brain of conscious fish. Tests with the vital dye, Sudan Black, confirmed that their mock EDF permeated all brain surfaces, including the ventral medulla and cerebral ventricles, and flowed into the cerebrospinal fluid (CSF). Manipulations of the EDF, such as making it hypoxic, hypercapnic, or acidotic, all failed to elicit changes in either the branchial or aerial ventilation of *Amia*. By contrast, a direct link between the CSF and neurons affecting the central respiratory drive in amphibians is suggested by experimental manipulation of the CSF in toads (Branco *et al.*, 1991; Smatresk and Smits, 1991). This again differs from mammal preparations where changes in CSF are not the primary stimulus for medullary chemoreceptors (Kiley *et al.*, 1985). Rather, ventilation is altered in response to a pH reduction of the extracellular fluid (ECF) within the medulla itself and these pH changes are more likely caused by blood borne acid-base fluctuations than by the CSF.

Cranial nerve ablation has also been used to test the role of peripheral chemoreceptive afferent pathways in aerial-ventilatory control. In most fishes, input from branchial O_2 receptors reaches the CNS via cranial nerves IX and X. Surgical sectioning or removal of these nerves offers a way of discriminating between branchial and other peripheral sensory stimuli eliciting air-breathing behavior. This is technically difficult and is seldom done without undesirable side effects that include altered behavior, metabolic depression, and stress (McKenzie *et al.*, 1991a; Hedrick and Jones, ms). Nevertheless, in addition to abolishing the stimulatory effects of aquatic hypoxia and external NaCN (see following) on aquatic ventilation, branchial denervation attenuated air-breathing behavior in *Protopterus* (Lahiri *et al.*, 1970), and stopped the hypoxic water induced air-breathing response of *Lepisosteus oculatus* (Smatresk, 1988, 1994). For *Amia*, McKenzie *et al.* (1991a) found that branchial denervation coupled with pseudobranch ablation entirely abolished air-breathing responses to hypoxia. In contrast, Hedrick and Jones (1993, ms) reported that branchial denervation depressed branchial ventilation but did not abolish the air-breathing reflex in hypoxic water. Thus, while the pattern for *Amia* is less defined, it is clear that peripheral sensors play a role in aerial respiratory control.

Ambient Effects on Ventilatory Control

Whereas Chapter 5 detailed the broad spectrum of factors affecting air-breathing behavior, this section focuses on the relationship of ambient respiratory gas levels (i.e., O_2 and CO_2) in both the aerial and aquatic media to ventilatory behavior.

Hypoxia

As detailed in Chapter 5, an increased air-breathing frequency in progressive aquatic hypoxia is a nearly uniform response of the aquatic air-breathing fishes. Although aquatic hypoxia does not affect the aerial tidal volume of most species (Chapters 3 and 5; Hedrick and Jones, 1993), such an effect was found in both *Lepisosteus* (Smatresk and Cameron, 1982a) and *Trichogaster* (Burggren, 1979).

How does hypoxia affect branchial ventilation? Most non-air-breathing fishes increase branchial ventilation and O_2 extraction in progressive aquatic hypoxia (Hughes, 1963, 1973b; Holeton, 1980; Randall, 1982). In the case of the air-breathing fishes, hypoxia will affect aquatic ventilation in a variety of ways depending upon respiratory partitioning. For the obligatory air breathers *Protopterus*, *Lepidosiren*, *Electrophorus*, and *Monopterus*, all of which have reduced gill surface area, hypoxia has little effect on an already low gill ventilation frequency. Among species having a larger gill area, some maintain a constant gill ventilation frequency in hypoxia (*Neoceratodus*, *Synbranchus*, *Lepisosteus*), while others (*Erpetoichthys*, *Misgurnus*, *Hoplosternum*, *Brochis*, *Piabucina*) reduce it slightly. Still other species having normally low rates of aerial O_2 utilization in normoxia (juvenile *Protopterus*, *Polypterus*, *Heteropneustes*, *Clarias*, *Umbra* and *Amia*) employ branchial hyperventilation down to the threshold PO_2 eliciting air breathing. This is the same pattern seen for the facultative air breathers *Ancistrus*, *Hypostomus*, and *Gymnotus*. Once air breathing is initiated, most species modulate branchial ventilation to some extent; gill ventilation frequency is usually the lowest just after the breath is taken and then gradually increases over the time air is held (see Chapter 5 and following).

The hypoxic responses of amphibious air breathers appear similar to aquatic air breathers in that ventilation increases until an O_2 threshold for emergence is reached. Data on the O_2 thresholds for air-breathing and emergence (Tables 6.1 and 6.2) provide comparative information on this, however, more studies are needed.

Internal versus External Oxygen Sensors

The aquatic hypoxia thresholds depicted in Tables 6.1 and 6.2 suggest the control of air breathing and emergence may lie with an external (water monitoring) O_2 chemoreceptor. However, it is undoubtedly true that progressive hypoxia also activates internal (i.e., tissue or intravascular hypoxemia) sensors. The problem becomes one of discovering these different O_2 receptors and discriminating their hierarchy in respiratory control (Smatresk, 1988).

There have been several approaches to this question including studies of hypoxia-acclimation effects on air-breathing threshold, the selective administration of chemicals stimulating respiratory responses,

TABLE 6.1 Hypoxia Air-Breathing Thresholds for Some Aquatic Breathers[a]

Species	O_2 Threshold (torr)	°C	Reference
Neoceratodus forsteri	85[e] (50-115)	18	Johansen *et al.*, 1967
Amia calva	90	18	Johansen *et al.*, 1970a
Hoplerythrinus unitaeniatus	81 (2–127)[d]	27–30	Stevens and Holeton, 1978b
	20–80	25	Mattias *et al.*, 1996
Clarias batrachus	40–60[c]	25	Jordan, 1976
Hypostomus plecostomus	21	24	Gee, 1976
	60	25–27	Graham and Baird, 1982
	79[g]	25–27	
	35	25	Perna and Fernandes, 1996
H. regani	60	20,25,30	Fernandes *et al.*, 1995[i]
Ancistrus chagresi	35	24	Gee, 1976
	33[b,f]	25–27	Graham and Baird, 1982
	60[g]	25–27	
Loricaria uracantha	32	24	Gee, 1976
Chaetostoma fishcheri	57	24	Gee, 1976
Sturisoma citurense	25	24	Gee, 1976
Rhinelepis strigosa	22	25	Takasusuki *et al.*, in review
Heteropneustes fossilis	100	24	Hughes and Singh, 1971
Gymnotus carapo	60	29	Liem *et al.*, 1984
Pygidium striatum	41	24	Gee, 1976
Umbra limi	12–48[b,f]	5–30	Gee, 1980
Synbranchus marmoratus	33–69[c,f]	25	Graham and Baird, 1984
	30	25	Bicudo and Johansen, 1979
	54[h]	25	

[a]Parenthetical values are ranges.
[b]Temperature dependent.
[c]Varies inversely with body size.
[d]A continuous breather but threshold data reported.
[e]Varies directly with size.
[f]No effect of hypoxia acclimation.
[g]O_2 threshold changed by hypercapnia (Table 6.3).
[h]Hypoxia acclimation effect.
[i]Also a personal communication for the threshold data.

TABLE 6.2 Progressive Hypoxia Emergence Thresholds for Some Amphibious and Marine Intertidal Fishes

Species	O_2 Threshold (torr)	°C	Reference
Mnierpes macrocephalus	21	27	Graham, 1970
Blennius pholis	25	20	Davenport and Woolmington, 1981
Taurulus bubalis	6.2[a]	20	Davenport and Woolmington, 1981
Helcogramma medium	18(1.5)	15	Innes and Wells, 1985
Oligocottus snyderi	22–25	12–15	Yoshiyama *et al.*, 1995
O. maculosus	22–25	12–15	Yoshiyama *et al.*, 1995
Clinocottus globiceps	22–25	12–15	Yoshiyama *et al.*, 1995
Ascelichthys rhodorus	22–25	12–15	Yoshiyama *et al.*, 1995

[a]Strong temperature effect; 0.6 torr at 5 C, 24.1 at 28 C.

surgical ablation of afferent sensory inputs, and exposure of air-breathing fishes to different ambient gas mixtures.

If internal O_2 sensors elicit air breathing, then it could be expected, assuming that hypoxia acclimation can increase air-breathing efficacy (Chapters 5 and 7), that this treatment might also shift the ambient O_2 threshold for facultative air breathing. Studies with *Ancistrus*, however, failed to show a change in air-breathing threshold with acclimation, thus suggesting an external O_2 receptor is the primary trigger for air breathing (Graham and Baird, 1982). Comparable experiments with *Umbra* (Gee, 1980) and *Synbranchus* (Graham and Baird, 1984) also revealed no hypoxia acclimation effect on air-breathing threshold. Similarly, Fernandes and co-workers (1995, and pers. comm.) report that temperatures between 20-30 C do not affect the air-breathing threshold of *Hypostomus regani*.

Application of abruptly acting stimuli such as nicotine (a cholinergic receptor agonist) or NaCN to discreet sites has also been used to approach the question of which O_2 receptors trigger air breathing. This work confirms the presence of both external and internal O_2 receptors. The application of a small quantity of nicotine under the opercula of *Protopterus* stimulated both gill and lung ventilation (Johansen and Lenfant, 1968). This response was also elicited by injection of a small bolus of cyanide into the anterior branchial circulation (Lahiri *et al.*, 1970).

Specific differences correlated with aerial respiratory dependence are also evident in the external and internal NaCN responses of *Amia* and *Lepisosteus*. In the case of *Amia*, addition of NaCN to its gill ventilatory flow stream or the injection of this compound into its dorsal aorta causes an increase in gill ventilation frequency but does not elicit air breathing (McKenzie *et al.*, 1991a). By contrast, both of these treatments stimulate aerial and aquatic respiration in *L. osseus* (Smatresk *et al.*, 1986).

Selective NaCN injections were used by Smatresk *et al.* (1986) to localize internal O_2 receptors in *L. osseus*. As opposed to injections into the dorsal aorta, injections into the ventral aorta and conus arteriosus elicited stronger and faster ventilatory responses, suggesting that the O_2 receptors are positioned in the head or branchial circulation. Extending these techniques further, Smatresk and co-workers were able to detail the interaction of external and internal O_2 receptors regulating the bimodal breathing of *Lepisosteus*. Internal receptors monitor blood PO_2 at a central vascular site downstream from the ABO (i.e., between the heart and dorsal aorta) and stimulate gill ventilation and air-breathing frequency as required by metabolic demand. External sensors monitor ambient O_2 and regulate aquatic ventilation accordingly. Separate tests in which the ABO was ventilated with air, N_2, and O_2 demonstrated convincingly that, irrespective of tissue O_2 levels, the external O_2 sensor shuts off gill ventilation in hypoxia and turns it on in normoxia (Smatresk *et al.*, 1986).

Although a somewhat unnatural situation, experiments with hypoxic gas media have provided information about respiratory control. *Amia* in normoxic water, breathing air containing only 8% O_2, greatly increased air-breathing frequency (Hedrick and Jones, 1993). The air-breath duration of *Monopterus albus* was significantly reduced in fish breathing gas mixtures containing 16% O_2 or less, and breaths containing 1.5% O_2 were rejected almost immediately (Graham *et al.*, 1995a).

Hyperoxia

Aquatic hyperoxia occurs naturally in some air-breathing fish habitats (Chapter 1). High partial pressures of O_2 increase this molecule's diffusive flux into the body, and because fish respiration is driven mainly by O_2, hyperoxia usually reduces branchial-ventilation frequency and amplitude and lessens air-breathing frequency (Johansen, 1970; Smatresk, 1988). Virtually all species that have been tested reduce air ventilation frequency in hyperoxic water (Chapter 5), and hyperoxic fish may become largely unresponsive to aquatic stimuli such as hypercapnia and other agents (McKenzie *et al.*, 1991a,b) that would otherwise initiate air breathing. Reduced aquatic ventilation in hyperoxia can lead to a slight respiratory acidosis which is tolerated and ultimately compensated (Smatresk and Cameron, 1982a; Smatresk, 1988).

Although an unnatural condition, aerial hyperoxia has also been shown to lower the air-breathing frequency of almost all species so tested (e.g., *Protopterus, Erpetoichthys, Lepisosteus, Electrophorus, Monopterus, Synbranchus, Trichogaster*, and others, Chapter 5). One exception in this regard is *Amia*, which because it loses vast amounts of ABO-O_2 transbranchially, switches almost exclusively to type II breathing (i.e., inhaling but not exhaling, Chapter 2) to sustain buoyancy (Hedrick and Katz, 1991; Hedrick and Jones, 1993; Hedrick *et al.*, 1994).

Aquatic Hypercapnia

The potential importance of CO_2 to respiratory control was first suggested by Willmer (1934). Believing CO_2 to be a more critical element in respiratory control than O_2, Willmer interpreted the nearly complete cessation of yarrow branchial ventilation at PCO_2s below 10 torr and above 25-30 torr (Figure 6.1) as a mechanism for regulating plasma and tissue CO_2. He observed (page 296):

> [the] lack of CO_2 is a very important factor causing the fish to come to the surface to breathe air, and it seems reasonable to conclude that the 'respiratory center' of this fish is controlled by the CO_2 content of the blood which must neither be too high nor too low.

In spite of Willmer's conviction, little evidence has been compiled showing the role of CO_2 in respiratory control. Moreover, the ways that aquatic hypercapnia affects bimodal breathing are not easily summarized because many of the observations are cursory. Most tests were done by slowly bubbling CO_2 into water which often resulted in unrealistically high ambient levels (Figure 1.3) or rates of CO_2 change that were too rapid to permit full assessment of the response. This technique introduces other variables for which there are few control data. At elevated tensions, CO_2 readily diffuses across epithelial surfaces into the body, and once there, interpretations of a direct "hypercapnia effect" are obscured by the complex chemistry of CO_2 in plasma, its relation to blood pH and PCO_2 (Equation 6.1), its potential broad spectrum effects on respiratory control, and its influence on hemoglobin-O_2 affinity (Chapter 7). Johansen (1966) observed that PCO_2 causes erratic air-breathing and behavior in *Synbranchus*, and there is considerable evidence that elevated CO_2 can have a narcotic effect on fishes or may act as an anesthetic for many species. Thus, some of the putative "CO_2 effects on bimodal breathing" may actually reflect planes of neural dysfunction.

Owing to Willmer's influence, there have been numerous efforts to differentiate between low and high ambient CO_2 effects. Dehadrai (1962a for *Notopterus*), Todd (1971, *Gillichthys*), and others presented graphical interpretations similar to Willmer's. Shelton *et al.* (1986) have reviewed most of this work.

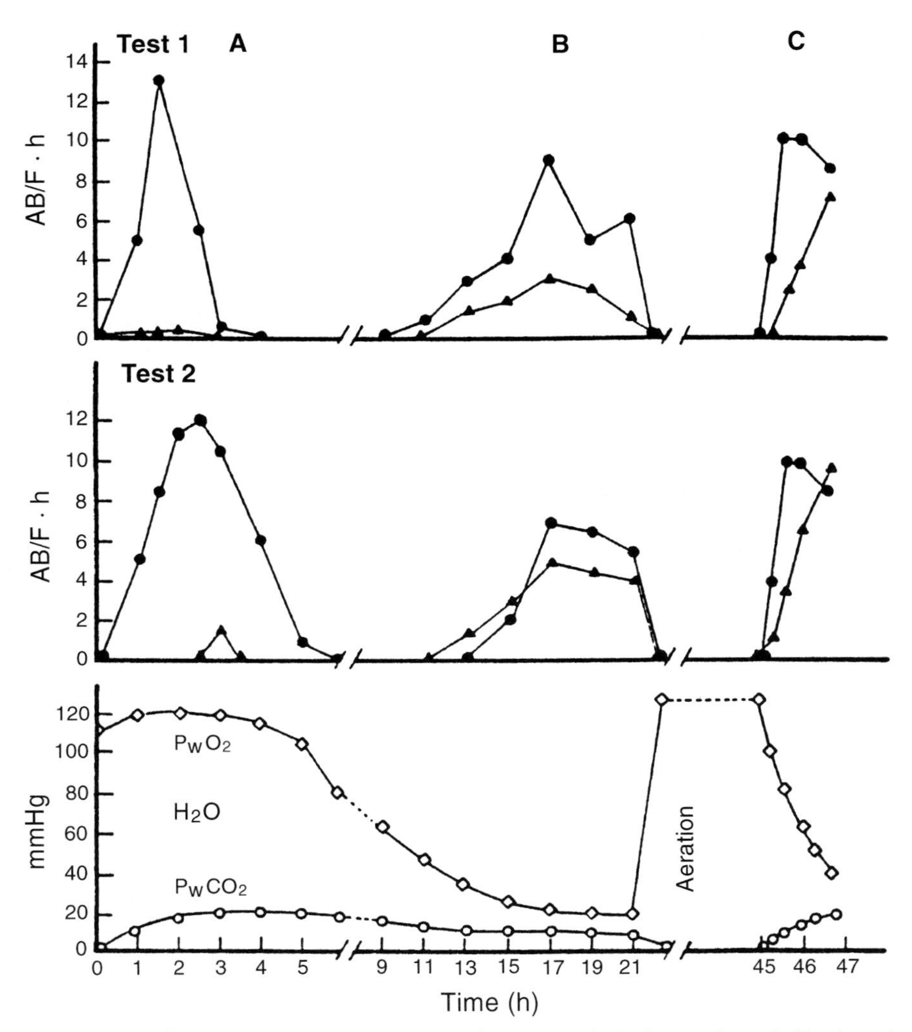

FIGURE 6.3 Air-breathing frequencies of groups of *Ancistrus chagresi* (dots) and *Hypostomus plecostomus* (triangles) during two 48 h tests in which fish were exposed to progressive hypercapnia while in normoxic water (A), followed by the gradual onset of hypoxia while in hypercapnic water (B); and, following 24 h in aerated water, abrupt exposure to simultaneous hypoxia and hypercapnia (C). Bottom panel shows O_2 and CO_2 tensions during the entire test period. (From Graham and Baird, 1982. Reprinted by permission of the Company of Biologists, Ltd., from the *Journal of Experimental Biology,* volume 96, pp. 53–67.)

Hypercapnia effects usually combine with those of hypoxia to trigger air breathing at a higher PO_2 (Table 6.1) and increase air-breathing frequency (Chapter 5). Hypercapnia can override the normal effect of aquatic hyperoxia in diminishing air-breathing frequency. The threshold levels of hypercapnia that either elicit air breathing or increase its rate have been determined for several species (Table 6.3). A rise in aquatic CO_2 to 3% increases the branchial ventilation of some species (*Neoceratodus*, *Anabas*, and *Protopterus*); higher levels depress aquatic ventilation in most. Aquatic hypercapnia thresholds affecting air breathing have been determined for a number of species (Table 6.3).

On the other hand, increases in aquatic CO_2 are without effect on the air breathing of *Electrophorus* and *Monopterus*. McMahon and Burggren (1987) determined that neither the branchial nor aerial ventilation of *Misgurnus* were affected by aquatic hypercapnia and concluded that CO_2 has only a minor respiratory drive. Todd (1972) found little evidence for a CO_2 effect in *Gillichthys*. Johansen (1966) found no effect of a high PCO_2 or low aquatic pH on the aquatic respiratory pattern of *Synbranchus*, which did, however, ultimately switch to air breathing.

TABLE 6.3 Aquatic CO_2 Threshold Levels Stimulating Either the Onset of Facultative Air Breathing or an Increased Frequency of Continuous Air Breathing

Species	O_2 Threshold (torr)	°C	Reference
Facultative Air Breathers			
Neoceratodus forsteri	9	18	Johansen *et al.*, 1967
Amia calva	14	18	Johansen *et al.*, 1970a
Heteropneustes fossilis	12	25	Hughes and Singh, 1971
Hypostomus plecostomus sp.	13[a] 11[b]	25–27	Graham and Baird, 1982
Ancistrus chagresi	9[a] 8[b]	25–27	Graham and Baird, 1982
Continuous Air Breathers			
Protopterus sp.	7[e]	25	Jesse *et al.*, 1967
Lepisosteus osseus	7–9	18–25	Rahn *et al.*, 1971
Notopterus sp.	5–7	—	Dehadrai, 1962a
Piabucina festae	20	24–26	Graham *et al.*, 1977
Anabas testudineus	14[c]	25	Hughes and Singh, 1970b
Trichogaster trichopterus	>2[d]	27	Burggren, 1979
Electrophorus electricus	10	—	Johansen *et al.*, 1968b
Monopterus cuchia	30	30	Lomholt and Johansen, 1974

[a]In normoxic water.
[b]In hypoxic water.
[c]PCO_2s between 14 and 28 torr had similar effect.
[d]Air-breathing frequency increases linearly from 0 to 45 torr.
[e]Juveniles.

Very few studies have documented the long-term effects of hypercapnia on air breathing. The air-breathing frequency and tidal volume of *Monopterus cuchia* were initially affected by an increased CO_2, but this response was neither prolonged nor consistent over time (Lomholt and Johansen, 1974). For *Ancistrus* and *Hypostomus*, an acutely applied hypercapnia of 20 torr evoked a short period of air breathing, but after 2–6 hours all air breathing ceased (Figure 6.3). A decline in a CO_2-elicited air-breathing response also occurred in *Piabucina* after a few hours (Graham *et al.*, 1977). Although not widely studied, this type of CO_2 compensation probably involves adjustments for respiratory acidosis that include elevation of extra- and intracellular bicarbonate, changes in other ion concentrations, adjustments of Hb–O_2 affinity, and other mechanisms (Cameron and Iwama, 1989; Chapter 7). *Lepisosteus oculatus*, for example, shows nearly complete blood pH compensation after 24 hour exposure to hypercarbic (1% CO_2) water (Smatresk and Cameron, 1982b).

More "invasive" methods of hypercarbic and other acid-base disturbances have been attempted in order to probe for CO_2 or pH receptors, be they peripheral or central, affecting air-breathing. As mentioned, Hedrick *et al.* (1991) found no evidence that EDF CO_2 or acidity affected the respiration of *Amia*. McKenzie *et al.* (1991b) found that infusion of ammonium carbonate (0.2 M, 2.5 ml/kg) into *Amia* in normoxic water significantly increased arterial PCO_2 and total CO_2 content. This treatment did not affect blood pH nor did it elicit air breathing. By contrast, infusion of HCl (10^{-5} M, 0.5 ml/kg) into normoxic *Amia* had a series of effects (reduced arterial pH, elevated PCO_2 but not total CO_2, and reduced total blood O_2). This treatment caused the massive release of the catecholamines epinephrine and norepinephrine, raised heart rate and dorsal-aortic pressure, and increased both aquatic ventilation and air breathing. However, when this acid infusion was given to hyperoxic *Amia*, these same cardiovascular parameters and blood-pH profiles were affected, but there was no effect on aquatic ventilation and neither air breathing nor catechol release were elicited. Thus, the blood acid-base disturbances caused by acid infusion appear to trigger air breathing only if they co-occur with reductions in blood O_2 to below normoxic levels.

Intravascular Chemosensor Feedback and Bimodal Control

Because bimodal breathing is cyclic, there are regular oscillations in ABO and blood respiratory gas levels and blood pH. These parameters are similarly affected by emergence. It could therefore be expected that chemoreceptors monitoring blood respiratory properties might function in bimodal respiratory control (Figure 6.2). Although studies have shown the effects of an air breath and emergence on blood respiratory properties, none have demonstrated a cause and effect relationship between either blood respiratory parameters or their oscillations and the triggering of ABO or branchial ventilation, or a return to water. Johansen and Lenfant (1968) failed to detect any pattern for the air breathing of *Protopterus* and variables such as arterial or venous PO_2 and pH. They wrote (page 463):

> the variability [in respiration] was large enough to suggest that the actual level of arterial blood gas tensions or intrapulmonary gas tension played no role in setting the breathing rhythm. It appears that the fish relies more on tolerance to large variations in gas tension than on regulations against them.

This state of knowledge has not changed in the past quarter century. Studies examining aspects of blood respiratory gas and acid-base properties in relation to bimodal control have been done on at least twelve genera (*Neoceratodus, Lepidosiren, Protopterus, Lepisosteus, Amia, Arapaima, Electrophorus, Hoplerythrinus, Channa, Monopterus, Synbranchus, Blennius* [*Lipophrys*], Chapter 7). In some instances, an altered air ventilation pattern can be linked to modification of a blood-respiratory property. For example, a post exercise pH reduction is associated with more air breathing in *Lepisosteus* (Smatresk, 1994). In the majority of cases, however, no pattern exists between breathing and blood or ABO parameters. There is no correlation between lung PO_2 and air ventilation in *Protopterus;* although, adding N_2 to the organ increases the mean arterial O_2 tensions at which air breaths are taken. The abrupt cessation of branchial ventilation in aquatic hypercapnia by *Protopterus* took place well in advance of any appearance of additional CO_2 in the arterial circulation (Johansen and Lenfant, 1968). For *Amia*, branchial apnea associated with the onset of air breathing causes a slight reduction in arterial pH (-0.06 units); however, this was rapidly compensated and did not recur with each subsequent breath. Neither the onset of air breathing nor intermittent air breaths in progressively hypoxic water altered the trend for declining arterial and venous O_2 tensions in this fish (Johansen *et al.*, 1970a, figure 8).

Experimental emersion causes an increased blood PCO_2 and total CO_2 and a reduction of pH in most air-breathing fishes so investigated (viz., *Protopterus, Amia, Channa, Monopterus, Synbranchus*) which also responded to air exposure by increasing air-gulping frequency. This, however, did not compensate for the pH reduction which was only relieved with immersion or, in the case of *Channa*, with auxiliary ABO ventilation (Ishimatsu and Itazawa, 1983b). Similarly, the onset of air breathing in *Synbranchus* resulted in a dramatic rise in blood PCO_2 and a drop in pH. While extracellular pH was uncompensated, the intracellular space was compensated by a bicarbonate influx (Heisler, 1977, 1982).

STUDIES OF ABO RECEPTORS

Most ABO receptor studies have examined behavioral responses to changes in air-phase O_2 level or to manipulations of ABO gas quality and volume. These approaches are affected by experimental design and by characteristics of the species under study. For example, air-breathing responses to alterations in respirometer gas-phase or ABO-O_2 contents would be expected to differ between obligatory versus nonobligatory air breathers and would also be affected by aquatic O_2 tension.

Mechanoreceptors

Although data are sparse, ABO mechanoreceptors appear to affect air ventilation by transducing information about the rate or extent of organ wall deformation. Recording afferent activity in the ABO branch of the vagus in *Protopterus, Lepisosteus*, and *Amia*, has played an important role in demonstrating the presence of mechanoreceptors responsive to both ABO deflation and inflation (and chemosensory components, see following) (Smatresk, 1988). These preparations have identified two classes of ABO mechanoreceptors: Slowly adapting receptors (SAR), and rapidly adapting receptors (RAR). SAR impulse frequencies change in proportion to ABO volume or respond tonically to step changes in volume. These receptors are most likely positioned in relatively noncompliant areas of the ABO. RARs respond to dynamic volume changes with quick firing bursts but go silent as volume stabilizes; they are likely located in more compliant regions of the organ (Smatresk and Azizi, 1987). Both SAR and RAR occur in *Protopterus, Lepidosiren*, and *Lepisosteus* (DeLaney *et al.*, 1983; Smatresk and

Azizi, 1987); however, only SAR have been found in *Amia* (Milsom and Jones, 1985). There appear to be no major differences in the RARs and SARs of the air-breathing Actinopterygii and Sarcopterygii. Graphs showing SAR activity in relation to absolute pressure during stepped volume changes reveal a deflation hysteresis suggesting that the ABO mechanoreceptors of *Protopterus*, *Amia*, and *Lepisosteus* are responsive to wall tension and not transmural pressure (Milsom and Jones, 1985; Smatresk and Azizi, 1987).

Milsom (1989a,b, 1990, 1991) stressed the importance of mechanoreceptor input for the coordination of motor control over the respiratory muscles and structures along the aerial respiratory pathway. Postulation exists that feedback control of ABO wall tension (i.e., by transduction of volume or pressure) may occur in *Lepisosteus* because vagal efferents regulate the tone of both smooth and striated muscle in the organ's wall (Smatresk and Azizi, 1987). Several workers have suggested that the pulmonary mechanoreceptors of lungfish constitute the evolutionary precursors of the mammalian Hering-Breuer reflex, a pulmonary stretch receptor-mediated reflex affecting respiratory frequency (Johansen, 1970; Jones and Milsom 1982; DeLaney *et al.*, 1983; Pack *et al.*, 1990, 1992).

Pack and colleagues (1990) found that increasing lung volume or pressure in anesthetized or decerebrate *Protopterus* also prolonged breath duration. In most air-breathing species, experimental deflation of the ABO triggers air gulping, while ABO inflation lengthens air-breath duration. However, because ABO inflation and deflation affect buoyancy, caution must be exercised in conclusions about the exclusivity of a bimodal control function. Indeed, the opinion of several groups has been that the primitive function of ABO mechanoreceptors was to aid in buoyancy regulation (DeLaney *et al.*, 1983; Pack *et al.*, 1990). Non-air-breathing fishes respond to gas bladder inflation (or decreases in atmospheric pressure) through the "gasspuckreflex," in which gas is voided to correct buoyancy (Fange, 1976, 1983; Graham *et al.*, 1978b). These fishes also gulp air in response to gas bladder deflation (compression). The type II air-breathing pattern of *Amia* (Hedrick and Jones, 1993; Hedrick *et al.*, 1994) demonstrates that air ventilation does occur for buoyancy purposes. Future experiments need to distinguish the respiratory versus buoyancy relevant inputs of ABO mechanoreception.

Finally, a few additional comments about buoyancy are needed. Several groups have suggested that a low aerial respiratory exchange ratio (R, see Chapter 5) and the ensuing steady deflation of the ABO, could in itself trigger air ventilation (Gee and Graham, 1978; Milsom and Jones, 1985) as a result of a temporal decline in buoyancy. Observations on some of the fishes subject to such a buoyancy change support this by showing they often breathe and restore ABO volume long before ABO-O_2 declines to a limiting level, indicating that buoyancy control can take precedence over O_2 acquisition (Gee and Graham, 1978; Hedrick and Jones, 1993; Hedrick *et al.*, 1994). On the other hand, an elevated ABO-O_2 tension in *Lepisosteus* overrides the normal reflex effects of ABO deflation and inflation on air breathing and on branchial ventilation (Smatresk, 1994).

Chemoreceptors

Testing for the presence of ABO chemosensors is difficult because the experimental techniques that are used often affect organ volume or pressure and may thus activate mechanoreceptors. It is also hard to discriminate between the activity of chemoreceptors within the ABO and those located remotely (i.e., in the efferent vasculature).

Evidence for ABO-O_2 chemoreceptors influencing air ventilation has been obtained for *Monopterus albus* through experiments in which step changes in the O_2 content of the air phase over the fish were accomplished while it was submerged (Graham *et al.*, 1995a). However, because *Monopterus* uses its buccal and branchial chambers as an ABO, it is very likely that these O_2 receptors are of branchial origin. The presence of O_2 chemoreceptors in other types of ABOs has not been demonstrated and their role in ventilation can only be speculated (Smatresk, 1988).

Experiments with ABO-O_2 responses show that a hypoxic air phase reduces breath-hold duration whereas hyperoxia most often lengthens it (Shelton and Croghan, 1988). Genera shown to increase air-breath duration in hyperoxia include *Protopterus*, *Lepidosiren*, *Erpetoichthys*, *Lepisosteus*, *Electrophorus*, *Ancistrus*, *Hypostomus*, *Clarias*, *Monopterus*, *Trichogaster* and others (Chapter 5). Analogous results are obtained when ABO-O_2 level is manually altered by injection or withdrawal of gas (Smatresk, 1988; Pack *et al.*, 1990; Graham *et al.*, 1995a). With minor exceptions (*Anabas*, Hughes and Singh, 1970b), species having well-developed gills are less affected by a N_2 gas phase or by injection of this gas into the ABO (*Neoceratodus*, Johansen *et al.*, 1967; *Lepisosteus*, Smatresk and Cameron, 1982c). By contrast, injection of N_2 into the lung of *Protopterus* doubles its air-breathing frequency (Johansen and Lenfant, 1968).

As previously indicated, alterations in the O_2 content of the ambient gas phase demonstrated an

O_2-mediated ventilatory response for *Monopterus albus* (Graham *et al.*, 1995a). For *M. cuchia* Lomholt and Johansen (1974) found that exposure to hypoxic gas mixtures (4 and 8% O_2) increased both air-gulp volume and frequency. Similar studies with *M. albus* (Graham *et al.*, 1995a) did not reveal changes in tidal volume (note that the ABOs of *M. cuchia* and *M. albus* do differ in that the former has pharyngeal sacs, Chapter 3), but did show that this fish voided extremely hypoxic and anoxic breaths within a few seconds of inspiration. The presence of an ABO chemosensor is suggested by the rapidity of the voiding reflex which was about two times faster than would be expected if stimulation had occurred via remote, downstream vascular receptors (Eclancher, 1975; Eclancher and Dejours, 1975).

ABO chemosensors sensitive to CO_2 were studied by vagal efferent recordings. These show that SARs in the lungs of *Protopterus* and *Lepidosiren* are inhibited by CO_2 (DeLaney *et al.*, 1983) as are those in *Lepisosteus oculatus* (Smatresk and Azizi, 1987). However, none of the SARs found in *Amia* by Milsom and Jones (1985) were CO_2 sensitive and no other CO_2 sensitive ABO receptors have been identified. Tests with lungfish (DeLaney *et al.*, 1983) showed that rapid elevation of ambient CO_2 from 0 to 6% immediately reduced the impulse rate activity of a CO_2 sensitive SAR; however, the return to zero CO_2 quickly restored activity. Smatresk and Azizi (1987) found that SAR CO_2 sensitivity in *Lepisosteus* occurred at concentrations between 6 and 10% (i.e., a PCO_2 range of 44–74 torr), well in excess of physiological levels of this gas in the ABO. These workers were of the opinion that a general pH susceptibility of the receptor cells, rather than a specific CO_2 sensory modality, elicited the observed "CO_2 response."

CARDIORESPIRATORY REFLEXES

Reviews of fish cardiorespiratory integration were done by Taylor (1992) and Burleson *et al.* (1992) and this section focuses on the integration of these processes with air breathing.

Vagal motor and sensory connections between the medulla, gills, ABO, heart, and other areas establish the potential for integrated cardiovascular and respiratory actions that enhance bimodal respiratory efficacy (Shelton *et al.*, 1986; Smatresk, 1988). Accordingly, the taking (or release) of an air breath, or an amphibious excursion may trigger simultaneous changes in branchial ventilation, cardiac output, and the pattern of blood flow (Figure 6.4).

Branchial Ventilation

For the majority of aquatic air-breathing fishes, ventilation frequency and amplitude are greatest just before the air breath, become maximally attenuated just after the breath, and then gradually increase during the interbreath interval (IBI). This pattern has been observed during volitional air breathing and has been induced by ABO inflation and deflation in: *Protopterus*, *Lepidosiren*, *Erpetoichthys*, *Lepisosteus*, *Amia*, *Ancistrus*, *Gymnotus*, *Anabas*, *Piabucina* and others (Smatresk, 1988).

Because most species have "in-series" ABO and systemic circulations, the reduction of branchial ventilation when a breath is taken lessens the potential for transbranchial O_2 loss (Chapter 4). Notably, even the lungfishes *Protopterus* and *Lepidosiren*, which have partial separation of systemic and pulmonary circulations, show this response. Thus, most of the reported exceptions to this general scheme appear to reflect experimental protocols rather than an alternative ventilatory control pattern. On the other hand, results indicating that *Neoceratodus* lacks this response (Lenfant and Johansen, 1967) may reflect a more primitive stage for its bimodal respiratory control and should be investigated. Species such as *Monopterus*, *Synbranchus*, *Electrophorus*, and *Hypopomus*, hold air in their mouths and must therefore cease aquatic ventilation during air breathing. Finally, when not air breathing, species like *Synbranchus* and *Periophthalmus* exhibit an intermittent aquatic ventilatory pattern; branchial apnea is affected by temperature, aquatic oxygen level, and body size (Graham *et al.*, 1987; Bridges *et al.*, 1995).

Emersion also affects branchial ventilation. Although patterns vary, most species make lower frequency, higher amplitude ventilations out of water. Upon emerging, *Mnierpes* flushes all water from its buccopharynx and then ventilates slowly for the extent of its exposure. During prolonged (several minutes) exposure, this fish opens its mouth and flares its opercula (Graham, 1970, 1973). After ventilating briefly when first exposed, *Blennius* seals its opercula and tidally ventilates its branchial chamber through its mouth until returned to water (Pelster *et al.*, 1988). Large *Sicyases* empty water from the buccopharynx, take an air gulp, and seal both the mouth and opercula for variable periods (Ebeling *et al.*, 1970; Gordon *et al.*, 1970). *Entomacrodus* also holds air gulps in a sealed branchial chamber for various periods (Graham *et al.*, 1985). Similar observations have been made for *Andamia* (Rao and Hora, 1938) and *Periophthalmus* (Stebbins and Kalk, 1961), although there is debate over whether or not mudskippers do in fact carry water in their branchial chambers (Graham, 1976a; and Chapter 2).

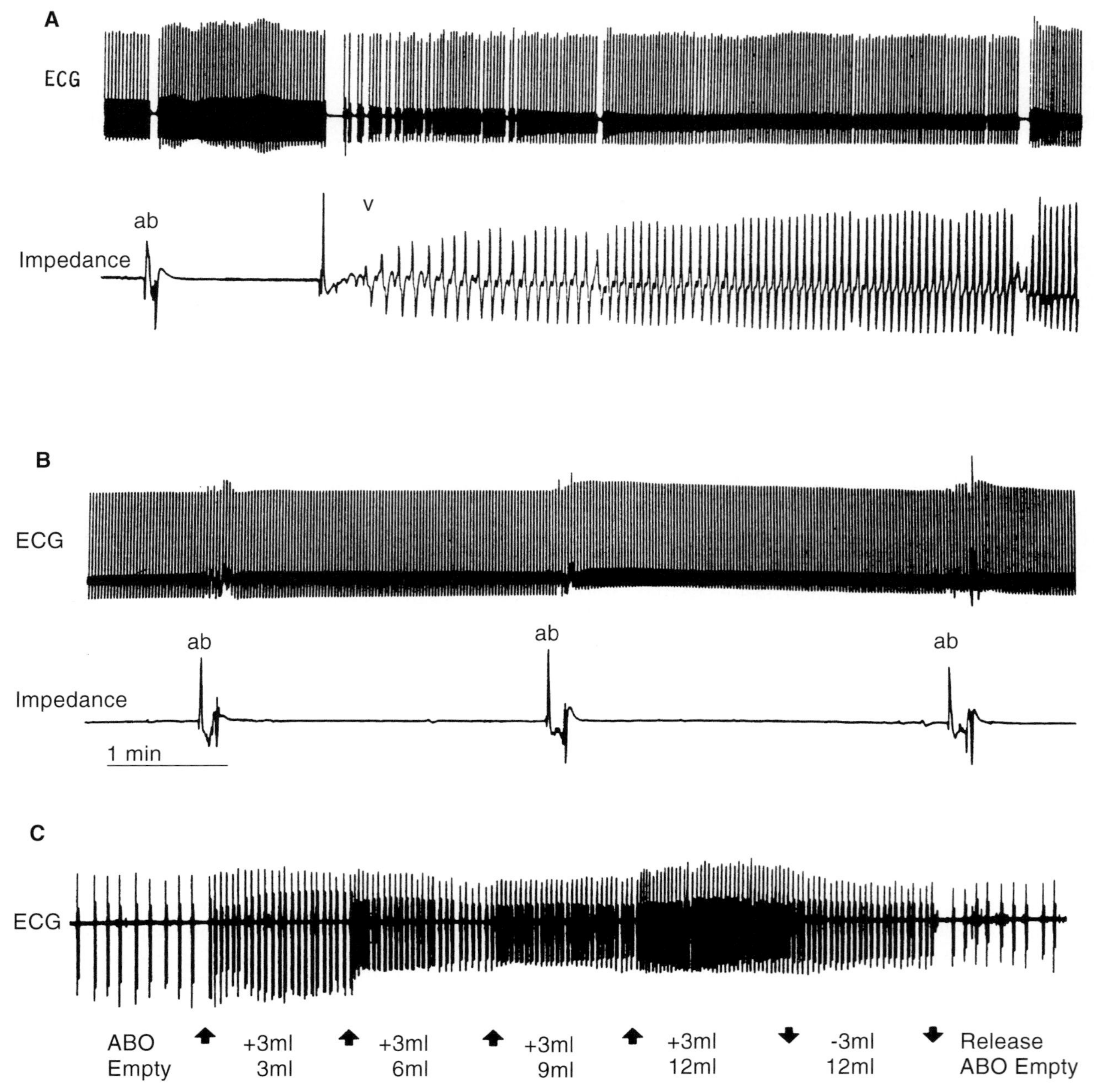

FIGURE 6.4 ECG and branchial chamber impedance records for a 390 g *Synbranchus marmoratus*. The 1 min time scale shown under trace **B** applies to both **A** and **B. A,** The effect on heart rate of a spontaneous transition from air-breathing to gill ventilation. The impedance trace shows a typical air-breath signal (ab) and the associated cardiac acceleration. Cyclic aquatic gill ventilation (v) begins about 80 s after the air breath and results in a lower level of tachycardia than followed the air breath. **B,** ECG and branchial impedance records showing air breathing (ab) following the injection of 1 mg/kg atropine. **C,** Effect on heart rate of four manual injections (3 ml each) into the ABO followed by the withdrawal of 3 ml and then the spontaneous release of all ABO gas by the fish. Numbers below the trace indicate injected and cumulative ABO volumes at each step. This fish had an ABO volume of about 12 ml. (From Graham *et al.*, 1995a. Reprinted by permission of the Company of Biologists, Ltd., from the *Journal of Experimental Biology,* volume 198, pp. 1455–1467.)

Regulation of Vascular Tone

Coordination of air breathing and vasomotor activity is well documented and is important in redistributing blood flow for optimal gas exchange. In most studies, it has only been possible to monitor air-breath effects on either central vessel blood velocity or regional blood flow. However, as detailed in Chapter 4, more exacting work with lungfishes and with *Electrophorus* has quantified changes in cardiac output to the systemic and ABO vascular beds taking place over the entire air-breath cycle.

In all cases, air intake initiates increased blood flow to the ABO which gradually subsides as ABO-O_2 falls. Investigations have demonstrated that adrenergic (vasodilatory) and cholinergic (vasoconstricting) vasomotor mechanisms alter branchial, ABO, and systemic resistances. The pulmonary artery vasomotor segment (PAVS) of lungfish is critically important in the phasic regulation of pulmonary blood flow (Chapter 4). For *Protopterus*, Laurent *et al.* (1978) showed that acetylcholine constricts the pulmonary artery but opens the ductus arteriosus and thus optimizes pulmonary and systemic flow in relation to the respiratory cycle. Both *in vitro* (Johansen and Reite, 1967) and *in vivo* (Johansen *et al.*, 1968a) studies demonstrated the antagonistic actions of cholinergic and adrenergic stimuli on branchial and pulmonary resistances in African lungfish. For *Hoplerythrinus*, Smith and Gannon (1978) found that cholinergic stimulation of the gills resulted in the selective perfusion of arches 3 and 4, which would direct more flow into the coeliac artery and the ABO (Chapter 4). Increased branchial ventilation in water increases the heart rate and blood flow of *Periophthalmus barbarus* (Aguilar and Graham, 1995; Bridges *et al.*, 1995) and both *Synbranchus* and *Monopterus* (Lomholt and Johansen, 1974, 1976; Graham *et al.*, 1995a). The remarkable finding of a diving bradycardia in *Periophthalmodon* (Garey, 1962) has not been observed in other mudskippers; neither *Periophthalmus sobrinus* nor *P. barbarus* exhibited a change in heart rate with transfer between water and air (Gordon *et al.*, 1969; Aguilar and Graham, 1995).

Cardiac Regulation

In most aquatic air-breathing fishes, ABO deflation elicits bradycardia and reduces ABO perfusion while spontaneous or experimental ABO inflation initiates tachycardia, increases ABO perfusion, and reduces aquatic ventilation. Some of the naturally amphibious fishes are not subject to an emergence bradycardia, an otherwise normal piscine reflex (Table 6.4).

TABLE 6.4 Cardiac Rate Responses to Emergence among Air-Breathing Fishes

Species	°C	Heart rate (bpm)		Reference
		Water	Air	
Anguilla vulgaris	15	45	25	Berg and Steen, 1965
	7	22–24	15	
Sicyases sanguineus	14	60–80	40–50	Gordon *et al.*, 1970
Periophthalmodon australis	—	46	108	Garey, 1962
Periophthalmus sobrinus	24	70–125	no change	Gordon *et al.*, 1969
Blennius pholis	12.5	82 ± 2	82 ± 3[a]	Pelster *et al.*, 1988
Anabas testudineus	25	40 (20–70)[b]	30 (15–70)	Singh and Hughes, 1973
Clarias batrachus	25	30-39	15	Jordan, 1976

[a]Values are one SD; heart rate over first several seconds in air was about 40/min.
[b]Range.

As in nearly all vertebrates, heart activity reflects the balance between the inhibitory influences of vagal (cholinergic) inputs and stimulatory (adrenergic) effects of either sympathetic nerve fibers or circulating catecholamines (Taylor, 1992). Vagal inhibition can affect both chrono- and inotropic cardiac action and may be elicited by nociceptor (stress) stimuli, hypoxia, or input from both mechano- and chemoreceptors (Figure 6.4). Excitatory heart stimulation may take place via the down regulation of vagal tone, an increase in central-vessel blood pool (i.e., Starling's law), the arrival of blood-borne catecholamines, or the direct adrenergic nerve (sympathetic innervation) action on the heart (Laurent *et al.*, 1983).

A marked, however little appreciated, difference in the cardiac responses to air breathing is seen for the lungfishes and to some extent in *Lepisosteus* (Table 6.5). Although air-breath-initiated changes in pulmonary and systemic blood flow are well documented for lungfishes (Chapter 4), the role played by air-breath tachycardia seems to have been overstated. The literature contains references to lungfish air-breathing tachycardia; however, available data indicate otherwise (Table 6.5). This difference may be the result of a phylogenetic consequence attributable to the patterns of adrenergic (sympathetic) cardiac innervation existing among the air-breathing fishes; lungfish lack this innervation entirely, and there is only limited sympathetic innervation into the heart of *Lepisosteus* (Laurent *et al.*, 1983). Nearly all teleosts have a sympathetic car-

TABLE 6.5 Cardiac Responses to Air Intake in Aquatic Air-Breathing Fishes

Species	°C	Heart rate (bpm) Pre-air-breath	Post-air-breath	Reference
Neoceratodus forsteri	18	14	18	Johansen *et al.*, 1967
Protopterus aethiopicus	25	34(6)[a]	34(5)	Johansen and Lenfant, 1968
	–	33[b,c]	36	Johansen *et al.*, 1968a
	25	36	36	Szidon *et al.*, 1969
Lepidosiren paradoxa	27	29[d]	32	Axelsson *et al.*, 1989
Lepisosteus	18–21	35	37	Smatresk and Cameron, 1982a
oculatus	–	60	70	Smatresk, 1988
Amia calva	–	33	34	Johansen *et al.*, 1970a
Arapaima gigas	26–30	34(11)	34(10)	Farrell, 1978
Electrophorus electricus	28	28	66	Johansen *et al.*, 1968
	–	14	30	Johansen *et al.*, 1970b
Hoplerythrinus unitaeniatus	26–30	61(12)	82(23)	Farrell, 1978
Ancistrus chagresi	25	110	150	Graham, 1983
Clarias batrachus	25	30	39	Jordan, 1976
Monopterus cuchia	20	16	34	Lomholt and Johansen, 1976
	28–30	12	70	
M. albus	24–27	18–25	36–42	Graham *et al.*, 1995a
Synbranchus marmoratus	20–22	6	27	Johansen, 1966
	–	21	35	Johansen *et al.*, 1970b
	24–27	3–10	40	Roberts and Graham, 1985
Anabas testudineus	25	25	62	Singh and Hughes, 1973

[a]Parenthetical values are SE.
[b]Data obtained by manual inflation with 20 cc air.
[c]Lung deflation reduced heart rate from 18 to 12.
[d]Resting heart rate. Rates following vasoactive drug administration were: atropine 32, propranolol 25.

diac innervation. The hearts of teleosts, lungfish, *Lepisosteus*, and *Amia* can also receive circulating catecholamines from concentrations of chromaffin cells located in the great veins (Youson, 1976; Laurent *et al.*, 1983; Taylor, 1992).

Cardiac stimulation via sympathetic innervation is the exact analog of the control mechanism of higher vertebrates and has the advantages of immediate, forceful, and restricted cardiac control. In the absence of this innervation, both lungfishes and *Lepisosteus* would need to apply one of two mechanisms in order to synchronize air breathing with an increased cardiac activity. The first is a dependence upon the regulated and repeated release of catechols from strategically placed chromaffin cells in the great veins draining into the heart. This mechanism would be dependent upon the production and storage of sufficient catechols and would also be subject to whatever effects these exerted throughout the circulation (e.g., competing with cholinergic effects on branchial vasoconstriction during air breathing). The second mechanism is dependence upon an air-breath-initiated withdrawal of vagal inhibition. While this would have the advantages of regularity and rapidity, without the adverse effects of a rising systemic catechol titre, it might not, on the other hand, provide a very large tachycardia. Axelsson *et al.* (1989) found that the heart rate of atropine (a cholinergic antagonist) treated *Lepidosiren* only increased from 29 to 32 bpm. This is the same increase observed during volitional air breathing (Table 6.5) and suggests that a minimal vagal tone contributes to bradycardia in this fish.

An important comparative question arising from these facts is, how does the presence or absence of adrenergic innervation, or its degree of expression, affect the cardiac responses to ABO inflation or air-breathing stimuli? Review of the data (Table 6.5) indicates that the scope of air-inspiration tachycardia is reduced in lungfish and *Lepisosteus* relative to most teleosts. This pattern needs further confirmation with comparative studies controlling for the many variables likely to influence pre- and post-air breath heart rate. Table 6.5 does, nevertheless, suggest a phyletic difference, based on innervation pattern, in the cardiac responses to air breathing. It further suggests that these differences have resulted in radically different vasomotor responses to ABO inflation. At one extreme is the highly integrated vasomotor control system of lungfish which regulates cardiac output to the branchial, pulmonary, and systemic circulations in phase with the air-breathing cycle (Johansen *et al.*, 1968a). In contrast to the actinopterygians, the absence of step changes in arterial pressure following an air breath may be essential for modulating, by peripheral resistance, the balance between the nearly, but not completely, separated pulmonary and systemic circulations of the lungfish.

Peripheral inputs can affect cardiac responses to air breathing. NaCN and branchial denervation studies documented the role of external O_2 receptors in mediating the cardiac responses of *Lepisosteus* and *Amia* (reviewed by McKenzie *et al.*, 1991a). In addition, cardiac-rate responses to natural air breathing or experimental manipulation of ABO volume have been documented for a number of species (Roberts and Graham, 1985; Graham *et al.*, 1995a; Figure 6.4).

Little is known about how ABO gas quality affects cardiac responses. In species having ABO chemoreceptors, direct gas quality information could potentially influence the magnitude of inspiration tachycardia as well as modulate heart rate over the breath-hold period. Recent experiments with *Monopterus albus* (Graham *et al.*, 1995a) demonstrated the presence of ABO-O_2 chemoreceptors affecting heart action. Inspiration of gas containing either 1.5 or 3% O_2 caused an immediate (1–2 seconds) behavioral response and air-breath tachycardia did not develop. Many of the breaths taken in 1.5% O_2 were released before 10 heart beats occurred (11–27 seconds), and exposure to pure N_2 greatly slowed the heart and in most cases the gas was not held long enough to record more than two or three heart beats. Although *Monopterus* is unlikely to encounter such hypoxic air in its natural environment, these findings demonstrate the activity of an O_2-sensitive receptor that is capable of overriding mechanoreceptor and other stimuli affecting tachycardia. The probable function of this receptor would be to signal the end of the effective gas exchange period of the breath and to reduce heart activity accordingly. In species lacking ABO-O_2 receptors, chemosensor coordination of cardiac action in relation to ABO-gas content would fall to afferent information originating from more remote cells in the vascular tissue or myocardium.

Surprising data for the electric eel (*Electrophorus*) suggest ABO-gas quality does not affect its heart activity. Johansen *et al.* (1968b) found no differences in the extent of tachycardia elicited by the injection of either pure O_2 or N_2 gas into the ABO of this fish. This result is unexpected because most fishes possess external O_2 receptors in their branchial arches and because the electric eel uses its buccal chamber as an ABO and retains a reduced, although functional, branchial epithelium. While branchial tissue atrophy might explain the lack of a specific response to either N_2 or O_2 in the *Electrophorus* ABO, the Johansen *et al.* (1968b) study bears repeating because their conclusions were based on only a few records made on specimens recovering from anesthesia. The latter effect may have "knocked out" normal peripheral chemosensory responses and the displayed records were too brief to determine whether centrally mediated receptors in either the blood or myocardium were subsequently sensitized.

SUMMARY AND OVERVIEW

The aquatically respiring fishes and the mammals occur at the opposite extremes of the spectrum of vertebrate respiratory modality. Nevertheless, these two groups are similar in respiring unimodally and being, for the most part, continuous breathers. These similarities may extend into the endogenously activated control regions of the CNS such as the CRG and CPG. It is emphasized, however, that the functionality of these two regions in the fish brain is not as well documented as in mammals. Also, there are numerous exceptions, for both mammals and fishes, to the generalization of both groups as "continuous ventilators." Finally, and in spite of the apparent similarities, there are major differences in the organization of sensory inputs; in fishes the primary chemosensory stimulus for ventilation is from peripheral O_2 receptors, whereas the mammalian respiratory drive has its basis in either or both pH and CO_2 sensors located in the ventral medulla. There can be no doubt that the shift from peripheral to central control and the replacement of a hypoxic drive with a hypercapnic drive, are closely associated with the evolutionary transition of the tetrapod lineage to exclusively aerial ventilation.

Investigations of fish bimodal control have sought to define "evolutionary points of origin" of the aerial respiratory control features of the higher vertebrates. This approach, however, has not been particularly fruitful and a commonly expressed conclusion has been that in terms of the relative importance of peripheral and central control elements, "there is little to indicate that the bimodal fishes occupy an intermediate position in the evolutionary progression from aquatic to aerial respiration" (Shelton *et al.*, 1986; Shelton and Croghan, 1988).

It may therefore be more rewarding to "turn this comparison on its head" and view bimodality as the starting point for the evolutionary tracks leading to unimodal aquatic and aerial respiration in fishes and mammals. This perspective suggests that selection for auxiliary aerial respiration, as well as the associated changes in the CPG needed for air gulping and for bimodal control, occurred early in the evolutionary history of the bony fishes and long before the derivation of the modern fish and tetrapod lines. Thus, while certain features of the control mechanisms of aquatically respiring fishes and the exclusively air-ventilating mammals may be relics of their common ancestry, the mechanisms in both groups strongly reflect evolutionary convergence for unimodal ventilatory control in media having vastly different physical properties.

The extant air-breathing fishes, on the other hand provide us with a time-encapsulated comparative window to the progressive stages of bimodal control and specialization, a subject that is examined in more detail in Chapter 9.

Comparisons among closely related groups reveal how different dependencies upon air breathing have modulated the relative importance of aquatic and aerial stimuli for bimodal ventilation. Among the lungfishes, *Neoceratodus* is more responsive to water-borne stimuli (e.g., hypoxia stimulates air breathing, hypercapnia shuts off branchial ventilation) than either *Protopterus* or *Lepidosiren,* which are more responsive to ventilatory stimuli arriving via the air phase. Alterations in lung O_2, N_2, and CO_2 all affect the air breathing of *Protopterus;* however, aquatic hypoxia does not affect its air-breathing or branchial respiration rates, and aquatic hypercapnia stops branchial respiration completely and elevates air-breathing frequency. There is evidence for the use of separate aerial and aquatic CPGs in *Protopterus,* but not for the role of either central pH or CO_2 receptors in lungfishes or any other air-breathing fish.

Among the actinopterygians, extensive studies enable definition and comparison of the peripheral control elements in *Amia* and *Lepisosteus*. No other species have been as thoroughly investigated. There is ample documentation for the effects on bimodal ventilation of peripheral inputs to the CPG from chemosensors (primarily O_2, but CO_2 and pH effects are seen) located in the branchial stream, intravascularly, and possibly in the ABO. Sectioning of the vagus does not affect branchial ventilation pattern (indicating that there is CRG activation of the branchial pump) but does affect reflex responses to hypoxia and hypoxemia such as branchial hyperventilation and bradycardia. Mechanoreceptor input can also affect bimodality. This can stimulate or stop air breathing, alter branchial ventilation, and affect both cardiac and branchial activity. Further, both *Amia* and *Lepisosteus,* as well as several other species, exhibit the "gasspuck-reflex," a gas-bladder deflation stimulated by excessive buoyancy. The weight of the experimental evidence indicates that the mechanisms of respiratory control in *Amia* and *Lepisosteus* are largely similar; the few differences between them, in general, reflect the greater dependence of *Amia* upon aquatic respiration (Chapter 5). For *Amia*, aquatic hypoxia stimulates the release of catechols which augment branchial ventilation. In *Lepisosteus*, by contrast, hypoxia more readily activates air breathing and in turn attenuates gill ventilation. External NaCN stimulation activates a branchial response in *Amia*, but an aerial respiratory response in *Lepisosteus*.

For the bimodal fishes in general, the effects of aquatic hypoxia in eliciting facultative air breathing and influencing air-breath duration are well-documented, as is the effect of breath duration on end-expired ABO-O_2 level (Chapter 5). Both aquatic air breathing and amphibious behavior are affected by physiological variables such as blood O_2 storage and metabolic rate, by behavioral factors including social facilitation and predator presence, and by environmental conditions (Figure 6.2). On the other hand, both intra- and interspecific comparisons do not enable the identification of the peripheral receptors eliciting these responses. There is a weak or no correlation between end-expired ABO-O_2 level and air ventilation. There is not a strong correlation between air ventilation and blood O_2 or CO_2 levels. Also, the blood acid-base parameters pH, PCO_2, and HCO_3^- do not appear regulated by air breathing; the small and slow oscillations taking place in these over the breath cycle appear to be a result of air ventilation rather than a stimulus for it. These facts suggest that neither intravascular O_2 nor acid-base parameters have a dominant role in bimodal fish respiratory control. In fact, aquatic bimodal fish seem better adapted for tolerating large internal variations in blood respiratory properties than for regulating ventilation and metabolism in a manner to preclude them.

Heart activity, vascular tone, and branchial ventilation are all integrated with air breathing through vagal and other sensory and motor neurons interconnecting the medulla, the gills, the heart, the ABO, and other structures affected by either aerial respiration or emergence. In aquatic air breathers, an air breath usually elicits tachycardia as well as branchial apnea. However, there are not instantaneous or fixed species-specific relationships between air intake and changes in these cardiorespiratory parameters. The degree of air-breathing tachycardia is also determined by the degree of separation of the ABO and systemic circulations, by the poising of adrenergic and cholinergic tone, and by whether or not there is sympathetic cardiac innervation. Tachycardia may also be influenced by the volume of inspired air and its O_2 content, the duration of the preceding IBI, the heart rate prior to inspiration, and by environmental and behavioral factors. Over the time air is held, tachycardia diminishes while branchial ventilatory frequency increases. These phasic changes in heart and ventilatory activity, along with integrated shifts in vascular resistance, optimize gas exchange efficiency by maximizing flow to the ABO just after air inspiration when the O_2 gradient is highest. The subsequent rise in branchial ventilation and perfusion occurs as the O_2 gradient (both in the ABO and transbranchially) is diminished. Amphibious air breathing elicits changes in branchial chamber ventilation and heart frequency; however, we lack data needed to define general patterns.

CHAPTER

7

Blood Respiratory Properties

High hemoglobin content would therefore be doubly advantageous to the fish in swamp waters. Other things being equal, it would mean a greater supply of oxygen in the body, and a greater buffering power of the blood to deal with excess CO_2, so that visits to the surface would need to be less frequent.

E.N. Willmer,
Journal of Experimental Biology XI, 1934

INTRODUCTION

This chapter examines the influence of aerial gas exchange on the blood-respiratory properties of air-breathing fishes. Lupu (1925) was one of the first biologists to investigate the respiratory properties of the blood of an air-breathing fish (*Misgurnus*). Willmer (1934) drew attention to the unique problems posed for blood-respiratory function by the combined effects of hypercapnia and hypoxia in tropical swamps and included several air-breathing fishes in his studies. Investigations of O_2 binding by hemoglobin (Hb), and other respiratory characteristics of air-breathing fish blood have been carried out for a number of species. Early syntheses of blood properties in relation to air breathing are found in Lahiri *et al.* (1970), Johansen and Lenfant (1972), and Lenfant and Johansen (1972). The volumes resulting from the 1976 R.V. *Alpha Helix* expedition to the Amazon (*Canadian Journal of Zoology*, 1978, 66 [4]); *Comparative Biochemistry and Physiology*, 1979, 62A,[1]) also contain a diversity of works on blood-respiratory function and Hb properties. Morris and Bridges (1994) emphasized both ecological and evolutionary aspects in their survey of Hb function in the bimodally breathing fishes. A volume by Val and de Almeida-Val (1995) describes the blood-respiratory properties of Amazonian air-breathing fishes.

Respiratory Properties

With few exceptions, fishes have a tetrameric Hb contained within a nucleated erythrocyte (Boutelier and Ferguson, 1989; Satchell, 1971, 1991). As reviewed by several workers (Wintrobe, 1934; Graham, 1985), interspecific differences have been documented in such commonly measured hematological properties as red cell size, hematocrit (Hmct), total Hb concentration [Hb], and the Hb content per cell (termed the mean corpuscular hemoglobin concentration, MCHC = Hb/Hmct · 100). Intraspecific variations in these parameters can occur as a result of season, temperature, salinity, diet, reproductive condition, and other factors. Hemoglobin plays an important role in the transport of both O_2 and CO_2 and, because of the allosteric interaction of each of these ligands with the binding site of the other, the sequential transport (i.e., O_2 in and CO_2 out) of these respiratory gases is a highly integrated feature of Hb function (Perry and Wood, 1989).

O_2 Transport

Oxygen transport involves the reversible, cooperative binding of O_2 to hemoglobin and is thus determined by both the concentration of blood Hb and its affinity for O_2 (Boutelier and Ferguson, 1989). When fully saturated a molecule of Hb binds four molecules of O_2. However, the percentage of Hb saturation depends upon plasma PO_2, plasma and red cell pH, temperature, and PCO_2. The relationship between PO_2 and Hb saturation is described by an O_2 dissociation curve (Figure 7.1). Under comparable conditions of pH, PCO_2, and temperature, intra- and interspecific comparisons of Hb–O_2 affinity are made with respect to the P_{50}, the PO_2 at which Hb is 50% saturated. The shape of the dissociation curve is influenced by physical factors and the binding cooperativity taking place between individual subunits of the Hb tetramer. High cooperativity means that the binding of O_2 to one subunit greatly enhances binding by neighboring subunits and in such cases the dissociation curve has a strongly sigmoidal shape. The index of cooperativity is described by the Hill coefficient, n which is about 2 for most vertebrate hemoglobins (Cameron, 1989).

Both a lower pH and higher PCO_2 "shift the dissociation curve to the right," and lessen affinity (Figure 7.1). The reduction in Hb–O_2 affinity caused by a lower pH, termed the Bohr effect, increases the quantity of O_2 that can be unloaded in tissue capillaries and is thus vital for normal O_2 delivery. At the gills, however, aquatic conditions such as increased temperature, elevated CO_2, and low pH can reduce Hb–O_2 affinity and limit O_2 uptake (Willmer, 1934; Lenfant and Johansen, 1972; Powers *et al.*, 1979a,b; Jensen, 1991). A stronger effect of CO_2 and pH is described by the Root effect (Figure 7.1) which shifts the curve further to the right and reduces the total amount of O_2 that can be bound to Hb. Krogh and Leitch (1919) were the first to show that fishes inhabiting O_2-poor waters usually have greater O_2 carrying capacities and higher Hb–O_2 affinities than do those residing in air-saturated waters.

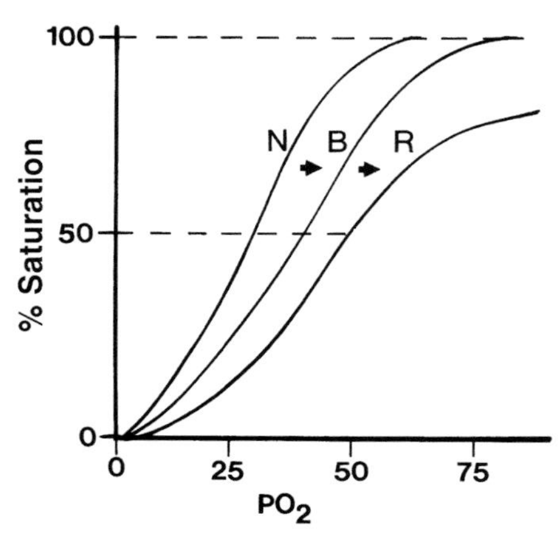

FIGURE 7.1 Generalized oxygen dissociation curve relating the percent saturation of hemoglobin to oxygen partial pressure (PO_2, torr). The dashed lines indicate the 50 and 100% saturation levels. Curve N denotes normal conditions prevailing at the gas-exchange surface (i.e., normal pH, low CO_2, and high PO_2). Curve B (Bohr effect) shows the right shift in Hb–O_2 affinity taking place in tissue capillaries as a result of increased respiratory CO_2 and slightly lowered pH. Curve R, signifying the Root effect, shows the effect of a more acid pH in decreasing affinity and preventing 100% saturation.

CO_2 Transport

Respiratory CO_2 is transported in reversible chemical combinations formed in both fish blood plasma and red cells (Figure 7.2). In the plasma, CO_2 combines with water to form carbonic acid (H_2CO_3) which dissociates into protons and bicarbonate (HCO_3^-). Because red cells contain carbonic anhydrase, CO_2 hydration proceeds more rapidly there than in the plasma. The protons formed in red cells bind with amino acid residues on the globin protein of the O_2-dissociated (reduced) Hb while bicarbonate diffuses back to the plasma in equal (charge balance) exchange with plasma chloride (chloride shift, Figure 7.2). As O_2 is unloaded, Hb has an increased capacity to bind protons and this is termed the Haldane effect. From a conceptual standpoint the Haldane effect can be seen as a reversed Bohr effect. Whereas the Bohr effect describes how CO_2 affects Hb–O_2 affinity, the Haldane effect describes the effect of O_2 on the ability of Hb to transport CO_2. The strength of the Haldane effect varies considerably among fishes (Jensen, 1991); the reduced or absent Haldane effect in some air breathers (e.g., *Protopterus*, Lahiri *et al.*, 1970) has been linked to differences in the sites of aerial (O_2) and aquatic (CO_2) gas exchange.

Some respiratory CO_2 is transported as carbamate which forms on the terminal amine groups of the reduced Hb molecule (Farmer, 1979; Albers, 1985). The quantity of carbamate formed in fish Hb is, however, extremely small (Jensen, 1991). The loading of CO_2 onto Hb in tissue capillaries takes place sequentially; the process commences with the rise in blood CO_2 and a drop in pH which, through the Bohr effect, unloads O_2 and frees Hb to bind CO_2 (Figure 7.2; Albers, 1970, 1985; Jensen, 1991). CO_2 transported in this manner is termed oxylabile CO_2 to signify that it is carried in association with Hb and will be given up in the presence of O_2 (Farmer, 1979). Exchange steps reverse in the gills where respiratory CO_2 is released primarily as a gas, but also as bicarbonate which may be incorporated into ion exchange processes associated with acid-base balance (Heisler, 1986; Cameron and Iwama, 1989; Jensen, 1991). In the case of air-breathing fishes, ABOs derived from the gills may contain carbonic anhydrase and thus mobilize and release CO_2 into the air phase. As was shown in Chapter 5, most fishes have low rates of aerial CO_2 release.

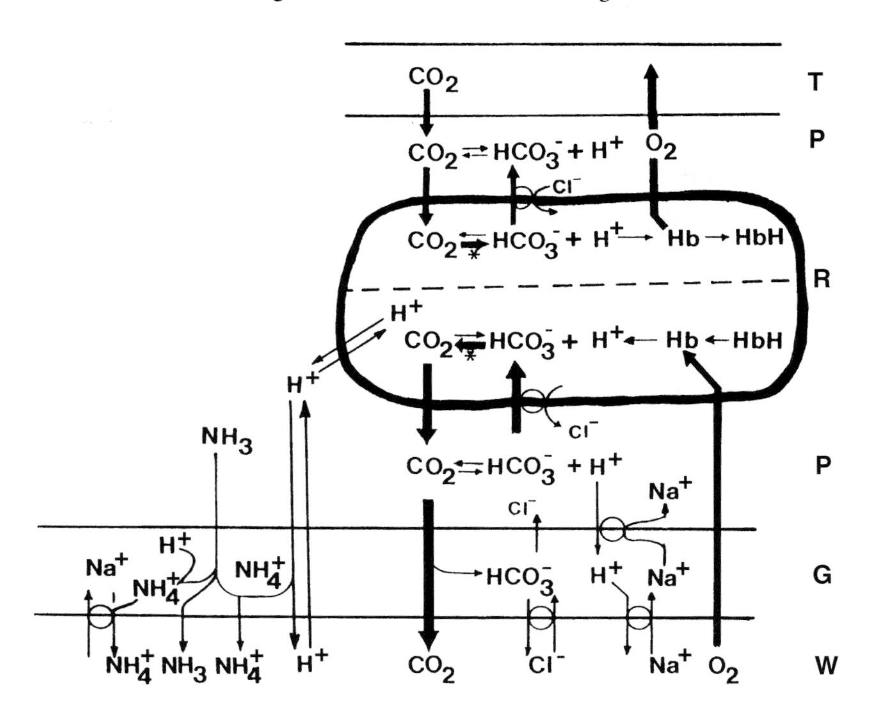

FIGURE 7.2 Diagram of the movements of O_2, CO_2, ions, and other substances between water and the tissues of a fish. Dashed line through the red cell (R) subdivides diagram into upper and lower sections. The lower section shows net movements of substances taking place in the gills (G), where CO_2 would move from R and through the plasma (P) into water (W). Upper diagram depicts exchanges taking place in tissues (T) where CO_2 moves towards R. The presence of carbonic anhydrase (*) in the red cell catalyzes CO_2 hydration. This enzyme is poised in the opposite direction in the gills. Note how the presence of O_2 affects proton binding with Hb and that protons may combine with NH_3 or have a charge balance function. (Adapted from Cameron [1989] and Heisler [1989].)

HEMOGLOBIN FUNCTION IN AIR-BREATHING FISHES

Oxygen Transport

Several characteristics of the air-breathing fishes place special importance on the Hb–O_2 binding properties of their blood. Because of the circulation patterns in some species (Chapter 4), there is the problem of the unsaturation of efferent ABO blood caused by venous mixing. There is also the potential for transbranchial loss of O_2 acquired in the ABO. In addition, because O_2 is more plentiful in air, a fish may need less ventilation to supply its metabolic demands. This can cause CO_2 to build up which is termed respiratory acidosis (Chapter 5; Johansen, 1970; Howell, 1970). In the case of amphibious air breathers, air-ventilatory capacity can also be limited by mechanical inefficiencies and the risk of dehydration (Graham, 1976a; Liem, 1987). Aquatic air breathers may also experience a respiratory acidosis because of their generally reduced gill surfaces and the tendency to limit aquatic ventilation while holding an air breath (Johansen, 1970; Lahiri *et al.*, 1970; Wood *et al.*, 1979). Acidosis can be induced by the often acidic and hypercarbic habitat of many air-breathing fishes. Because an elevated CO_2 and lowered pH reduce Hb–O_2 affinity, air-breathing fishes may increase their blood buffer reserve and increase Hb–O_2 affinity to ensure normal O_2 loading in the respiratory organs and normal respiratory CO_2 transport (Graham, 1983, 1985; Jensen, 1991). Non-air-breathing species typically respond to acute aquatic hypoxia by increasing both their [Hb] and Hb–O_2 affinity (Krogh and Leitch, 1919; Wood and Johansen, 1972; Powers, 1980). As will now be detailed, some air-breathing fishes also do this, but others do not.

Hemoglobin and Hematocrit

Table 7.1 summarizes the [Hb] and hematocrit (Hmct) data for the air-breathing fishes. In a number of species these two blood parameters are as high, or higher, than in many mammals and birds (i.e., about

TABLE 7.1 Comparative Values of Blood Hemoglobin Concentration [Hb] (g%) and Hmct (%) for Air-Breathing Fishes[a]

Species	[Hb]	Hmct	Reference
Lepidosiren paradoxa	7.0	39.8	19
	6.5	28	15
Protopterus aethiopicus	7.4	27.4[b]	6
Protopterus sp.	7.8	30.1	15
P. annectens	6.9	37.6	2
Neoceratodus forsteri	7.0	35.0	15
Polypterus senegalus	10-14	30-43	38
Erpetoichthys calabaricus	7.5	21.8	3
Lepisosteus osseus	9.0	—	21
L. oculatus	7.0–8.0[e]	—	32
Atractosteus tristoechus	9.1	29.0	31
Amia calva	5.8	22.8	17
	7.7	26.4	39
Arapaima gigas	7.5	30.8	18
Megalops sp.	9.8	—	29
Lepidocephalus guntea	17.0–19.0[j]	50–62[j]	43
	14.3–16.2[k]	50–59[k]	43
Erythrinus erythrinus	8.2–10.5[c]	34.0–42.5[c]	19
Hoplerythrinus unitaeniatus	6.3–12.3[c]	41.0–42.5[c]	19
Piabucina festae	12.7[l]	44.3	10,11
Clarias batrachus	7.6	41.0	33
	12.6	36.2	30
C. gariepinus	5.8	28.9	13
C. lazera	9.7	39.4	2
Heteropneustes fossilis	9.2	45.2	33
	11.6	43.8	30
	9.5	—	29
	12.0–17.6[d]	32.0–42.0[b,d]	20
	4.5–7.2[e,i]	—	27
Hoplosternum littorale	13.9[e]	—	42
	14.4	49.4[f]	19
	16.8	62[q]	19
	9.1–9.2	32–37	28
H. thoractatum	9.2	35–39	11
Ancistrus chagresi (spinosus)	4.6–6.4[d,l]	23.4–31.2	11
A. cirrhosus	4.5	20	19
Hypostomus plecostomus	5.1–10.3[d,l,m]	24.9–32.8	11
H. regani	8.6	29.7	9
	10.0	34[p]	9
Hypostomus sp.	8.6[e,f]	31.5	40
	6.8[e,g]	31.1	40
Pterygoplichthys sp.	8.5[e,f]	28.7	40
	6.8[e,g]	23.4	40
	11.4	43.3	19
P. multiradiatus	9.2[l]	29.8[l]	36
Loricaria (Rineloricaria) uracantha	4.9[n]	21.2–23.5	11
Electrophorus electricus	11.2	41.0	16
	12.7	45.3	19
Neochanna burrowsius	6.9	24.4	41
Helcogramma medium	3.06	11.6	14
Mnierpes macrocephalus	8.5	—	12
Blennius pholis	3.8	18.3	5
Dormitator latifrons	15.5	39.0	35
	8.1–10.5[d,l]	33.9–41.2	11
Gillichthys mirabilis	5.0	—	34
Periophthalmus sp.	15.8	—	29
P. sobrinus	14.3	47.4	23
Boleophthalmus	14.8	—	37
boddarti	16.7	51.3	23
Anabas testudineus	10.0–19.8[i]	—	8
	18.1[j],19.8[k]	—	4
Trichogaster trichopterus	10.4	27.3	45
T. leeri	8.4	—	12
T. pectoralis	13.2	37.6	26
Osphronemus olfax	17.8[j]	48.6	25
	15.9[k]	—	25
Channa punctatus	12.1	28.2	33
	10–12	44–46	7
C. maculata	7.9	32.7	44
Channa sp.	12.0–20.0[i]	60–70	29
Synbranchus marmoratus	17.3[e,f]	48.8	40
	16.2[e,g]	53.5	40
	13.0	47.0	21
	13.8	48.3	19
	10.4–11.6[n,o]	34.5–40.3	11
	12.1 ± 0.7	—	1
Monopterus cuchia	14.2	[h]	33
	12.8–15.5[c]	40.0–52.0[c]	22
	18–24[j]	40–67[j]	24
	14–26[k]	39–71[k]	24
	17.8–20.8	55–57	30

[a]Classification and phylogenetic ordering of species follows Chapter 2.
[b]Packed cell volume.
[c]Range of values reported.
[d]Seasonal changes.
[e]Back-calculated from O_2 capacity.
[f]From normoxic water.
[g]After 4–7 days in hypoxic water.
[h]Reported Hmct = ≥ 100.
[i]Direct size effect.
[j]Males.
[k]Females.
[l]Increased in hypoxia (Table 7.5B).
[m]Varies among habitats.
[n]No seasonal change.
[o]Panama population.
[p]After 4 h in air.
[q]After 13 h in air.
[1]Amaral and Carneiro, 1982; [2]Babiker, 1979; [3]Beitinger *et al.*, 1985; [4]Banerjee, 1966; [5]Bridges, 1988; [6]DeLaney *et al.*, 1976; [7]Dheer *et al.*, 1987; [8]Dube and Munshi, 1973; [9]Favaretto *et al.*, 1981; [10]Graham, 1976b; [11]Graham, 1985; [12]Graham, unpublished; [13]Hattingh, 1972; [14]Innes and Wells, 1985; [15]Johansen, 1970; [16]Johansen *et al.*, 1968b; [17]Johansen *et al.*, 1970a; [18]Johansen *et al.*, 1978a; [19]Johansen *et al.*, 1978b; [20]Joshi and Tandon, 1977; [21]Lenfant and Johansen, 1972; [22]Lomholt and Johansen, 1976; [23]Manickam and Natarajan, 1985; [24]Mishra *et al.*, 1977b; [25]Natarajan, 1980; [26]Natarajan and Rajulu, 1982; [27]Pandey *et al.*, 1971; [28]Perez, 1979; [29]Pradhan, 1961; [30]Singh, *et al.*, 1976; [31]Siret *et al.*, 1976; [32]Smatresk and Cameron, 1982a; [33]Srivastava, 1968; [34]Todd, 1971; [35]Todd, 1972; [36]Val *et al.*, 1990; [37]Vivekanandan and Pandian, 1979; [38]Vokac *et al.*, 1972; [39]Weber *et al.*, 1976a; [40]Weber *et al.*, 1979; [41]Wells *et al.*, 1984; [42]Willmer, 1934; [43]Yadav *et al.*, 1978; [44]Yu and Woo, 1987b; [45]Zanjani *et al.*, 1969.

TABLE 7.2 Hemoglobin Concentrations (g%) of Air-Breathing Fishes from the Finca Tocumen and Rio Frijoles, Panama during the Wet and Dry Seasons and Following 14–21 Days in Aquatic Hypoxia (PO_2<25 torr) in the Laboratory (T = 24 – 26 C)[a]

Habitat	Species	Wet season		Dry season		Laboratory hypoxia
Rio Frijoles	*Hypostomus plecostomus*	5.1 ± 0.35 (29)	*[b]	5.9 ± 0.34 (20)	*	7.9 ± 0.37 (13)
	Loricaria[c] *uracantha*	4.9 ± 0.50 (19)		4.8 ± 0.36 (20)		4.9 ± 0.62 (11)
	Ancistrus chagresi[c]	4.6 ± 0.30 (13)	*	5.4 ± 0.26 (21)	*	6.4 ± 0.42 (23)
	Synbranchus marmoratus	10.4 ± 0.48 (6)		11.1 ± 0.96 (6)		11.6 ± 0.74 (7)
Finca Tocumen	*Hoplosternum thoracatum*	9.3 ± 0.38 (13)		9.1 ± 1.04 (9)		9.2 ± 0.23 (14)
	Piabucina festae	12.7 ± 0.64 (14)		12.4 ± 1.06 (7)	*	14.6 ± 0.56 (4)
	Dormitator latifrons	8.1 ± 0.26 (23)	*	9.4 ± 0.64 (9)	*	10.5 ± 0.47 (33)
	Hypostomus plecostomus	9.6 ± 0.66 (5)		10.3 ± 0.40 (9)		—

[a]Rio Frijoles seasonal data are based on monthly fish samples. [Hb] values are mean ± 2(SE)(*n*).
[b]Asterisks indicate significant differences between adjacent comparison groups (Graham, 1985).
[c]See Table 7.1 for revised nomenclature.

15 g% Hb and Hmct > 40%; Johansen *et al.*, 1978b; Graham, 1976b, 1985). This elevation has been attributed to aquatic environmental conditions (pH, O_2, and CO_2), the physiological characteristics of a species including its aerobic requirements, and factors such as its extent of reliance upon air breathing (i.e., obligate versus facultative), the relative size of its ABO, and the blood-flow pattern to that organ (Johansen, 1970; Todd, 1972; Satchell, 1976; Johansen *et al.*, 1978a; Graham, 1985).

A long-standing problem with most air-breathing fish [Hb] and Hmct data is that the published values are often obtained in connection with other physiological studies and, as such are not in all cases based on standardized techniques, are usually derived from a small number of specimens, and can be quite variable both intra- and interspecifically (Table 7.1). For selected species, several studies document the importance of factors such as nutrition, season, reproductive condition and gender, and pollutants on these two parameters (Mishra *et al.*, 1977a,b; Mahajan and Dheer, 1979; Dheer *et al.*, 1987; Munshi and Hughes, 1992). Both macro- and microscale differences in habitat conditions within the broad distribution range of some species can also affect these parameters (Graham and Baird, 1984; Graham, 1985). Historically, however, few studies have examined the nature of relationships between air breathing *per se* and the [Hb] and Hmct of the air-breathing fishes (Johansen, 1970).

The plasticity of the Hb–Hmct relationship in air-breathing fishes has been examined through techniques such as altering the ambient aquatic conditions of a fish, denying it access to the air phase, or air exposure (Johansen *et al.*, 1978b; Weber *et al.*, 1979; Graham, 1983; Babiker, 1984b; Val *et al.*, 1990). The "natural" field experiments presented by seasonal changes and by differing extremes of physical conditions prevailing in various habitats containing the same species have also provided new information (Graham, 1985; Val *et al.*, 1990).

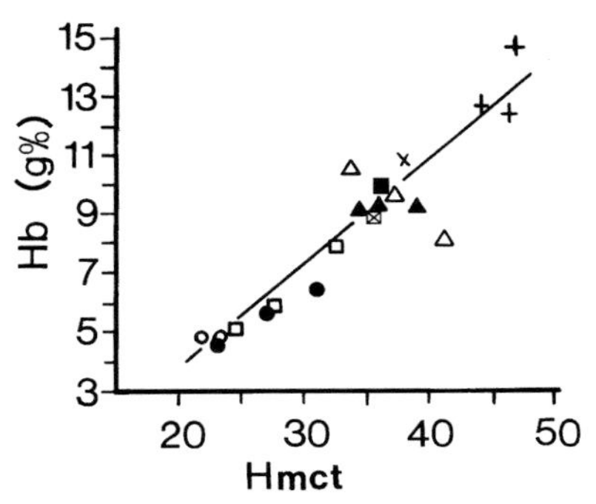

FIGURE 7.3 Hb–Hmct relationships for seven air-breathing fishes occurring in different habitats in Panama. *Hypostomus*, Rio Frijoles (open box); Rio Mandinga (box with ×); Finca Tocumen (filled box); *Loricaria* (open circle); *Ancistrus* (filled circle); *Hoplosternum* (filled triangle); *Piabucina* (+); *Synbranchus* (×); *Dormitator* (open triangles). Regression equation Hb = -3.76 + 0.36 (Hmct), r = 0.93, n = 20, $p < 0.001$. Note three values for *Dormitator latifrons* suggest changes in Hb affect MCHC (see text). (From *Env. Biol. of Fishes*, volume 12[4], pp. 291–301, figure 3. Graham, 1985, with kind permission of Kluwer Academic Publishers.)

Work in Panama has shown that some air-breathing fishes alter [Hb] and Hmct both seasonally and in laboratory hypoxia (Table 7.2). As in most vertebrates (Altland, 1971; Johansen *et al.*, 1978b) a linear relationship exists between [Hb] and Hmct in these fishes (Figure 7.3). This linearity indicates that the observed [Hb] increases take place mainly through polycythemia; MCHC is thus conserved in most, but not all, species. Table 7.2 also shows that the Rio Frijoles, Panama populations of the armored catfishes (Loricariidae) *Ancistrus chagresi* (now *A. spinosus*) and *Hypostomus plecostomus* increased their [Hb] and Hmct by 12 to 17% during the dry season (December to

May), the time of year when the environmental requirement for facultative air breathing, if it materializes at all, is most likely. These data were obtained from monthly samples of fish in the same locality of the stream. In the year when this work was done the Rio Frijoles did not stop flowing and remained normoxic. The change in blood-respiratory properties therefore suggests a genetically mediated dry season response that better enables these fishes to withstand aquatic hypoxia and carry out facultative air breathing, should this be required.

Studies in Panama further revealed that [Hb] in different populations of *Hypostomus plecostomus* was inversely related to habitat O_2 concentration (Table 7.3A). Populations of this species dwelling in the nearly anoxic swamp at Finca Tocumen had the highest [Hb], followed by those from the Rio Mandinga (seasonally stagnant), while the lowest [Hb] was found for the population in the Rio Frijoles (where stagnation seldom occurs). Transfer of Finca Tocumen *Hypostomus* to normoxic laboratory tanks reduced their [Hb] somewhat, but not to the levels of the Rio Frijoles fish (Table 7.3B). This suggests the presence of fixed intraspecific differences in total [Hb] that may have evolved in isolated populations and in response to different regimes of aquatic hypoxia and air-breathing requirement.

In some species, however, blood [Hb] and Hmct appear to be independent of aquatic conditions. In *Hoplosternum thoracatum*, for example, there was no effect of 14 days exposure to aquatic hypoxia and air breathing on either [Hb] (9 g%) or Hmct (37%) (Table 7.2). This finding parallels that of Perez (1979) who determined that limiting *Hoplosternum littorale* to aquatic respiration in normoxic water did not affect the relative proportion of either of its two Hb types or its total [Hb] (9 g%), but did significantly increase Hmct (from 33 to 37%, i.e., MCHC was reduced).

Hemoglobin-Oxygen Affinity

Molecular Aspects

Inter- and intraspecific differences in air-breathing fish Hb–O_2 affinity have several bases. There may be differences in the primary structure of the Hb molecule itself, in red cell constituents that affect Hb quartenary structure (allosteric effectors), or in the respiratory behavior (Chapter 5) of the fish. Environmental differences can affect blood gas tensions and plasma and red cell pH, which determine the Hb molecule's microenvironment (Howell, 1970; Powers, 1980; Graham, 1983).

TABLE 7.3A Temperature and Oxygen Levels in Different Habitats in Panama in Relation to Various Blood Parameters in *Hypostomus plecostomus* and *Synbranchus marmoratus*

Habitat[a]	Rio Frijoles	Rio Mandinga	Finca Tocumen	Ocelot Pond
Temperature Range (°C)	25–27	26–28	26–28	24–32
Oxygen range (torr)	130–147	40–90	5–30	< 5–332
Hypostomus plecostomus				
Hb (g%)	5.79 ± 0.33 (6)	9.05 ± 0.22 (8)	9.92 ± 0.49 (6)	—
Hmct (%)	27.4 ± 4.20 (6)	35.1 ± 2.86 (8)	36.4 ± 4.90 (6)	—
NTP (μM/100 ml)	173.12 ± 4.16 (6)	—	211.50 ± 31.0 (6)	—
NTP/Hb	1.94	—	1.39	—
Synbranchus marmoratus				
Hb (g%)	10.7 ± 0.70 (12)	10.8 ± 0.42 (6)	10.7 ± 0.55 (5)	11.3 ± 0.36 (7)
Hmct (%)	40.2 (3)	38.2 (3)	34.5 (3)	40.3 ± 4.8 (7)

[a]Habitat oxygen and temperature conditions between December, 1975 and July, 1976.

TABLE 7.3B The Effect of Exposure to Normoxic Water (PO_2 = 150 torr) in the Laboratory on the Mean [Hb] of Two Groups of *Hypostomus* from Finca Tocumen[a]

Group	I	II
Mean [Hb] at collection (date)	9.56 ± 0.65 (5) (1 March)	10.57 ± 0.39 (6) (15 April)
Mean [Hb] after exposure to normoxic water (date)	7.65 ± 0.75 (4) (9 April)	9.92 ± 0.38 (6) (27 April)

[a]Values are mean ± 2(SE) (*n*) (data from Graham, 1985).

Hemoglobin. The Hb–O_2 affinity of a particular species might be an intrinsic property of its Hb brought about by selected amino acid substitutions in subunit tetramers (Johansen *et al.*, 1978a; Powers, 1980). In addition, many fishes possess and circulate multiple hemoglobins, each with different O_2 affinities or responses to variables that affect affinity (Weber *et al.*, 1976a,b; Garlick *et al.*, 1979; Perez, 1980). Some fishes also have the capacity to alter the percentage of Hb components to favor O_2 binding and transport under the prevailing ambient and red cell environmental conditions. In this regard, however, there do not seem to be notable differences between air-breathing and non-air-breathing fishes. The average number of major Hb components found among the 15 air-breathing genera surveyed by Fyhn *et al.* (1979) was 2.4 with a range of from one (*Lepidosiren*, *Arapaima*, *Electrophorus*, *Gymnotus* and some loricariids) to six (*Synbranchus*). Several genera are polymorphic for their dominant Hb component (Fyhn *et al.*, 1979; Farmer *et al.*, 1979). Weber *et al.* (1976a) identified five Hb components in *Amia*. Both Garlick *et al.* (1979) and Perez (1979) reported two types in *Hoplosternum*, and Val *et al.* (1990) found at least four different Hbs in *Pterygoplichthys*. Finally, shifts in the relative amounts of the four different Hb subunits of *Protopterus amphibius* take place during estivation (Johansen *et al.*, 1976).

Hb–O_2 affinity. Various erythrocyte cofactors bind to Hb and affect O_2 affinity by altering quarternary structure and heme–heme interactions. Among fishes, the most important of these are the nucleoside triphosphate compounds (NTP), of which adenosine and guanosine triphosphate (ATP, GTP) are the principal types (Weber *et al.*, 1979; Powers, 1980; Boutelier and Ferguson, 1989). These substances, also termed organophosphates, preferentially bind to the deoxy-state Hb and lower Hb–O_2 affinity. Although both ATP and GTP influence Hb–O_2 affinity, GTP is regarded as the more potent modifier. Other organophosphates affect Hb. The lungfishes *Protopterus* and *Lepidosiren* both have, in addition to ATP and GTP, significant levels of uridine di- and triphosphates (Bartlett, 1978a; Isaacks *et al.*, 1978a,b). Inisotol pentaphosphate (IP_5), the phosphate modulator prevalent in birds, has also been found in these two lungfishes, in *Erpetoichthys* (Beitinger *et al.*, 1985), and appears to be the principal Hb modulator in *Arapaima* (Bartlett, 1978b, 1980).

As seen in Table 7.4, there is considerable intra- and interspecific variation for air-breathing fish red cell organophosphate levels. Values range between about 200 and 2500 μM, and there is often little consistency among estimates for the same species by different workers. There are no clear comparative trends for the amount or type of NTP in different species or for their quantitative relationship to air-breathing (Bartlett, 1978a,b,c, 1980; Babiker, 1984b; Val *et al.*, 1990). The initial generalization that GTP is more prevalent in air breathers (Isaacks *et al.*, 1978a,b; Johansen *et al.*, 1978b) has not been sustained, although, as will be seen below, the relative and total amount of GTP will change with ontogeny and in response to air breathing and hypoxia. Finally, there is considerable interspecific variation in the ratio of total phosphate to Hb which determines O_2 affinity (Bridges and Morris, 1989; Morris and Bridges, 1994).

Table 7.4 shows Babiker's (1984b) erythrocytic phosphate data for *Protopterus*, *Polypterus*, and *Clarias*. This data set, which offers the advantage of having been gathered by the same investigator, demonstrates both

TABLE 7.4 Red Cell Organophosphate Levels in Some Air-Breathing Fishes[a]

Species	Total NTP (μM/100 ml)	% GTP	Reference
Lepidosiren paradoxa	464	66	4
	206	30	5
Protopterus aethiopicus	1470[b]	33	3
P. annectens			
juveniles	835	23	6
adults	732	40	6
adult, estivating	500	21	6
Polypterus senegalus			
juveniles	461	9	6
adults	409	23	6
Arapaima gigas	905[b]	47	2
	570	46	4
Hoplerythrinus unitaeniatus	381	41	4
Clarias lazera			
juveniles	543	10	6
adults	560	15	6
Pterygoplichthys sp.	2500	60	1
	419	80	4
in water	251	44	5
in air 28 h	335	50	5
Hoplosternum littorale			
in water	240[b]	65	5
in air 13 h	307	69	5
Electrophorus electricus	342	81	4

[a]Quantities defined are total nucleoside triphosphate levels (NTP = ATP + GTP) and the percentage of this amount that is glutamine triphosphate (GTP).

[b]Median values.

[1]Bartlett, 1978a; [2]Bartlett, 1978b; [3]Bartlett, 1978c; [4]Isaacks *et al.*, 1978a,b; [5]Johansen *et al.*, 1978b; [6]Babiker, 1984b.

the range of organophosphate differences that can exist among species as well as the effect of air-breathing ontogeny on NTP concentration. (Chapters 2 and 5 review aspects on the respiratory physiology of these three genera.) Compared to juveniles (which are not obligatory air-breathers), adult *Protopterus annectens* have a slightly reduced total red-cell phosphate concentration and a greater percentage of GTP. Adult *Polypterus senegalus* also have a lower total phosphate and a larger percentage of GTP than juveniles; however, the absolute differences are much less than for *Protopterus*. Relatively small ontogenetic increases in the total organophosphate level and in the percentage of GTP also take place in *Clarias lazera*.

In response to hypoxia, most fishes decrease the molar ratio of organophosphate to Hb which increases Hb–O_2 affinity (Wood and Johansen, 1972; Jensen, 1991). The ratio of red cell organophosphate concentration to Hb (NTP/Hb) can therefore be used to index O_2 affinity (Morris and Bridges, 1994). Table 7.5 presents data from Babiker (1984b) and Val *et al.* (1990) detailing the effects of hypoxia exposure on red cell NTP. The relatively mild level of hypoxia used by Babiker (3.6 mg/l, about 50% saturation) did not markedly alter the red cell ATP levels of *Protopterus*, *Polypterus*, or *Clarias*; nonetheless, total NTP levels in hypoxia were elevated because of significant increases (relative to normoxia) in GTP. The increase in NTP suggests the Hb–O_2 affinity of these fishes was reduced. Unfortunately, Babiker (1984b) did not measure affinities, nor did he provide companion Hb data for these fishes. Without Hb data we do not know how the rise in GTP with hypoxia acclimation affected NTP/Hb. This information would have proven

TABLE 7.5A Hypoxia Effects on Red-Cell ATP and GTP Concentration in Three Air-Breathing Fish Species[a]

		Control 6.6 mg O_2/l 28–32 °C	Hypoxia (12 days) 3.6 mg O_2/l 28–32 °C
Protopterus annectens	Total (ATP + GTP)	725(μM/100 ml)	900
(40–55 g)	% GTP	34	51
Polypterus senegalus	Total	560	685
	% GTP	20	37
Clarias lazera	Total	440	620
	% GTP	11	39

[a]Babiker, 1984b.

TABLE 7.5B Ambient Oxygen Effects on the Red-Cell Phosphate and Other Blood Respiratory Properties of *Pterygoplichthys multiradiatus*[a]

	Normoxia 7.3 mg O_2/l 25–27 °C	Hypoxia 1.3 mg O_2/l 25–27 °C	Natural 0.5 mg O_2/l 27 °C
Blood Parameters			
Hmct	29.8	30.4	36.0
Hb	9.2	11.4	10.9
MCHC	31.1	37.3	31.1
Total NTP (μM/100 ml)	1010	629	440
%GTP	35	8	7
GTP/ATP(%)	59	15	18
NTP/Hb	7.16	3.59	2.63
P_{50} (@pH = 7.4)	4.7	3.8	3.5

[a]Val *et al.*, 1990.

extremely valuable in light of the different air-breathing physiologies of these three genera.

Information about the effect of hypoxia on NTP/Hb is available from studies by Val *et al.* (1990) on *Pterygoplichthys* (Table 7.5B). This catfish was obtained from hypoxic, dry season pools in Brazil, and Val and co-workers did blood analyses on fresh-caught individual fish as well as on those that were maintained in hypoxic and normoxic laboratory tanks. As seen in Table 7.5, the natural (habitat) and hypoxia-acclimated (laboratory) *Pterygoplichthys* had a lower total NTP and a lower relative amount of GTP than did the normoxic group. The hypoxia and natural groups also had a higher [Hb], and their correspondingly lower NTP/Hb correlates directly with the (slightly) higher Hb–O_2 affinities measured for them. The data of Val *et al.* (1990) thus confirm the NTP/Hb ratio as an index of affinity. This ratio is very important for air-breathing fishes, because it is a predictor of other properties affecting O_2 transport including the magnitude of the Bohr effect (Johansen *et al.*, 1976; Farmer, 1979; Bridges *et al.*, 1984) and the influence of catecholamines and other substances affecting red cell pH and ion concentration; all of which indirectly affect Hb–O_2 affinity (Mangum *et al.*, 1978; Weber *et al.*, 1979; Nikinmaa *et al.*, 1986; Jensen, 1991; Nikinmaa, 1990).

Comparative Oxygen Dissociation Curves

The results for *Pterygoplichthys* indicate that, in a pattern similar to the non-air-breathing fishes (Krogh and Leitch, 1919), this facultative air-breather responded to aquatic hypoxia with a left shift of its O_2 dissociation curve. This finding has important comparative implications in light of the general perception that the onset of air breathing is most often accompanied by a right shift in Hb–O_2 affinity.

An early link between comparative respiratory physiology and vertebrate evolution developed around the idea that, because O_2 is more plentiful and always accessible in air, the evolutionary transition from aquatic to aerial respiration would be generally characterized by an increase in metabolism and a decrease in Hb–O_2 affinity (McCutcheon and Hall, 1937; Lahiri *et al.*, 1970; Lenfant and Johansen, 1972; Johansen and Lenfant, 1972; Morris and Bridges, 1994). Evidence for evolutionary shifts in Hb affinity with changes in respiratory mode and O_2 availability can be inferred at least in the broadest sense from the natural occurrence of such patterns among extant species. For the bullfrog (*Rana*), metamorphosis from water-ventilating polliwogs to air-breathing subadults is accompanied by a decline in Hb–O_2 affinity (Riggs, 1951). A similar pattern occurs in several other amphibians (*Ambystoma*, *Amphiuma*, *Bufo*, McCutcheon and Hall, 1937; Lenfant and Johansen, 1967), and the theme of a right-shifted curve among air-breathing fishes has persisted for a long time (Johansen, 1970; Frey and van Aardt, 1995).

Interspecific comparisons. How has the evolutionary transition to a high, and relatively constant, supply of aerial O_2 affected the Hb–O_2 affinity of the air-breathing fishes? Is there an evolutionary trend towards reduced Hb–O_2 affinity with a greater dependence upon air breathing? Comparison of the Hb–O_2 binding properties of over 20 species of air-breathing fishes (Table 7.6) reveals P_{50} values ranging from 3 to as high as 35 torr and Bohr coefficients ($d\log P_{50}/d\text{pH}$) extending from -0.13 to -1.81. This variation doubtlessly reflects specific differences in behavior, physiology, habitat, and air breathing. However, other factors are also at play. For example, the blood samples on which these data are based were obtained from specimens in a variety of experimental conditions (i.e., from in water to a range of air-exposure periods including an estivating lungfish), and a number of different measurement techniques and conditions were employed. Thus, the comparative approach in this case hinges on our ability to make selected comparisons in which neither species diversity nor analytical methods vary too much.

Powers *et al.* (1979a) compared the Hb–O_2 dissociation curves of 42 species of air-breathing and non-air-breathing Amazon fishes under standardized conditions of temperature and PCO_2, but not pH. Their summary data (Figure 7.4) showed that P_{50} values for air-breathing fishes, which ranged from about 3 to 14 torr (30 C, CO_2 = 0 torr), were completely encompassed by values for non-air breathers, suggesting no fundamental effect of air breathing on P_{50}.

An analysis of selected literature values by Morris and Bridges (1994) did point to a right shift in affinity with air breathing; however, this effect is not large. The mean P_{50} (± SD) values reported for the three respiratory categories defined by these workers were: Aquatic respiration 11 ± 7 torr, bimodal 13.5 ± 6, and obligatory air 20 ± 10.

Broad species surveys, however, are not meaningful because of the bias introduced by phyletic differences among the species and specific differences in activity, habitat, and air-breathing efficacy. Comparisons of dissociation curves need to be made within phyletically similar groups for which specific differences in air breathing are known. Data of this type exist only for the lungfishes, the osteoglossids, and

TABLE 7.6 Comparative P_{50} Values and Bohr Coefficients of Some Air-Breathing Fishes

Species	P_{50} (torr)	°C	pH	Bohr Coefficient ($d\log P_{50}/dpH$)	Reference
Lepidosiren paradoxa	8.0	29	7.4	-0.31	12
Protopterus aethiopicus	32.0	25	7.4	-0.35	19
	19.0	25	7.5	-0.47	13
P. amphibius	34.5	26	7.2	-0.25[b,w]	10
	10.0	26	7.6	-0.55[m,w]	10
Neoceratodus forsteri	12.5	18	7.6	-0.62	14
Polypterus senegalus	23.5	30	7.7[f]	-0.43	21
Erpetoichthys calabaricus	17.9	25	7.6	-0.29[a,f]	1
	24.6	25	7.0	[a,p]	1
	29.4	25	6.8	[a,s]	1
	19	25	7.7	-0.25[t,f]	1
	23.5	25	—	[t,p]	1
	31.3	25	6.9	[t,s]	1
Lepisosteus oculatus	24.1	20	7.8[o]	-0.5[e]	18
	30.4	20[f]	—	—	18
Amia calva	24	27	7.6	-1.8	9
	8	16	7.6	—	9
Arapaima gigas	21	29	7.4	-0.30	11
	35.4	30	7.4	-0.60	3
	—	—	—	-0.57	17
Erythrinus erythrinus	7.1	29	7.4	—	12
Hoplerythrinus unitaeniatus	12.1	29	7.4	-0.64	12
	—	—	—	-0.75	17
Hoplosternum littorale	11.1	29	7.4	-0.39[a]	12
	14.2	29	7.4	-0.31[u]	12
	12[h]	28	—	—	24
	19[i]	28	—	—	24
	7.8	30	7.7[i]	—	4
	17.8	30	6.6[p]	—	4
	12.8	20	—	—[i,q]	16
	15.0	30	—	—[i,q]	16
	17.0	40	—	—[i,q]	16
	16.5	20	—	—[p,q]	16
	19	30	—	—[p,q]	16
	24	40	—	—[p,q]	16
Ancistrus chagresi	17	25	7.4[c]	—	5
	10	25	7.4[d]	—	5
Hypostomus sp.	22.0	30	7.4	-0.32	22
Pterygoplichthys multiradiatus	3.5	27	7.4	-0.13[e]	20
	4.7	27	7.4	-0.13	20
	3.8	27	7.4	-0.16	20
Pterygoplichthys sp.	10.4	29	7.4	-0.31[b]	12
	8.5	29	7.4	-0.36[v]	12
Electrophorus electricus	10.7	28	7.6	-0.68	8
	9.7	29	7.4	-0.43	12
Neochanna burrowsius	11	12	7.8	-0.6[k]	23
	18	12	7.4	-0.8[j]	23
Helcogramma medium	19.0	15	7.7	-0.85	6
Blennius pholis	8–24	12.5	7.9	-0.6 to -1.1[n]	2
	12–29	25	7.9	-0.5 to -1.2[n]	2
Channa maculata	7.6	25	7.6	-0.70	25
	14.5	25	7.2	—	25
Synbranchus marmoratus	7.0	29	7.4	-0.45[k]	12
	11.0	28	7.4	-0.23[l]	12
Synbranchus sp.	7.2	25	7.6	-0.40	7
Monopterus cuchia	7.8	30	7.6	-0.57	15

(continues)

[a]Water breathing only.
[b]Bimodal breathing.
[c]Normoxia acclimated.
[d]Hypoxia acclimated.
[e]Fresh caught fish from hypoxic habitat.
[f]Equilibrated with1% CO_2.
[g]Equilibrated with 2% CO_2.
[h]No CO_2.
[i]3.3% CO_2.
[j]Exposed to air for three weeks.
[k]Kept in water.
[l]44 h in air.
[m]Estivating.
[n]Strongly dependent on P/Hb.
[o]*In vivo* pH.
[p]Equilibrated with 5 or 5.6% CO_2.
[q]No effects of forced aerial (18 h) or aquatic respiration on any blood parameter.
[r]Median values, four experimental treatments.
[s]Equilibrated with 9% CO_2.
[t]In air 4 h.
[u]In air 13 h.
[v]In air 18 h.
[w]Mean values, equilibrated with 3–8% CO_2.
[1]Beitinger *et al.*, 1985; [2]Bridges, *et al.*, 1984; [3]Galdames-Portus *et al.*, 1979; [4]Garlick *et al.*, 1979; [5]Graham, 1983; [6]Innes and Wells, 1985; [7]Johansen, 1970; [8]Johansen *et al.*, 1968b; [9]Johansen *et al.*, 1970a; [10]Johansen *et al.*, 1976; [11]Johansen *et al.*, 1978a; [12]Johansen *et al.*, 1978b; [13]Lenfant and Johansen, 1968; [14]Lenfant *et al.*, 1966; [15]Lomholt and Johansen, 1976; [16]Perez, 1980; [17]Powers *et al.*, 1979a; [18]Smatresk and Cameron, 1982a; [19]Swan and Hall, 1966; [20]Val *et al.*, 1990; [21]Vokac *et al.*, 1972; [22]Weber *et al.*, 1979; [23]Wells *et al.*, 1984; [24]Willmer, 1934; [25]Yu and Woo, 1987b.

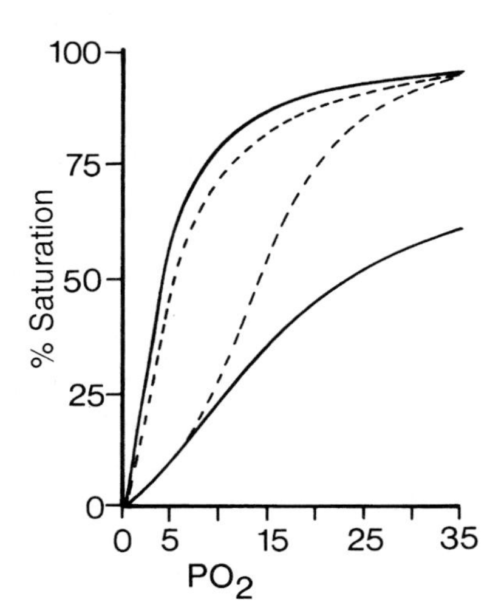

FIGURE 7.4 Comparative oxygen dissociation curves for some air-breathing and non-air-breathing Amazon fishes. Curves for the air-breathers fell at or within the dashed lines; those for the non-air-breathers occur at or within the two solid lines. Measurements were made at zero PCO_2 and 30 C. PO_2 units are torr. (Reprinted from *Comparative Biochemistry and Physiology*, volume 62A, Powers, *et al.* A comparative study of the oxygen equilibra of blood from 40 genera of Amazonian fishes, pp. 67–85, 1979, with kind permission of Elsevier Science-NL, Sara Burgerhartstraat 25, 1055 KV Amsterdam, The Netherlands.)

TABLE 7.7 Hematological Properties of Some Closely Related Air-Breathing and Non-Air-Breathing Fishes[a]

(A) Blood Parameters

	Mode	[Hb] (g%)	Hmct (%)	P_{50} (torr)	(°C/pH)
Ceratodontidae					
Neoceratodus forsteri	C,O	7.2	31	12.5	(18/7.6)
(Lenfant *et al.*, 1966)					
Lepidosirenidae					
Lepidosiren paradoxa	C,O	7.0	40	8.0	(29/7.4)
(Johansen *et al.*, 1978b)					
Protopteridae[b]					
Protopterus aethiopicus	C,O				
(Lenfant and Johansen, 1968)		6.4	25	19.0	(25/7.5)
(Swan and Hall, 1966)		6.96	25	32.0	(25/7.4)
Osteoglossidae					
Osteoglossum bicirrhosum	W	7.5–9.7	28.0	6.1	(28/7.4)
Arapaima gigas	C,O	7.5	30.8	21.0	(29/7.4)
(Johansen *et al.*, 1978a)					
Erythrinidae					
Hoplias malabaricus	W	7.2	27.5	6.1	(29/7.4)
Erythrinus erythrinus	C,F	8.2–10.5	34–43	7.1	(29/7.4)
Hoplerythrinus unitaeniatus	C,F	6.3–12.3	41–43	12.1	(29/7.4)
(Johansen *et al.*, 1978b)					
Lebiasinidae					
Piabucina panamensis	W	8.9	31	—	—
P. festae	C,F	12.7	44	—	—
(Graham, 1976b; 1985; and unpublished)					

(B) CO_2 and Temperature Effects on P_{50}[c]

	Mode	°C	P_{50} (torr) @ 0% CO_2	P_{50} (torr) @ 5.6% CO_2
Erythrinidae				
Hoplias malabaricus	W	30	4.8	16.4
Hoplerythrinus unitaeniatus	C,F	30	11.4	36.2
Osteoglossidae				
Osteoglossum bicirrhosum	W	30	9.4	25.0
Arapaima gigas	C,O	20	7.2	13.6
		30	14.8	37.3

[a]Respiratory mode of each species: Water-ventilating exclusively (W), Continuous air-breather (C), Facultative air breather (F), Obligatory air breather (O).

[b]Additional data are on Table 7.6.

[c]Powers et al., 1979a.

erythrinids (Table 7.7), and they provide only minimal support for the hypothesis of an evolutionary trend towards a reduced Hb–O_2 affinity with air breathing.

For the lungfishes, differences in analytical methods and in the temperature at which affinity was determined limit direct comparisons of P_{50} in relation to air breathing (Table 7.7). The P_{50} of the facultatively air-breathing *Neoceratodus* is 12.5 torr at 18 C. Values for the obligatory air-breathing *Protopterus* (in water) range from 19 to 32 torr at 25 C, whereas the value for *Lepidosiren*, also an obligatory air breather, is 8 torr at 29 C. Because warmer temperature raises P_{50}, the expected P_{50} of *Neoceratodus* at 25 C (about 22 torr, see Lenfant *et al.*, 1966, figure 4) would approach the values of *Protopterus* and would be much higher than *Lepidosiren*. Thus, while additional comparative studies are clearly needed, the data do not support the idea of a right shift in the dissociation curve with increased air-breathing reliance among the lungfishes.

For the osteoglossids, the obligate air breather *Arapaima gigas* has a much lower Hb–O_2 affinity than the non-air breather *Osteoglossum bicirrhosum* (Table 7.7A and B). These affinity differences are demonstrable at different temperatures, even in stripped hemoglobins, indicating they have their basis in subunit amino acid sequences. For the erythrinids (Table 7.7), *Hoplias*, a non-air breather, has a higher Hb–O_2 affinity than the air-breathing *Hoplerythrinus* but about the same affinity as the air-breathing *Erythrinus* (Johansen *et al.*, 1978b; Powers *et al.*, 1979a).

Table 7.7 lists the only Hb–O_2 comparisons of closely related air breathers and non-air breathers made to date. The pattern of a higher Hb–O_2 affinity for the non-air breather holds in one of the three groups. Additional species-pair comparisons (e.g., for *Piabucina*, see Table 7.7, and among the lungfishes) are needed to further clarify the relationships between air breathing and Hb–O_2 affinity. From the whole blood data now at hand, it appears that, in the case of *Arapaima* and *Osteoglossum*, the right-shifted curve of the air breather signifies a relaxation of molecular affinity made possible in part by access to aerial O_2. This evolved concomitantly with air-breathing specializations, such as increased branchial shunting (*Arapaima* is an obligatory air breather), which reduced the potential for O_2 loss, and adaptations for CO_2 retention (see following). It remains unknown if, and how, the P_{50} of this air-breathing fish might change in response to different levels of hypoxia. Does *Arapaima* alter its P_{50} when exposed to different environmental conditions of hypoxia and hypercapnia? Preliminary data for two gobies, the facultatively air-breathing *Gillichthys mirabilis*, and the amphibious air breathing *Periophthalmus barbarus*, indicate that *Gillichthys* has a lower Hb–O_2 affinity (Aguilar and Graham, 1995), and thus does not support the putative evolutionary pattern of a right shift in affinity with increased O_2 access.

Intraspecific variations in Hb–O_2 affinity. The potentials for ABO blood unsaturation and transbranchial O_2 loss can limit the air-breathing efficiency of some fishes. In light of this, the high [Hb] of some species and the rise in [Hb] observed for other species in hypoxia (*Ancistrus*, *Hypostomus*, *Piabucina*, Tables 7.1–7.3) can be viewed as mechanisms that conserve the net rate of tissue O_2 supply (Johansen *et al.*, 1978b;

Graham, 1985; Jensen, 1991). This alone may not be sufficient because the risk of branchial O_2 loss is particularly accentuated in some facultative and continuous air breathers with gill surfaces perfectly adequate for respiration in normoxic water. These fish usually occur in habitats where the PO_2 varies in space and time (Willmer, 1934; Kramer *et al.*, 1978), and air breathing is not continually required. A tropical pool or swamp, for example, may be highly oxygenated during daylight hours but nocturnally hypoxic (Figure 1.4; Graham, 1985).

In response to such a range of conditions, the facultative air breather benefits from the capacity to selectively control its gill perfusion and ventilation during the air-breath cycle and may also shunt blood past its gills (Chapters 4–6; Graham, 1983). These fish would be aided by an increase in O_2 affinity and, as shown for *Pterygoplichthys* (Table 7.5B), experiments employing hypoxia-induced facultative air breathing show that many of these fishes actually shift their O_2 dissociation curves to the left (Graham, 1983; Val *et al.*, 1990), or such a shift is suggested by reductions in red cell NTP/Hb (Weber *et al.*, 1979).

Following fourteen days of hypoxia (PO_2 < 25 torr) acclimation and facultative air breathing, *Ancistrus chagresi* had a 40% rise in [Hb] and, through a reduction in erythrocyte NTP/Hb, had a left-shifted dissociation curve (P_{50} reduced from 17 to 10 torr, Table 7.6). Weber *et al.* (1979) found that brief (48 hours) acclimation to hypoxia (PO_2 < 25 torr) lowered the NTP/Hb of two other facultatively air-breathing loricariids (*Hypostomus* sp. and *Pterygoplichthys* sp.), but did not affect the NTP/Hb of *Synbranchus marmoratus*. These workers also found that hypoxia acclimation "left shifted" the P_{50} of *Hypostomus* (from 19 to 14 torr, pH 7.65, 30 C). These results are consistent with field observations in Panama (Graham, 1985) where it was found that *Hypostomus plecostomus* occurring in less oxygenated habitats had a lower NTP/Hb, but that *Synbranchus* from differently oxygenated habitats invariably had the same [Hb]. However, unlike *A. chagresi* (Table 7.6) and *H. plecostomus* (Table 7.3), the 48 hour hypoxia acclimation period given by Weber *et al.* (1979) was apparently too brief to bring about [Hb] changes in the *Hypostomus* they studied.

Therefore, among the facultative air-breathers, a left shift in Hb–O_2 affinity, and not the right-shift predicted by the general evolutionary trend for vertebrates, takes place with the onset of air breathing. By allowing a "tighter grip" on O_2, an affinity increase lessens the transbranchial loss of this gas and compensates for the effects of CO_2 retention on affinity (Weber *et al.*, 1979). This, together with the rise in [Hb], increases air-breathing effectiveness. However, the unloading of O_2 and gas transport at the tissue level must inevitably have a reduced efficiency because of the affinity increase, and this would need to be countered by increased Bohr and Root effects (see following).

Species in which blood properties are not affected by aquatic conditions and air breathing. As documented, the blood properties of certain species appear to be largely unaffected by various experimental manipulations that alter respiratory partitioning. Perez (1980), for example, found no changes in a series of blood–respiratory properties (e.g., Hb, Hmct, NTP/Hb, P_{50}, pH, and others) of *Hoplosternum littorale* following exposure to air for 18 hours, forced air breathing (hypoxic water), or restricted air access. A similar response was seen in *Synbranchus* in which Hb–O_2 affinity is high but, along with [Hb], remained unchanged as a result of hypoxia or increased air breathing (Powers *et al.*, 1979a; Weber *et al.*, 1979; Graham, 1985).

When *Synbranchus* holds air in its buccal and branchial cavities, it cannot ventilate aquatically and its mode of air breathing supplies O_2 directly to the systemic circulation (Chapter 4). With this flow pattern, there can neither be venous O_2 dilution nor transbranchial O_2 loss. It follows, therefore, that 48 hours of hypoxia acclimation would not alter the NTP/Hb ratio (= affinity) of this fish (Weber *et al.*, 1979) and that, in contrast to observations for *Hypostomus* (Table 7.3), *Synbranchus* from differently oxygenated habitats in Panama would have the same [Hb] and not alter [Hb] in laboratory hypoxia. Subsequent sections discuss the probable O_2 storage function of Hb in *Synbranchus* and other species and the effect of air exposure on Hb–O_2 affinity. The independence of *Hoplosternum* and *Synbranchus*, insofar as hypoxia effects on blood respiratory properties is concerned, illustrates a fundamental principle for the air-breathing fishes: The presence of any mechanism permitting the effective isolation of the aerial and aquatic respiratory systems, whether it is the circulatory pattern or a respiratory mode that excludes simultaneous action of both systems, lessens the need for adaptive changes in either P_{50} or total [Hb].

Carbon Dioxide Transport and Respiratory Acidosis

CO_2 is highly soluble in water and aquatically ventilating fishes typically have low plasma PCO_2 and bicarbonate values (Rahn, 1966a,b; Howell, 1970). With a restriction in aquatic ventilation, plasma CO_2 and bicarbonate levels can become elevated in many air-breathing fishes (Lahiri *et al.*, 1970; Rahn and

Garey, 1973; Mangum *et al.*, 1978; Heisler, 1982, 1986), a condition considered to be an integral aspect of the evolution of air breathing in vertebrates (Rahn, 1966a,b; Gans, 1970; Howell, 1970; Johansen and Lenfant, 1972; Ultsch, 1987, 1996).

Several air breathers (mudskippers, African lungfish, *Channa*) are reported to have high plasma bicarbonate levels; however, the presence of an elevated CO_2 has not been confirmed. Although reduced ventilation results in a high blood CO_2 in some species (Lahiri *et al.*, 1970; Garey and Rahn, 1970), it has not been unequivocally established that this, in all instances, was strictly a respiratory acidosis due to branchial hypoventilation or hypoperfusion, because aquatic PCO_2 was not recorded. Verification that an elevated CO_2 is due to restricted gill exchange during air breathing is therefore needed for many species, and this requires simultaneous determination of blood and aquatic PCO_2, preferably over a range of ambient PCO_2 levels. This is because in many hypoxic swamps aquatic PCO_2 is often elevated (Carter, 1931; Wood and Lenfant, 1976; Kramer *et al.*, 1978), and a fish can, with sufficient ventilation, equilibrate this PCO_2 with its plasma. Thus, the blood CO_2 of an air-breathing fish may reflect the ambient CO_2 and not alteration of the respiratory mechanism (Heisler, 1986).

Regardless of its source, a rise in plasma CO_2 during air breathing reduces blood pH, which in turn lowers Hb–O_2 affinity (Bohr effect, Table 7.6) and perhaps even blood O_2 capacity (Root effect). While pH regulation during the carriage of oxylabile CO_2 from tissues to the respiratory organ is important, the larger question is how long-term exposure to increased PCO_2 during air breathing (owing to hypoventilation, gill hypoperfusion, aquatic hypercapnia, or all of these), and its effect on pH are compensated to permit continuation of normal respiratory gas transport? Do air-breathing fishes have an increased blood-buffer capacity or perhaps lower Bohr and Root effects? Do properties of their different Hb components enhance transport of respiratory CO_2 over a greater range of blood CO_2 tensions?

Blood buffering capacity. The buffering capacity of blood is described by the ratio of the increase in plasma bicarbonate ion to the decline in blood pH ($dHCO_3^-/d$pH). This characteristic is determined by *in vitro* measurements of the pH of whole blood that has been equilibrated with gas containing different amounts of CO_2 (Albers, 1970, 1985). The principal buffer agents in blood are Hb, the plasma proteins, and plasma bicarbonate. Hb and, to a lesser extent, plasma proteins are effective buffers because they have ionizable groups with pK's that are in the range of blood pH. The buffering capacity of fish blood is not large compared to most other vertebrates (values range from about -10 to -20 mM HCO_3^-/l pH) and appears to be poised mainly towards pH compensation associated with the transport of O_2 and oxylabile CO_2. Fishes appear unable to correct for large pH variations caused by changes in temperature or metabolic state (Heisler, 1986) or the consequences of air exposure (Lenfant *et al.*, 1966). There appear to be no unique buffer properties that distinguish the blood of air-breathing fish from non-air breathers.

Analyses of pH regulation in air-breathing fishes have focused primarily on the effects of hypercapnia, the result of either air exposure or branchial hypoventilation during aquatic air breathing. Elevated blood PCO_2 values have been recorded for a number of air-breathing fishes during air exposure or aquatic air breathing (Lenfant *et al.*, 1966; Lahiri *et al.*, 1970; Rahn and Garey, 1973; DeLaney *et al.*, 1977; Johansen *et al.*, 1978b; Ishimatsu and Itazawa, 1983b; Ultsch, 1996). The Henderson-Hasselbalch equation was introduced in Chapter 6 to define the quantitative relationships between blood bicarbonate, PCO_2, and pH:

$$\text{pH} = \text{pK}_1 + \log([HCO_3^-] / \alpha CO_2 \cdot PCO_2) \qquad (7.1)$$

This equation shows that blood pH will remain constant as long as the ratio of HCO_3^- to PCO_2 does not change (i.e., if temperature is stable, both pK_1 and α remain unchanged). In the event of change, the equation predicts a linear relationship between pH and bicarbonate concentration. This linearity, termed the blood buffer value or constant, is usually shown in the context of a Davenport diagram (Figure 7.5) which graphs the dependency of plasma $[HCO_3^-]$ on pH in relation to PCO_2 isobars. Buffer capacity is indicated by the slope of this line; a steeper line signifies more bicarbonate production per unit pH change, the usual characteristic of blood containing more hemoglobin or plasma proteins (i.e., more attachment sites for protons, Albers, 1970).

The steps in pH compensation in response to hypercapnia are illustrated in Figure 7.5 which shows data for the Australian lungfish (*Neoceratodus*) during a 39-minute air exposure (Lenfant *et al.*, 1966). In the first nine minutes of air exposure the plasma PCO_2 of this fish increased from about 3 to 9.5 torr; however, because there was a proportionate increase in bicarbonate (from about 3 to 10 mM), pH was nominally affected. Over the next 30 minutes, PCO_2 approached 20 torr, and in the absence of additional bicarbonate formation, pH fell dramatically to about 7.3. At this point the fish was returned to water and, as the figure shows, the reversal sequence for the pH, PCO_2, and bicarbonate excesses almost exactly mirrored their onset in air.

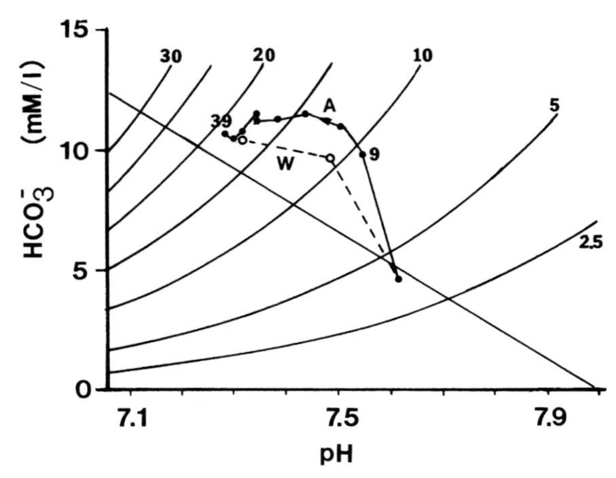

FIGURE 7.5 Davenport diagram showing blood pH in relation to HCO_3^- and PCO_2 (see text). Solid line is the *in vitro* buffer line for oxygenated whole blood of *Neoceratodus*. Data points are *in vivo* values for *Neoceratodus* during a 39-min air exposure (*A*, solid circles and line) and following its return to water (*W*, dashed line, open circles). Time span for the water recovery not indicated. (Reprinted from *Respiratory Physiology,* volume 2, Lenfant, *et al.* Respiratory properties of blood and pattern of gas exchange in the lungfish, *Neoceratodus forsteri* [Krefft], pp. 1–21, 1966, with kind permission of Elsevier Science-NL, Sara Burgerhartstraat 25, 1055 KV Amsterdam, The Netherlands.)

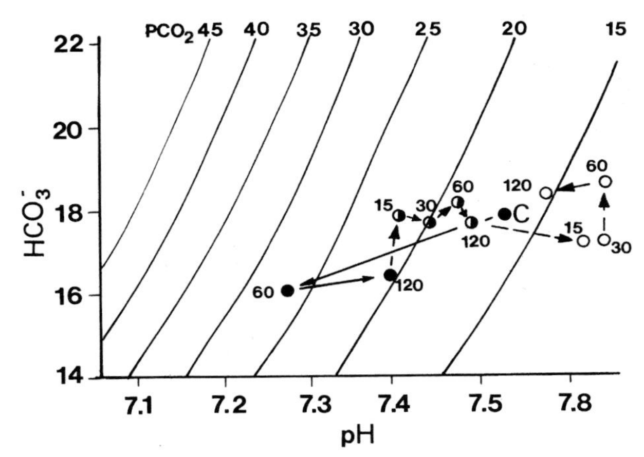

FIGURE 7.6 Davenport diagram showing acid-base changes in the blood plasma of *Channa argus* during a 2 h exposure to air (closed circles), during 2 h of artificial ventilation (half-closed circles), and then in water for 2 h (open circles). *C* indicates control or starting point and values next to each symbol indicate elapsed time since the onset of the treatment. Data are for fish No. 91 which was representative of four of the five fish tested in this manner by Ishimatsu and Itazawa. (T = 25 C, HCO_3^- units as in Figure 7.5.) (Reprinted from *Comparative Biochemistry and Physiology,* volume 74A, Ishimatsu and Itazawa. Blood oxygen levels and acid-base status following air exposure in an air-breathing fish *Channa argus,* pp. 787–793, 1983b, with kind permission of Elsevier Science–NL, Sara Burgerhartstraat 25, 1055 KV Amsterdam, The Netherlands.)

Although *Neoceratodus* is not amphibious, this data set illustrates both the timing and the scope of emersion-hypercapnia effects on blood acid-base parameters affecting respiration. Because *Neoceratodus* is not naturally amphibious and does not estivate (as does the African *Protopterus*), questions about its long-term adaptation to hypercapnic acidosis are not particularly relevant to its natural history. Such questions are important for species that naturally emerge from water, because pH compensation would be needed to sustain normal physiological and biochemical functions. It is possible, using data in Figure 7.5 and equation 7.1, to estimate the amount of bicarbonate required to fully restore the blood pH (7.28) of the air-exposed *Neoceratodus* to its pre-emersion level (7.62), even with a PCO_2 of 20 torr. Assuming a pK_1 of 6.1 and rearranging equation 7.1,

$$\log [HCO_3^-] = (7.62 - 6.10) + \log [20 \cdot 0.028]$$
$$[HCO_3^-] = 18.5 \text{ mM/l} \qquad (7.2)$$

Thus, complete pH compensation, which is rarely seen in fishes, including the air breathers, would require a sixfold increase in bicarbonate over the initial amounts recorded for this fish in water (Figure 7.5). Relevant to this calculation are data for *Protopterus* exposed to air in order to induce estivation (DeLaney *et al.,* 1977). These workers found the blood was hypercapnic and acidotic for about two weeks, after which pH was gradually restored by an increased bicarbonate.

In nearly all studies, emersion has resulted in a general hypercapnic-blood acidosis. The switch to air breathing in *Synbranchus* led to a rise in blood PCO_2 and a drop in pH, but no change in bicarbonate (Heisler, 1982). For air-exposed *Amia*, PCO_2 increased, pH fell, but total CO_2 was apparently unchanged (Daxboeck *et al.,* 1981). In the case of *Channa argus* a four-hour exposure to air lowered blood pH and increased PCO_2 but did not affect bicarbonate and, just as in *Neoceratodus*, these effects in *Channa* were reversed within one hour of immersion (Ishimatsu and Itazawa, 1983b). As detailed in Chapter 3, the aerial-ventilatory mechanism of *Channa* is critically dependent upon the piston-like action of water in the branchial chamber. Ishimatsu and Itazawa (1983b) were able to demonstrate this dependence by showing that artificially ventilated *C. argus* could reverse most symptoms of respiratory acidosis induced by air exposure. Figure 7.6 shows changes in blood pH, bicarbonate, and PCO_2 brought on by a two-hour period of air exposure which was followed first by a two-hour period of artificial ventilation, and then by the return of the fish to water. Artificial ventilation lowered the volitional breathing rate of *Channa* and caused a decrease in plasma PCO_2 and an elevated plasma pH. Artificial ventilation did not affect plasma HCO_3^- nor did it alter either arterial PO_2 or total blood O_2 concentration.

Just as a limited ventilation in air affected the acid-base status of *Channa*, studies with the aquatic *Hypostomus* by Wood *et al.* (1979) demonstrated that blood buffer capacity can be affected by facultative air breathing. Whole blood buffer curves for control and 48 hour hypoxia-acclimated *Hypostomus* (Figure 7.7) reveal that the control group (30 C) has a buffer curve typical of aquatically ventilating fish (i.e., with a low PCO_2 and HCO_3^-) and an *in vivo* pH of 7.4. The hypoxia-acclimated fish have the same blood pH but a much higher PCO_2 (20 torr) and HCO_3^- (13 mM). The *in vitro* buffer lines of the blood from these two groups, determined by equilibration with air and with air containing 5.6% CO_2, are parallel. This means that CO_2 retention resulting from reduced ventilation during facultative air breathing (although ambient CO_2 tension was not reported) was completely compensated by an increased buffer reserve (Wood *et al.*, 1979). It can be speculated that a longer period of hypoxia acclimation may have, through an increased [Hb], affected the slope of the buffer line.

The Bohr Effect

Carter (1931) initially hypothesized that, in view of the large and variable quantities of CO_2 present in the habitats and blood of air-breathing fishes, they would likely have a Hb that lacked a Root effect and had a minimal Bohr effect. Some literature on this subject pointed to a reduced Bohr effect among air breathers (Willmer, 1934; Fish, 1956; Johansen, 1970; Lahiri *et al.*, 1970; Farmer *et al.*, 1979).

Early studies emphasized a reduced Bohr effect in species with increased air-breathing dependence (e.g., *Protopterus* -0.35 versus *Neoceratodus* -0.62, Lenfant *et al.*, 1966). Molecular studies further affirm that the Hb components of some species may have a reduced and even a reversed Bohr effect (Weber *et al.*, 1976b; Farmer, 1979; Garlick *et al.*, 1979). However, many species also show a Root effect (Farmer *et al.*, 1979), and the large body of comparative data now available for air-breathing and other species (Table 7.6, also see Johansen *et al.*, 1978a; Mangum *et al.*, 1978; Powers *et al.*, 1979a; Morris and Bridges, 1994) indicate no relationship between air-breathing and the magnitude of the Bohr effect. Nor is such a pattern seen in comparisons of the closely related air-breathing and non-air-breathing pairs: *Osteoglossum* and *Arapaima*, -0.41 versus -0.30; *Hoplias* and *Erythrinus*, -0.40 versus -0.64 (data of Johansen *et al.*, 1978a,b; also see Powers *et al.*, 1979a; and Table 7.6).

A facet of the air-breathing adaptation question that has received little attention is the apparent change in the Bohr coefficient with hypoxia acclimation and the onset of facultative air breathing. Figure 7.8 shows that the whole blood Bohr coefficient of hypoxia-acclimated *Hypostomus* is significantly lessened relative to that of control fish (Weber *et al.*, 1979). Thus, even if an air-breathing fish could not buffer its blood sufficiently to precisely regulate pH, it could lessen the effects of pH

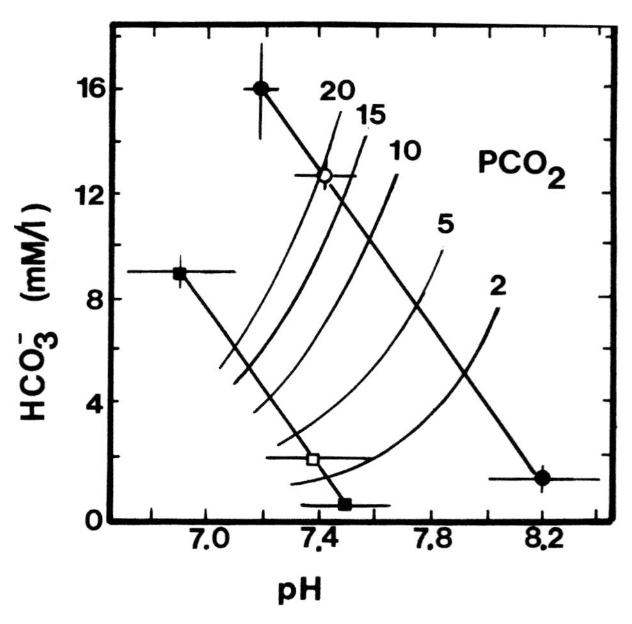

FIGURE 7.7 Davenport diagram showing oxygenated blood buffer curves for *Hypostomus* acclimated to hypoxic water for 48 h (circles) and a control group (squares). *In vivo* data for each group indicated by the appropriate open symbol (see text). (Data are from *Comparative Biochemistry and Physiology*, volume 62A, Wood, *et al.* Effects on air-breathing on acid-base balance in the catfish, *Hypostomus* sp., pp. 185–187, 1979, with kind permission of Elsevier Science-NL, Sara Burgerhartstraat 25, 1055 KV Amsterdam, The Netherlands.)

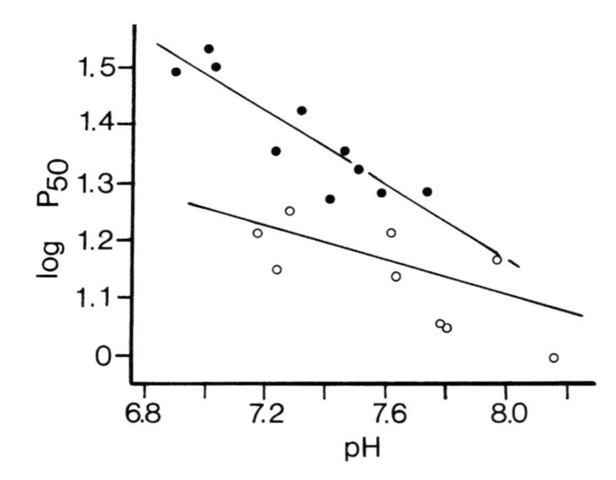

FIGURE 7.8 Whole blood Bohr coefficient data for *Hypostomus* acclimated to hypoxia (open circles) and a control group (filled circles). Regression equations: Hypoxia log P_{50} = 2.5 – 0.18 (pH) (n = 9, r = -0.72). Normoxia log P_{50} = 3.7 – 0.32 (pH) (n = 10, r = -0.89). (Data are from *Comparative Biochemistry and Physiology*, volume 62A, Weber *et al.* Acclimation to hypoxic water in facultative air-breathing fish: Blood oxygen affinity and allasteric effectors, pp. 125–129, with kind permission of Elsevier Science–NL, Sara Burgerhartstraat 25, 1055 KV Amsterdam, The Netherlands.)

and PCO_2 on Hb–O_2 affinity. As previously detailed, Weber *et al.* (1979) demonstrated an increased Hb–O_2 affinity for the hypoxia-treated groups of *Hypostomus*. These workers suggest that lowered red cell phosphate concentrations, by affecting both Hb–O_2 affinity and red cell pH, play a major role in reducing the Bohr effect (also see Johansen *et al.*, 1976; Bridges, 1988; and Jensen, 1991). The rise in [Hb] in some species and the use of different Hb subunits may be important in achieving this response.

Apart from the effect of air breathing on the Bohr coefficient of *Hypostomus*, a strong case cannot be made for the presence of any functional differences, insofar as CO_2 retention is concerned, in the blood properties of air-breathing and non-air-breathing fishes. There are no known mechanisms to conserve Hb–O_2 affinity during respiratory acidosis. Thus, the air-breathing fishes appear, for the most part, to endure the effects of a diminished O_2 capacity, although some can potentially compensate for this by increasing [Hb] and transporting more O_2 (Farmer, 1979; Farmer *et al.*, 1979; Graham, 1983, 1985).

OTHER BLOOD FUNCTIONS

O_2 Storage

The high [Hb] of air-breathing fishes may also serve for the storage of O_2 for times when air is not held. Table 7.1 shows that high [Hb] values occur among species (*Electrophorus*, *Synbranchus*, *Monopterus*, mudskippers) that hold air breaths in their buccal and branchial chambers. This type of air breathing precludes aquatic ventilation and mouth use, and because these fishes must at some time feed and have at least minimal branchial flows in order to carry out ancillary exchange processes (Stiffler *et al.*, 1986), they obviously do not spend their entire lives holding air in their mouths. Most of these fishes do, nevertheless, have diminished gill surfaces (Chapter 3; Carter and Beadle, 1931; Rosen and Greenwood, 1976) and some may not ventilate frequently or efficiently in hypoxic water (Farber and Rahn, 1970; Roberts and Graham, 1985; Graham *et al.*, 1987).

Studies with *Synbranchus* have demonstrated the probable O_2-storage function of its Hb and show how this specialization is integrated with variables such as the average duration of an air breath, ABO volume, O_2 utilization, and $\dot{V}O_2$ to ensure that while air is held, enough O_2 is loaded into the blood to supply needs during its inter-air-breath interval (IBI). This is illustrated for a 200 g fish using data contained in Table 7.8. A generalized hourly respiratory pattern for this fish in hypoxic water would consist of four, approximately 15 minute periods of alternating air breaths and IBIs. There are usually no, or very few, water ventilations during the IBI which is followed by another air breath. As detailed in Table 7.8, the amount of O_2 utilized from an air breath is 1.24 ml, and at two breaths/hour, this totals 2.48 ml O_2/h. Because little or no O_2 uptake occurs during the IBI, part of the O_2 removed from the ABO must be stored for use during the subsequent IBI. Thus, over an hour the amount of O_2 used during each 15 minute period is 2.48/4 = 0.62 ml.

With an estimated blood volume of about 8 ml and a [Hb] of 11.1 g% (Table 7.2), this fish has about 0.9 g of circulating Hb. Assuming an Hb–O_2 combining coefficient of 1.35 ml O_2/gHb, the saturated Hb of *Synbranchus* holds 1.2 ml O_2. With a mean Hmct of 40% (Tables 7.1–7.3), the approximately 4.8 ml of blood plasma in this fish would hold 0.03 ml of O_2 at saturation, giving a total blood O_2 capacity of 1.23 ml.

Table 7.8 shows that the calculated O_2 capacity of this fish approximates its measured O_2 uptake rate during a 15-minute air breath. It could therefore be expected that blood O_2 content of this fish would cycle between 100% saturation (i.e., late in the air-breath holding period) to about 50% saturation late in the IBI. Large swings in blood O_2 commonly occur in intermittent air breathers (Shelton and Boutelier, 1982) and were documented for *Synbranchus* by Johansen (1966, 1970). Oxygen delivery is not affected by these swings because *Synbranchus* has a high Hb–O_2 affinity that is only slightly affected by equili-

TABLE 7.8 An Hour in the Life of a 200 g Air-Breathing *Synbranchus marmoratus*[a,b]

Breath cycle					
Cycle	Air held	IBI	Air held	IBI	
Duration (min)	15.0[c]	15.0	15.0[c]	15.0	
Aerial O_2 utilization					**Total**
By interval	1.24[d]	0	1.24	0	2.48
Four period integral	0.62	0.62	0.62	0.62	2.48
Blood O_2 capacity					
	1.23 ml[e]				

[a]Indicating how air-breath duration, ABO Volume, O_2 utilization, metabolic rate, and blood O_2 capacity are integrated to permit O_2 storage for possible use during the inter-air-breath interval (IBI).

[b]Data are from Graham and Baird (1984), T = 25 C.

[c]15.7 min was rounded to 15 min to simplify calculations.

[d]From Chapter 3, Figure 3.26, a 200 g fish has an average air-breath volume of 7.4 ml · 0.21 = 1.55 ml O_2. Mean O_2 utilization in 15 min = 80%. Air-breath O_2 extraction in 15 min = 0.8 · 1.55 = 1.24 ml.

[e]8 ml total volume, 60% plasma, 40% Hmct; [Hb] = 11.1 g% = 0.89 g Hb in 8 ml. HbO_2 = 0.89 · 1.35 ml O_2/g Hb = 1.20 ml O_2. Plasma O_2 = 4.8 ml = 0.03 ml O_2, total O_2 capacity = 1.23 ml.

bration with 5.6% CO_2 (P_{50} = 9.4 torr at zero CO_2 versus 11.3 torr in 5.6% CO_2, 30 C; Powers *et al.*, 1979a). This means that most of the ABO O_2 can be loaded and then off-loaded in the tissues without interference from accumulating CO_2. The O_2 storage function for Hb in *Synbranchus* may have utility in its water-ventilation cycle which is characterized by an average of 15 minutes of eupnea and 7 minutes of apnea (Graham *et al.*, 1987).

Effects of Air Exposure and Estivation

Air exposure effects on blood respiratory properties have been studied in a number of air-breathing species and some results are detailed. In these tests the duration of air exposure has ranged from minutes to years (i.e., estivating lungfish). Observations have included exposure effects on various properties of erythrocytes, from their number and size to their ability to bind and carry O_2. The objective of much of this research was to determine how blood properties would be affected by the evolutionary transition from water to land.

Favaretto *et al.* (1981) found that keeping *Hypostomus regani* in air for four hours elevated both Hmct and [Hb] (Table 7.1). These increases were attributable to the release of immature blood cells which, while constituting about 2.6% of the red cell population of fish in water, amounted to 12.6% of the red cells in four-hour-air-exposed fish. For air-exposed and control *Hypostomus*, these workers found no differences in total blood or plasma volume, in extracellular (interstitial) volume, or in total body water content and could rule out dehydration as a causal factor in the observed increase in Hb and Hmct. Their data did, however, indicate that red cell volume declined in air which could have been a consequence of CO_2-associated shifts in water and ion content.

Beitinger *et al.* (1985) reported no effects of a four-hour air exposure on red-cell morphology, on most major features related to respiratory gas transport in *Erpetoichthys* including P_{50}, pH, buffer capacity (Table 7.6), or on either the type or total amount of NTPs. These workers did note a slight increase in the Hill coefficient (i.e., 2.0 for control and 2.53 for fish in air). Nonetheless, their overall conclusion was that *E. calabaricus* possesses no blood-respiratory mechanisms to facilitate exclusively aerial respiration. For *Hoplosternum littorale*, Perez (1980) similarly determined that 18 hours of air exposure had no effect on many of the very same blood-respiratory properties studied by Beitinger *et al.* (1985), including the Hill coefficient.

Johansen *et al.* (1978b) also studied air-exposed *Hoplosternum littorale*; however, unlike Perez (1980), these workers reported that 13 hours in air caused a radical hemoconcentration (increased Hmct and Hb, Table 7.1). The NTP/Hb of this fish was elevated, and somewhat paradoxically in view of the latter, Hb–O_2 affinity decreased (Table 7.6). It is not known what factors account for the different findings of these two studies on air-exposed *Hoplosternum*. The pronounced hemoconcentration suggests the fish studied by Johansen *et al.* (1978b) became more dehydrated than those in the Perez study. The findings of Johansen *et al.* (1978b) for *Pterygoplichthys* paralleled their *Hoplosternum* results: Fish in air for 28 hours had an elevated Hb and Hmct, whereas NTP/Hb was unchanged. A P_{50} was not reported for 28 hour fish; however, the P_{50} measured after 18 hours in air was less than that of fish in water (Table 7.6).

The final air-exposure test done by Johansen *et al.* (1978b) was with *Synbranchus*; several fish were exposed for as long as 21–35 hours and one was kept in air for 44 hours. These workers found a dramatic hemoconcentration (i.e., [Hb] rose from 13 to 15 g% while Hmct went from 48 to 59%), however, MCHC did not change. Air exposure caused a drop in plasma pH (7.8 to 7.4) and a doubling of the red cell phosphate concentration. The Bohr effect was shifted from -0.45 in water to -0.23 after 21–44 hours in air. Air exposure decreased Hb–O_2 affinity; the P_{50} of *Synbranchus* in water was 5 torr, but 11 torr for a fish in air for 44 hours.

All of these changes were interpreted by Johansen *et al.* (1978b) as an adaptive compensatory mechanism for long-term air exposure. However, based on the results reported for *Hoplosternum* and *Pterogoplichthys* by this team, it is possible that dehydration played a role in the observed changes. It is unclear why a proficient air breather like *Synbranchus* should lower its Hb–O_2 affinity during air exposure when obligatory air breathing in hypoxic water affects neither its NTP:Hb ratio (Weber *et al.*, 1979) nor its Hmct (Heisler, 1982). *Synbranchus* routinely makes terrestrial sojourns and naturally endures long periods of air exposure in humid mud burrows. Studies of fish confined to moist burrows without access to free water revealed little effect of either comparable, or much longer, exposure times on [Hb] and Hmct, and neither of these parameters changed in burrow-dwelling fish returned to water (Graham unpublished). Johansen *et al.* (1978b) did not detail their air-exposure techniques; however, the possibility that their fish were dehydrated cannot be excluded. More importantly, some of the characteristics they interpreted as long-term adaptations to air exposure for *Synbranchus* are handled quite differently in estivating lungfish (Weber *et al.*, 1976b).

The blood of *Protopterus amphibius* estivating in cocoons for 28–30 months differed in several respects from that of fish in water (DeLaney *et al.*, 1976; Johansen *et al.*, 1976). The initial exposure caused a relatively rapid hemoconcentration (i.e., increased [Hb], Hmct, and plasma electrolytes), owing to the loss of plasma. There was a change in the relative amounts of the four Hb subunits in this fish (Weber *et al.*, 1976b). After about 30 days there was a large release of neutrophils which is reminiscent of the findings of Favaretto *et al.* (1981) for *Hypostomus*.

The blood characteristics of estivating *P. amphibius* bear some similarity to those in 44-hour-air-exposed *Synbranchus*. There was about a 50% hemoconcentration and a conserved MCHC. However, *Protopterus*, unlike *Synbranchus*, had a strongly elevated plasma bicarbonate, and its total red cell phosphate levels (ATP and GTP, but not IP_5 were measured) were reduced by 50% relative to controls, which elevated Hb–O_2 affinity. Whereas the P_{50} of *P. amphibius* in water was about 35 torr, the value for estivating fish was 10 torr (Table 7.6). In addition, estivating fish had a larger Bohr effect than control fish (-0.5 versus -0.25). These changes in affinity and Bohr effect are the opposite of those reported for 44-hour *Synbranchus*. Johansen *et al.* (1976) concluded that the dramatic increase in Hb–O_2 affinity during estivation is an adaptive response to the confinement in cocoons, which are generally regarded as hypoxic (and are thought to restrict ventilation) (Fishman, *et al.*, 1987). A recent study by Lomholt (1993), however, has not provided support for a link between lung hypoxia and an affinity increase. Lomholt found that cocoon dwelling *P. amphibius* use a series of small tidal ventilations rather than single, large volume breaths taken by free living lungfish. In agreement with the more frequent breath pattern, Lomholt (1993) measured high expired lung PO_2s (100–110 torr) for cocoon fish and not the hypoxic levels postulated by Johansen *et al.* (1976).

SUMMARY AND OVERVIEW

Several air-breathing fish species have blood Hb concentrations approaching those of birds and mammals. Other species increase [Hb], usually by polycythemia, when their requirement for air breathing is heightened by aquatic hypoxia. To a large extent, both the Hb concentration and Hb–O_2 affinity of air-breathing fish blood are functionally linked to circulatory specializations for conserving aerially-obtained O_2 and preventing mixing between the ABO efferent and systemic bloods. An elevated blood O_2 capacity compensates for the forced unsaturation of O_2-rich efferent ABO blood when it mixes with O_2-poor systemic blood in the venous circulation. A high [Hb] can serve both an O_2 storage and a blood-buffering function in air-breathing fishes.

Research has focused on determining the extent that the evolution of air breathing has resulted in a reduction in hemoglobin-oxygen affinity. Although the pattern of reduced Hb–O_2 affinity is seen among closely related species of non-air-breathing and air-breathing osteoglossids (*Osteoglossum* versus *Arapaima*) and erythrinids (*Hoplias* and *Hoplerythrinus*), comparisons among a diversity of air-breathing and non-air-breathing species fail to verify this as a "universal" evolutionary pattern. The pattern is not, for example, seen among the three lungfish genera (*Neoceratodus*, *Lepidosiren*, and *Protopterus*) which have different air-breathing requirements. Nor is it apparent in a comparison of *Erythrinus* and *Hoplias*. Moreover, other and quite different patterns have been found. Facultative air breathers (*Ancistrus*, *Hypostomus*, and *Pteroglopichthys*), increase Hb–O_2 affinity during aerial respiration. This increase not only reduces the potential for the transbranchial loss of O_2 to hypoxic water, it compensates for the effect of increased CO_2 on Hb–O_2 affinity. Another pattern is found among species such as *Synbranchus* that hold air breaths on or near their gills. These fishes are not subject to transbranchial O_2 loss, have a high Hb–O_2 affinity, and change neither [Hb] nor Hmct as a result of exposure to hypoxic water. These various patterns all suggest that the degree of branchial isolation "enjoyed" by a species, that is, its degree of independence from gill respiration, is a major determinant of Hb–O_2 affinity during air breathing. More research is required to determine how aquatic hypoxia and other conditions affect the relative O_2 affinity of the different air-breathing species. Notably missing at the present time are detailed studies that experimentally evaluate both the environmental and evolutionary factors influencing the blood-respiratory properties of the Asian air breathing fishes (e.g., the mudskippers, the anabantoids, *Nototopterus*, and many others).

CO_2 retention due to hypoventilation, exposure to hypercarbic water, or both, is another general consequence of air breathing. Initial predictions that air-breathing fish Hb would be adapted for this condition through molecular specializations that reduced Bohr and Root effects have not been sustained empirically, and there appear to be neither qualitative nor quantitative differences in the blood-buffer capacities of air- and non-air-breathing fishes. Nonetheless, a limited data set for *Hypostomus* reveals that, following its transition to facultative air breathing, this fish can compensate for a decreased blood-plasma pH in hypercarbia by increasing HCO_3^- concentration and lessening its Bohr effect.

CHAPTER

8

Metabolic Adaptations

Though it is not invariable, the pressure of natural selection is frequently of such intensity as to seem to operate not through its subtler modes but by the very threat of death, by calling forth new adaptations as the only alternative to sudden and complete extinction.

H.W. Smith, *From Fish To Philosopher,* 1959

INTRODUCTION

In the same way that Chapter 7 considered the effects of aerial O_2 on blood respiratory properties, the objectives of this chapter are to determine the extent to which energy metabolism, enzyme kinetics, organ function, and system physiology have been affected by air breathing and the relatively routine access to aerial O_2 it permits. The perspective taken is shown in Table 8.1 which contrasts several metabolic and organ functions in different vertebrate classes to illustrate the involvement of separate organ systems in the basic physiological functions of respiration, ion and water regulation, acid-base balance, and nitrogen excretion. This table shows that major shifts in the division of labor and in the integration of organ systems have coincided with the evolutionary transitions from aquatic to aerial respiration and from aquatic to terrestrial life. In a freshwater fish, for example, four of these key metabolic processes reside principally within the gills and only one in the kidney. In mammals, gas exchange and respiratory acid-base regulation are lung functions; whereas, ion and water regulation, nitrogen excretion, and metabolic acid-base regulation are the province of the kidney. Although fishes do excrete some organic acids via their urinary system, this is not integrated with respiration as are the respiratory and metabolic components of acid-base regulation in mammals. In larval amphibians, excretion, osmoregulation, and respiration are branchially mediated; the post-metamorphic frog occupies an intermediate position between fish and mammals in terms of kidney function and, depending upon its habitat, auxiliary organs such as the skin and urinary bladder may be involved in respiratory and osmotic functions.

This chapter seeks evidence for changes in basic metabolic and physiological functions that have occurred with the transition to air breathing in fishes. Has, for example, the availability of aerial O_2 affected oxygen delivery to the tissues or tissue aerobic and anaerobic potentials? What are the implications of the reduction in gill perfusion associated with air breathing (Chapter 5) for the normal branchial functions of CO_2 removal, osmotic regulation, acid-base balance, and nitrogen excretion? How do routine amphibious excursions affect the nitrogen excretion pattern? In addition to being intrinsically interesting, these questions contribute to an understanding of the biochemistry and physiology of the first vertebrate air breathers and the first amphibious tetrapods (Stiffler *et al.*, 1986).

OXYGEN ACCESS AND INTERMEDIARY METABOLISM

Table 8.2 presents comparative $\dot{V}O_2$ data for three pairs of closely related air-breathing and non-air-

TABLE 8.1 Comparative Metabolic and Functional Properties of Organs in Different Vertebrate Groups

Physiological process	Functional organ[a] in specified group			
	Mammal	Amphibian: Tadpole	Amphibian: Adult	Fish
Gas exchange				
O_2/CO_2	L	G/S	L/S	G/S
Ion regulation	K	G	K,S,B	G
Water regulation	K	K	K	K
Acid-base balance	K,L	G?	K	G
Nitrogen excretion	K	G	K	G

[a]L, lung; K, kidney; G, gill; S, skin; B, urinary bladder.

TABLE 8.2 Comparative Total $\dot{V}O_2$ Data for Some Closely Related Air-Breathing and Non-Air-Breathing Fish Species in Three Families[a]

Species	°C	$\dot{V}O_2$ (ml/kg h)	Reference
Osteoglossidae			
Osteoglossum bicirrhosum	28–30	71.4	4
Arapaima gigas[a]	28–30	157.8	4
Erythrinidae			
Hoplias malabaricus	27–30	49.7	1
	25	80–100	6
H. microlepis	25	73.2	2
Hoplerythrinus unitaeniatus[a]	27–30	59.1	1
	27–30	174	5
	25	80–100	6
Lebiasinidae			
Piabucina panamensis	25	75.7	3
P. festae[a]	25	135	3

[a]The air-breathing species.
[1]Cameron and Wood, 1978; [2]Dickson and Graham, 1986; [3]Graham *et al.*, 1978b; [4]Stevens and Holeton, 1978a; [5]Stevens and Holeton, 1978b; [6]Mattias *et al.*, 1996.

breathing fishes. In most cases, comparisons were made by the same investigators employing the same methodologies. The table shows that in two cases the air-breather has a considerably higher $\dot{V}O_2$ than the non-air breather. In the third case (Erythrinidae), such a difference appears in some, but not all, data sets. The finding of an elevated $\dot{V}O_2$ associated with air breathing is consistent with early hypotheses about the physiological effects of air access (McCutcheon and Hall, 1937; and see Chapter 7). In addition, these metabolic differences must carry implications for the behavioral ecology and ecological energetics of air breathing (reviewed by Kramer 1983b, 1988). These $\dot{V}O_2$ differences may be expressed in the biochemistry and intermediary metabolism of air-breathing fishes; it is the nature and extent of this expression that is the subject of this section.

This review begins with consideration of anaerobic and aerobic metabolism in air-breathing fishes. This subject was previously treated for the Amazonian air-breathing fishes by Hochachka (1980) and updated accounts are found in de Almeida-Val and Hochachka (1995) and in Val and de Almeida-Val (1995). Volumes such as Hochachka and Somero (1984) should be consulted for elucidation of the biochemical principles underlying much of the following discussion.

Aerobic and Anaerobic Metabolism

In vertebrates the initial steps of both aerobic and anerobic metabolism proceed along the same pathway in which a metabolic fuel (some amino acids and proteins, glucose, glycogen) is converted to pyruvate (Figure 8.1). In anaerobic metabolism the enzyme lactate dehydrogenase (LDH) catalyzes the conversion of pyruvate to lactate which accumulates in the cell. Even though lactate formation has a favorable free-energy change (-25 kJ/mole), this process cannot be sustained indefinitely. The capacity of lactate conversion to briefly prolong glycolysis resides in the regeneration of nicotine adenine dinucleotide (NAD^+) from its reduced form ($NADH/H^+$). This regeneration is of key importance because the quite small amount of NAD^+ contained in the cell is used in a preceding glycolytic step (2 glyceraldehyde 3-phosphate ——> 2 1,3-bisphosphoglyceric acid).

In aerobic metabolism, pyruvate or lipid (free fatty acids) is converted into acetyl-CoA which is channeled to the citric acid cycle, the products of which are CO_2, $NADH/H^+$, and reduced flavin dinucleotide ($FADH_2$). These steps take place in the mitochondria where NAD^+ and FAD are regenerated; O_2 serves as an electron acceptor and combines with the H^+ to form water, and a large quantity of ATP is synthesized.

Biochemical Correlates of Aerial O_2 Access

Osteoglossids As a Case Study

Comparative metabolic data were obtained for the non-air-breathing *Osteoglossum bicirrhosum* and the air-breathing *Arapaima gigas* during the 1976 *Alpha Helix* Amazon expedition. Ultrastructural, biochemical, and enzymatic indicators point to a relatively depressed metabolic activity in both these genera compared to other species. For example, the activity of brain citrate

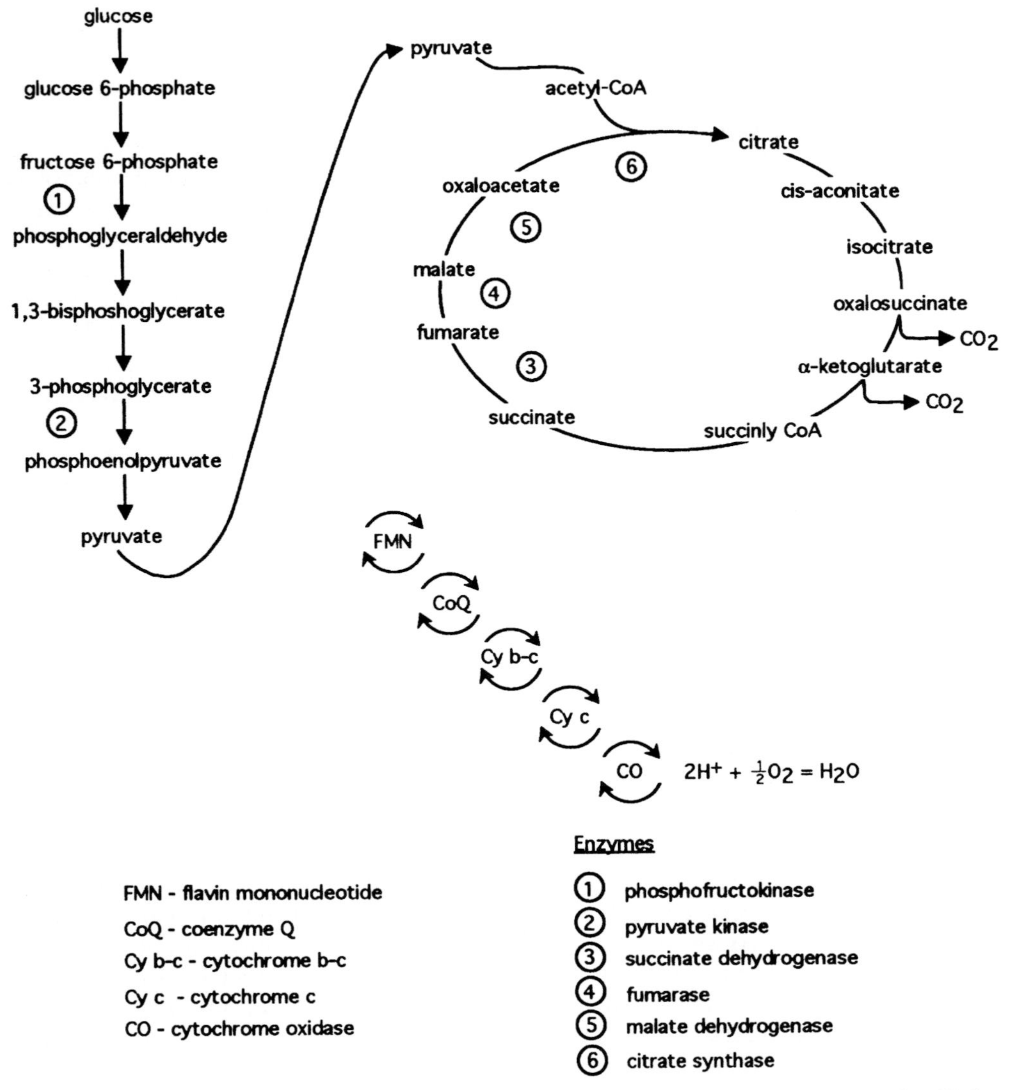

FIGURE 8.1 Generalized depiction of the anaerobic (glycolytic) and aerobic (citric acid or Krebs cycle) metabolic pathways for the breakdown of glucose and for the coupled oxidative phosphorylation via electron transfer. Key anaerobic and aerobic enzymes mentioned in the text are indicated.

synthase (CS, a mitochondrial marker enzyme that regulates carbon flow into the citric acid cycle) in *Arapaima* and *Osteoglossum* is well below that of most other fishes and vertebrates in general (Hochachka *et al.*, 1978b). Because both genera have low brain CS activity, this appears to be an osteoglossid character and may therefore be independent of respiratory adaptation *per se*, and both genera have a high anaerobic tolerance. Evidence for an air-breathing effect on tissue oxidative capacity is seen in several tissues.

Heart. Differences in the heart tissues of *Arapaima* and *Osteoglossum* correlate with the differences in oxygen-use rate shown in Table 8.2. Hochachka *et al.* (1978b) found that the heart of *Arapaima* is much larger than that of *Osteoglossum* and is surrounded by lipid stores. The cells making up the innermost heart layer (spongy myocardium) of *Arapaima* contain numerous, regionally differentiated mitochondria that appear specialized for O_2 utilization and the exchange of nutrients with blood. Although glycogen is present in these cells, they also contain an abundant store of intracellular triglyceride which is the primary metabolic fuel.

By contrast, the spongy myocardium of *Osteoglossum* contains two cell types. Type I cells are similar to those in the spongy layer of *Arapaima* and have numerous mitochondria, a store of glycogen, and are

TABLE 8.3 Comparative Activities of Selected Heart-Tissue Enzymes[a,b,c]

Enzyme and role	Enzyme activity	
	Osteoglossum	*Arapaima*
Anaerobic metabolism		
PK	144	52
LDH-B (low pyruvate)	409	307
LDH-A (high pyruvate)	205–243	195–325
Aerobic metabolism		
CS	11	10
GDH	2.9	3.4
Other functions[d]		
PFK	10.7	5.7
FDPase	0.25	0.54
Aldolase	11.1	10.9

[a]Non-air-breathing *Osteoglossum bicirrhosum* and the air breather *Arapaima gigas*.
[b]Activity units are μM/min g tissue (wet weight) at 25 C.
[c]Hochachka *et al.*, 1978b.
[d]Note: PFK and aldolase both function in glycolysis. The Aldolase reaction is reversible and it can also function in gluconeogenesis.

aerobic. Type II cells, which are about twice as abundant as type I cells, contain fewer mitochondria and have large glycogen stores indicating they are less aerobic than either the type I cells or the heart cells of *Arapaima* (Hochachka *et al.*, 1978b). Thus, the greater oxidative capacity of *Arapaima* heart tissue is indicated by larger populations of aerobic myocardial cells and a store of triglycerides and other lipids. The activities of three glycolytic enzymes PFK (phosphofructokinase), PK (pyruvate kinase), and LDH (lactate dehydrogenase) are all lower in the heart cells of *Arapaima* than in *Osteoglossum*. Moreover, the differences may be more extreme than indicated by data in Table 8.3 which were obtained using homogenates of the type I and II cells.

LDH isomers and the role of the heart in lactate metabolism. In most lower vertebrates the bulk of the lactate that is formed by various tissues during anaerobic glycolysis enters the circulation and is transported to organs such as the heart, kidney, and liver for further processing as a source of energy or carbon. The first processing step is the conversion of lactate to pyruvate which can then have several fates: Conversion to acetyl-CoA and oxidation to CO_2 via the Krebs cycle (Figures 8.1 and 8.2), conversion to alanine which enters the intracellular free amino acid pool, or utilization as a substrate for gluconeogenesis which restores tissue glycogen levels (Dickson and Graham, 1986).

The relative activities of the LDH-A and LDH-B isomers indicate the extent to which a tissue is poised for anaerobic glycolysis or the oxidation of lactate (French and Hochachka, 1978). The LDH-A isomer (also known as M_4LDH or the pyruvate reductase [PR] isomer) is commonly found in anaerobically poised tissue such as white muscle. LDH-A is not inhibited by high pyruvate concentrations and thus continues to reduce pyruvate to lactate, an essential requirement for a burst of muscular activity. By contrast, the LDH-B isomer (also known as H_4LDH or the lactate oxidase isomer) occurs in more oxidative tissues (heart, red muscle) where it functions to initiate lactate clearance by oxidizing lactate to pyruvate. In light of this enzyme's function, it is not surprising that LDH-B is inhibited by the moderate accumulation of its own end product—pyruvate.

Table 8.3 compares the relative activities of LDH in the heart tissues of *Osteoglossum* and *Arapaima*. Both species have a high LDH activity and although LDH-A is present, LDH-B is the prodominant type. Thus, in addition to having an anaerobic capacity, the heart tissues of both *Osteoglossum* and *Arapaima* can function in the clearance of lactate circulated to the heart from other parts of the body (Hochachka *et al.*, 1978b, Hochachka, 1980).

Red and white muscle. The myotomal muscles of both *Arapaima* and *Osteoglossum* are largely adapted for anaerobic function (Hochachka *et al.*, 1978a). A number of differences do, however, correlate with the aerial O_2 dependence of *Arapaima*. This fish has both red (i.e., oxidative) and white (glycolytic) myotomal fibers, whereas *Osteoglossum* has only white muscle (Hochachka *et al.*, 1978a). Although *Arapaima* red muscle has a greater mitochondrial abundance and higher levels of aerobic enzymes than its white muscle, the levels of the glycolytic enzymes PFK, PK, LDH, as well as aldolase (this enzyme cleaves and forms carbon bonds and, by keeping fructose diphosphate levels low, facilitates glycolysis and pyruvate formation) are as high in red muscle as in white muscle (Table 8.4). This is in stark contrast to the general pattern seen in most fishes (i.e., white > > red) and indicates that red muscle also has considerable anaerobic potential.

The much greater anaerobic capacity of *Osteoglossum* white muscle is indicated by its higher PK (5×) and LDH (3×) levels relative to *Arapaima* (Table 8.4). Another indicator is the higher ratio of fructose-1,6 diphosphatase (FDPase, catalyzes the hydrolysis of fructose diphosphate in the glucogenic direction) to PFK (0.24 in *Osteoglossum* versus 0.06 in *Arapaima*) which, together with a higher aldolase activity, would

GLUCONEOGENSIS

LACTATE –(1)–> PYRUVATE –(2)–> Oxaloacetate –(3)–> PEP ->->-> FDP –(4)–> F6P ->-> Glucose

TRANSAMINATION

PYRUVATE –(5)–> alanine

OXIDATION VIA CITRIC ACID CYCLE

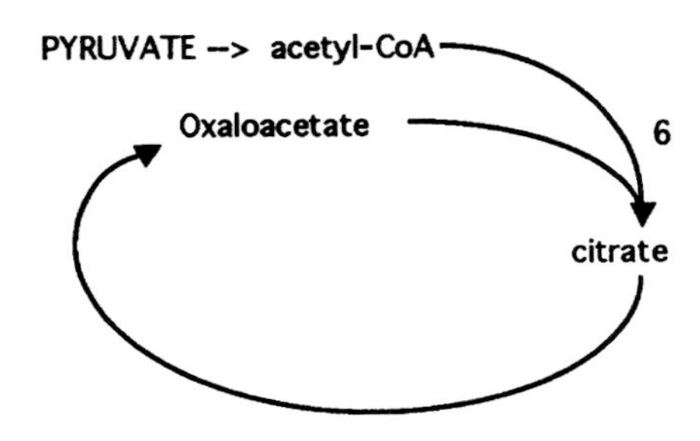

SUBSTRATES

- PEP phosphoenolpyruvate
- FDP fructose - 1,6 -biphosphate
- F6P fructose - 6 - phosphate

ENZYMES

1. LDH (Lactate oxidase)
2. pyruvate carboxylase
3. PEP carboxykinase
4. FDPase
5. glutamate - alanine transaminase
6. citrate synthase

FIGURE 8.2 The metabolic fates of lactate. See text for detail.

TABLE 8.4 Comparative Activities (μM/minute g, 25 C) of Selected Muscle Enzymes[a,b]

Enzyme and role	*Osteoglossum* white muscle	*Arapaima* White muscle	*Arapaima* Red muscle
Anaerobic metabolism			
PK	536	103	134
LDH-B (low pyruvate)	760	260	263
LDH-A (high pyruvate)	702	182	181
Aerobic metabolism			
CS	0.95	1.71	3.3
GDH	0.56	1.3	3.1
GOT	5.5	11.2	54.4
Other functions			
PFK	9	12	9
FDPase	2	0.7	1
FDPase/PFK	0.24	0.06	—
Aldolase	69	20	21

[a] *Osteoglossum* is a non-air-breather; *Arapaima* is an air breather.
[b] Hochachka *et al.*, 1978a.

sustain a greater rate of glycolysis (Hochachka *et al.*, 1978a) or possibly enhance gluconeogenesis. By contrast, the activities of the aerobic enzymes CS, GDH (glutamate dehydrogenase, reversibly catalyzes ammonia or glutamate formation) and GOT (aspartate aminotransferase, catalyzes oxaloacetate formation; [note, this enzyme is also abbreviated as AST]) in *Arapaima* white muscle are between two to three-times higher than in *Osteoglossum* white muscle (Table 8.4).

Kidney and gills. The premise offered at the beginning of this chapter was that, for some groups, the effects of air breathing on gill ventilation may have resulted in selection for the transference of ancillary branchial functions to other organs. Morphological and biochemical comparisons of *Arapaima* and *Osteoglossum* (Table 8.5) indicate that the kidney of *Arapaima* has partially assumed functions that would be exclusively branchial in most species and suggests that air breathing is the causal factor in this transition (Fields *et al.*, 1978a; Hochachka *et al.*, 1978c; Hulbert *et al.*, 1978a,b,c; Storey *et al.*, 1978).

TABLE 8.5 Comparative Activities (μM/minute g, 25 C) of Selected Enzymes from the Kidney and Gills of *Osteoglossum* and *Arapaima*[a]

Enzyme and role	Kidneys		Gills	
	Osteoglossum	*Arapaima*	*Osteoglossum*	*Arapaima*
Anaerobic metabolism				
PK	205	31	50	30
LDH–B (low pyruvate)	141	332	69	52
LDH–A (high pyruvate)	102	313	50	37
Aerobic metabolism				
CS	1.3	0.9	1.4	2
GDH	4	23	2	3.2
GOT	20	90	8.4	14.7
Other functions				
PFK	2	0.68	1.5	0.6
FDPase	1.6	3.6	0.37	0.71
FDPase/PFK	0.8	5.3	0.25	1.18
Aldolase	1.3	7.8	3.2	3.5
Ion activated ATP–ases				
Na^+/K^+	0.01	0.04	1.14	0.03
Na^+/NH_4^+	0	0.06	0	trace

[a]Hochachka *et al.*, 1978c; Hulbert *et al.*, 1978a.

Reflecting its air-breathing specializations, the gills of *Arapaima* have about a 50–70% smaller mass than those of *Osteoglossum*. This smaller size, combined with structural specializations for branchial shunting (Chapter 4), may reduce *Arapaima* gill efficiency for CO_2 and nitrogen excretion and ion regulation. Enzyme profile data for gills correlate with the different O_2-use rates of *Osteoglossum* and *Arapaima*. Although the gills of both genera have an oxidative capacity, this appears greater in *Arapiama* which has higher activities of CS, MDH (malate dehydrogenase, converts malate to oxalaoacetate in the Krebs cycle), GOT, and GDH. Higher PK, PFK, and LDH activities and a lower FDPase:PFK (Table 8.5) are indicative of a greater anaerobic potential in *Osteoglossum* gills. The higher proportion of LDH-B in both fishes indicates that lactate arriving from other tissues can be processed (see preceding). On the other hand, a lower FDPase:PFK ratio and low aldolase activity in *Osteoglossum* implies a smaller role for gluconeogenesis.

A predominating role for gills in Na^+ regulation is the expected pattern for fishes, and the gills of both *Osteoglossum* and *Arapaima* have about the same density of chloride cells (Hulbert *et al.*, 1978c). However, Na^+/K^+-ATPase, the enzyme involved in branchial Na^+ transfer, is about five times more active in the gills of *Osteoglossum* than in *Arapaima* (Table 8.5). Further, because gill mass is two to three times less in *Arapaima*, this translates into a 12-fold activity difference in favor of the non-air breather.

A nearly inverse pattern is seen for the kidneys of these fishes. Hulbert *et al.* (1978a) reported that the kidney of *Osteoglossum* has about one-tenth the mass of its gills and, given its 14 times lower Na^+/K^+-ATPase activity (Table 8.5), this translates into nearly 150-fold reduction in Na^+ regulation potential for the kidney (see Hochachka *et al.*, 1978c). The kidney of *Arapaima* has a larger relative mass than that of *Osteoglossum*; the just cited works by Hulbert *et al.* (page 804) and Hochachka *et al.* (page 828) contain unpublished observations of a 3.5× larger kidney in *Arapaima*. Thus, although the relative activity of kidney Na^+/K^+-ATPase in *Arapaima* is about the same as for gills (Table 8.5), the much larger kidney size increases its role in Na^+ regulation to a level approaching that of the gills. Similarly, the much greater activity of GDH in the *Arapaima* kidney, combined with a larger organ mass, signify a considerable importance for the kidney's role in nitrogen excretion and acid-base balance (Hochachka *et al.*, 1978c).

Anaerobic and aerobic enzyme profiles of the kidney indicate a greater glycolytic capacity for *Osteoglossum* and a greater oxidative capacity in *Arapaima* (Table 8.5). In mammals the liver is the primary organ for lactate clearance; however, Hochachka *et al.* (1978c) point out that the high levels of LDH and the high percentage of lactate oxidase (LDH-B) in the osteoglossid kidney suggest it is the dominant site for lactate conversion in osteoglossids, especially *Arapaima*. Because this reconversion requires oxygen, the kidney's position inside the gas bladder (Chapters 3 and 4) may especially favor this activity.

Kidney gluconeogenesis in *Arapaima* may be related to the high GDH activities measured in this organ by Hochachka *et al.* (1978c). Storey *et al.* (1978) determined that GDH kinetics in *Arapaima* kidney favor the oxidative deamination of glutamate, which forms 2-ketoglutarate (2KGA), NH_3, and H^+. Glucogenic enzymes in the kidney could form glucose from 2KGA and, assuming the proton by products are excreted in the urine, GDH would thus promote one facet of a mammal-like acid-base function in this kidney. In addition, the presence of Na^+/NH_4^+-ATPase in *Arapaima* kidney tissue suggests a mechanism for the urinary release of ammonia (Hochachka *et al.*, 1978c).

The general enzyme profiles of *Arapaima* and *Osteoglossum*, together with the details surrounding the reciprocity of renal and branchial function in *Arapaima*, offer a classic portrait of how routine aerial-O_2 access and the anatomical and physiological changes enhancing this access have influenced tissue intermediary metabolism as well as organ and system function. However, as will become evident in subsequent sections, visiting this same theme with other groups of closely related air-breathing and non-air-breathing species pairs does not always lead to the same set of satisfying results. Furthermore, it is important to emphasize that even though these tissue and enzyme studies are especially appealing from a theoretical standpoint, no comparative studies of integrated, whole organism bimodal and renal function have ever been carried out for these osteoglossids or any other species and this needs to be done.

The Erythrinid Paradox

Driedzic *et al.* (1978) compared the energy metabolism of the heart of the air-breathing *Hoplerythrinus unitaeniatus* and the non-air-breathing *Hoplias malabaricus*, determining that the energy charge (the proportion of high energy phosphate in the adenylate pool, Hochachka and Somero, 1984; Vetter *et al.*, 1986) of these two fishes was not different. *Hoplias* had much greater stores of glycogen in its liver, heart, and gills; *Hoplerythrinus* had much greater lipid stores around its heart, in its liver, and surrounding its viscera.

The implication of this finding, that glycogen was the preferred metabolic fuel of the non-air breather's tissues, was confirmed with experiments on isolated, perfused hearts. Not only did the hearts of *Hoplias* beat longer in anoxia than those of *Hoplerythrinus*, heart activity in the former could be sustained by adding glucose to the perfusate. Oxygen deprivation experiments, in which buffered cyanide was perfused, further showed that the heart of *Hoplias* was much more anoxia resistant than that of *Hoplerythrinus*; the heart of the former used both perfused glycogen and myocardial glycogen stores to sustain anaerobic function. Driedzic *et al.* (1978) concluded that, under normal conditions, the heart of *Hoplias* has a much greater anaerobic potential than that of *Hoplerythrinus* and that, in severe anoxia, this heart's larger quantity of glycogen stores ensures a prolonged anaerobic function.

Although differences in diet and food availability of these two genera might account for the observed differences in fat deposition, Driedzic *et al.* (1978) did tentatively conclude that the presence of larger fat stores in the tissues of *Hoplerythrinus* was indicative of a greater dependence upon free fatty acids as a metabolic fuel, particularly in the heart. This aerobic metabolic option would be consistent with the continued oxygen access provided by air breathing.

Paradoxically, virtually all other comparative studies of *Hoplias* and *Hoplerythrinus* tissue metabolic capacity, including some whole-organism data, have proved negative insofar as a correlation with air breathing is concerned. Principal among these results are enzyme profiles for muscle (Hochachka *et al.*, 1978a; Fields *et al.*, 1978b) and gill (Hulbert *et al.*, 1978b) and measurements of $\dot{V}O_2$, renal function, and acid-base balance (Cameron and Wood, 1978). Specifically, the glycolytic enzyme activities in both red and white muscle are higher for *Hoplerythrinus*, and both CS and GDH activities are twofold higher in *Hoplias* (Table 8.6). Despite being an air breather with a reduced gill surface area (Cameron and Wood, 1978) and, according to some data sets, having a higher $\dot{V}O_2$ (Table 8.2), the tissues of *Hoplerythrinus* appear to be more glycolytic and less oxidative than those of *Hoplias*.

There is not an easy explanation for this set of facts. The premise of this chapter, that air breathing could obviate a major reliance upon metabolic adaptations for hypoxia and permit greater aerobic activity, is simply not supported by data for erythrinid tissues other than the heart. Very little can be done about this except to point out that these metabolic differences do not seem consistent with the behavior patterns of these fishes (i.e., *Hoplias* is a slow moving, stalking, quick strike and primarily piscine predator; the air-breathing erythrinids are more on-the-move feeders and consume a greater variety of prey [Graham, pers. obs.]). The metabolic differences are not in concert with some of the data on whole-organism metabolism (e.g., Table 8.2; Stevens and Holeton, 1978b; Dickson and Graham, 1986; Mattias *et al.*, 1996) and the heart studies of Driedzic *et al.* (1978) which support the idea that *Hoplias* has a greater anaerobic capacity. It is therefore quite important to revisit the *Hoplias–Hoplerythrinus* comparison. Hochachka *et al.* (1978a) mention

TABLE 8.6 Comparative Activities (μM/minute g, 25 C) of Selected Enzymes from the Gills[a] and Red and White Muscle[b] of the Non-Air-Breathing Erythrinid *Hoplias malabaricus* and the Air-Breathing *Hoplerythrinus unitaeniatus*

	Gills		White Muscle		Red Muscle	
Enzyme and role	*Hoplias*	*Hoplerythrinus*	*Hoplias*	*Hoplerythrinus*	*Hoplias*	*Hoplerythrinus*
Anaerobic metabolism						
PK	35	37	174	448	144	341
LDH-B (low)	126	94	576	1064	419	810
LDH-A (high)	66	67	508	911	280	692
Aerobic metabolism						
CS	2.2	0.35	2	1.3	3.7	1.4
GDH	4.4	1.5	1	0.75	1.7	0.8
Other functions						
PFK	2.1	2.7	5.6	3.6	2	4.1
FDPase	0.65	0.46	2.5	1.9	3.4	1.8
Aldolase	—	—	27.3	16.2	13.8	13.7
Ion activated ATPase						
Na^+/K^+	0.03	0.04				

[a]Hulbert *et al.*, 1978b.
[b]Hochachka *et al.*, 1978a.

(pages 744, 746) preservation and loss problems with some of their erythrinid material. Cameron and Wood (1978) cite numerous limiting factors in their study (sample size, animal condition, experiment duration), all of which would warrant further erythrinid comparisons. Finally, the white muscle enzyme profile data obtained for *Hoplias microlepis* in this laboratory (Dickson and Graham, 1986) show much lower levels of CS, MDH, PK, and LDH than were reported for *H. malabaricus* by Hochachka *et al.* (1978a). The data for *H. microlepis* also show that hypoxia acclimation can affect both oxidative and glycolytic enzyme levels. The lessons of Chapter 7 were that hypoxia acclimation, habitat, and season can affect the blood-respiratory properties of some species. Long-term comparisons of *Hoplias* with the air-breathing erythrinids may therefore be required to resolve the present enigma.

Other Air-Breathing Fishes

Lungfishes. In spite of proficient air breathing, the metabolic organization of the African (*Protopterus*) and South American (*Lepidosiren*) lungfishes is poised primarily for anaerobiosis, and these animals are capable of entering a state of metabolic depression in response to acute hypoxia exposure or during seasonal estivation (Lahiri *et al.*, 1970; Hochachka and Hulbert, 1978; Hochachka, 1980; Dunn *et al.*, 1981, 1983). These two lungfishes have large stores of glycogen in their hearts, livers, and both red and white muscle can use this energy substrate either aerobically or anaerobically. In fact, these stores are so large that Hochachka and Hulbert (1978) used the term "glycogen seas" to describe their appearance in red muscle and "glycogen rosettes" to characterize their presence in both liver and white muscle. These lungfishes have large fat stores (illustrated in Hochachka and Somero, 1984, figure 7-4) which, along with glycogen and protein, are utilized in energy metabolism.

Enzyme profiles indicate that, although the African and South American lungfish are capable of oxidizing a range of substrates, they, like the osteoglossids, have a low metabolic turnover relative to most fishes. As expected, lungfish red muscle contains greater numbers of mitochondria than white muscle; however, the metabolism of both *Protopterus* and *Lepidosiren* seems poised primarily for anaerobiosis (Tables 8.7 and 8.8). Fish that are forcibly submerged or given nitrogen to breathe will concentrate lactate in their muscles and release this when air is again available (Lahiri *et al.*, 1970). Both the hearts and brains of these fishes have a high glycolytic potential, and the heart, liver, and kidney can all participate in gluconeogenesis.

The combination of air breathing and a primarily glycolytic muscle metabolism has suggested certain similarities between lungfishes and diving mammals with respect to both the need to sustain anaerobic glycolysis while away from the surface and to deal with the clearance of lactate following the resumption of respiration (Hochachka, 1980). Dunn *et al.* (1983) tested the anaerobic capacity of *P. aethiopicus* through "forced laboratory dives" of 12 hours in duration. These forced dives did not have the same effect on lungfish as they have on mammals. First, lungfish cardiac output was not redistributed as it is in forcibly

TABLE 8.7 Activities (μM/minute g, 25 C) of Selected Metabolic Enzymes from *Protopterus aethiopicus*[a]

Enzyme and role	Tissue			
	Heart	Muscle	Liver	Brain
Anaerobic metabolism				
PK	101	99	14	103
LDH-B (low pyruvate)	594	256	249	149
LDH-A (high pyruvate)	403	150	168	117
Aerobic metabolism				
CS	18	0.84	2.44	2.58
AAT[b]	83.4	3.7	29.1	—
HOACD[c]	17.4	1.97	32.9	—
Other functions				
Aldolase	16	27.8	6.7	—

[a]Dunn *et al.*, 1983.
[b]Aspartate aminotransferase.
[c]Beta-Hydroxyacyl–CoA dehydrogenase.

submerged mammals. Second, metabolite assays done on forcibly submerged lungfish revealed that both brain and heart can function anaerobically and that liver remains active in glucose synthesis, which is also different from mammalian divers (Dunn *et al.*, 1983). Thus, although superficially appealing, parallels between diving seals and "diving lungfish" can only be drawn so far because of the many radical differences between these organisms. Indeed, the survival of lungfish held without access to air for 12 hours in the Dunn *et al.* (1983) studies was possible because of the experimenters' use of aerated water which, owing to the ability of the fish to respire aquatically, probably mitigated many of the anaerobic stressors that would be quickly faced by submerged mammals.

The apparent paradox of air breathing and a primarily anaerobic muscle metabolism has been linked to burrowing and estivation (Hochachka, 1980). The connection to burrowing is logical; burrowing is an energetically intense undertaking done most often below the water surface and thus without air access. The low O_2 tensions that prevail in the cocoons of *Protopterus* and the difficulties imposed by ventilating within a cocoon (Chapters 2 and 7) have fueled ideas that a heightened anaerobic capacity relates to estivation. Although this is appealing, it was shown in Chapter 7 that cocoon fish have a reduced metabolic rate and that respiration within the cocoon may not be as limiting as was previously thought. In addition, several features of estivation metabolism appear to obviate anaerobic capacity. The metabolic changes associated with estivation, for example, occur gradually and appear induced by seasonal factors such as food supply and habitat drying. Among the changes taking place in estivation are the endogenous production of a metabolic depression factor (Swan *et al.*, 1968), a switch to a primarily fat dependent oxidative maintenance metabolism (this spares glycogen reserves for the tissues—brain, red cells, kidney—requiring this substrate [Hochachka and Somero, 1984]), and a switch to ureotelism (see Nitrogen Metabolism, page 234). Depending upon environmental conditions at the time of capture, metabolic characteristics related to estivating may or may not be apparent in a lungfish. Further, not all African lungfishes form cocoons, and this behavior has never been reported for *Lepidosiren* (Chapter 2).

It has been suggested that lungfish muscle-O_2 requirements could be low because these fishes are slow moving and primarily searching, opportunistic feeders without need of a sustained swimming performance. This is a reasonable explanation; however, it should be noted that *Protopterus* occurs as deep as 20 m in Lake Victoria, a habitat that may necessitate greater mobility (Greenwood, 1987). If the extent of muscle utilization is a valid explanation for the prevailing low metabolic demand of lungfish muscle, then the Australian *Neoceratodus*, with its more open-water behavior and less dependence upon air breathing, would be an ideal comparative subject to test this hypothesis.

Lungfish and *Synbranchus*: Habitat effects on heart metabolic adaptations. The natural histories of the South American lungfish (*Lepidosiren*) and *Synbranchus* have numerous parallels including air breathing, burrowing, and the ability to survive moist burrow confinement during the dry season. As might be expected from these similarities, the hearts of these two fishes have been shown to have metabolic specializations for elevated oxidative and glycolytic capacities (French and Hochachka, 1978; Hochachka, 1980). In addition to large glycogen stores and numerous mitochondria, both hearts have intracellular stores of fatty acids. As shown in Table 8.8, the hearts of

TABLE 8.8 Comparative Activities (μM/minute g, 25 C) of Selected Heart Enzymes of *Lepidosiren* and *Synbranchus*[a]

Enzyme and role	*Lepidosiren*	*Synbranchus*
Anaerobic metabolism		
LDH-B (low pyruvate)	1177	871
LDH-A (high pyruvate)	464	385
Aerobic metabolism		
CS	11.5	13.5
GDH	2.9	8.9

[a]Hochachka and Hulbert, 1978.

Lepidosiren and *Synbranchus* have high levels of both oxidative and glycolytic enzymes. Most notable are the high activities of LDH-A and LDH-B which are greater than in most other fishes, including the tunas, and three to five times greater than in most mammal hearts (Hochachka and Hulbert, 1978; Hochachka, 1980)!

A first interpretation, based on both high LDH and glycogen stores, is that these hearts have elevated anaerobic capacities that sustain activity during burrowing. While this is probably the case, the air-breathing capabilities of these fishes and their specialized coronary circulation patterns (Chapter 4) would seem to favor sustained O_2 delivery to the heart whenever air breathing was possible. Related to this, it is important to note that LDH-B predominates in both of these hearts, which means they are strongly capable of lactate oxidation. Thus, in *Lepidosiren* and *Synbranchus*, the sustained aerobic capacity of the heart enables this organ to function in the initial processing of lactate arriving from less aerobic tissues such as the muscle. An analogous situation exists in the heart of *Arapaima* (see preceding) which may also use its kidney for this purpose.

Mudskippers. Mudskippers have a high metabolic rate, are accomplished bimodal and exclusively air breathers, and spend a large percentage of their lives out of water. These fishes therefore have unlimited environmental O_2 access most of the time. However, mudskipper behavior often involves bursts of intense, and probably anaerobic, terrestrial locomotion (Hillman and Withers, 1987); these fish are also confined to hypoxic and, perhaps even, anoxic burrows at certain times (e.g., during high tide, while guarding eggs, Ip *et al.*, 1991).

A range of metabolic-performance capacities have been attributed to mudskippers. Bandurski *et al.* (1968) reported that *Periophthalmodon australis* could survive up to 40 minutes in a nitrogen atmosphere and that both nitrogen-exposed fish and exercised fish developed large concentrations of lactate in their muscles and brains. By contrast, Gordon (1978) reported that *Periophthalmus vulgaris* has very limited anaerobic capacity; it had a short survival time (5–10 minutes) when acutely exposed to anoxic water and only a limited survival capacity (1–2 hours) in gradual hypoxia. These fish could not sustain burst activity for very long and did not form much lactate after the burst was completed.

Comparisons of muscle and liver enzyme activities by Siau and Ip (1987) suggested that both *B. boddarti* and *Periophthalmodon schlosseri* have a well-developed capacity for anaerobic glycolysis. These workers found no differences between the two species in the following enzymes: PEPCK (PEP carboxylase, a gluconeogenic enzyme), GOT (= AST), MDH, PK, ALT (alanine aminotransferase), LDH. Nor did exposure of fish to a range of salinities or to air affect tissue-enzyme profiles in a manner relating to the different aerial-exposure patterns and respiratory capacities of *Boleophthalmus* (which spends less time in air) and *Periophthalmus*. Siau and Ip did, however, find differences between liver and muscle tissues. Relative to the liver, the muscle of both species has higher PK and LDH activities and higher ratios of LDH:MH (indicating muscle's greater capacity to sustain lactate production than Krebs cycle activity) and PK:PEPCK (reflecting muscle poising toward lactate oxidation rather than gluconeogenesis). The liver tissue of both species has higher GOT (AST):MDH ratios than most other fish species, suggesting that the malate-aspartate shuttle may be important in oxidizing NADH formed by glycolysis. Finally, this study showed that exposure to air for three days resulted in increased aspartate levels (which may relate to nitrogen excretion, see following and Chew and Ip [1987]) in the muscle of both species, but did not affect the levels of muscle lactate, malate, or alanine.

Chew and co-workers (Chew *et al.*, 1990; Ip *et al.*, 1991; Chew and Ip, 1992a,b,c, 1993) examined the capacity of mudskippers to tolerate ambient hypoxia, the result of burrow confinement, and functional hypoxia, which was induced by bursts of forced anaerobic activity. They found that both *Boleophthalmus boddarti* and *Periophthalmus chrysospilos* tolerated six hours' exposure to hypoxic water by entering a state of metabolic arrest. The oxygen levels inducing this arrest are very low (less than one percent saturation), however, *P. chrysospilos* enters this state at a higher oxygen tension than *B. boddarti* which may reflect its lower natural resistance to hypoxia as a result of spending more time in air. Measurements with *B. boddarti* do show that acute exposure to nearly anoxic water results in an immediate threefold reduction in $\dot{V}O_2$ (Chew and Ip, 1992a).

Biochemical studies reveal that exposure of *B. boddarti* and *P. chrysospilos* to hypoxia resulted in virtually no changes in muscle glycogen or total energy charge (creatine phosphate, ATP, ADP, and AMP). Muscle lactate levels and the quantities of all the intermediary metabolites that were measured (citrate, malate, pyruvate, succinate, alanine, aspartate) were conserved. Further, the total excess $\dot{V}O_2$ measured for these fish during recovery from ambient hypoxia amounted to only a small fraction of the rate of control fish over the same period. Anoxia experiments on *B. boddarti* (done by exposing fish to pure N_2 gas for 40 minutes), caused slightly greater reductions of muscle adenylates and creatine phosphate and larger increases in both lactate and succinate, than were recorded during hypoxia, however, none of the other variables were affected.

Siau and Ip (1987) suggested that the PK:PEPCK ratios in *P. chrysospilos* and *B. boddarti* might favor the production of oxaloacetate and then succinate which might be a source of anaerobic energy production. Certain invertebrates can sustain anaerobic ATP production by converting the glycolytic intermediate phosphoenolpyruvate (PEP) to succinate; very little is known about this mechanism in vertebrates (Hochachka and Somero, 1984). Although anoxia and functional hypoxia (see following) did elevate succinate levels in *B. boddarti*, Chew and Ip (1992b) concluded that the small quantity that was formed had little impact on energy production.

The induction of functional hypoxia, done by chasing fish for 5–10 minutes until they were exhausted, caused a significant depletion of muscle glycogen stores and a reduced energy charge, and a sixfold increase in lactate in *P. chrysospilos* (Ip *et al.*, 1991). Similar changes occurred in *B. boddarti* which increased its muscle lactate by 20-fold; however, of the six intermediary metabolites listed, only aspartate and citrate remained unchanged in this species following functional hypoxia. Experiments showing a relatively high tolerance to cyanide poisoning established that *Boleophthalmus* has a high resistance to anoxia, the capacity to detoxify cyanide to thiocyanate, and a high cytochrome oxidase activity. In most other respects the mitochondrial metabolism of this species is largely similar to other fishes (Chew and Ip, 1992c; 1993).

Channa. The metabolic effects of acute exposure to air and hypoxic water have been studied in different snakehead species. Ramaswamy (1983) reported that exposure of *Channa gachua* to air for periods of 5 and 10 hours depleted body glycogen reserves by 73% and 92% relative to control fish. Although liver glycogen was reduced by air exposure, a significant redistribution of this metabolite also took place; glycogen levels increased in both heart (+40% after 5 hours, +82% after 10 hours) and muscle (+50%, +13%).

Yu and Woo (1987a) compared the metabolic responses of *C. maculata* to acute and chronic aquatic hypoxia. Acute (1 hour exposure) tests were carried out at aquatic oxygen levels ranging from 50 to 10 torr, and biochemical assays were done on plasma, liver, white muscle, heart, and brain. Relative to the control group (130 torr), the largest effects were elicited by exposure to 30 and 10 torr O_2. Among the tissues examined, the following parameters were significantly increased in acute hypoxia: *Plasma*, glucose, lactate, glycerol; *liver*, water, lactate, amino acids; *muscle*, lipid; *heart*, alanine; *brain*, lactate, succinate. The parameters found to decrease in acute hypoxia exposure were: *Liver*, lipid; *heart*, glycogen, succinate.

Prolonged (three days) exposure to hypoxia resulted in a greater utilization of lipid and glycogen fuel reserves. Accordingly, total glycogen was depleted and there was an increase in plasma glycerol and an elevated liver LDH activity. However, none of the metabolite changes seen in the acute hypoxia fishes were evident in fish exposed to hypoxia for three days. Yu and Woo (1987a) concluded that, in spite of its air-breathing capacity, *C. maculata* was partially dependent upon anaerobic metabolism during acute hypoxia exposure. By contrast, during the course of 3 days' hypoxia acclimation, this fish was able to elevate is respiratory efficacy and supply its total metabolic requirements by oxidative metabolism. However, the increased metabolic cost of air breathing in hypoxia required the tapping of glycogen and lipid reserves.

Clarias. Johnston *et al.* (1983) studied the metabolic adjustments of hypoxia-acclimated *C. mossambicus*. Relative to fish in aerated (114 torr PO_2) water, catfish held for 11–13 days in hypoxia (18 torr) had a higher air-breathing frequency and a higher metabolic rate. This species does not have a large glycogen store (280 mg/100g of tissue), and the hypoxia-acclimated fish had slightly lower amounts of glycogen in their livers, but higher amounts of glycogen in both their red (+78%) and white (+76%) muscle fibers. Hypoxia acclimation additionally resulted in a 36% reduction in the mean cross-sectional area of the white muscle fibers. Neither the number of muscle fibers per capillary nor the mitochondrial volume densities of the white and red muscle fibers were affected by hypoxia.

Based on the low amounts of stored glycogen and the absence of changes in either the muscle fiber:capillary ratio or mitochondrial volume, Johnston *et al.* (1983) concluded that the need for anaerobic metabolism in *C. mossambicus* was largely mitigated by its ability to acquire oxygen by air breathing. The potential limit to oxygen supply imposed by an aquatic hypoxia of 18 torr was compensated by an increased air-breathing frequency and by acclimatory processes that increased aquatic-respiration proficiency.

Amia. Singer and Ballantyne (1991) profiled the metabolic organization of *Amia calva*. Although their study did not specifically focus on air-breathing specializations, they observed that, relative to teleosts, *Amia* had lower aerobic (CS) and glycolytic (PFK, LDH) enzyme activities in its red muscle, liver, and heart. Regarding air breathing, Singer and Ballantyne (1991) concluded that the ability of *Amia* to tolerate hypoxic water is more related to air breathing than to biochemical specialization. This conclusion echoes that reached for *Clarias mossambicus* by Johnston *et al.* (1983).

NITROGEN METABOLISM

Because fishes excrete nitrogenous waste as ammonia via their gills, it is important to ask how air breathing might affect this process. Data presented above indicating that the kidney of *Arapaima* may function in ammonia excretion is one example of a compensatory mechanism for a reduction in branchial ventilation associated with aquatic air breathing or the reduced availability of water during routine terrestrial excursions. Could air-breathing fishes be expected to have an increased tolerance for ammonia, an alternate way of disposing of this substance, or might they use mechanisms to detoxify ammonia (e.g., urea or amino acid synthesis) and store it until such time as it can be excreted?

Comparative Overview

The cellular breakdown of nitrogen-containing substances (e.g., structural and dietary proteins, nucleotides, adenylates) inevitably results in deamination and the formation of ammonia. In its gaseous (NH_3) form, ammonia is highly soluble in tissues and plasma. However, this molecule also has a dissociation constant between 9 and 10 which means that, at normal cytosolic pH levels of 6–7, it strongly attracts protons and thus readily forms ammonium ion (NH_4^+). Thus, while ammonia can be considered a pH buffer (i.e., the binding and excretion of H^+ eliminates acid build-up), its strong proton affinity renders ammonia toxic and requires that it be kept dilute and promptly excreted, or that it be converted into less toxic compounds such as urea or uric acid (Hochachka and Somero, 1984; Walsh and Henry, 1991).

Organisms can be categorized on the basis of their predominant nitrogen excretory product, and this has led to comparative insight into the range of environmental, evolutionary, and physiological factors influencing the overall process (Smith, 1959; Baldwin, 1964; Huggins *et al.*, 1969; Hochachka and Somero, 1984). There are three basic patterns in vertebrates: Ammonotelism (ammonia excretion), ureotelism (urea), and uricotelism (uric acid). Uricotelism is not known to occur in fishes and is not considered further.

Ammonotelism is the primary method utilized by the fishes, by both larval and aquatic amphibians, and by most aquatic invertebrates. In fishes, ammonia diffuses from tissues to the plasma and is then transported to the gills where it rapidly diffuses into water. Ammonotelism is possible because ammonia is highly soluble (about 10,000 times more than CO_2) in plasma and water. However, because ammonia is toxic, it must be kept in a dilute solution, and ammonotelism requires a nearly unlimited access to water.

If environmental water availability is limited or if the water contains toxins such as ammonia, catabolically formed ammonia can be detoxified by urea synthesis (ureotelism). Urea formed in this way may be voided via either the urine or gills. Ureotelism combines the advantage of conserving water with the disadvantage of requiring metabolic energy (2.5 ATP per N) (Mommsen and Walsh, 1991, 1992). Ureotelic animals include most adult amphibians, some reptiles, most mammals and, as will now be detailed, certain fishes.

Fish Nitrogen Metabolism

The principles of fish nitrogen metabolism were formulated in classic comparative studies by Homer Smith (1930, 1931, 1935a, 1959), and there are several recent reviews (Campbell, 1973; Hochachka and Somero, 1984; Griffith, 1991; Walsh and Henry, 1991; Mommsen and Walsh, 1991, 1992; Wood, 1993). The intricate association between environmental water availability and lower-vertebrate nitrogen metabolism has influenced theories about the habitat and evolution of the early vertebrates and has contributed to ideas about causal factors for the origin of vertebrate terrestriality (Baldwin, 1964; Hochachka and Somero, 1984; Griffith, 1991).

Ammonotelism

Ammonia forms in mitochondria, diffuses through the cytoplasm to the extracellular space, and is transported to the gills. Ammonotelism requires no specific biochemical steps once deamination (NH_2 removal or cleavage) occurs, only the presence of water for dilution. Whereas diffusion is the primary mechanism for the branchial release of ammonia to the ventilatory water stream, other factors can complicate the rate of ammonia flux (Wood, 1993). First, in some species and in different tissues, ammonia may be converted to a carrier molecule (e.g., glutamine or alanine or other molecule) prior to release into the plasma. When the carrier arrives at the gills, ammonia would be split off and excreted. However, the carrier molecule may not reach the gills, and the various synthetic and degradative processes taking place in other tissues, for example, the liver, may recycle it into a protein, a deaminated fat or carbohydrate, or oxidize it in the Krebs cycle (Mommsen and Walsh, 1992).

Another variable affecting ammonotelism is the relative proportions of NH_3 and NH_4^+ excreted branchially which is influenced by biochemical factors, by fish metabolic and osmotic state, and by ambient water conditions (Wood, 1993). Ammonium, for example, can be involved in the stoichiometric exchanges for Na^+ or H^+ and thus affect acid-base balance (Figure 7.2).

Ureotelism

Urea is the predominant nitrogen excretory form in marine elasmobranchs and the coelacanth (*Latimeria chalumnae*), which concentrate sufficient amounts of this solute in their plasma and tissues to be isosmotic with seawater. Estivating African lungfish and a few other species, including some air breathers, also form large amounts of urea (Smith, 1930; Campbell, 1973; Hochachka and Somero, 1984; Saha and Ratha, 1989; Mommsen and Walsh, 1989, 1991; Rozemeijer and Plaut, 1993). The urea formed by many of these fishes is synthesized by the ornithine urea cycle (OUC, Figure 8.3) which combines ammonia (or other nitrogen donors such as glutamine or aspartate) with bicarbonate to form urea (Hochachka and Somero,

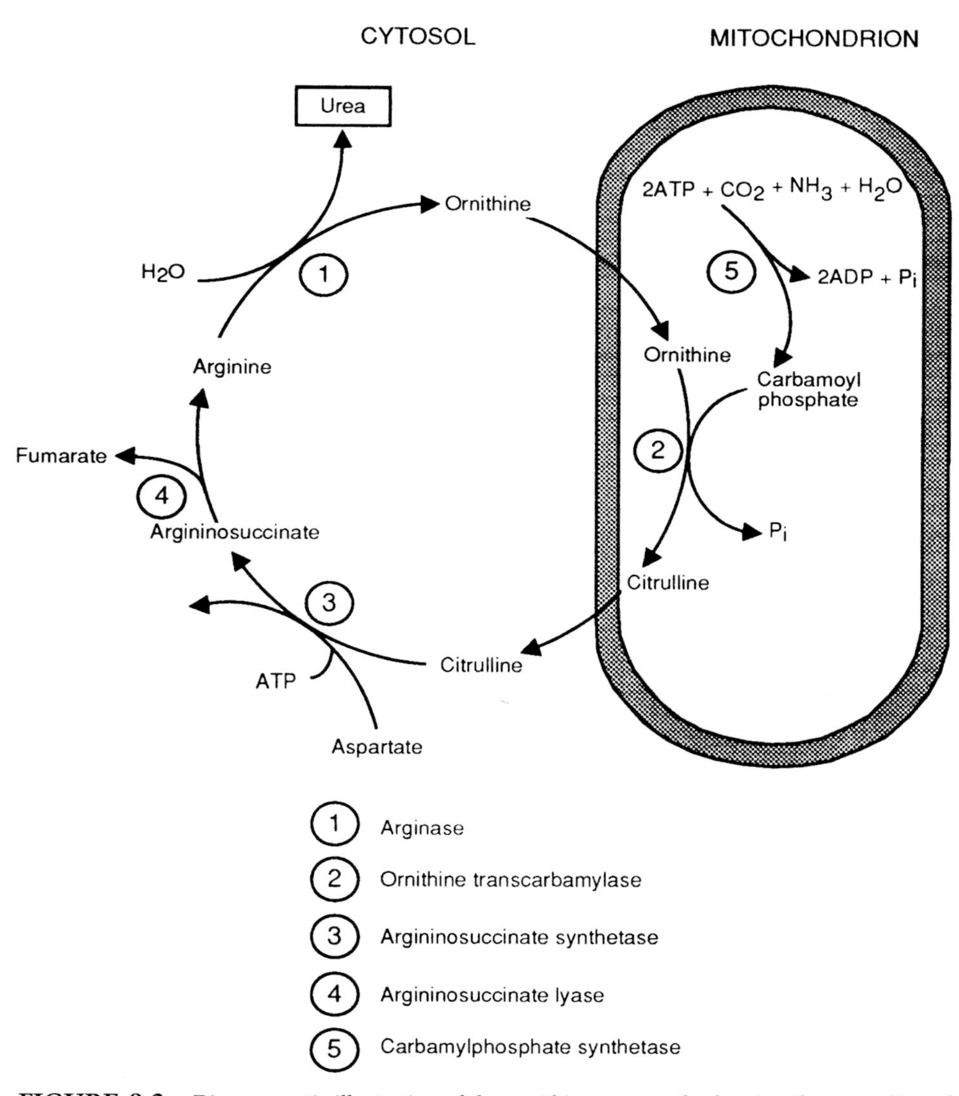

FIGURE 8.3 Diagrammatic illustration of the ornithine–urea cycle showing the cytosolic and mitochondrial steps. See text for detail.

1984). Whereas fishes also make urea through the degradation of purines, and by the hydrolysis of the amino acid arginine (a major constituent of most fish diets), the OUC is the major source of urea in ureotelic fishes (Campbell, 1973). The OUC is the major pathway for nitrogen excretion in terrestrial vertebrates (Hochachka and Somero, 1984; Gordon and Olson, 1994). Urea has the physiological advantages of being relatively inert, of requiring less water, and of being easy to store.

The presence of the OUC in species such as the African and South American lungfish and tetrapods has historically provided the rationale for an evolutionary link between the OUC and the invasion of land by the vertebrates. This is because the OUC is ideally suited for solving the problems of hypercapnia, desiccation, and nitrogen excretion imposed by air breathing and terrestriality. Urea formation by an amphibious fish would enable it to conserve water and avoid both hypercapnia and ammonia toxicity. Moreover, Atkinson and co-workers (Atkinson and Bourke, 1987; Atkinson, 1992) have suggested that, in the case of the terrestrial vertebrates, bicarbonate consumption by urea synthesis is an important feature of acid-base balance. Might such a mechanism be operating in the air-breathing fishes, and could CO_2 accumulation actually mitigate the effects of ammonia?

Urea and Fish Air Breathing: Natural History, Physiology, and Biochemistry

During the 1960s comparative interest in the relationship between nitrogen metabolism and the evolutionary transition from water to land grew significantly. The amphibians provided the striking example of a metamorphosis-induced transition from ammonotelism to ureotelism due to increased activities of OUC enzymes in the liver. It was discovered that amphibian ornithine-urea activity could be stepped up in response to osmotic stress (Gordon *et al.*, 1961; McBean and Goldstein, 1967), natural or induced estivation (Balinsky *et al.*, 1967; Janssens and Cohen, 1968a,b), or increased ambient ammonia levels (Janssens, 1972). Among the fishes, OUC utility in osmoregulation had already been demonstrated for both the elasmobranchs and *Latimeria* and additional evidence was accumulating for the role of urea formation during lungfish estivation (i.e., while in a burrow and wrapped in a cocoon [Smith, 1959; Gordon *et al.*, 1961; Brown and Brown, 1967]).

Lungfish Estivation and Urea

All of the lungfishes are ammonotelic in water and have tissue and plasma urea and ammonia levels typical of other freshwater fishes. However, both the African (*Protopterus*) and South American (*Lepidosiren*) lungfishes potentially use the OUC to form and store urea during estivation (Smith, 1930, 1935a; Janssens, 1964; Janssens and Cohen, 1966, 1968a,b). The African and South American lungfishes have a greater OUC capacity than their non-estivating Australian counterpart, *Neoceratodus* (Goldstein *et al.*, 1967). *Protopterus* and *Lepidosiren* can also form urea via purine catabolism (Brown *et al.*, 1966; Forster and Goldstein, 1966; Janssens and Cohen, 1968a,b; Carlisky and Barrio, 1972; Campbell, 1973).

In the case of estivating African lungfish, Huggins *et al.* (1969) made a distinction between ureogenesis—the presence of the OUC enzymes and a low rate of urea formation, and ureotelism—urea as the primary nitrogen excretion product. The metabolic rate of estivating *Protopterus* declines to below 20% of resting rate in water (Smith, 1930, 1935a; Janssens, 1964; Chapter 5) and with this comes a dramatic reduction in the rate of ammonia production. While urea becomes concentrated during estivation, the actual rate of urea synthesis is about the same for an estivating and a non-estivating African lungfish, which means that there is little, if any, up-regulation of urea synthesizing capacity (Janssens and Cohen, 1968a,b; Huggins *et al.*, 1969).

Thus, for the African lungfish, the combined absence of water and food and a need for exclusive air breathing triggers a metabolic-rate reduction rather than a transition to an elevated rate of urea production. This is fundamentally different from the changes in ornithine-urea enzymes seen in various amphibians exposed to a habitat water shortage.

Urea Metabolism: Other Fishes

The mid-1960s also saw a concerted effort to extend principles learned about ammonia detoxification by urea synthesis to the amphibious fishes. Malcolm Gordon and co-workers did the first experiments testing for the induction of ureogenesis. They compared rates of urea and ammonia excretion in fishes that were first held in water for a control period, exposed to air for a set time, and then returned to water.

These tests revolved around measuring the content, by mass, of ammonia-N and urea-N in water samples removed from the holding chamber at regular intervals during the control and post-emersion periods. The underlying assumption of this approach was that, while nitrogen metabolism (ammonia production) continues during air exposure, nitrogenous waste elimination is interrupted because the branchial system is not functional and because fluid economy precludes urine excretion. The post-emersion excretion pattern would indicate changes in nitrogen metabolism taking place during air exposure. Gordon *et al.* (1969, 1970, 1978) reported that air exposure followed by immersion could increase urea excretion in both mudskippers (*Periophthalmus sobrinus* and *P. cantonensis*) and the Chilean clingfish (*Sicyases*).

As compelling as these findings were, however, a clear relationship for urea and the OUC in air-breathing teleost nitrogen metabolism has not been demonstrated, and there is marked variation in the apparent patterns of nitrogen excretion in different species and questions regarding the presence and efficacy of the OUC.

Nitrogen Excretion

Investigations of the types of nitrogen excretion products that are formed and their formation rate have been conducted on 17 air-breathing species (Table 8.9). Comparison of pre- and post-emersion ammonia and urea-nitrogen excretion rates has been the most common experimental method. The techniques have ranged from transferring fish from water to air (humidified chambers) and then back to water, to placing them on a grid over water, which could be sampled for fecal and urinary N releases at regular intervals (Davenport and Sayer, 1986; Sayer, 1988; Sayer and Blackstock, 1988), to catheterizing the urinary papilla to collect urinary outflow during air exposure (Stiffler *et al.*, 1986). Most fishes in water, including non-air breathers, are strongly ammonotelic and excrete very small amounts of urea. Although a post-emersion increase in the proportion of urea excretion has been shown for several species, the results are highly variable (Table 8.9).

TABLE 8.9 Percentage of Total Nitrogen Excretion Made Up by Urea in Some Air-Breathing Fishes[a]

Species	% Urea	Reference
Clarias batrachus	8	Saha and Ratha, 1989
C. mossambicus	13	Eddy *et al.*, 1980
After 4 h in air	36	
Heteropneustes fossilis	9	Saha and Ratha, 1989
	12	Saha and Ratha, 1990
In 25 mM NH_4Cl	23	
In 75 mM NH_4Cl	27	
Sicyases sanguineus		
Water control	58	Gordon *et al.*, 1970
After 20 h in air	86	
After 35 h in air	97	
Alticus kirki	55	Rosemeijer and Plaut, 1993
After 5 h in air	57	
After 24 h in air	66	
Blennius pholis	18	Davenport and Sayer, 1986
Pholis gunnellus		
Water control	19	Kormanik and Evans, 1988
After 24 h in air	13	
Periophthalmus sobrinus		
Water control	42	Gordon *et al.*, 1968
After 12 h in air	50	
P. cantonensis	54	Gordon *et al.*, 1978
	20	Morii *et al.*, 1978
Water control	20	Iwata *et al.*, 1981
After 72 h in air	52[b]	
P. expeditionium	33	Gregory, 1977
P. gracilis	14	Gregory, 1977
Boleophthalmus pectinirostris	14	Morii *et al.*, 1978
Scartelaos histophorus	19	Gregory, 1977
Channa gachua		
Water control	58	Ramaswamy and Reddy, 1983
After 5 h in air	60	
After 10 h in air	78	
Channa punctatus	3	Ramaswamy and Reddy, 1989
	3	Saha and Ratha, 1989
Anabas testudineus		
Water control	43	Ramaswamy and Reddy, 1989
After 5 h in air	50	
After 10 h in air	56	
Water control	4	Saha and Ratha, 1989
Monopterus cuchia	9	Saha and Ratha, 1989

[a]Unless otherwise detailed, values are for control fish that were kept in water.

[b]Total N excretion reduced to 20–30% of fish in water.

Mudskippers. Mudskipper nitrogen excretion has been reviewed by Clayton (1993) and most of the experimental results are shown in Table 8.9. The existing data reflect considerable variability, and there is no correspondence between the natural amphibious-exposure time of different species and their methods of handling nitrogen excretion. Subsequent work has failed to verify the initial reports (Gordon *et al.*, 1965, 1969, 1978) that mudskippers switched from ammonotelism to ureotelism during forced emergence and that stored urea was excreted following the return to water (Morii *et al.*, 1978; Iwata *et al.*, 1981; Iwata, 1988). Gregory (1977) did determine that urea made up as much as 33% of the total nitrogen flux from *Periophthalmus* and *Scartelaos*. However, studies with *Boleophthalmus pectinorostris* and with *P. cantonensis* (a species studied by Gordon *et al.*, 1978) showed that both species continued to excrete ammonia and urea while exposed to air, but at lower rates than when in water (Morii *et al.*, 1978, 1979; Morii, 1979). This implies that nitrogen storage occurs during air exposure, a fact confirmed with tests on fish returned to water. However, these and other tests (Ip *et al.*, 1990) also revealed that ammonia and not urea was the major excretory product; ammonia does accumulate in several tissue compartments but conversion of ammonia to urea did not occur (Morii, 1979). Urinary tract blockage experiments and tests using a subdivided holding chamber confirmed that the gills, skin, and urinary tract were all involved in nitrogen excretion (Morii *et al.*, 1978).

Other species. Three air-breathing species, *Sicyases sanguineus* (Gordon *et al.*, 1970), *Channa gachua* (Ramaswamy and Reddy, 1983), and *Alticus kirki* (Rozemeijer and Plaut, 1993) appear to be ureotelic in water (i.e., more than 50% of their nitrogen excretion takes the form of urea), and the percentage of urea excreted by these fishes increased with air-exposure time (Table 8.9). Although urea only accounted for 43% of the total nitrogen excreted by *Anabas testudineus* in water, this quantity climbed to 50% or higher with air exposures of 5 and 10 hours. The findings of increased urea excretion for both *Anabas* and *Channa* were also reflected, qualitatively, in corresponding increases in blood and liver ammonia and urea levels during the emergence period (Ramaswamy and Reddy, 1983).

Rozemeijer and Plaut (1993) reported that urea constitutes the major nitrogenous excretion product of *Alticus* in water and, although the percentage of urea increased with air exposure, the total quantity of nitrogen excretion became less. These workers found that well over 50% of the nitrogen excreted by exposed fish was actually released to the air as a gas and that a substantial quantity became trapped in exuded mucus (similar to *Blennius pholis*, see following).

For *Pholis gunnellus* held in air for 24 hours, Kormanik and Evans (1988) found elevated rates of ammonia and urea excretion but no difference from the pre-emersion urea:ammonia ratio for up to three hours after the return to water. However, over the ensuing 21 hours, ammonia release returned to its pre-emersion rate while urea excretion continued to increase, suggesting that urea had been stored during emersion. Davenport and Sayer (1986) found no effect of air exposure on the urea-excretion rate of *Blennius pholis* which is similar to findings for *Periophthalmus* and *Boleophthalmus* (Morii *et al.*, 1979). Studies with *Blennius*, moreover, demonstrated that much of the nitrogen excreted by this fish exited via the skin and mucus (Sayer and Blackstock, 1988).

In summary, what can be said about the whole animal pre- and post-air-exposure tests? Table 8.9 shows an extreme variability among the different species. Some of this variability can be attributed to methodology. The experimental emersion periods are unequal for the different species and have often been unrealistically high. In addition, variables such as temperature and pre-experimental diet were often uncontrolled. Wood (1993) observed that the methods used to measure levels of plasma, urinary, and aquatic ammonia and urea have been subject to errors arising from confusion over units (ammonia versus ammonia-N, NH_3 versus total ammonia, and μmoles versus mmoles), analytical techniques (i.e., the failure to deproteinate leads to overestimates), and sampling methods such as not freezing samples or not making determinations immediately (i.e., bacteria in the test water could have rapidly altered the chemical form of the excreted nitrogen). Sufficient consideration may not have always been given to the volume of water in the experimental chamber used for the pre- and post-emersion experiments. If too little water volume was used, accumulated ammonia levels might have become too high and inhibited ammonotelism. If the volume was too large then the results might have been affected by the limits of analytical resolution.

Finally, we have little basis for understanding the gross apparent differences in the urea excretion patterns of closely related and ecologically similar species. One example is the extreme differences in the urea-formation rates of *Alticus* and *Blennius* (both are blennies). Whereas *Channa gachua* is ureotelic, *C. punctatus* is ammonotelic (Table 8.9; Roy and Das, 1989), and Saha *et al.* (1994) report this species lacks the full complement of OUC enzymes.

Comparative Enzymology of Nitrogen Metabolism

Differences in the biochemistry of the ornithine urea cycle separate the mammals and fishes and there are also differences among the fishes. The major difference is in CPS, carbomyl phosphate synthetase, the mitochondrially formed enzyme that converts glutamine into carbamoylphosphate (Figure 8.3). The CPS of fishes is glutamine- and N-acetylglutamate-dependent and is designated as CPS III. Lungfishes and tetrapods have CPS I which is inactive in the absence of ammonia and N-acetylglutamate. Another difference is arginase, the last enzyme of the O–U cycle (Figure 8.3) which is mitochondrial in fish liver but cytosolic in lungfishes and tetrapods. This means that, in most fishes, arginine formed in the cytosol must be transported into the mitochondria where urea is formed. In lungfishes and most other vertebrates, however, ornithine must be transported into the mitochondria from the cytosol, and citrulline must be transported back into the cytosol where urea forms. Although arginase is common in fish liver, its activity tends to be scaled up in air-breathing and potentially ureogenic species (Singh and Singh, 1986; Saha and Ratha, 1989). A significant question surrounding teleost ureogenesis is uncertainty about the source of the urea that was formed. Was this the result of an up-regulation of the OUC or did it come from another urea-forming pathway? Although a functional OUC was found in estivating African lungfish (preceding), the initial picture for actinopterygian nitrogen metabolism was that many of these fishes lacked the full complement of OUC enzymes (Brown and Cohen, 1960). Cohen (1966) suggested that although the OUC was probably a universal characteristic of the early vertebrates, genes regulating expression of some of the OUC enzymes were likely lost in the actinopterygians because their initial evolution in freshwater had eliminated the selection pressure for retaining the energetically expensive urea-production capacity.

As reviewed by Huggins *et al.* (1969), several reports indicated that most actinopterygians, particularly the modern teleosts, lacked the genes needed to synthesize the OUC enzymes, or, if these enzymes were present, they were in such low amounts that the quantitative importance of urea formation to total nitrogen flux could be doubted (Table 8.10). Consistent with this, Gregory (1977) determined that the activities of the OUC enzymes in three amphibious fishes *Scartelaos histophorus*, *P. expeditionium*, and *P. gracilis* were not sufficient to sustain significant ureotelism. Thus, the urea measured in these fishes, and probably most teleosts, would have had to originate from either purines or arginine and not as a direct result of the OUC. Further, the work by Iwata *et al.* (1981) showed that ammonia accumulation in *P. cantonensis* during air exposure causes an up-regulation of muscle GDH. This enzyme catalyzes the conversion of ammonia and alpha-ketoglutarate to glutamate (Das *et al.*, 1991), which can be transaminated into different, non-essential free amino acids (FAA). FAAs stored in body muscle would serve the needs of ammonia detoxification and nitrogen storage and can be used in osmotic adaptation for variable habitat salinity (Iwata *et al.*, 1981; Iwata, 1988). Whereas FAA accumulation offers an osmotic advantage to euryhaline species such as most gobies, a comparison of *P. cantonensis* and the non-amphibious goby *Tridentiger obscurus* demonstrated that the muscle GDH kinetics of *P. cantonensis* made it more proficient at converting ammonia to glutamate, suggesting an adaptation for the added stress of amphibious life (Iwata and Kakuta, 1983).

TABLE 8.10 Relative Activities (μM of product/hour g) of Five Ornithine–Urea Cycle Enzymes from the Livers of a Shark (*Scyliorhinus canicula*), the Lungfish (*Protopterus aethiopicus*), Six Teleost Fishes, and the Frog (*Rana temporaria*)[a] and Four Air-Breathing Fishes[b]

Species	Enzymes[c]				
	CPS	OTC	ASS	ASL	ARG
Shark					
S. canicula	1	620	75	25	1856
Lungfish					
P. aethiopicus	22	1133	10	12	4023
Teleosts					
Anguilla anguilla	0.16	6	1.6	1.8	4565
Salmo salar	0.03	4	0.6	0.7	1117
Rutilus rutilus	0.11	1	0.3	0.8	329
Tinca vulgaris	0.19	2.4	3.6	1.3	366
Gadus callarias	0.13	1	4	1	245
Mullus barbatus	0.24	179	3.4	1.7	714
Frog					
R. temporaria	365	9532	51	121	19787
Air-breathing fishes[b]					
Heteropneustes fossilis	4.6	252	29	27	7699
Clarias batrachus	2.4	312	13	22	6654
Amphipnous (Monopterus) cuchia	1.2	93	23	16	597
Anabas testudineus	2.8	113	20	13	3580

[a]Huggins *et al.*, 1969.
[b]Saha and Ratha, 1989.
[c]CPS, carbamyl phosphate synthetase; OTC, ornithine transcarbamylase; ASS, argininosuccinate synthetase; ASL, argininosuccinate lyase; ARG, arginase.

Although most recent accounts (e.g., Wood, 1993) hold to the view that the OUC enzymes are not functional in most teleosts, trace levels of all the OUC enzymes are found in many species (Huggins *et al.*, 1969). It is further recognized that use of this cycle for ammonia detoxification may be important during ontogeny (Dépêche *et al.*, 1979; Mommsen and Walsh, 1991), and recent work has documented enhanced urea production in such ecologically and phylogenetically diverse fishes as *Oreochromis* (Randall *et al.*, 1989) and *Opsanus* (Mommsen and Walsh, 1991).

There is new evidence linking OUC enzyme activity and air breathing (Table 8.10). Singh and Singh (1986) reported greater liver arginase activities for air breathers (*Clarias*, *Heteropneustes*, *Channa punctatus*, *C. striatus*) than for non-air breathers (*Labeo*, *Catla*, *Rita*). Saha and Ratha (1987, 1989) reported active OUC enzymes in several air-breathing teleosts and other species. They reported that *Heteropneustes*, along with *Clarias batrachus*, *Anabas testudinus*, and *Monopterus* [= *Amphipnous*] *cuchia*, all have the full complement of OUC enzymes in both liver and kidney tissues; however, another air breather, *Channa punctatus* does not. These workers also found that, when *Heteropneustes* was exposed to either aquatic hyper-ammonia or air, its tissue urea concentration was increased as was the rate of urea excretion (Saha and Ratha, 1990). Paradoxically, some of the species reported to have the OUC enzymes are not ureotelic; urea, in fact, amounts to a quite small percentage of the total nitrogen excreted by *Heteropneustes* (even in 75 mM ammonium chloride), *Clarias*, and *Monopterus* (Table 8.9). No other studies have reviewed both tissue urea and ammonia levels and the enzyme quantities and, for this reason, we lack insight into how air breathing is integrated with nitrogen metabolism and, whether or not, urea production is regulated in response to tissue ammonia concentration.

SUMMARY AND OVERVIEW

In "*Living Without Oxygen*," Peter Hochachka (1980) begins his chapter on air-breathing fishes by drawing parallels between them and the mammalian divers. He focuses on the limits to oxygen access imposed by "diving" in these fishes and emphasizes the relatively pronounced anaerobic capacities of air breathers such as *Arapaima* and *Protopterus* (Dunn *et al.*, 1983). Although equating air-breathing fishes and diving mammals enabled Hochachka (1980) to contrast different states of oxygen access among vertebrates, this perspective is largely paradoxical from the standpoint of air-breathing fish biology. As emphasized in Chapter 2 and elsewhere in this book, the overwhelming impact of air breathing is obviation of the problem of environmental hypoxia for many species, and in most cases this specialization actually enables fishes to survive in and exploit hypoxic aquatic environments, thus filling ecological niches not open to non-air-breathing species. In addition, all air-breathing fishes retain functional gills or have some other means of aquatic respiration and invariably utilize aquatic O_2 whenever it is accessible (Chapters 5 and 7). Accordingly, it seems more reasonable to argue that air-breathing fishes present a radical departure from the classic "diving-mammal paradigm," and, with the exception of periods of intense physical exertion or long-term burrow confinement, most air-breathing fish tissues would seem to have unrestricted O_2 access. Nevertheless Hochachka does make a powerful point and it remains somewhat enigmatic as to why some of the air-breathing fishes "having access to a nearly infinite O_2 reservoir," have not exploited aerobic metabolism to a greater extent.

This chapter has examined the air-breathing fish biochemical data for indications of how aerial respiration has altered the patterns of normal piscine function among the various species. Biochemical investigations of *Amia* and *Clarias* both reached the conclusion that air breathing obviated cellular mechanisms for ambient hypoxia adaptation. As emphasized in the preceding chapters and demonstrated in Table 8.2, there is good reason to suspect that the continued aerial O_2 access "enjoyed" by air-breathing fishes should translate into demonstrable changes in the aerobic poising of their intermediary metabolism and organ function. Indeed, strong evidence for this scenario is seen in a comparison of the osteoglossids *Arapaima*, an air breather, and *Osteologlossum*, a non-air breather. Data for these fishes demonstrate that the air breather has a much higher metabolic rate. There is also evidence for the transfer of some level of branchial-excretory function to the kidney as well as the increased capacity of the kidney and the heart to oxidize lactate. While integrated, whole organism studies are needed, most of the enzymatic and biochemical data presently available suggest that the kidney of *Arapaima* has assumed some role in acid-base regulation (proton excretion) and in the co-transport of Na^+ and NH_4^+, which are strictly branchial functions in most fishes.

Although there is the suggestion that the air-breathing erythrinid *Holplerythrinus* has evolved to a higher aerobic state than its non-air-breathing counterpart *Hoplias*, the evidence is equivocal. On one hand, some data indicate a higher $\dot{V}O_2$ for *Hoplerythrinus*, and studies of its heart verify that this tissue is more aerobically poised than that of *Hoplias*. However, studies of red and white muscle tissues provide no basis for the assertion that *Hoplias* has less of an oxidative capacity than *Hoplerythrinus*. In fact, the opposite argument is supported. It appears that the enhanced oxidative capacity imparted by air breathing is limited to the heart and perhaps other visceral organs of *Hoplerythrinus*. If this is true, then *Hoplerythrinus* would be similar to the lungfish and probably *Synbranchus* and *Arapaima*. The muscle tissues of all of these fishes are characterized by having a fairly low oxidative capacity which means that anaerobic bursting and lactate production are inevitable. It may be that air breathing sustains a higher standard metabolic rate in the heart and other organs and thus enables the somewhat rapid restoration of expended metabolites and the clearance of lactate.

The available enzymic and biochemical data for the mudskippers *Boleophthalmus* and *Periophthalmus* indicate their burst activity is also largely anaerobic which is again similar to the other air-breathing fishes. Thus, whereas access to aerial oxygen can ensure survival and the occupation of ecological niches that are beyond the scope of non-air-breathing fishes, the limitation to aerobic scope in these animals, as in most ectotherms, is imposed by their limited capacity to store and transfer oxygen. Even though there is a plentiful supply of oxygen in and around the ABO, there is limited oxygen storage capacity in tissues or blood and the energy requirement of bursting muscle quickly exceeds the capacity of the circulation to deliver the oxygen required for aerobic energy production.

The data for mudskippers also indicate that these fishes may endure ambient (burrow) hypoxia by entering into a period of metabolic arrest. A suspended metabolic state couples energetic advantages (i.e., conserving metabolic substrate and avoiding the debilitating energetic costs of post-hypoxia lactate clearance) with those of safety (i.e., remaining in burrows at high tide when large predatory fish enter the area). However, it is not known how metabolic arrest is induced, for how long it can be sustained, or how mudskippers that remain in burrows for reasons other than shelter (i.e., to care for developing eggs) deal with the metabolic costs of such endeavors in the face of ambient hypoxia. In this regard it is important to learn more about the relative activites of the LDH isomers in different tissues of mudskippers, particularly the heart. Whereas a number of fishes and ectotherms have the ability to enter into arrested metabolic states that require little oxygen (Storey and Storey, 1990), this most often occurs gradually, and much of the reduction in metabolic demand is related to cold temperature (Ultsch, 1989). Mudskippers, however, appear to abruptly reduce metabolic demand and must do this at usually tropical temperatures (20 C or warmer).

On the other hand, there are practically no data on burrow oxygen levels. It may be that mudskippers build burrows that have natural air traps or that induce water percolation and thus mitigate the hypoxia problem. In any event, strong parallels exist between the metabolic responses of mudskippers and those of the burrowing lungfish and synbranchids. These are in harmony with Hochachka's diving metaphor and Scholander's original ideas about the diving reflex itself—it minimizes aerobic costs until oxygen is again available.

The patterns of nitrogen excretion in relation to air breathing remain somewhat obscure. The first principles are that fish normally excrete nitrogen in the form of branchial ammonia and that air breathing, which interrupts or reduces branchial function, precludes this. For a fish periodically exposed to air, the interruption of gill function leads to the build-up of CO_2 and ammonia. Would respiratory acidosis progress to the point of becoming a metabolic limitation? Or, would the increase in ammonia reach toxic levels before acidosis became a problem? Studies have suggested the transition to ureogenesis in some species or the proclivity of air-breathing fishes to utilize urea excretion. However, the documentation of this is uneven, and additional data and a broader synthesis are needed.

If the theme of the progressive transfer of branchial functions to other organs holds for nitrogen metabolism among the amphibious air-breathing fishes, then it is theoretically possible that the build-up of CO_2 and ammonia is countered by the formation of urea and that the associated flux of protons from acid to a base is further used to manipulate extracellular pH by affecting bicarbonate. In this manner nitrogen metabolism would continue to exert an influence on ventilation, even though the gills are less involved.

CHAPTER

9

Synthesis

Whereas most biologists vary a few factors under controlled conditions and observe the effects of each deviation, the evolutionary biologist observes the results already obtained, as learned from studies of natural history, and tries to infer the factors that operated in the past. Where the experimental biologist predicts the outcome of experiments, the evolutionary biologist retrodicts the experiments already performed by Nature; he teases science out of history.

E.O. Wilson, *Naturalist*, 1994

INTRODUCTION

This chapter has two objectives. The first is to unify the information presented in the preceding chapters into a biological synthesis of fish air breathing. This synthesis, the main features of which are illustrated in Figure 9.1, integrates the physical factors affecting air breathing with its many and diverse biological facets. Fish air breathing is an ancient trait, and a rich diversity of bimodal respiratory specialization is found throughout this group's entire phyletic succession. Air breathing has evolved independently in many fishes and there are numerous instances where, in addition to serving a respiratory function, bimodal breathing has pervaded other facets of a species' natural history such as, and including, reproductive behavior and buoyancy (Graham, 1994).

The second objective of this chapter is to explore the evolutionary relationships between air breathing and the origin of tetrapods. The first air-breathing vertebrates were fishes, and some of these were ancestral to the tetrapods. The sequence leading from fish air breathing to tetrapod origins and the diverse physical and biological factors affecting this transition will be examined.

THE NATURAL HISTORY OF FISH AIR BREATHING

The air-breathing fishes are a diverse assemblage and this section examines air-breathing in relation to natural history. Factors to be compared include habitat, geographical and micro-distributions, patterns of ecological radiation, the different types of air breathers, and relationships between air breathing and variables such as body size, feeding behavior, and reproductive behavior. The data for the 49 known fish families with air-breathing species (Table 9.1) form the basis for most of this analysis.

Habitat

In terms of general habitat, the 49 families with air-breathing fishes are distributed as follows:

i. Exclusively in freshwater, 32
ii. In brackish to freshwaters, 6
iii. In marine to brackish waters, 2
iv. From marine to freshwaters, 3
v. Only in marine waters, 6.

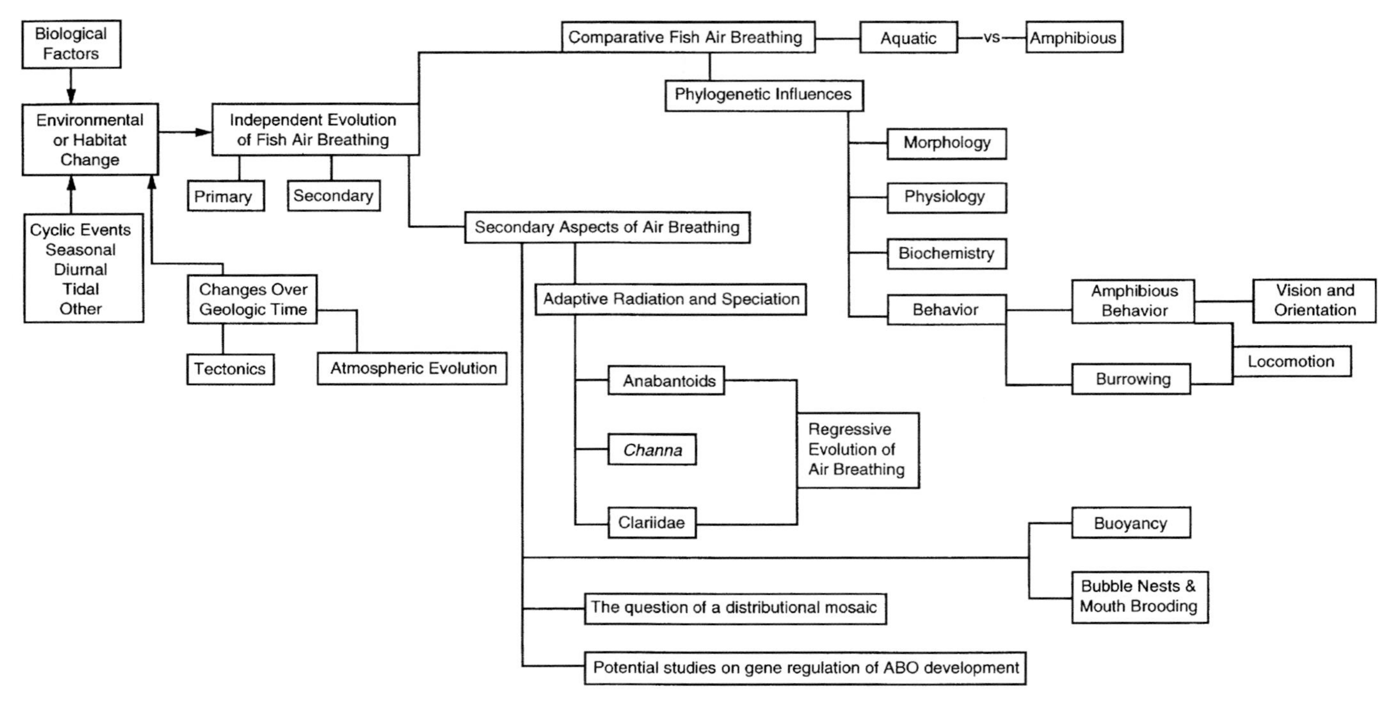

FIGURE 9.1 Flow scheme showing the relationships between physical and biological factors in the origin of fish air breathing and the effect of this adaptation on the diversification and radiation of the various species. (Reprinted from *American Zoologist,* volume 34, pp. 229–237, figure 1. Graham, 1994. Used with permission of the American Society of Zoologists.)

TABLE 9.1 List of Fish Families Containing Air-Breathing Species Together with Summary Information about the Species Diversity, Ecology, and Respiratory Adaptations in Each

Order Family	No. genera	No. species	Habitat	Vertical distribution	Feeding	ABO	Respiratory pattern
Ceratodontiformes							
Ceratodontidae	1	1	F	B	C	+	A–F
Lepidosireniformes							
Lepidosirenidae	1	1	F	B	C	+	A–C,Am–S
Protopteridae	1	4	F	B	C	+	A–C,Am–S
Polypteriformes							
Polypteridae	2	11	F	B/M	C	+	A–C,Am–V
Lepisosteiformes							
Lepisosteidae	2	7	B–F	M	C	+	A–C
Amiiformes							
Amiidae	1	1	F	M	C	+	A–C
Osteoglossiformes							
Osteoglossidae	2	2	F	M	C/P	+	A–C(?)
Pantodontidae	1	1	F	M	C	+	A–C
Notopteridae	3	5	F	M	C	+	A–C
Gymnarchidae	1	1	F	M	C	+	A–?
Elopiformes							
Megalopidae	1	2	M–F	M	C	+	A–C
Anguilliformes							
Anguillidae	1	1	F	B	C	–(+?)	Am–V
Gonorynchiformes							
Phractolaemidae	1	1	F	B/M	C/H	+	A–?
Cypriniformes							
Cobitididae	4	7	F	B	C	+	A–C+F

(continues)

TABLE 9.1 (*continued*)

Order Family	No. genera	No. species	Habitat	Vertical distribution	Feeding	ABO	Respiratory pattern
Characiformes							
Erythrinidae	2	2	F	M	C	+	A–C
Lebiasinidae	2	2	F	M	C	+	A–C
Siluriformes							
Pangasiidae	1	4	F	B	H	+	A–C
Clariidae	3	44	F	B	C/S	+	A–C+F,Am–V+S
Heteropneustidae	1	2	F	B/M	C/S	+	A–C,Am–V+S
Aspredinidae	1	2	F	B	C/S	+	A–F
Trichomycteridae	2	2	F	B	C	+	A–F
Callichthyidae	4	131	F	M	C	+	A–C
Loricariidae	10	14	F	B	H	+	A–F
Gymnotiformes							
Hypopomidae	1	3	F	B	C	–(+?)	A–F
Gymnotidae	1	1	F	M	C	+	A–F
Electrophoridae	1	1	F	B	C	+	A–C
Salmoniformes							
Umbridae	2	5	F	M	C	+	A–F
Lepidogalaxiidae	1	1	F	B	C	–	A–F,Am–S
Galaxiidae	3	10	F	B	C	–	Am–V
Gobiesociformes							
Gobiesocidae	5	7	M–F	B	C	–	Am–S
Cyprinodontiformes							
Aplocheilidae	1	5	F	M	C	–	Am–V
Cyprinodontidae	1	4	M–F	M	C	–	Am–V+S
Scorpaeniformes							
Cottidae	2	4	M	B	C	–	Am–V
Perciformes							
Stichaeidae	4	5	M	B	C	–	Am–S
Pholididae	3	5	M	B	C	–	Am–S
Tripterygiidae	1	1	M	B	C	–	Am–V
Labrisomidae	2	2	M	B	C	–	Am–V
Blenniidae	7	32	M	B	C/H	–	Am–V
Eleotridae	2	2	B–F	M	C	–	A–F
Gobiidae	15	40	M–B	B	C	+/–	A–F,Am–V
Gobioididae	1	1	M–B	B	C	–	A–F,Am–S
Mastacembelidae	2	3	B–F	B	C	–	Am–S
Anabantidae	3	24	B–F	M	C	+	A–C,Am–V+S
Belontiidae	12	44	B–F	M	C/O	+	A–C(Am–V?)
Helostomatidae	1	1	F	M	H/P	+	A–C
Osphronemidae	1	1	F	M	H	+	A–C
Luciocephalidae	1	1	F	M	C	+	A–C
Channidae	1	12	F	M	C	+	A–C,Am–S
Synbranchidae	3	14	B–F	B	C	+	A–C+F,Am–V+S

Habitat: Freshwater, F; Brackish, B; Marine, M.
Vertical distribution: Benthic, B; Midwater or surface, M.
Feeding: Carnivore, C; Planktivore, P; Herbivore, H; Scavenger, S; Omnivore, O.
ABO: Present, +; Absent, –.
Respiratory pattern: Aquatic, A – Continuous, C or Facultative, F; Amphibious, Am – Volitional, V or Stranded, S.

Distribution within the Water Column

Because proximity to surface air or the water's edge is an integral aspect of air breathing, most air-breathing fishes frequent relatively shallow depths, at least during the air-breathing phases of their life history (e.g., freshwater *Anguilla*). The air-breathing fishes can, however, be separated on the basis of their distribution within a shallow water column (Table 9.1). Most, and nearly all species within at least 29 families, are demersal and found on, or in, the substrate and are classified as benthic. Included among these are many bottom feeders (catfishes), burrowers (mastacembelids, synbranchids, gobies), and the amphibious marine fishes.

In contrast, most of the species in the remaining 20 families are pelagic, meaning they occur up in the water column or near the surface (Table 9.1). In many of these fishes, air breathing is closely tied to buoyancy and the ABO contributes importantly to static lift. Phasic changes in ABO volume during ventilation and gas transfer have been shown to affect buoyancy in *Neoceratodus*, *Piabucina*, *Hoplosternum*, and *Umbra* and, under certain conditions, buoyancy control may take precedence over the respiratory needs of these fishes. Both *Lepisosteus* and *Amia* are known to regulate ABO volume for purposes of buoyancy control. Conversely, ABO volume is more or less fixed by bony structures in some groups (e.g., all the anabantoids, *Channa*, and to a large extent, the Clariidae), which may be less subject to buoyancy deviations resulting from variations in ABO volume and respiratory gas flux rates.

The Types of Air-Breathing Fishes

As detailed in Chapter 1, air-breathing fishes are classified on the basis of whether they are amphibious or aquatic air breathers. Aquatic air breathers are in turn divisible into either facultative or continuous, with the latter being further separated into the obligatory and non-obligatory air breathers. Table 9.2 (A and B) summarizes the respiratory patterns found among the 49 air-breathing fish families. There are 26 families with species known to use only aquatic air breathing, 12 that use amphibious air breathing, and 11 in which both patterns are present. Considering the 37 (i.e., 26 + 11) families containing aquatic air breathers, about 21 of these have species that air-breathe continuously, 11 are facultative air breathers, 3 contain both continuous and facultative forms, and there are 2 families (Gymnarchidae and Phractolaemidae) for which these details are completely unknown (see Table 9.2D).

TABLE 9.2 Summary of the Types of Air-Breathing Occurring among the 49 Families, with a List of Known and Unknown Obligatory Air-Breathing Genera

A. *Types of Air Breathing*

Amphibious	Aquatic	Both
12	26	11

B. *Divisions of Aquatic Air Breathing* (n = 37)

Continuous	Facultative	Both	Unknown
21	11	3	2

C. *Genera Having Obligatory Air-Breathing Species* (n = number of known obligatory species in genus)

Family	Genus	(n)
Lepidosirenidae	*Lepidosiren*	(1)
Protopteridae	*Protopterus*	(4)
Osteoglossidae	*Arapaima*	(1)
Megalopidae	*Megalops*	(1)
Pantodontidae	*Pantodon*	(1)
Pangasiidae	*Pangasius*	(1)
Notopteridae	*Notopterus*[a]	(1)
Clariidae	*Clarias*[a]	(3)
Callichthyidae	*Hoplosternum*[a]	(1)
Electrophoridae	*Electrophorus*	(1)
Osphronemidae	*Osphronemus*	(1)
Anabantidae	*Anabas*	(1)
Belontiidae	*Trichopterus*[a]	(2)
	Colisa[a]	(1)
Synbranchidae	*Monopterus*[a]	(2)
Channidae	*Channa*[b]	(12)
		34

D. *Genera with Unknown Obligatory/Non-Obligatory Status*

Family	Genus
Osteoglossidae	*Heterotis*
Gymnarchidae	*Gymnarchus*
Phractolaemidae	*Phractolaemus*
Notopteridae	*Xenomystus*
	Papyrocranus
Pangasiidae	*Pangasius*[a]
Anabantidae	*Ctenopoma*
	Sandelia

[a]Some species.
[b]Obligatory for all species not established.

The Occurrence of Obligatory Air Breathing

Found among the continuous air breathers are at least 16 genera with obligatory air-breathing species (Table 9.2C). Surprisingly, it appears that 6 of these genera contain both obligatory and non-obligatory species. The total number of known obligate air-breathing species is 34.

Species for Which Nothing Is Known

Table 9.2D lists genera for which additional air-breathing information, including obligatory status, is needed. First and most important, virtually nothing is known about the air breathing of *Gymnarchus*, *Phractolaemus*, *Heterotis*, *Papyrocranus*, and *Xenomystus*. Knowledge of these primitive teleostean species (all of which occur in Africa and have well-developed ABOs) would significantly broaden our comparative perspective. Especially useful would be the contrast between *Heterotis*, a plankton feeder, and its primarily piscivorous and obligatory air-breathing South American counterpart *Arapaima*. The potential comparison between the African (*Xenomystus*, *Papyrocranus*) and Asian notopterids is important. Based on ABO structure, I expect some of these fishes are obligatory air breathers.

Additional data on the obligatory status of *Pangasius* would be useful in view of the large size and diversity of its numerous species, the variable obligatory aerial-breathing responses reported for *P. sutchii* (Browman and Kramer, 1985), and the variability in ABO structure reported by Roberts and Vidthayanon (1991). Studies with species of the African anabantid genera *Ctenopoma* and *Sandelia* are important because they appear less dependent on air than their Asian counterpart *Anabas*. *Sandelia* in particular has a small ABO surface area and is likely a facultative air breather. Documentation of this would thus establish a pattern of regressive ABO development like that seen in the Clariidae.

The Occurrence of Amphibious Air Breathing

Defined as the gas exchange of fishes during exposure to air, amphibious air breathing occurs during prolonged seasonal exposure in drying mud, during brief stranding by receding waters, and during volitional (behaviorally directed) emergence. Table 9.1 lists 24 families containing amphibious air-breathing species. These are uniformly dispersed in both fresh and marine habitats. Amphibious air breathing is the exclusive aerial mode of species in 12 of these 24 families. The root causes for amphibious air breathing among the 12 primarily amphibious air-breathing families are:

i. Volitional behavior, 8
ii. Stranding, 4.

Among the 12 families containing both amphibious and aquatic air-breathing species, the causes for amphibious exposure are:

i. Volitional behavior, 3
ii. Stranding, 5
iii. Both, 4.

Table 9.3 lists fishes involved in the various types of amphibious air breathing. The 16 genera capable of long-term exposure out of water are all freshwater and include forms that can reduce their metabolism or estivate. The 14 genera that routinely endure short-term exposure are all intertidal. Among the 26 genera listed as making volitional terrestrial sojourns, 17 (a minimum of 75 species) occur in marine to brackish water, littoral habitats, and nine (about 25 species) occur in freshwaters. In addition to these fishes, sections of Chapter 2 cite anecdotal accounts of terrestriality in several genera including *Heteropneustes*, *Hoplosternum*, *Hoplerythrinus*, *Callichthys*, and *Hypopomus*. While these are not discounted, it is not possible in all cases to distinguish between terrestrial migrations and the consequences of stranding following seasonal flood dispersion, and more data are needed.

TABLE 9.3 Categories of Amphibious Air-Breathing Fishes

A. Genera Capable of Prolonged (Seasonal) Mud Exposure	
Protopterus	*Lepidosiren*
Clarias	*Heteropneustes*
Callichthys	*Anguilla*
Neochanna	*Lepidogalaxias*
Mastacembelus	*Macrognathus*
Anabas	*Ctenopoma*[a]
Channa[a]	*Synbranchus*
Monopterus (*Amphipnous*)	*Ophisternon*
B. Genera Experiencing Periodic (Tidal) Aerial Exposure	
Sicyases	*Phallerodiscus*
Gobiesox	*Tomicodon*
Lepadogaster	*Anoplarchus*
Cebidichthys	*Xiphister*
Xererpes	*Pholis*
Apodichthys	*Pseudapocryptes*
Apocryptes	*Taenioides*
C. Genera Capable of Volitional (= Purposeful) Terrestrial Sojourns	
Erpetoichthys	*Anguilla*[a]
Clarias[a]	*Galaxias*
Rivulus	*Fundulus*[a]
Clinocottus	*Mnierpes*
Dialommus	*Andamia*
Alticus	*Salarias*
Blennius	*Coryphoblennius*
Istiblennius	*Entomacrodus*
Periophthalmus	*Boleophthalmus*
Periophthalmodon	*Scartelaos*
Gobionellus	*Anabas*
Ctenopoma[a]	*Monopterus* (*Amphipnous*)
Synbranchus	*Ophisternon*

[a]Some species.

Zoogeography

Air-breathing fishes extend from the tropics to high latitudes in habitats ranging from ditches to the open ocean. Several genera occur in the temperate zone (*Lepisosteus*, *Amia*, *Misgurnus*, some *Channa*, the cottids, stichaeids and pholids, and some anabantoids and blennies) and some occur in high latitudes (*Umbra*, *Galaxias*, *Neochanna*, and *Sicyases*), even within the Arctic circle (*Dallia*).

As is generally observed for the freshwater fishes, the occurrence of primitive air-breathing groups (e.g., lungfishes, osteoglossids, and both the Ginglymodi and Halecostomi when their fossils are taken into account) on different continents reflects an origin that predates the separation of Pangaea (about 200 mybp). Whitmore (1981) has suggested continental drift as a possible explanation for the Asian occurrence of one osteoglossid, *Scleropages formosa*.

A number of freshwater teleosts including the air-breathing families Notopteridae, Clariidae, Mastacembelidae, Anabantidae, and the Synbranchidae are presently found in both Africa and Asia (Lowe-McConnell, 1987), but are not continuously distributed in the intervening areas between these two continents. One scenario suggested by this pattern is that climatic conditions, which were at one time favorable for the broad radiation of these groups, have subsequently changed to the extent that once continuous distributions are now disjunct. Another possibility is that the primitive stocks of these groups evolved in Gondwana prior to its subdivision into Africa, South America, Antarctica, Madagascar, and India (i.e., in the Late Cretaceous–Early Tertiary Periods, 65–75 mybp, Windley, 1984) and that continental drift established the African and Indian-Asian distributions.

This latter scenario seems consistent with the evolutionary history and distribution of the Anabantoids. According to Liem (1963a) ancestral anabantoids first appeared in the Upper Cretaceous or Paleocene (60–80 mybp, i.e., before or during the break-up of Gondwana), and it has been shown that the most primitive family, Anabantidae, occurs in both Africa (*Sandelia* and *Ctenopoma*) and India (*Anabas*). The remaining four families (Belontiidae, Osphronemidae, Helostomatidae, Luciocephalidae) are more recently derived and occur only in Asia, a distribution that might be expected if their origin occurred after the Indian-African split (according to Windley [1984], the Indian continent collided with Asia between 55 and 40 mybp).

The Ecological Radiations of Certain Groups

In several cases the diversification of air-breathing fishes reflects the successful adaptive radiation of a particular air-breathing morphotype. This important secondary aspect of air breathing (Figure 9.1) is exemplified by the Callicthyidae and Clariidae and the five anabantoid families. Each of these groups has undergone extensive diversification (Table 9.1), and they constitute the three most diverse taxa in which air breathing is present in nearly all species within the group. When considered together, these families form a large fraction of the entire air-breathing fish species diversity (i.e., the seven families are 14% of the 49 air-breathing fish families; the 25 genera are 16%, the 143 species are 39%).

There is almost no doubt that the presence of air breathing has in itself provided some of the impetus for diversification within these groups. Air breathing enables populations to expand into habitats where, because of adverse habitat conditions, ecological success is partly assured through limited competition from non-air-breathing species. In this manner, formerly contiguous populations may also become reproductively isolated and begin genetic divergence into separate species. The diversity of *Channa* (12+ species) may reflect this speciation process.

Anabantoid air breathing contains many lessons regarding adaptive radiation, and is an important area for continued study because of this group's diversity and the variations seen in the degree of dependence on air breathing. The five anabantoid families have all undergone significant modifications in dentition, jaw structure, and diet, as well as notable alterations in body shape and size (Liem, 1963a). These functional changes occurred concomitantly with adjustments in the sites and surface areas of the muscular attachments to the skull which, while affecting the position and volume of the suprabranchial chambers, has not apparently altered ABO respiratory function (Liem, 1963a; Beadle, 1981; Lauder and Liem, 1981, 1983a, 1989b, 1990). In addition to respiration, the suprabranchial chambers of some anabantoids function in both sound production and reception, and contribute to buoyancy (Liem, 1963a; Schuster, 1989).

Furthermore, elements of air breathing have pervaded many facets of anabantoid life, such as conspecific behavioral displays in courtship and territoriality, the reception and production of sound, bubble nest building, and parental care, including mouth brooding (Peters, 1978). Bubble nests, for example, are constructed using expired air laced with mucus. Nothing is known about relationships between air breathing and mouth-brooding behavior (*Betta*, *Macropodus*). Bader's (1937) work sheds important light on the role of the environment in ABO ontogeny. The anabantoids are thus a window to the manner in which the adaptive radiation of a group has been influenced by, or alternatively, has taken place in spite of, the requirement to sustain a mechanism for aerial gas exchange.

Another facet of adaptive radiation is evident in the clariids, some of which have undergone regressive ABO development (Figure 9.1). In these instances the strong selection pressures establishing air breathing have been lessened in novel environmental settings. Similarly, fishes of the anabantid genus *Sandelia* and the clariid genus *Dinotopterus* are, in terms of ABO development, highly derived from the obligatory air breathing members of their families (*Anabas* and *Clarias*). In both cases ABO regression appears strongly tied to radiation into habitats containing more O_2. Finally, the example provided by the air-breathing and non-air-breathing species of *Piabucina* illustrates that selection pressures for air breathing can vary in space and time: *P. festae* and *P. panamensis* are closely related and even sympatric in certain areas.

Nevertheless, selection for air breathing occurred in one but not the other.

Mosaic Distributions: Air Breathers and Non-Air Breathers

Another secondary aspect of air-breathing fish biology is the phenomenon of the distributional mosaic (Figure 9.1). In many habitats there is considerable overlap in the distributions of air-breathing and non-air-breathing fishes. A frequent observation of many workers was that the small drying pools they sampled contained both air breathers and non-air breathers (Kramer and Graham, 1976; Kramer *et al.*, 1978). Carter and Beadle (1931), for example, only found eight air-breathing species among the 20 fish species collected from a drying swamp pool in the Paraguayan Chaco. Similarly, air breathers constituted only a small fraction of the fishes removed from a dry season mudhole in Venezuela by Beebe (1945).

Observations such as these prompted several writers to question the efficacy of air breathing. If air breathing was so advantageous, why then were not more of the species trapped in drying pools capable of using this specialization? Could it be that other adaptive responses, for example, ASR (aquatic surface respiration, see Chapters 1 and 2 and Kramer, 1983a) were equally effective at ensuring survival? A number of species have elaborate structures (Braum, 1983; Saint-Paul and Soares, 1988) and behaviors related to ASR, and Kramer (1983a) argued that under certain circumstances air breathing is not necessarily an advantage. In a drying pool, for example, it is the air breathers, and not the other fishes, that must periodically surface for air and would thereby be more subject to attack by avian or shore-based predators. (Behavior such as synchronous air breathing is one mechanism to counter this [Kramer and Graham, 1976, and see following]).

Notwithstanding such a disadvantage, in my opinion neither anaerobiosis nor ASR would offer the same long-term survival prospects of air breathing. Anaerobiosis cannot be sustained indefinitely (Chapter 8) and ASR requires more or less continuous surface contact which increases vulnerability to certain predators, is energetically expensive, and may preclude feeding (Kramer and Mehegen, 1981; Poulin *et al.*, 1987). Thus, the apparent distributional mosaic suggested by the concentration of air-breathing and other fishes in drying mudholes may say more about the large-scale crowding and concentrating effects of the dry season than about the relative advantages of air breathing in this situation. In any case the distribution of air-breathing and non-air-breathing species at a particular place and time may reflect the history of habitat severity; strong overlaps in the two types would indicate the absence, for several years, of the severe conditions of hypoxia and near drying that would eliminate non-air breathers from habitats where only air-breathers can survive.

Feeding Behavior

Most air-breathing fishes are carnivorous. Many aquatic air breathers take prey at or near the water surface (anabantoids, *Arapaima*, *Pantodon*, *Piabucina*), and emergent feeding occurs in several amphibious species (*Clarias*, *Synbranchus*, *Monopterus*, *Periophthalmus*). The few exceptions to carnivory include the planktivorous *Heterotis* (this is probably facultative) and several primarily herbivorous fishes (*Phractolaemus*, *Pangasius*, *Helostoma*, *Osphronemus*, the Loricariidae, and some blennies).

Studies examining the relation of air breathing to feeding tactics, such as stealth, have not been carried out. When *Monopterus* and *Synbranchus* are holding air in their ABOs, their heads are positively buoyant, pointed vertically, and held motionless. Observations of these fish in my laboratory suggest that this posture may be an effortless, cryptic, stealth behavior that allows the fish to position their mouths near the water surface and close to small prey that may swim by.

Body Size

Air-breathing fishes encompass a range of body sizes. At the large end is *Arapaima*, one of the largest freshwater fish in the world. Some gars (*Atractosteus*) grow to nearly 2 m as does *Electrophorus*. Freshwater species reaching a maximum length between 1 and 2 m include the three lungfish genera, certain clariids, and some of the pangasiids (some of which are as large as *Arapaima*). *Clarias gariepinus* (200 cm, 60 kg) is the largest freshwater fish in South Africa (Bruton, 1979). Fishes reaching about one meter at maximum length include the polypterids, most gars, *Amia*, *Heterotis*, *Anguilla*, most of the clariids, some *Channa*, and the Synbranchidae. At the other extreme are many small anabantoid genera which reach a maximum length of about 3 cm.

Studies of developing fishes suggest that 10 mm total length is about the minimum body size at which aquatic air breathing commences. Critical determining factors include the relative body mass needed to develop the swimming thrust required to overcome water-surface tension (i.e., a small larva risks becoming glued to the surface) and the buoyancy of ingested air. A number of studies document that some aquatic air breathers (*Anabas*, *Colisa*, *Macropodus*, and others) begin doing so before their ABOs are developed and it appears in some cases that air breathing may be needed to promote the expression of genes regulating the development of this organ (Bader, 1937; Liem, 1981; Prasad and Prasad, 1985; Chapter 3).

Among the intertidal fishes, *Sicyases* is probably the largest (to about 4 kg), whereas, most other species, particularly the agile ones, are less than 20 cm in length, relatively slender, and have a mass of 60 g (*Boleophthalmus*) or much less.

Aspects of Parental Care

There are no general patterns relating fish air breathing and parental care. Air-breathing fishes engaging in offspring care include a number of the primitive species (*Protopterus*, *Lepidosiren*, *Notopterus*, *Arapaima*, *Heterotis*, *Gymnarchus*), *Hoplosternum*, the synbranchids, and several anabantoids and loricariids. Families with air-breathing fishes in which parental care is exhibited (although not necessarily by the air-breathing species) include the Erythrinidae, Lebiasinidae, Clariidae, Aspredinidae, Heteropneustidae, Channidae, and in the marine environment, some of the Labrisomidae, Gobiidae, and Blenniidae (Breder and Rosen, 1966).

Air breathing appears to play a role in nest oxygenation (through filamentous fins) for *Lepidosiren* and possibly *Synbranchus* (via skin exudation); it has also been suggested that the cirri on the heads of male loricariids serve this function. All of these observations require additional details. Air ventilation is crucial to bubble-nest formation by species of *Betta*, *Hoplosternum*, and other callichthyids. In addition, there are mouth-brooding species of *Macropodus* and *Betta* as well as *Channa orientalis*, and it is reported that male *Arapaima* shelter swarms of new hatchlings in their mouths. How brooding or sheltering affects the mechanics and physiology of air-breathing remains uninvestigated. Finally, Kramer (1978c) has reviewed terrestrial spawning in fishes, most of which remain unstudied insofar as air-breathing and emergence tolerance are concerned.

FACTORS LEADING TO THE EVOLUTION OF AIR BREATHING

The habitat and behavioral data contained in Table 9.1 suggest that two major factors, hypoxia and emergence, have influenced the origin of fish air breathing.

Hypoxia in Freshwater Species

The presence or potential occurrence of aquatic hypoxia is the single environmental feature linking the habitats of all freshwater air-breathing fishes. In most cases, seasonal hypoxia, the result of reduced rainfall, leads to habitat shrinkage, crowding, stagnation, and severe hypoxia. Also, some air-breathing fishes reside in chronically hypoxic habitats (Chapter 1). Because no other environmental factor is so widespread, hypoxia stands alone as the major influencing factor in the origin of air breathing among freshwater fishes (Inger, 1957; Johansen, 1970; Graham *et al.*, 1978a).

This does not imply that hypoxia has been the exclusive causal factor in the origin of freshwater air breathing. Hora (1935) suggested that air breathing in certain Indian species (e.g., the cobitids *Lepidocephalichthys* and *Acanthophthalmus* and possibly *Amblyceps* [see Chapter 2]) may be precipitated by changes in water flow (i.e., from torrential to slow-flowing streams) rather than aquatic hypoxia. This idea has never been investigated. As reviewed in Chapter 2, hypoxia and habitat drying appear to be the primary causes for air breathing in the mastacembelids. Similarly, the potential for dry season stranding is a primary force in the amphibious excursions of some rivulines. Even though both mastacembelids and the rivulines can tolerate dry season exposure, they lack the ability to air breathe aquatically. Seasonal tolerance of aerial exposure in mud is not, on the other hand, common to all air-breathing fishes, nor is it exclusive to them. For example, non-air-breathers such as carp (*Puntius* and *Cyprinus*; Day, 1877) and tilapia (*Sarotherodon*; Donnelly, 1973; Bruton, 1979) also tolerate seasonal mud exposure.

A number of the freshwater air-breathing genera are also capable of terrestrial excursions (Table 9.3C); however, in view of the large number of freshwater air breathers that do not emerge (including the lungfishes), the argument that emergence was the driving force for the development of air breathing has virtually no basis. A review of Chapter 2 suggests that a major reason for terrestrial sojourns by freshwater air-breathing fishes is to search out new habitable areas. Habitat expansion is triggered by the onset of the rainy season in both *Anabas* and *Clarias* (Chapter 2, Inger, 1957). Amphibious excursions also occur in response to habitat drying, crowding, or a lack of food (Liem, 1987; Sayer and Davenport, 1991).

Emergence and Marine Species

Many reviewers have considered the diverse factors leading to amphibious air breathing among the

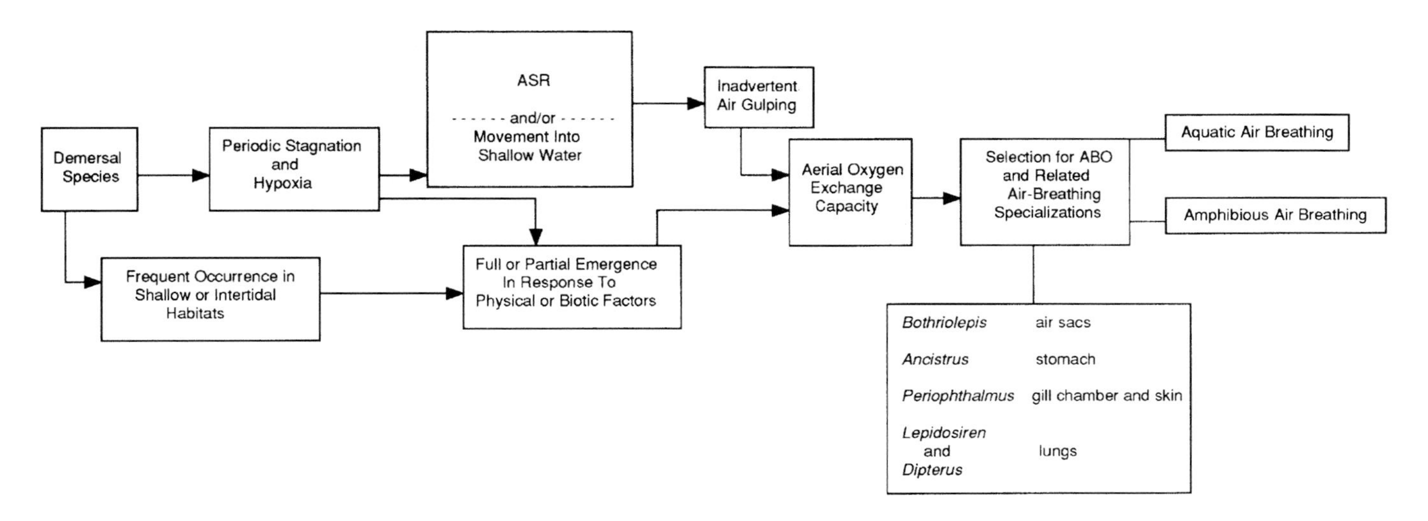

FIGURE 9.2 Role of the habitat variables periodic hypoxia and emergence in the evolution of amphibious and aquatic air-breathing specializations among fishes.

marine–brackish water littoral fishes (Gibson, 1969, 1982; Graham, 1976a; Horn and Gibson, 1988; Sayer and Davenport 1991). In some cases it appears that aquatic hypoxia has led to both aquatic (i.e., air gulping by *Pseudapocryptes* and *Taenioides*) and amphibious (*Helcogramma, Taurulus, Blennius*) air breathing. Similarly, the presence of air breathing in the pelagic *Megalops* is linked to its ontogeny in hypoxic freshwaters. Nevertheless, and for the most part, hypoxia as a causal factor for air breathing in these habitats takes second place to the role of emersion.

Chapter 2 detailed various causal factors for emergence which include tidal (e.g., among pholids, stichaeds, and probably most clingfish) as well as wave-to-wave strandings. Recent work on cottids by Karen Martin (1991, 1993, 1995, 1996) has shown how both tidal strandings and the potential for aquatic hypoxia can lead to specializations for exposure resistance and amphibious air breathing. Much more to the point, however, are the numerous species that emerge to actively exploit the ecological boundary between the littoral and terrestrial environments. These amphibious littoral fishes include blennies (*Entomacrodus, Andamia, Alticus*), the rockskippers (*Mnierpes, Dialommus*), some moderately terrestrial gobies (*Gobionellus*), and the mudskippers (*Periophthalmus, Boleophthalmus, Periophthalmodon*, and *Scartelaos*).

Other Mechanisms of Habitat Adaptation Possibly Leading to Air Breathing

The effects of hypoxia and emersion on air breathing have been influenced by whether a species was pelagic or demersal. Figure 9.2 proposes likely scenarios leading to natural selection for air breathing in demersal species. Many shallow water fish behaviorally respond to hypoxia by moving into very shallow water where they initiate ASR or emerge part or all of their body (Breder, 1941; Wright and Raymond, 1978).

Depending upon the factors involved, these responses, when coupled with the regular occurrence of aquatic hypoxia, could have triggered selection for air breathing. Selection, for example, would occur when such responses enabled a species to survive in an otherwise uninhabitable location and thus gain access to additional ecological resources. By the same token, if a fish utilizing ASR in hypoxic water inadvertently aspirates air (Burggren, 1982) and by chance gains from the net oxygenation of either its tissues or the ASR stream (e.g., if this lowered its energetic costs for ventilation or eliminated the reliance on surface contact), then selection for air breathing would have also been possible (Gans, 1970). For the intertidal fishes, on the other hand, oxygen probably played a minor role in the evolution of amphibious behavior; exploitation of ecological resources occurring at or just above the water line appears to have been a stronger selective factor.

Figure 9.3 suggests alternative scenarios for the evolution of air breathing from ASR. Many species utilizing ASR elevate their buoyancy for this purpose (Gee and Gee, 1995). Buoyancy increases may entail hyperinflation of the physosclistous gas bladder (*Dormitator*) or, in species lacking this organ (e.g., most littoral and benthic fishes), the inhalation of air bubbles that are held in the mouth. In the case of

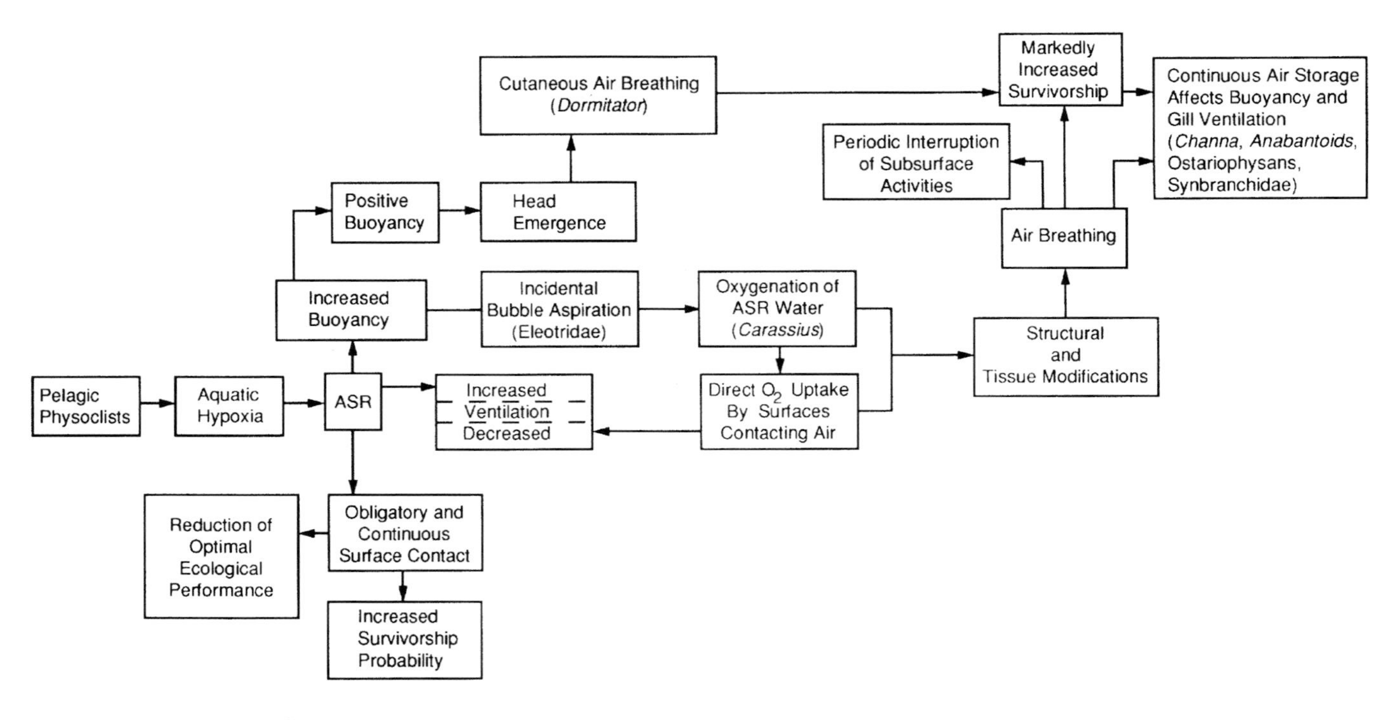

FIGURE 9.3 Possible mechanisms and steps leading to the derivation of air-breathing in open water fishes, beginning with ASR (aquatic surface respiration) and related buoyancy adaptations in physoclistous fishes.

Dormitator, positive buoyancy emerges the anterio-dorsal skin surface which, through natural selection, has become modified for aerial gas exchange (Todd, 1973). For gobies, most of which lack a gas bladder, Gee and Gee (1995) presented histological evidence for changes in the dorsal-buccal epithelium (i.e., the surface contacting air inhaled for purposes of buoyancy augmentation, see Chapter 2) of several genera. In these cases, natural selection has exploited a mechanism allowing O_2 absorption across a body surface brought into contact with air. The air breathing of *Gillicthys* may have originated in this manner. For *Dormitator* and the air-gulping gobies, the transition from ASR to facultative air breathing has the advantage of reducing the ventilatory costs of ASR.

An Estimate of the Number of Independent Origins of Fish Air Breathing

When the entire diversity of air-breathing fishes is considered, there is hardly anything that binds all of them together. Other than correspondence between the adaptive advantages conferred by the capacity to exchange respiratory gases aerially, the species in this group do not have structural, physiological, or behavioral air-breathing traits in common; the expression of this specialization in the groups is as diverse as the groups themselves.

In view of this, there can be no doubt that air breathing has evolved independently many times; the phylogenetic analysis presented in Chapter 2 permits estimation of the frequency with which this has occurred. First, from discussions in Chapter 1 (and see Figure 1.1), air breathing is considered to be symplesiomorphic, that is a shared primitive trait, in all bony fishes. This symplesiomorphy, based on the common presence of a lung and lung-like gas bladder (see Chapter 3) for aerial respiration, extended from the Sarcopterygii to the Neopterygii and perhaps even to some of the Euteleostei. Referring to Figure 2.1, this means that all families, from the Ceratodontidae to the Megalopidae, and perhaps even the Phractolaemidae, may share a common air-breathing ancestral linkage.

In contrast, it appears, from the diversity of ABO structure and behavior seen among most euteleosts and the neoteleosts, that ancient selection pressures for air breathing were interrupted (i.e., replaced by other selective pressures that altered lung and gas-bladder function) during their evolution. Air breathing thus appears to have been reinvented by many euteleosts, as evidenced by the appearance of novel ABO types, and is thus largely autapomorphic (i.e., an independently evolved specialization unique to each group) in most euteleosts and neoteleosts. Using the lack of air-breathing synapomorphy (i.e., the absence of air-breathing in an ancestral group) as an index of the independent origin of this specialization in various genera, it is estimated, based on Figure 2.1 and Table 9.1, that air breathing has independently evolved among the fishes at least 38 times and perhaps as many as 67 times.

AIR-BREATHING SPECIALIZATIONS

Specializations for air breathing are unequally distributed among fishes.

Air-Breathing Organs

Development

There is an apparent lack of morphological specializations for air breathing among mastacembelids, pholids, stichaeds, and some of the blennies and clingfishes, all of which appear to respire aerially by default, that is, when stranded by receding waters.

Although morphologic specialization for aerial respiration is not seen in all species, it is not rare among the more than 370 species of air-breathing fishes. Considering the degree of modification found in the branchial structures of gobies (i.e., mudskippers and *Pseudapocryptes*) to be the lower limit of "functional ABO development," the numbers are as follows (Table 9.1):

i. Total number of the 49 air-breathing families containing species with an ABO, 33 (67%)
ii. The number of the 125 air-breathing genera possessing an ABO, 73 (58%)
iii. The number of the 375 species with an ABO, 248 (66%).

ABO Types

Classifications arising from ABO-type usually place fishes in four to six groups, largely without reference to phylogeny (Carter, 1957). As detailed in Chapter 3, these include:

i. Derivatives of the digestive tube, including lungs, gas bladder, pneumatic duct, stomach, and intestine
ii. Organs formed in the head region, including the buccal, pharyngeal, branchial, and opercular surfaces as well as pouches formed adjacent to the pharynx which are in some cases filled with branchial diverticulae
iii. Existing organs such as the gills or skin.

Phylogenetic and Morphological Influences on the ABO

ABO diversity has been affected by phylogeny and by morphological constraints (Chapter 3). The lung was a primitive fish characteristic present in both the sarcopterygians and actinopterygians (Liem, 1989a). Whereas tetrapod evolution is linked, in nearly all instances, to increased lung efficacy, the evolutionary history of the actinopterygian lung followed a different course; it was gradually transformed from a respiratory gas bladder into a non-respiratory physostomous gas bladder, and finally to a physoclistous gas bladder primarily specialized for sound reception, buoyancy control, or both (Figure 3.2B).

The ostariophysan order Siluriformes (catfishes) is indicative of the manner in which the evolutionary canalization of gas bladder function proceeded to the point that, in cases where temporal climatic changes dictated facultative bimodal breathing in some habitats, this organ could no longer be acted upon by selection for air breathing. Only one siluriid, *Pangasius*, has a respiratory gas bladder (Browman and Kramer, 1985). All other air-breathing siluriids use secondarily derived ABOs that range in diversity and complexity from the stomach and intestine (loricariids and callicthyids) to the branchially derived dendrites and fans of *Clarias* and its allies and the extended pharyngeal pouches of *Heteropneustes*. Thus, and beginning with the Ostariophysi, adaptive specialization of the gas bladder for functions discordant with bimodal respiration appears to have led to the evolution of novel ABOs in all of the higher teleosts.

Examples detailing the integration of phylogeny and bimodal breathing morphology extend to the patterns of circulation (Chapter 4), and to the control of and mechanisms for ABO ventilation (Chapters 3 and 6), and are also relevant to the morphological specializations (i.e., fin placement) for terrestriality present in the amphibious gobies and blennies. Phylogeny influenced the patterns of ABO ontogeny and differentiation and may further combine with habitat conditions in expression of phenotypic parameters such as ABO and gill surface areas (Bader, 1937; Ebeling and Alpert, 1966).

Physiology

Comparative gas exchange, quantification of respiratory partitioning between air and water, and the factors regulating this have been well studied, as reviewed in Chapter 5. Major principles are that ABOs are primarily specialized for O_2 uptake and not for the release of CO_2 and that aquatic air-breathers remain largely dependent upon water for branchial CO_2

release. Amphibious air breathers are more adept at releasing CO_2 aerially and thus largely avoid respiratory hypercapnia during terrestrial sojourns (Graham, 1976a, 1994).

The diverse respiratory patterns found among the air-breathing fishes suggest the presence of different evolutionary stages of ventilatory control as well as differences in central and peripheral receptor interaction (Chapter 6). Although a variety of ventilatory control responses have been recorded for air-breathing fishes, none has attained the tetrapod proficiency level as exemplified by the absence of gills, exclusive reliance upon unimodal (aerial) respiration, and a central pH or CO_2 modulated (even partially) control. On the other hand, species such as *Andamia*, *Entomacrodus*, *Mnierpes*, and other amphibious air-breathing fishes, while still in possession of gills, appear to have evolved separate aquatic- and aerial-respiratory control responses, probably without either an intermediate aquatic air-breathing control stage or the stimulus of aquatic hypoxia.

Cardiorespiratory integration has been well documented (Smatresk, 1994; Graham *et al.*, 1995a). Among the aquatic air breathers tachy- and bradycardic reflexes correlate with ABO inflation and deflation and are stimulated by alterations in respiratory gas content and gill ventilation. Marked phylogenetic differences in adrenergic cardiac innervation exist: Lungfish lack this entirely, it is present in *Lepisosteus* and fully expressed in most teleosts. A survey of heart responses to ABO-volume change reveals that the scope of air-inspiration tachycardia is reduced in lungfish and *Lepisosteus* relative to teleosts (Chapter 6 and Graham *et al.*, 1995a).

The respiratory properties of the blood and the hemoglobin (Hb)–O_2 affinity of bimodal fishes were reviewed in Chapter 7. Whereas the evolutionary progression to a greater dependence on air breathing (as seen in tetrapods and exemplified by amphibian metamorphosis) results in a lowering of Hb–O_2 affinity, this trend is not strongly evident among the bimodal fishes. Rather, species forced to endure more extreme conditions of aquatic hypoxia (which requires more air breathing) often do so through adaptations augmenting aquatic respiration.

The small amount that is known about the comparative physiology of other systems in air-breathing fishes emphasizes the role of this adaptation in ensuring an aquatic existence rather than offering a track to terrestrial life. In bimodal fishes, for example, the gills remain critically important for CO_2 release, nitrogen excretion, and both acid-base balance and ion regulation. Gill dependence is a principal factor restricting fishes to an essentially aquatic existence; no bimodal fish has progressed in breathing efficacy to the point that the respiratory and other essential ancillary functions of the gills have been subsumed by other organs. As detailed in Chapter 8, there is a trend toward reciprocal kidney and gill functions in *Arapaima*. Nonetheless, progress along this trajectory is relatively small compared to the evolutionary changes in renal and respiratory physiology coincident with either amphibian metamorphosis or the evolution of tetrapods (Toews and Stiffler, 1989).

Biochemistry

The biochemistry of bimodal breathing largely hinges on the availability of water. Aquatic air-breathing fishes depend upon gill ventilation to flush CO_2 from their blood. With the exception of certain amphibious air breathers, most bimodal fishes exposed to air cannot release all of their CO_2 and thus become hypercapnic, which has repercussions for a number of systemic, cellular, and metabolic processes (Heisler, 1989). Even when not exposed to air, the acid-base equilibrium of a bimodal fish can be affected by the pH or PCO_2 of its ambient water (e.g., by seasonal temperature changes or the seasonal influx of leaf tannins at the onset of the wet season). It is also affected by the extent that gill ventilation is restricted (leading to the retention of respiratory CO_2) in order to limit transbranchial O_2 loss. A limited gill ventilation can influence biochemical processes by restricting ion transfer (Oduleye, 1977; Stiffler *et al.*, 1986; Olson, 1994; Graham, 1994).

Aspects of biochemical specialization touched on in this book, but remaining largely unexplored, include the apparent ability of some fishes (*Heteropneustes*, *Notopterus*, mudskippers) to arrest their metabolism while in hypoxic water without access to air. Ultsch (1989) has reviewed the ecological and physiological aspects of hibernation in fishes and other ectotherms. Several fishes including some air breathers are capable of enduring prolonged periods of estivation. However, most species can only accomplish this at cool temperatures which reduce metabolic demand (Storey and Storey, 1990). In light of this, the ability of certain air-breathing species to endure hypoxia and even anoxia is noteworthy because many of these occur in warm habitats.

Another apparent metabolic difference is the extent to which air-breathers are independent of air quality. Species that air-breathe under ice (*Umbra*) or in burrows (*Synbranchus*, *Periophthalmus*) may, for example, normally encounter hypoxic gas pockets; whereas, species respiring in open air would not. Finally there are the relative differences among closely related air-breathers and non-air breathers. Although the air-breathing *Arapaima* has a greater aerobic capacity than the non-air-breathing *Osteoglossum*, the aerobic capacity of both these genera is reduced relative to most other fishes. On the other hand, the air-breathing and non-air-breathing erythrinids display few differences insofar as the aerobic versus anaerobic poising of their tissues is concerned.

The branchial release of ammonia (a toxic end product of nitrogen metabolism) is the prevalent method of nitrogen excretion among fishes and is highly dependent upon water access (Griffith, 1991). Ammonia may thus concentrate in bimodal fish denied access to water. The biochemical evolutionary view is that ureotelism, the capacity to detoxify ammonia by the synthesis of urea through the ornithine-urea cycle (OUC), was present in the Sarcopterygii and inherited by tetrapods. The Actinopterygii, by contrast, have been thought to lack the OUC and remain dependent upon the branchial release of ammonia (ammonotelism). Ureotelism is essential during lungfish estivation, however, the absolute rate of urea production in estivating *Protopterus* is about the same as in awake fish; total metabolic rate has merely been lowered to a level where the intrinsic rate of ureogenesis keeps pace with nitrogenous waste production. What significance does actinopterygian bimodal breathing and, in particular, the independent appearance of amphibious air breathing in this group, hold for the biochemical evolution of ammonotelism and ureotelism? There are indications that some air-breathing teleosts are ureotelic and some have the full complement of OUC enzymes, however, generally divergent findings about the nitrogen metabolism of different species do not permit a clear answer to this question at this time (Griffith, 1991; Chapter 8).

Behavior

In addition to air breathing itself, several other complicated behavioral patterns have been noted for the air-breathing fishes. The feeding, swimming, and air-breathing activity of *Hoplosternum* (Boujard *et al.*, 1990) and *Amphipnous* (Patra *et al.*, 1978) and *Lepidocephalus* (Natarajan, 1984b) are strongly affected by diel cycle. Also, both the activity and air-breathing frequency of *Colisa*, *Anabas*, and *Amphipnous* were reduced during a solar eclipse (Pandey and Shukla, 1982). Clariids in shallow water exhibit socially facilitated prey herding and channel-excavating behaviors (D. Donnelly, 1966; B. Donnelly, 1973, 1978; Bruton, 1979). The latter appears to enable fish trapped in drying pools to get deeper into the substrate (i.e., below the water surface). Bruton (1979) also observed that when he approached a drying pool containing *Clarias*, fish would leave the water and climb a mud bank in order to take cover in a burrow (i.e., this burrow had been below the water surface at one time).

A striking example of a "higher level of air-breathing control" is provided by synchronous air breathing. This behavior, characterized by the closely timed and nearly simultaneous air gulping by a group, is considered a form of socially mediated temporal schooling that reduces the danger of predation during surfacing (Kramer and Graham, 1976). Synchronous air breathing has been described in varying degrees for *Polypterus*, *Lepisosteus*, *Heterotis*, *Notopterus*, *Clarias*, *Heteropneustes*, *Hoplosternum*, *Corydoras*, *Ancistrus*, *Hypostomus*, *Piabucina*, and *Colisa* as well as in young *Arapaima* and *Megalops*. Most recently, synchrony has also been observed in *Hoplerythrinus* by Mattias *et al.* (1996).

Even though amphibious air-breathing fishes are seldom far from water, bimodality does permit littoral-zone exploitation and is reflected in attendant morphological and behavioral adaptations for terrestriality. Amphibious fishes purposely engage in this behavior for reasons that include territoriality and courtship, the pursuit of prey, orientation and avoidance of turbulence (*Galaxias* and other species), escape from predators (however, mudskippers are occasionally bitten by mosquitos), even the shedding of ectoparasites (*Gillichthys*) and other reasons (Magnus, 1962, 1966; Graham, 1970; Brillet, 1984, 1986; Sayer and Davenport, 1991).

While emerged these fish have the capacity to orientate to water using known reference points and possibly sun-compass (Graham *et al.*, 1985; Berti *et al.*, 1994). Aerial vision poses a problem because of the refractive differences of air and water (Denny, 1993). Solutions include, for *Periophthalmus*, reduction of lens curvature to compensate for the added corneal refraction in air or, in *Mnierpes* and some blennies (*Entomacrodus*), flattening of the corneas to reduce aerial astigmatism and allow emmetropia in air without compromising aquatic accommodation (Graham and Rosenblatt, 1970; Graham, 1971; Herald and Herald, 1973; Stevens and Parsons, 1980; Land, 1987).

AIR BREATHING AND AQUACULTURE

Air-breathing fishes are ideal for aquaculture and at least six genera (*Clarias*, *Channa*, *Heteropneustes*, *Pangasius* and *Dormitator*) and probably many more are raised commercially (Santhanam *et al.*, 1987;

Chang, 1984). Santhanam *et al.* (1987) devote a full chapter to the special qualities of air-breathing fishes for aquaculture. These fishes are generally hardy and their lower requirement for aquatic respiration makes them more resistant to natural and anthropogenic toxins (Day, 1877; Jacob *et al.*, 1972; Natarajan, 1983), and less vulnerable to certain parasites (Ravindranath *et al.*, 1985). Air breathing translates into a less labor-intensive maintenance because the fish farmers can be less concerned about water quality. Not only can air breathers be raised in waters that are stagnant or have poorer turnover rates than could be used for non-air-breathers, less care needs to be given to the type, quality, and amount of feed that is provided. Feces (chicken, porcine, and human) supplemented with a mixture of grains, weeds, and slaughter-house waste constitute the diets of air-breathing fish cultured in India.

Air breathers can also be raised in greater densities than non-air breathers (i.e., two to three times greater than carp or trout). Both *Clarias* and *Heteropneustes* can be cultured *in vitro* or their fertilized eggs can be obtained in the field. *Channa* is more difficult to rear from eggs and sperm, however, large numbers of young fry are plentiful most of the year. In addition, *Channa*, *Pangasius*, and probably other fishes are rapidly growing (i.e., the growth rate of *Pangasius* is said to rival that of pelagic species such as tuna, a rare phenomenon for a freshwater species). Moreover, the flesh of *Channa* and *Clarias* brings a high price because of its texture, low fat content, and a low number of intramuscular bones. Air breathing also simplifies the problems of transport to market or from natural capture sites to culture ponds. Santhanam *et al.* (1987) do, however, caution that culture ponds need to be fenced to prevent the fishes from wandering away.

One scientifically useful "benefit" of mixed species aquaculture was the finding by Tarnchalanukit (1986) that *Clarias* and *Pangasius* had hybridized. These genera have vastly different ABOs (Chapter 3) and it would be most informative to examine hybrid ABO morphology and their air-breathing behavior and physiology.

FISH AIR BREATHING AND THE EVOLUTION OF TETRAPODS

Air-Breathing Ancestry

Speculations that air breathing was broadly distributed among Paleozoic fishes are based on climatology (Barrell, 1916), partly on early atmospheric O_2 levels (Bray, 1985), on the details of certain fossils (Wells and Dorr, 1985), and on the evolutionary relationships of the early fishes (Schaeffer, 1965). Groups implicated in these conjectures include the Silurian sharks and acanthodians, all of the Early Devonian Osteichthyes, and possibly even the Devonian antiarch placoderm *Bothriolepis* (Denison, 1941; Wells and Dorr, 1985). With the possible exception of the latter, there is little fossil evidence pertaining to the presence of auxiliary ABOs.

Concerning the bony fishes, the consensus view is that the Sarcopterygii and Actinopterygii diverged in the Late Silurian (Denison, 1956; Romer, 1966; Thomson, 1993; Long, 1995). The assumption of a homologous lung in these groups therefore places the origin of vertebrate air breathing in the Late Silurian or earlier. Nonetheless, there is some uncertainty as to the origin of the lung and this in turn engenders questions about the homology of the lung and gas bladder and about the physical and biotic factors that led to the evolution of these organs.

How Valid Is the Assumption of Lung Symplesiomorphy?

Chapters 2, 3, and 4, emphasize the homology of the lung and gas bladder. This appears warranted, assuming that sarcopterygians and actinopterygians have a common ancestor, and because of similarities in ABO structure and circulation pattern among the extant lungfishes (Sarcopterygii), and *Polypterus* and *Erpetoichthys* (Actinopterygii), and also because of factors such as the homologies of structures formed by the embryonic aortic arches and morphological similarities among lungs and physostomous gas bladders in the primitive Osteichthyes (Chapters 3 and 4).

The common ancestry of the sarcopterygians and actinopterygians is suggested by many features of their skull and jaw. However, in light of the possibility that the morphological similarities among extant species could reflect convergence, what is the nature of the fossil evidence for a common ancestor for lobe- and rayfinned-fishes?

Unfortunately there is no concrete fossil evidence for this. If there was a common ancestor, it was probably a Late Silurian acanthodian. The sparse fossil record does suggest an acanthodian ancestry for the Osteichthyes. There are also Early Devonian fossil traces of both actinopterygians and sarcopterygians. Even so, the earliest material (Middle Devonian) available for comparison shows these two fish groups already display an extreme divergence (Romer, 1966; Carroll, 1988). Whereas the common ancestry of both groups is widely assumed, it is possible that the Osteichthyes is a paraphyletic assemblage of several groups including the ray- and lobe-finned fishes.

Did the Lung Evolve in the Late Silurian?

The assumption of a common origin for sarcopterygians and acanthopterygians requires that there was a lunged common ancestor in the Late Silurian or earlier. Studies of new fossil sarcopterygian material from China (Yu, 1990; Thomson, 1993) indicate a rather abrupt diversification of the major lobe-fin lineages beginning in the Lower Devonian (408 mybp). This radiation included the early separation of the Dipnoi (lungfishes, the only extant sarcopterygian with a functional lung) from the Rhipidistians, the now extinct ancestral line that gave rise to the Porolepiformes, Osteolepiformes, *Panderichthys*, and the tetrapods (Long, 1995) (Figure 9.4). Whereas Campbell and Barwick (1988) have postulated the absence of lung breathing among some groups of primitive lungfishes, the pattern of divergence shown by Yu (1990) suggests that a functional lung was likely present in the all of the earliest lobe-fins (Thomson, 1969, 1993).

The Habitat in Which Air Breathing Evolved

Vertebrates evolved in the sea (Denison, 1956; Robertson, 1957) and expanded to freshwaters in the Late to Upper Silurian (425–408 mybp) where they continued to diversify and radiate (Bray, 1985; Halstead, 1985). It is thought that many of these early vertebrates were euryhaline and that for some, occurrence in freshwater may have been seasonal for purposes of feeding or reproduction (Bray, 1985; Halstead, 1985; Griffith, 1987). Accordingly, the Late Silurian common ancestor of the sarcopterygians and actinopterygians could have lived in salt or freshwater, or both.

There is meager fossil information on the habitat of the Early Devonian actinopterygians. The Middle and Late Devonian fossils of this group occur in both marine and freshwater deposits. Early Devonian sarcopterygian fossils occur primarily in marine deposits, but freshwater fossils predominate in both the Middle and Late Devonian (Campbell and Barwick, 1988; Thomson, 1993).

A number of workers have reviewed or suggested new ideas about the environmental and habitat conditions leading to the evolution of the vertebrate lung (Denison, 1956; Inger, 1957; Romer 1958a,b, 1966, 1967; Little 1983, 1990; Bray, 1985). The scenario most commonly described involves exposure of populations to chronic or seasonal hypoxia in warm lowland waters of tropical latitudes. The Paleozoic habitat conditions selecting for a lung resemble, in the minds of most writers, those currently encountered in modern-day tropical lowlands where seasonal drought commonly results in widespread stagnation, severe aquatic hypoxia or even anoxia, and in some cases the total evaporation of water (Schmalhausen, 1968; Graham *et al.*, 1978a). It is in these habitats that the greatest diversity of extant air-breathing fishes and ABOs is found. The generally warm and dry conditions of the Late Silurian would have further contributed to the intensity and duration of hypoxic conditions in these habitats.

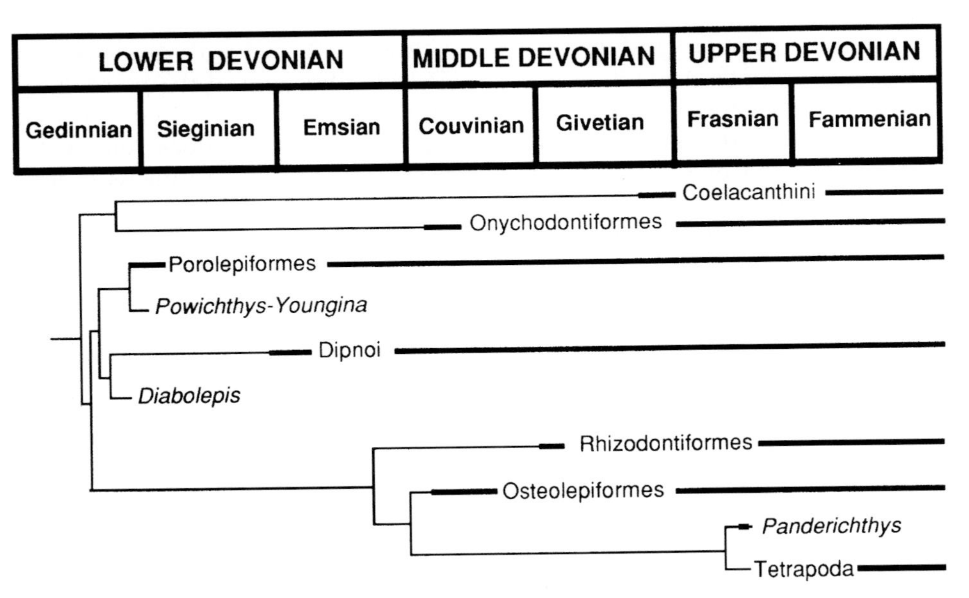

FIGURE 9.4 Cladogram for the sarcopterygian fishes. (Modified from Thomson, 1993. Reprinted by permission of *American Journal of Science*, volume 293, pp. A33–62, figure 13.)

From the foregoing, the air-breathing ancestor of the actinopterygian and sarcopterygian fishes probably inhabited freshwaters where it encountered hypoxia with sufficient regularity to lead to selection for the lung. Nevertheless, as discussed in Chapter 3, the early view, one that goes back to Darwin, was that buoyancy and not respiration was the source of selection pressure for this organ. Chapter 3 has presented arguments against the primacy of buoyancy in lung-gas bladder evolution. That discussion moreover, indicated there is scarcely any basis of support for the views (Burggren and Bemis, 1990) that the question of primacy is unresolvable and possibly moot, or that the organ could have arisen through "dual-purpose" selection. If it is assumed that buoyancy was the causal factor for this organ's evolution, then the habitat of the ancestral group would be open to reconsideration.

Whereas the primacy of the lung seems no longer in question, issues such as fossil distribution patterns may require consideration of alternate scenarios for its evolution. These scenarios include the possibility of a saltwater-origin of air breathing or its origin due to a factor entirely unrelated to aquatic hypoxia.

Did Vertebrate Air Breathing Evolve in Saltwater?

Packard (1974) argued that the lower dissolved oxygen content of saltwater could have been a factor in selection for the evolution of air breathing in isolated coastal seas of the Silurian. Thomson (1980, 1993) voiced general accord with this model for air breathing evolution, mainly because a preponderance of Middle and Late Silurian fish fossils occurred in both freshwater and marine habitats, and because many of the Early Devonian dipnoans were marine. Campbell and Barwick (1988) have, however, argued that the marine occurrence of the early lungfishes obviated air breathing. Thomson (1993), on the other hand, does not agree with this stand and describes these fishes as "at least partial lungbreathers" (page 55). Even though this controversy concerns Devonian lungfishes, the question itself is not far removed from the Silurian origin of air breathing. What therefore are the merits of Packard's hypothesis for the saltwater-origin of air breathing?

As was explained in Chapter 1, Henry's Law (equation 1.2) governing the concentration of any gas dissolved in water depends on two components, α the gas solubility coefficient (i.e., the factor affected by temperature and salinity) and the partial pressure (PO_2) of the gas (Dejours, 1994). The solubility of O_2 in water is inversely proportional to salinity; air-saturated freshwater thus contains between 25–30% more oxygen than does water with a salinity of 35‰. Packard hypothesized that, for fishes enclosed in the Silurian coastal seas, this solubility difference and seawater's greater density (i.e., denser water is harder to pump), acted in concert with an increased metabolic demand (due to a rise in water temperature), to elevate the metabolic costs for aquatic ventilation sufficiently to select for air breathing.

In a critique of this hypothesis, Graham *et al.* (1978a) calculated that the metabolic increment required to compensate the increased ventilatory costs in saltwater was quite small and probably within the scope of normal compensatory adjustments brought about through changes in gill area or geometry, or alterations in Hb–O_2 affinity. However, and more important, that critique also focused on a fundamental weakness of Packard's (1974) hypothesis, its emphasis on the physical factors (temperature and salinity) affecting the O_2 capacitance of water rather than the occurrence of a lowered O_2 partial pressure caused by an imbalance between habitat biological oxygen demand and the factors replenishing aquatic oxygen (photosynthesis and diffusion from the atmosphere). Biotic respiration in a freshwater swamp, for example, can easily deplete over 90% of the oxygen, lowering the PO_2 to or below 15 torr (Kramer and Graham, 1976). On the other hand, a sizable increase in water temperature and salinity (extremes probably too large to permit survival) only changes the O_2 content of sea water by 50% and does not affect PO_2 which remains at air-saturation (about 150 torr). As was shown in Chapter 1, it is the partial pressure of oxygen that determines diffusion rate and thus the respiratory utility of a medium (Graham *et al.*, 1978a).

Although Packard's emphasis on factors affecting O_2 solubility missed the point in terms of the critical environmental factor (hypoxia) that has selected for air breathing, this does not necessarily imply that coastal marine environments of the Silurian were not sufficiently hypoxic to induce selection for air breathing. Back bays, coastal lagoons and even enclosed Silurian seas could all have experienced episodic hypoxia. Hypoxia is also not rare in marine environments where it has occasionally caused mass mortalities (Brongersma-Sanders, 1957).

Nevertheless, owing to influences such as tidal flushing and wind-driven vertical mixing, hypoxic conditions are generally not as extreme, widespread, or prolonged in marine habitats as they can be in freshwaters (Richards, 1957; Graham *et al.*, 1978a; Holeton, 1980; Dunn, 1983). Nor are they as likely to occur regularly in the same locale or encompass an entire fauna. A coastal upwelling, red tide event, or other factor may, for example, occur every few years for periods ranging from a number of days to several weeks, and may span several hundred kilometers. A contrast between the effects of a hypoxic event of this magnitude and the annual occurrence of the dry season across most of the South American tropics, tropical Africa, and southern and central India indicates why the dry season has had a much greater impact on selection for hypoxia adaptation. In tropical lowland freshwater habitats, it is the regular occurrence of aquatic deoxygenation, brought on by seasonal drying and sustained by shading, high biological oxygen demand, and isolation, that, because it encompasses vast numbers of the endemic fauna, has selected for air breathing (Graham *et al.*, 1978a).

Was the Evolution of Air Breathing Linked to Changes in Metabolic Scope or Atmospheric Oxygen?

Bray (1985) noted that the evolution of vertebrate air breathing preceded the marked increases in atmospheric oxygen that occurred in the Late Devonian and Carboniferous. Beyond this, the possible link between paleoatmospheric O_2 levels and the evolution of air-breathing has not been considered.

Several different models of atmospheric O_2 evolution have been proposed and, although differing somewhat in detail, these show a general rise in O_2 throughout the Phanerozoic. Cloud's (1983) model (Figure 9.5), puts Silurian atmospheric O_2 at about 70% of present atmospheric level (PAL, 21%). This agrees with estimates by Berner and Canfield (1989, Figure 9.5) whose model also suggests that O_2 levels declined slightly in the Early Devonian, rose dramatically during the Late Devonian and Carboniferous to nearly 35% (about 1.7 times PAL) and fell in the Permian.

From Henry's law (equation 1.2), a reduction of atmospheric PO_2 would reduce aquatic O_2 content and thus might have placed a greater premium on aerial respiration. We do not know whether the ecological radiations of fishes in the Silurian and Early Devonian were associated with increased metabolic costs (e.g., possible causal factors would have included greater mobility, changes in jaw structure and thus feeding behavior, increased competition, higher energetic costs of osmoregulation in freshwater, etc.), or if these costs exceeded the oxygen-extraction capacity of the gills sufficiently to select for air breathing as a mechanism for increasing aerobic scope. Questions such as whether a limiting ambient O_2 tension led to air breathing need greater consideration and experimentation. As documented in Chapter 5, increased aerobic scope is an integral feature of air breathing in the Australian lungfish, *Neoceratodus*.

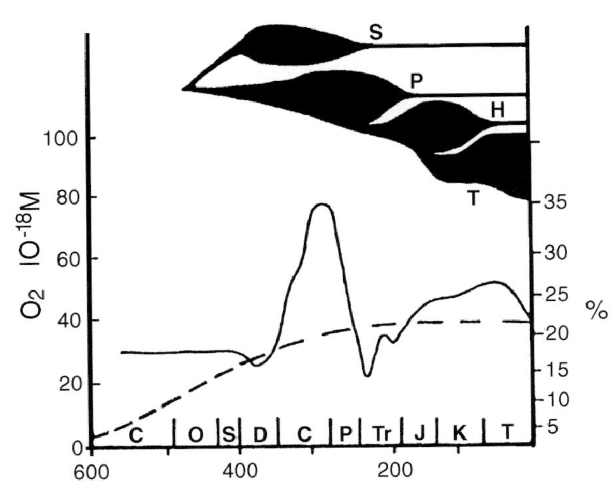

FIGURE 9.5 Estimates of atmospheric O_2 over geologic time based on models by Cloud, 1983 (dashed line) and Berner and Canfield, 1989 (solid line), and the evolutionary succession of the four major bony fish groups; Sarcopterygii (S), Polypteriformes (or Paleonisciformes, P), Holosteans (H, note that *Amia* and *Lepisosteus* are the only extant members of this group) and Teleosts (T). Abscissa shows geologic periods from Cambrian to Tertiary and millions of years before present. Vertical axes are molar amounts and percentage O_2 levels. (Reprinted from *American Zoologist*, volume 34, pp. 229–237, figure 2. Graham, 1994. Used with permission of the American Society of Zoologists.)

Figure 9.5 also shows the relationship between the atmospheric O_2 changes postulated by Berner and Canfield (1989) and the evolutionary transitions from the primitive to more advanced bony fish stocks, as indicated by the fossil record (Romer, 1966). Whereas the Permian O_2 decrease corresponds with the sarcopterygian decline and the paleonisciform expansion, none of the other transitions correlate with O_2 change. The evolutionary replacement of primitive by more derived bony fish grades was a complex process (Greenwood and Norman, 1963; Lauder and Liem, 1983a; Carroll, 1988) and not the result of solitary factors such as atmospheric O_2 concentration or bimodal breathing. Nevertheless, and with few exceptions, the surviving remnants (Figure 9.5) of these largely extinct Paleozoic and Mesozoic groups are bimodal breathers (e.g., lungfishes [*Neoceratodus*, *Protopterus*, *Lepidosiren*]; bichirs [*Polypterus*], the ropefish [*Erpetoichthys*], gars [*Lepisosteus*, *Atractosteus*], and the bowfin [*Amia*]). This could indicate that air breathing evolved independently in these groups as a consequence of their competitive exclusion to environmentally extreme refugia. On the other hand, this volume has repeatedly emphasized that the pattern of a lung or respiratory gas bladder in nearly all of the primitive air-breathing genera indicates that bimodal breathing is an ancient trait that first appeared among the Early Paleozoic fish stocks.

Tetrapod Evolution

Romer's Scenario

The factors leading to the evolution of tetrapods have been discussed by a number of authors (Schmalhausen, 1968; Panchen, 1980; Little, 1983, 1990; Gordon and Olson, 1994; Long, 1995). Alfred S. Romer was one of the first biologists to postulate the sequence of tetrapod evolution based on vertebrate relationships established by comparative anatomy and the fossil record. Romer had been influenced by Joseph Barrell (1916) who suggested the presence of semi-arid conditions throughout much of the Devonian (408–360 mybp), when the fluvial Old Red Sandstone deposits were formed, and suggested that the dry conditions could have spawned the evolution of amphibians. Several of Romer's works contain vignettes describing essential features of the evolutionary transition from fish to amphibian. The following is from *Vertebrate Paleontology* (Romer, 1966, page 86):

> But the amphibian, with his short and clumsy but effective limbs, could crawl out of the pool and walk (probably very slowly and painfully at first) and reach the next pool where water still remained and resume his normal existence as a water dweller.

In a less condensed form, Romer's (1958a,b, 1964, 1966, 1967, 1972 and elsewhere) account of tetrapod origins emphasized the monophyletic origin of amphibians and the inheritance of both lungs and lobe-fins from a rhipidistian ancestor. Among the qualities Romer attributed to the first amphibians were, a primary occurrence in water, the presence of a long, finned tail, major dependence on a bimodal respiration (lungs and internal gills), the presence of a thick, scaled armor that precluded cutaneous respiration, and polydactyly. Romer also stressed that, with tetrapods probably appearing in the Middle Devonian, the first amphibians underwent a considerable period of aquatic evolution prior to their Early Carboniferous (360–320 mybp) invasion of land.

Timing

An unfortunate feature of Romer's often quoted vignettes was their tendency to make vertebrate air breathing and tetrapod evolution part of the same piece of evolutionary history. In actuality, and this was clearly Romer's view, a large span of time separated the origin of air breathing in fishes, the first appearance of tetrapods, and their invasion of land. Assuming, for example, that air breathing had evolved by the Late Silurian (408 mybp), this preceded the Middle Devonian (380 mybp) appearance of tetrapods by some 28 million years and it was at least another 10 and perhaps 25 million years until largely terrestrial groups appeared (Thomson, 1993; Ahlberg and Milner, 1994; Daeschler *et al.*, 1994; Ahlberg, 1995; Carroll, 1995, 1996).

Climate and the Origin of the Tetrapod Limb

Although accepted in principle, several features of Romer's account have been criticized. It has been argued on various grounds that Romer placed too much emphasis (as in the preceding quotation) on the role of drought and dry Devonian climates in leading to the evolution of tetrapods. Romer envisioned the dry climate as a critical factor for the evolution of the tetrapod limb because it enabled relocation from a drying pool to one with more water,

> . . . [lobe-fins] would give their fortunate possessor the chance of crawling up or down the stream bed . . . and enable him to reach some surviving water body where he could resume a normal piscine existence. (Romer, 1964, page 57.)

Romer's scene explained the initial selection pressure for changes in limb structure. However, critics observed that this mechanism did not account for the extent of limb development that had to have taken place in aquatic amphibians between the Middle and Late Devonian. As detailed by Ahlberg and Milner (1994), the lobed, pectoral fin of the ancestral rhipidistian osteolepiform was completely transformed into the tetrapod forelimb, complete with its proximal (humerus) and distal (ulna, radius, digits) elements. Romer's view was that, with a selection process for limb evolution underway in the Devonian, other selection factors would lead to refinements for maneuverability in extremely shallow water and for terrestrial locomotion:

> Once this process had begun, it is easy to see how a land fauna might have eventually been built up. Instead of seeking water immediately, the amphibian might linger on the banks and devour stranded fish. Some types might gradually take to eating insects (primitive ones resembling cockroaches and dragonflies became abundant in the Pennsylvanian), and, finally, plants. (Romer, 1966, page 86.)

Regardless of whether or not movement from pool to pool was a critical factor, a cursory examination of the fins of mudskippers (Harris, 1960) and rockskippers (Graham, 1970; Zander, 1972a,b), and *Clarias* demonstrates how life in extremely shallow water and in conjunction with an expanding terrestrial biosphere could select for changes in fin structure over 10 or more million years (Inger, 1952). In addition to paddling, lobe-fins would assist in pushing the body away from, or lifting it over uneven substrate (e.g., climbing over submerged logs), and supporting the body during partial or full emergence (Eaton, 1960). Even though the early tetrapods may have never been out of water for very long, locomotion across exposed mud for short distances seems a likely requirement (and there are many fossil trackways, Campbell and

Barwick, 1988; Thomson, 1993). Further, and because these early tetrapods were very "fish-like," they would have probably swam in open water with limbs adducted, using the sinusoidal action of their body and caudal fin (Long, 1995). Lastly, other limb functions have also been suggested. Halstead (1968), for example, felt that stout forelimbs and pectoral girdles may have had a vital role in lifting the body and head weight off the trunk to enable inspiration.

The Ecological Radiation of Aquatic Tetrapods

Inger (1957, and see Thomson, 1993) suggested that Romer had over emphasized the selective effects of habitat drying and not adequately stressed the role of ecological radiations in evolution of the terrestrial tetrapods. Nonetheless, it is clear from the preceding quotation that Romer linked the evolution of a true terrestrial tetrapod fauna to the exploitation of terrestrial biosphere which did not begin to diversify until the Carboniferous. In response to Inger (1957), Romer (1958a, page 365) wrote:

> I heartily agree with . . . his major thesis that the development of vertebrate terrestrial life took place under favorable conditions of temperature, humidity, and other environmental factors. The seeming contrast is due to the fact that he has synapsed into a supposed single event two chapters in tetrapod history: (1) the development of limbs giving the potentiality of terrestrial existence, and (2) the utilization of these limbs for life on land. These two steps need not have been taken synchronously and, I believe, were separated in time by many millions of years.

Relationships

Subsequent paleontological research has largely supported essential features of Romer's scenario, particularly its timing, and has significantly improved our knowledge of the organisms involved. Through demonstration of morphological traits such as long tails with fin rays and internal gills (Coates and Clack, 1990, 1991; Thomson, 1991, 1993; Gordon and Olson, 1994; Zimmer, 1995), paleontological studies verify that some of the Late Devonian amphibians were primarily, if not exclusively aquatic.

It now seems rather certain that the tetrapods evolved from osteolepiform sarcopterygian fishes near the Middle Devonian or possibly earlier (Figs. 9.4 and 9.6). The early hypothesis that lungfishes were ancestral to the tetrapods was given new life by Rosen and co-workers (1981), but has now been completely rejected (Marshall and Schultze, 1992; Ahlberg and Milner, 1994). Fishes in the family Panderichthyidae (*Panderichthys*, *Elpistostege*, Figure 9.4) appear to be the ancestral group of the tetrapods (Schultze and Arsenault, 1985; Thomson, 1993; Ahlberg and Milner, 1994; Carroll, 1995, 1996; Long, 1995). Panderichthyids lived in the Upper Devonian (Figure 9.6) and were specialized for life in shallow, lentic waters. These fish, the fossils of which are found in freshwater deposits in North America and Europe (the tropical regions of Laurentia), share many morphological characters with the early tetrapods (Schultze and Arsenault, 1985; Long, 1995).

The Early Tetrapods

A total of eight fossil tetrapod genera are now known (Figure 9.6). These all occur in freshwater deposits and are found within an approximately 5 million year strata of the Upper Devonian (between the Upper Frasnian and Upper Famennian ages, 368–363 mybp). The oldest Upper Frasnian genera are *Elginerpeton* (Scotland) and *Obruchevichthys* (Latvia). (These two genera form the Elginerpetontidae in Figure 9.6.) Occurring in the Lower Famennian are *Metaxygnathus* (Australia) and *Hynerpeton* (North America), whereas *Ventastega* (Latvia), *Acanthostega* (Greenland), *Ichthyostega* (Greenland) and *Tulerpeton* (Russia) all occur in the Upper Famennian (Ahlberg, 1995; Carroll, 1995, 1996). The early tetrapods were thus broadly distributed across the tropical regions of Laurentia and Gondwana (Thompson, 1993; Daeschler *et al.*, 1994; Long, 1995).

As reviewed by Thomson (1993) and Long (1995), the exact time and site of the tetrapod origin remains unknown. However, this is important to know because, assuming evolutionary divergence had begun prior to the Middle Devonian, then it becomes uncertain as to whether the ancestral rhipidistians occurred in salt or freshwater and, assuming the latter, how the early tetrapods (all from freshwater deposits) became so widely dispersed.

Morphological comparisons reveal considerable differences among these eight tetrapods, reflect a mosaic pattern of radiation with respect to terrestriality and probably other features of their Upper Devonian habitats, and further suggest the possibility that there were several different tetrapod clades (Romer 1966 [reviews earlier literature]; Gordon and Olson, 1994; Ahlberg, 1995; Carroll, 1995, 1996). Analyses by Coates and Clack (1990) reveal differences in the forelimb mobility and strength of *Acanthostega* and *Ichthyostega*. The forelimb of *Ichthyostega* is quite robust, is permanently bent at the elbow, and could have probably supported the body out of water. By contrast, the forelimb of *Acanthostega* does not appear strong enough to have provided support on land, suggesting this animal was permanently aquatic (this is further suggested by the presence of both dorsal and ventral caudal fin rays). Recent work has documented additional variation in forelimb structure; *Tulerpeton* appears to have a stronger forelimb than either *Acanthostega* or

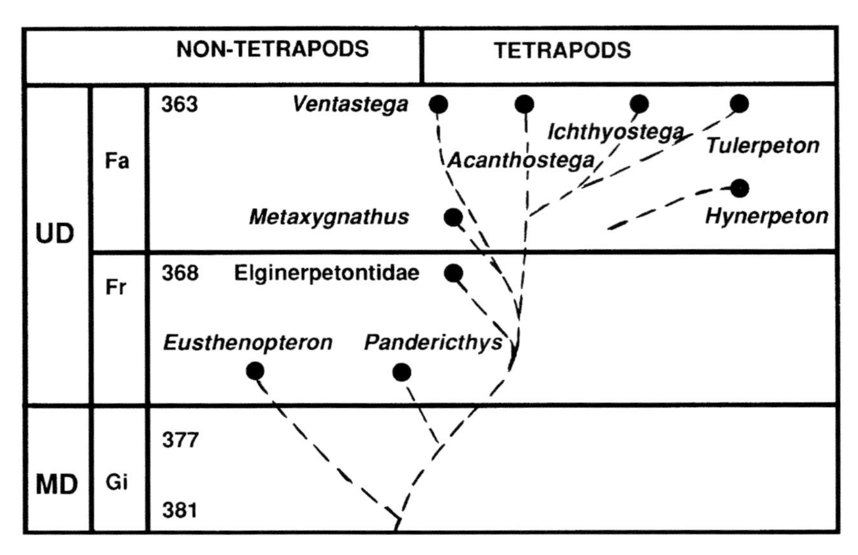

FIGURE 9.6 Postulated phyletic relationships among the known early tetrapod genera. Abbreviations: UD, Upper Devonian; MD, Middle Devonian; Fa, Fammenian; Fr, Frasnian; Gi, Givetian. Numbers are mybp. (Modified from Ahlberg, 1995 and Carroll, 1995. Reprinted with permission from *Nature,* volume 373, pp. 389–390. Copyright 1995. Macmillan Magazines Limited.)

Ichthyostega (Thomson, 1993). Daeschler *et al.* (1994) also report a robust forelimb in *Hynerpeton*.

Finally, polydactly (i.e., as many as 6 and 8 digits per limb) has been documented for several of the early tetrapod genera (Coates and Clack, 1990; Ahlberg and Milner, 1994). With respect to the number of digits on a limb Romer (1958b, page 92) observed:

> . . . some primitive fossil forms had five and there is some reason to believe that there may have been as many as seven in some primitive land types. . . . there may have been considerable variation among early types before higher tetrapods 'settled down' to the orthodox five toes.

Respiration

Notwithstanding the lesser emphasis given to tetrapod gills earlier, Romer (1972) observed that these were important for respiration in the early aquatic amphibians. Whereas lungs permitted early tetrapods to obtain oxygen from air and thus respire while confined in water that was too foul or hypoxic for gill ventilation, gills were essential for CO_2 removal, ion regulation, and nitrogen excretion. Coates and Clack (1991) reported internal gills in *Ichthyostega* and these also appear to be present in *Acanthostega*. On the other hand, Daeschler *et al.* (1994) concluded that the pectoral girdle structure of *Hynerpeton* did not suggest the presence of gills.

Various theories have been advanced concerning how the transition from bimodal to unimodal respiration could have affected the evolution of tetrapod gas exchange (Packard, 1976; Ultsch, 1987, 1996; Graham *et al.*, 1995b). Schmalhausen (1968) suggested that, because of the work associated with lifting the body to ventilate, the early tetrapods may have evolved accessory aerial respiratory surfaces such as open patches of skin in the mouth and on other protected surfaces. Romer (1972), however, emphasizes that the early amphibians were covered by thick scales that retarded water loss and impeded cutaneous gas exchange.

It would appear that, as the tetrapod emergence from water eliminated gill function, there would have been strong selection pressure for increasing pulmonary $\dot{V}O_2$ and $\dot{V}CO_2$ and for finding alternative mechanisms for carrying out functions that had been the province of branchial tissue. Graham *et al.* (1995b) suggested that a Late Devonian hyperoxic atmosphere could have improved lung respiratory efficiency and favored terrestriality by lessening the ratio of evaporative water loss to oxygen uptake and reducing the threat of desiccation during aerial respiration. As was shown for the air-breathing fishes, the exploitable avenues for achieving a greater aerial $\dot{V}CO_2$ could have included changes in ventilation, changes in the levels of carbonic anhydrase, and in hemoglobin respiratory and buffer capacities (see Chapters 5–8). It is also possible that an intrinsically higher plasma PCO_2 and HCO_3^- may have been a necessary consequence of this transition, just as in certain air-breathing fishes (*Electrophorus*, Garey and Rahn, 1970; Ultsch, 1987, 1996; Chapters 5 and 7).

Graham *et al.* (1995b) also suggested that changes in the Paleozoic atmosphere may have partially mitigated the CO_2 elimination problem for early tetrapods. Berner's (1993) model for atmospheric carbon dioxide, for example, indicates that this gas underwent largely reciprocal changes to those determined for oxygen. From a value of about 0.36% in the Late Devonian, atmospheric CO_2 dropped by a factor of 10 to about 0.035% in the Mid Carboniferous, and then increased in the Late Permian. Just as an elevated atmospheric O_2 would heighten its diffusion gradient across the lung surface into the body, the decline in CO_2 would increase the gradient for its diffusion from the body to the atmosphere. Even though atmospheric changes may have modulated the CO_2 problem somewhat, the total elimination of branchial function with the emergence of tetrapods was no doubt an important selective force in the evolution of a mechanism for acid-base regulation that was dependent upon both the renal and pulmonary systems (Chapter 8, Packard, 1976; Ultsch, 1987, 1996).

Finally, and as was indicated earlier regarding tetrapod lung function, several workers linked pectoral fin development to the need for body support during lung ventilation. Related to this, there are differing opinions as to whether the primary function of the extensive thoracic ribs, with their limb-girdle connections, of the early tetrapods was to support the viscera during emersion or to enable aspiratory breathing (Gans, 1970; Romer, 1972; Carroll, 1988; Thomson, 1993).

In summary, a review of Romer's paradigm for tetrapod evolution shows that it remains essentially intact and serves to emphasize the vast separation in time of three of its central features: The evolution of air breathing, the appearance of tetrapods, and the tetrapod land invasion.

SUMMARY AND OVERVIEW

Air breathing first evolved in the fishes and, over the 400 million year history of this group, this capacity has persisted in certain lineages and has been re-acquired by others. The great significance of air breathing for the fishes is that it ensures a heightened survival capacity for species that naturally occur, or are temporarily trapped, in hypoxic waters. Rather than having "tendencies toward invading land," most of these fishes remain tied to an exclusively aquatic existence. Air breathing permits a fish to have a high degree of immunity from ambient water O_2 tension as well as some independence from dissolved toxins, both natural and anthropogenic. The lesser degree of aquatic-habitat reliance afforded by this capacity allows air-breathing fishes to access habitats and ecological resources inaccessible to non-air-breathers, thus reducing competitive and predatory pressures on both adult and juvenile populations and promoting ecological radiation. Evidence that ecological radiation has occurred is found in groups such as the anabantoids, clariids, and callichthyids which are diverse and in which bimodal breathing has been exquisitely integrated into several features of their natural history.

To many biologists "the evolution of vertebrate air breathing," and "the evolutionary transition to land," mean about the same thing and conjure up images (Figure 1.2) of Paleozoic landscapes under assault by primitive fish-like tetrapods. The concept of a close link between bimodal respiration and the origin of the tetrapods is bolstered by comparative anatomical and fossil evidence and by dynamic transitional phenomena such as metamorphosis.

However, the evolution of vertebrate air breathing, the evolution of tetrapods, and their invasion of land were discreet historical events separated by tens of millions of years (Romer, 1966). The vertebrate lung is thought to have evolved during the Silurian Period in fishes ancestral to the ray- and lobe-finned Osteichthyes. It was in the Upper Devonian that tetrapods were derived from the lobe-finned fishes. The earliest tetrapods were aquatic, bimodal breathers, and they underwent a 10 million year or longer period of evolution in aquatic habitats before gaining a foothold on land in the Carboniferous. The occurrence of the first tetrapods in habitats that may have been subject to periodic drought and hypoxia, together with the competitive and predatory pressures exerted by more open water fishes probably led to increased natural selection for specializations needed to live in very shallow water and at the land-water interface.

As emphasized by Romer (1966) and many others, the radiation of tetrapods onto land occurred in concert with the evolution of the terrestrial biosphere and the opportunities for ecological exploitation this afforded. In this regard loose parallels exist between the first tetrapods and the extant amphibious fishes. It is not difficult to view the latter as a small piscine terrestrial invasion force that, were it not for the presence of a highly diversified and "ecologically hostile" terrestrial fauna already in place, could potentially mount one front of another terrestrial invasion. Eiseley (1957, page 54) wrote:

> The world is fixed, we say: fish in the sea, birds in the air. But in the mangrove swamps by the Niger, fish climb trees and ogle uneasy naturalists who try unsuccessfully to chase them back to the water. There are still things coming ashore.

References

Abel, D.C., Koenig, C.C. and Davis, W.P. 1987. Emersion in the mangrove forest fish *Rivulus marmoratus*: A unique response to hydrogen sulfide. Environ. Biol. Fishes 18:67–72.

Abel, E.F. 1973. Zur Öko-Ethologie des amphibisch lebenden Fisches *Alticus saliens* (Forster) und von *Entomacrodus vermiculatus* (Val.) (*Blennioidea, Salariidae*), unter besonderer Berücksichtigung des Fortpflanzungsverhaltens. Sitzungsber. Osterreich Akad. Wiss. (Naturw. Kl., Abt. I.) 181:137–153.

Agar, W.E. 1908. On the appearance of vascular filaments on the pectoral fin of *Lepidosiren paradoxa*. Anat. Anz. 33:27–30.

Aguilar, N.M. and Graham, J.B. 1995. The respiratory physiology of air-breathing gobies. Physiol. Zool. 68:72.

Ahlberg, P.E. 1995. *Elginerpeton pancheni* and the earliest tetrapod clade. Nature 373:420–425.

Ahlberg, P.E. and Milner, A.R. 1994. The origin and early diversification of tetrapods. Nature 368:507–514.

Al-Kadhomiy, N.K. and Hughes, G.M. 1988. Histological study of different regions of the skin and gills in the mudskipper, *Boleophthalmus boddarti* with respect to their respiratory function. J. Mar. Biol. Assn. U.K. 68:413–422.

Albers, C. 1970. Acid-base balance. In: Hoar, W.S. and Randall, D.J. (eds.) *Fish Physiology Vol. IV*. Academic Press, New York, pp. 173–208.

Albers, C. 1985. Gas transport properties of fish blood. In: Gilles, R. (ed.) *Circulation, Respiration, and Metabolism*. Springer-Verlag, New York, pp. 82–90.

Allen, G.R. 1982. *A Field Guide to Inland Fishes of Western Australia*. W. Australia Mus., Perth, 86 pp.

Allen, G.R. and Berra, T.M. 1989. Life history aspects of the west Australian salamanderfish, *Lepidogalaxias salamandroides*. Mees. Rec. West. Australian Mus. 14:253–267.

Altland, P.D. 1971. Erythrocyte and hemoglobin values: Vertebrates other than man. In: Altman, P.L. and Dittmer, D.S. (eds.) *Respiration and Circulation*. Fed. Am. Soc. Exp. Biol., Bethesda, pp. 151–153.

Amaral, A.D. and Carneiro, N.M. 1982. Estudo comparativo de alguns parametros sanguineos dos peixes: *Pimelodus maculatus* (Mandi) e *Symbranchus marmoratus* (Mucum). An. Da Acad. Brasileira Ciênc. 54:768–769.

Andresen, J.H., Ishimatsu, A., Johansen, K. and Glass M.L. 1987. An angiocardiographic analysis of the central circulation in the air breathing teleost *Channa argus*. Acta Zool. 68:165–171.

Ar, A. and Zacks, D. 1989. Alterations in the bimodal gas exchange of the African catfish *Clarias lazera*. In: Paganelli, C.V. and Farhi, L.E. (eds.). *Physiological Function in Special Environments*. Springer-Verlag, NY, pp. 172–190.

Arunachalam, S., Vivekanadan, E. and Pandian, T.J. 1976. Food intake, conversion, and swimming activity in the air-breathing catfish *Heteropneustes fossilis*. Hydrobiologia 51:213–217.

Assheton, R. 1907. The development of *Gymnarchus niloticus*. In: Kerr, J.G. (ed.) *The Work of John Samuel Budgett*. Cambridge Univ. Press, London, pp. 293–421, Pl. I–XXI.

Atkinson, D.E. 1992. Functional roles of urea synthesis in vertebrates. Physiol. Zool. 65:243–267.

Atkinson, D.E. and Bourke, E. 1987. Metabolic aspects of the regulation of systemic pH. Am. J. Physiol. 252:F947–F956.

Axelsson, M., Abe, A.S., Bicudo, J.E.P.W. and Nilsson, S. 1989. On the cardiac control in the South American lungfish, *Lepidosiren paradoxa*. Comp. Biochem. Physiol. 93A:561–565.

Babak, E. and Dedek, B. 1907. Vergleichende Untersuchungen über die Darmatmung der Cobitidinen und Betrachtung über die Phylogenese derselben. Biol. Centr. 27:697–703.

Babcock, L.L. 1951. *The Tarpon: A Description of the Fish With Some Hints on Its Capture*. Louis L. Babcock, Buffalo, NY, 157 pp.

Babiker, M.M. 1979. Respiratory behaviour, oxygen consumption and relative dependence on aerial respiration in the African lungfish (*Protopterus annectens*, Owen) and an air-breathing teleost (*Clarias lazera*, C.). Hydrobiologia 65:177–187.

Babiker, M.M. 1984a. Development of dependence on aerial respiration in *Polypterus senegalus* (Cuvier). Hydrobiologia 110:351–363.

Babiker, M.M. 1984b. Adaptive respiratory significance of organophosphates (ATP & GTP) in air-breathing fishes. Hydrobiologia 110:339–349.

Backhouse, G.N. and Vanner, R.W. 1978. Observations on the biology of the dwarf galaxiid *Galixiella pusilla* (Mack) (Pisces; Galaxiidae). Victorian Nat. 95:128–132.

Bader, R. 1937. Bau, Entwicklung und Funktion des akzessorischen Atmungsorgans der Labyrinthfische. Z. Wiss. Zool. 149:323–401.

Baldwin, E. 1964. *An Introduction to Comparative Biochemistry*. Cambridge University Press, Cambridge, 179 pp.

Balinsky, J.B., Choritz, E.L., Coe, G.L. and van der Schans, G.S. 1967. Amino acid metabolism and urea synthesis in naturally aestivating *Xenopus laevis*. Comp. Biochem. Physiol. 22:59–68.

Ballantyne, F.M. 1927. Air-bladder and lungs: A contribution to the morphology of the air-bladder of fish. Trans. R. Soc., Edinburgh 55:371–394.

Bandurski, R.S., Bradstreet, E. and Scholander, P.F. 1968. Metabolic changes in the mudskipper during asphyxia or exercise. Comp. Biochem. Physiol. 24:271–274.

Banerjee, T.K. and Mittal, A.K. 1975. Histochemistry and the functional organization of the skin of a "live-fish" *Clarias batrachus* (Linn.) (Clariidae, Pisces). Mikroskopie 31:333–349.

Banerjee, V. 1966. A note on the haematological study of *Anabas testudineus* (Bloch.). Sci. Cult. 32:326–327.

Barnard, K.H. 1943. Revision of the indigenous freshwater fishes of the S.W. Cape Region. Ann. S. Afr. Mus. 36:101–262.

Barrell, J. 1916. Influence of Silurian-Devonian climates on the rise of air-breathing vertebrates. Bull. Geol. Soc. Am. 27:387–436.

Bartlett, G.R. 1978a. Phosphates in red cells of two lungfish: The South American *Lepidosiren paradoxa*, and the African, *Protopterus aethiopicus*. Can. J. Zool. 56:882–886.

Bartlett, G.R. 1978b. Phosphates in red cells of two South American osteoglossids *Arapaima gigas* and *Osteoglossum bicirrhosum*. Can. J. Zool. 56:878–881.

Bartlett, G.R. 1978c. Water-soluble phosphates of fish red cells. Can. J. Zool. 56:870–877.

Bartlett, G.R. 1980. Phosphate compounds in vertebrate red blood cells. Am. Zool. 20:103–114.

Barton, M. 1985. Response of two species of amphibious stichaeoid fishes to temperature fluctuations in an intertidal habitat. Hydrobiologia 120:151–157.

Baumert, M. 1853. Chemische Untersuchungen über die Respiration des Schlammpeizgers (*Cobitis fossilis*). Ann. Chemie und Pharmacie 88:1–56.

Baylis, J.R. 1982. Unusual escape response by two cyprinodontiform fishes, and a bluegill predator's counter-strategy. Copeia 1982:455–456.

Beadle, L.C. 1981. *The Inland Waters of Tropical Africa: An Introduction to Tropical Limnology*. Longman, London, 2nd ed., 365 pp.

Beebe, W. 1945. Vertebrate fauna of a tropical dry season mud-hole. Zoologica 30:81–87.

Beitinger, T.L. and Pettit, M.J. 1984. Comparison of low oxygen avoidance in a bimodal breather, *Erpetoichthys calabaricus* and an obligate water breather, *Percina caprodes*. Environ. Biol. Fishes 11:235–240.

Beitinger, T.L., Pettit, M.J. and Hutchinson, V.H. 1985. Oxygen transfer characteristics of the blood of reedfish, *Erpetoichthys calabaricus*. Comp. Biochem. Physiol. 82A:553–558.

Bemis, W.E. and Northcutt, R.G. 1992. Skin and blood vessels of the snout of the Australian lungfish, *Neoceratodus forsteri*, and their significance for interpreting the cosmine of Devonian lungfishes. Acta Zool. 73:115–139.

Bemis, W.E., Burggren, W.W. and Kemp, N.E. (eds.). 1987. *The Biology and Evolution of Lungfishes*. Alan R.Liss, Inc., New York, 383 pp.

Berg, T. and Steen, J.B. 1965. Physiological mechanisms for aerial respiration in the eel. Comp. Biochem. Physiol. 15:469–484.

Berner, R.A. 1993. Paleozoic atmospheric CO_2: Importance of solar radiation and plant evolution. Science 261:68–70.

Berner, R.A. and Canfield, D.E. 1989. A new model for atmospheric oxygen over Phanerozoic time. Am. J. Sci. 289:333–361.

Berra, T.M. and Allen, G.R. 1989. Burrowing, emergence, behavior, and functional morphology of the Australian salamanderfish, *Lepidogalaxias salamandroides*. Fisheries 14:2–10.

Berra, T.M., Sever, D.M. and Allen, G.R. 1989. Gross histological morphology of the swimbladder and lack of accessory respiratory structures in *Lepidogalaxias salamandroides*, an aestivating fish from western Australia. Copeia 1989:850–856.

Berti, R., Colombini, I, Chelazzi, L. and Ercolini, A. 1994. Directional orientation in Kenyan populations of *Periophthalmus sobrinus* Eggert: Experimental analysis of the operating mechanisms. J. Exp. Mar. Biol. Ecol. 181:135–141.

Bertin, L. 1958. Organes de la respiration aérienne. Traité de Zoologie VIII:1363–1398.

Bevalander, G. 1934. The gills of *Amia calva* specialized for respiration in an oxygen deficient habitat. Copeia 1934:123–127.

Bevan, D.J. and Kramer, D.L. 1986. The effect of swimming depth on respiratory behavior of the honey gourami, *Colisa chuna* (Pisces, Belontiidae). Can. J. Zool. 64:1893–1896.

Bevan, D.J. and Kramer, D.L. 1987a. The respiratory behavior of an air-breathing catfish, *Clarias macrocephalus* (Clariidae). Can. J. Zool. 65:348–353.

Bevan, D.J. and Kramer, D.L. 1987b. Lack of a depth effect on the growth of the air-breathing catfish, *Clarias macrocephalus*. Can. J. Fish. Aquat. Sci. 45:1507–1510.

Bicudo, J.E.P.W. and Johansen, K. 1979. Respiratory gas exchange in the air-breathing fish, *Synbranchus marmoratus*. Environ. Biol. Fishes 4:55–64.

Bischof, G. 1818. Untersuchung der Luft, welche die Fischart *Cobitis fossilis* von sich giebt. Schweiggers J. Chemie und Physik 22:78–92.

Bishop, I.R. and Foxon, G.E.H. 1968. The mechanism of breathing in the South American lungfish, *Lepidosiren paradoxa*; a radiological study. J. Zool., London 154:263–271.

Biswas, N., Ojha, J. and Munshi, J.S.D. 1979. Bimodal oxygen uptake in relation to body weight of the amphibious mudskipper *Boleophthalmus boddaerti* (Pall). Indian J. Exp. Biol. 17:752–756.

Biswas, N., Ojha, J. and Munshi, J.S.D. 1981. Morphometrics of the respiratory organs of an estuarine goby, *Boleophthalmus boddaerti*. Japan. J. Ichthyol. 27:316–326.

Black, V.S. 1945. Gas exchange in the swimbladder of the mudminnow *Umbra limii* (Kritland). Proc. Nova Scotia Inst. Sci. 21:1–22.

Blaese, M.G. 1980. Morphologische und physiologische Untersuchungen zur Atmung und Verdauung des Panzerwelses *Brochis coeruleus*. Diplomarbeit, Mainz.

Boake, P. 1865. On the air breathing fish of Ceylon. J. Ceylon Br. Asiat. Soc. Great Britain & Ireland, pp. 128–142.

Böker, H. 1933. Über einige neue Organe bei luftatmenden Fischen und im Uterus der Anakonda. Anat. Anz. 76:148–155.

Boujard, T., Keith, P. and Luquet, P. 1990. Diel cycle in *Hoplosternum littorale* (Teleostei): Evidence for synchronization of locomotor, air breathing and feeding activity by circadian alternation of light and dark. J. Fish Biol. 36:133–140.

Boutilier, R.G. and Ferguson, R.A. 1989. Nucleated red cell function: Metabolism and pH regulation. Can. J. Zool. 67:2986–2993.

Brainerd, E.L. 1994. The evolution of lung-gill bimodal breathing and the homology of vertebrate respiratory pumps. Am. Zool. 34:289–299.

Brainerd, E.L., Liem, K.F. and Samper, C.T. 1989. Air ventilation by recoil aspiration in polypterid fishes. Science 246:1593–1595.

Branco, L.G., Glass, M. and Hoffmann, A. 1991. Central chemoreceptor drive to breathing in unanesthetized toads, *Bufo paracnemis*. Respir. Physiol. 87:195–204.

Braum, E. 1983. Beobachtungen uber eine reversible Lippenextension und ihre Rolle bei der Notatmung von *Brycon* spec. (Pisces, Characidae) und *Colossoma macropomum* (Pisces, Serrasalmidae). Amazoniana 7:355–374.

Bray, A.A. 1985. The evolution of the terrestrial vertebrates: Environmental and physiological considerations. Phil. Trans. R. Soc., London 309:289–322.

Breder, Jr., C.M. 1927. The fishes of the Rio Chucunaque drainage, eastern Panama. Bull. Am. Mus. Nat. Hist. 57:91–176.

Breder, Jr., C.M. 1941. Respiratory behavior in fishes not especially modified for breathing air under conditions of depleted oxygen. Zoologica 26:243–244.

Breder, Jr., C.M. and Rosen, D.E. 1966. *Modes of Reproduction in Fishes*. Nat. Hist. Press, Garden City, NY, 941 pp.

Bridges, C.R. 1988. Respiratory adaptations in intertidal fish. Am. Zool. 28:79–96.

Bridges, C.R. 1993a. Adaptation of vertebrates to the intertidal environment. In: Bicudo, J.E.P.W. (ed.) *The Vertebrate Gas Transport Cascade—Adaptations to Environment and Mode of Life*. CRC Press, Boca Raton, pp. 12–22.

Bridges, C.R. 1993b. Ecophysiology of intertidal fish. In: Rankin, J.C. and Jensen, F.B. (eds.) *Fish Ecophysiology*. Chapman and Hall, London, pp. 375–400.

Bridges, C.R. and Morris, S. 1989. Respiratory pigments: Interactions between oxygen and carbon dioxide transport. Can. J. Zool. 67:2971–2985.

Bridges, C.R., Freitag, J. and Harder, V. 1995. Intermittent ventilation *in vivo* and *in situ* in intertidal fish—effects on circulation and the role of hypoxia, hyperoxia and air-breathing. Physiol. Zool. 68:90.

Bridges, C.R., Taylor, A.C., Morris, S.J. and Grieshaber, M.K. 1984. Ecophysiological adaptations in *Blennius pholis* (L.) blood to intertidal rockpool environments. J. Exp. Mar. Biol. Ecol. 77:151–167.

Brillet, C. 1975. Relations entre territoire et comportement agressif chez *Periophthalmus sobrinus* Eggert (Pisces, Periophthalmidae) au laboratoire et en milieu naturel. Z. Tierpsychol. 39:283–331.

Brillet, C. 1976. Structure du terrier, réproduction et comportement des jeunes chez le poisson amphibie *Periophthalmus sobrinus* Eggert. Rev. Écol. Terre Vie 30:465–483.

Brillet, C. 1984. Étude comparative de la parade nuptiale chez deux espèces de poissons amphibies sympatriques (Pisces, Periophthalmidae). Comptes Rendus des Séances de l'Académie des Sciences, Serie III. 298:347–350.

Brillet, C. 1986. Notes on the behaviour of the amphibious fish *Lophalticus kirkii* Günther (Pisces, Salariidae). Comparison with *Periophthalmus sobrinus* Eggert. Rev. Écol. Terre Vie 41:361–375.

Brongersma-Sanders, M. 1957. Mass mortality in the sea. In: Hedgpeth, J.W. (ed.) *Treatise on Marine Ecology and Paleoecology, Vol. 1. Ecology*. Nat'l Research Council, Nat'l Acad. Sci., Washington, D.C., pp. 941–1010.

Browman, M.W. and Kramer, D.L. 1985. *Pangasius sutchi* (Pangasiidae), an air-breathing catfish that uses the swimbladder as an accessory respiratory organ. Copeia 1985:994–998.

Brown, C.R., Gordon, M.S. and Chin, H.G. 1991. Field and laboratory observations on microhabitat selection in the amphibious Red Sea rockskipper fish, *Alticus kirki* (Family Blenniidae). Mar. Behav. Physiol. 19:1–13.

Brown, C.R., Gordon, M.S. and Martin, K.L.M. 1992. Aerial and aquatic oxygen uptake in the amphibious Red Sea rockskipper fish, *Alticus kirki* (Family Blenniidae). Copeia 1992:1007–1013.

Brown, G.W. and Cohen, P.P. 1960. Comparative biochemistry of urea synthesis: III. Activities of urea cycle enzymes in various higher and lower vertebrates. Biochem. J. 75:89–91.

Brown, Jr., G.W. and Brown, S.G. 1967. Urea and its formation in coelacanth liver. Science 155:570–573.

Brown, Jr., G.W., James, J., Henderson, R.J., Thomas, W.N., Robinson, R.O., Thompson, A.L., Brown, E. and Brown, S.G. 1966. Uricolytic enzymes in liver of the dipnoan *Protopterus aethiopicus*. Science 153:1653–1654.

Bruton, M.N. 1976. On the size reached by *Clarias gariepinus*. J. Limnol. Soc. S. Africa 2:57–58.

Bruton, M.N. 1979. The survival of habitat desiccation by air breathing clariid catfishes. Environ. Biol. Fishes 4:273–280.

Buckley, J.L. and Horn, M.H. 1980. Comparative respiratory physiology of two species of intertidal stichaeid fishes from California. Am. Zool. 20:1025 (Abst.).

Budgett, J.S. 1901. On some points in the anatomy of *Polypterus*. Trans. Zool. Soc. London 15:323–338.

Budgett, J.S. 1902. On the structure of the larval *Polypterus*. Trans. Zool. Soc. London 16:315–346.

Budgett, J.S. 1903. Notes on the spiracles of *Polypterus*. Proc. Zool. Soc., London 1903:10–11.

Bugge, J. 1961. The heart of the African lungfish, *Protopterus*. Vidensk. Medd. Dansk Naturhist. Foren. Kjobenhavn 123:193–210.

Burggren, W. and Doyle, M. 1986a. Ontogeny of heart rate regulation in the bullfrog, *Rana catesbeiana*. Am. J. Physiol. 251:231–239.

Burggren, W. and Doyle, M. 1986b. Ontogeny of regulation of gill and lung ventilation in the bullfrog, *Rana catesbeiana*. Respir. Physiol. 66:279–291.

Burggren, W. and Haswell, S. 1979. Aerial CO_2 excretion in the obligate air breathing fish, *Trichogaster trichopterus*, a role for carbonic anhydrase. J. Exp. Biol. 82:215–225.

Burggren, W.W. 1979. Bimodal gas exchange during variation in environmental oxygen and carbon dioxide in the air breathing fish *Trichogaster trichopterus*. J. Exp. Biol. 82:197–213.

Burggren, W.W. 1982. "Air gulping" improves blood oxygen transport during aquatic hypoxia in the goldfish, *Carassius auratus*. Physiol. Zool. 55:327–334.

Burggren, W.W. and Bemis, W.E. 1990. Studying physiological evolution: Paradigms and pitfalls. In: Nitecki, M.H. (ed.) *Evolutionary Innovations*. Univ. Chicago Press, Chicago, pp. 191–228.

Burggren, W.W. and Infantino, Jr., R.L. 1994. The respiratory transition from water to air breathing during amphibian metamorphosis. Am. Zool. 34:238–246.

Burggren, W.W. and Johansen, K. 1987. Circulation and respiration in lungfishes (Dipnoi). In: Bemis, W.E., Burggren, W.W. and Kemp, N.E. (eds.) *The Biology and Evolution of Lungfishes*. Alan R. Liss, Inc., New York, pp. 217–236.

Burggren, W.W. and Pinder, A.W. 1991. Ontogeny of cardiovascular and respiratory physiology in lower vertebrates. Ann. Rev. Physiol. 53:107–135.

Burleson, M.L., Smatresk, N.J. and Milsom, W.K. 1992. Afferent inputs associated with cardioventilatory control in fish. In: Hoar, W.S., Randall, D.J. and Farrell, A.P. (eds.) *Fish Physiology, Vol. XII, Part B*. Academic Press, Inc., San Diego, pp. 389–426.

Burne, R.H. 1896. On the aortic-arch system of *Saccobranchus fossilis*. J. Linn. Soc. Zool. 25:48–55.

Cala, P. 1987a. The fish fauna and aquatic milieu of the llanos of Colombia (Orinoco basin) with special regard to respiratory patterns of fishes inhabiting extreme hypoxic waters. Proc. V Congr. Europ. Ichthyol., Stockholm, pp. 117–126.

Cala, P. 1987b. Aerial respiration in the catfish, *Eremophilus mutisii* (Trichomycteridae, Siluriformes), in the Rio Bogota Basin, Colombia. J. Fish Biol. 31:301–303.

Cala, P., del Castillo, B. and Garzon, B. 1990. Air-breathing behavior of the Colombian catfish *Eremophilus mutisii* (Trichomycteridae, Siluriformes). Exp. Biol. 48:357–360.

Calugareanu, D. 1907. Die Darmatmung von *Cobitis fossilis*. I. Über den Bau des Mitteldarms. Pflüg. Arch. 118:42–51.

Cameron, J.N. 1989. *The Respiratory Physiology of Animals*. Oxford Univ. Press, New York, 353 pp.

Cameron, J.N. and Iwama, G.K. 1989. Compromises between ionic regulation and acid-base regulation in aquatic animals. Can. J. Zool. 67:3078–3084.

Cameron, J.N. and Wood, C.M. 1978. Renal function and acid-base regulation in two Amazonian erythrinid fishes: *Hoplias malabaricus*, a water breather, and *Hoplerythrinus unitaeniatus*, a facultative air breather. Can. J. Zool. 56:917–930.

Campbell, J.W. 1973. Nitrogen excretion. In: Prosser, C.L. (ed.) *Comparative Animal Physiology*, 3rd ed., W.B. Saunders, Philadelphia, pp. 279-316.

Campbell, K.S.W. and Barwick, R.E. 1987. Paleozoic lungfishes—a review. In: Bemis, W.E., Burggren, W.W. and Kemp, N.E. (eds.) *The Biology and Evolution of Lungfishes*. Alan R. Liss, Inc., New York, pp. 93–131.

Campbell, K.S.W. and Barwick, R.E. 1988. Geological and palaeontological information and phylogenetic hypotheses. Geological Magazine 125:207–227.

Carlisky, N.J. and Barrio, A. 1972. Nitrogen metabolism of the South American lungfish *Lepidosiren paradoxa*. Comp. Biochem. Physiol. 41B:857–873.

Carroll, R. 1995. Between fish and amphibian. Nature 373:389–390.

Carroll, R.L. 1988. *Vertebrate Paleontology and Evolution*. W.H. Freeman, New York, 698 pp.

Carroll, R.L. 1996. Revealing the patterns of macroevolution. Nature 383:19–20.

Carter, G.S. 1931. Aquatic and aerial respiration in animals. Biol. Rev. 6:1–35.

Carter, G.S. 1933. Ecology of tropical swamps. Nature 132:896–897.

Carter, G.S. 1935. Respiratory adaptations of the fishes of the forest waters, with descriptions of the accessory respiratory organs of *Electrophorus electricus* (Linn.) (= *Gymnotus electricus* Aucct.) and *Plecostomus plecostomus* (Linn). J. Linn. Soc. London, Zool. 39:219–233.

Carter, G.S. 1957. Air breathing. In: Brown, M. (ed.) *Physiology of Fishes*. Academic Press, New York, pp. 65–79.

Carter, G.S. and Beadle, L.C. 1930. Notes on the habits and development of *Lepidosiren paradoxa*. J. Linn. Soc. London, Zool. 37:197–203.

Carter, G.S. and Beadle, L.C. 1931. The fauna of the swamps of Paraguayan Chaco in relation to its environment. II. Respiratory adaptations in the fishes. J. Linn. Soc. London, Zool. 37:327–368.

Cech, Jr., J.J. 1990. Respirometry. In: Schreck, C.B. and Moyle, P.B. (eds.) *Methods for Fish Biology*, Am. Fish. Soc., Bethesda, pp. 335–362.

Chan, P.K.Y. 1990. Aerial and aquatic oxygen consumption rates of Hong Kong mudskippers. Asian Mar. Biol. 7:189–199.

Chang, B.D. 1984. Tolerances to salinity and air exposure of *Dormitator latifrons* (Pisces, Eleotridae). Rev. Biol. Trop. 23:155–157.

Chew, S.F. and Ip, Y.K. 1992a. Tolerance of the mudskipper, *Boleophthalmus boddaerti*, to a lack of oxygen. Zool. Sci. 9:227–230.

Chew, S.F. and Ip, Y.K. 1992b. Biochemical adaptations of the mudskipper *Boleophthalmus boddaerti* to a lack of oxygen. Mar. Biol. 112:567–571.

Chew, S.F. and Ip, Y.K. 1992c. Cyanide detoxification in the mudskipper, *Boleophthalmus boddaerti*. J. Exp. Zool. 261:1–8.

Chew, S.F. and Ip, Y.K. 1993. Respiration in the muscle mitochondria of the mudskipper, *Boleophthalmus boddaerti*. Comp. Biochem. Physiol. 104B:681–688.

Chew, S.F., Lim, A.L.L., Low, W.P., Lee, C.G.L., Chan, K.M. and Ip, Y.K. 1990. Can the mudskipper, *Periophthalmus chrysospilos*, tolerate acute environmental hypoxic exposure? Fish Physiol. Biochem. 8:221–227.

Clayton, D.A. 1993. Mudskippers. Oceanogr. Mar. Biol. Annu. Rev. 31:507–577.

Clayton, D.A. and Vaughan, T.C. 1986. Territorial acquisition in the mudskipper *Boleophthalmus boddarti* (Pisces: Gobiidae) on the mudflats of Kuwait. J. Zool., London 209:501–519.

Clayton, D.A. and Vaughan, T.C. 1988. Ethogram of *Boleophthalmus boddarti* (Teleostei: Gobiidae), a mudskipper found on the mudflats of Kuwait. J. Univ. of Kuwait (Sci.) 15:115–138.

Cloud, P. 1983. The biosphere. Sci. Am. 249:176–189.

Coates, M.I. and Clack, J.A. 1990. Polydactyly in the earliest known tetrapod limbs. Nature 346:66–69.

Coates, M.I. and Clack, J.A. 1991. Fish-like gills and breathing in the earliest known tetrapod. Nature 352:234–236.

Cohen, P.P. 1966. Biochemical aspects of metamorphosis: Transition from ammonotelism to ureotelism. In: *Harvey Lectures*, Series 60. Academic Press, New York, pp. 119–154.

Conant, E.B. 1987. Bibliography of lungfishes, 1811–1985. In: Bemis, W.E., Burggren, W.W. and Kemp, N.E. (eds.) *The Biology and Evolution of Lungfishes*. Alan R. Liss, Inc., New York, pp. 305–373.

Courtenay, W.R. and Stauffer, J.R. (eds.). 1984. *Distribution, Biology and Management of Exotic Fishes*. Johns Hopkins Univ. Press, Baltimore, 431 pp.

Crawford, R.H. 1971. Aquatic and aerial respiration in the bowfin, longnose gar and Alaska blackfish. Ph.D. Thesis, Univ. of Toronto, Toronto, Ontario, Canada, 202 pp.

Crawford, R.H. 1974. Structure of an air-breathing organ and the swim bladder in the Alaskan blackfish, *Dallia pectoralis* Bean. Can. J. Zool. 52:1221–1225.

Cruz-Höfling, M.A., Cruz-Landim, C. and Patelli, A.S. 1981. Morphological and histochemical observations on swimbladders of an obligate water-breathing and a facultative air-breathing fish from the Amazon. Zool. Jahrb. Anat. 105:1–12.

Cruz-Landim, C. and Cruz-Höfling, M.A. 1979. Diferenças ultraestruturais entre bexigas natatórias de peixes teleósteos de respiração aquática e respiração aérea facultativa. Acta Amazonica 9:317–323.

Cunningham, J.T. 1932. Experiments on the interchange of oxygen and carbon dioxide between the skin of *Lepidosiren* and the surrounding water, and the probable emission of oxygen by the male *Symbranchus*. Proc. Zool. Soc., London 2:876–887.

Cunningham, J.T. and Reid, D.M. 1932. Experimental researches on the emission of oxygen by the pelvic filaments of the male *Lepidosiren* with some experiments on *Symbranchus marmoratus*. Proc. R. Soc., London 110:234–248.

Cunningham, J.T. and Reid, D.M. 1933. Pelvic filaments of *Lepidosiren*. Nature 131:913.

Cuvier, G. and Valenciennes, A. 1831. *Histoire Naturelle des Poissons Vol. VII*. Paris. Réimpression 1969. A. Asher and Co. Amsterdam, 1–531 pp., Pl. 170–206.

Cuvier, G. and Valenciennes, A. 1837. *Histoire Naturelle des Poissons Vol. XII*. Paris. Réimpression 1969. A. Asher and Co. Amsterdam, 1–507 pp., Pl. 344–368.

Cuvier, G. and Valenciennes, A. 1840. *Histoire Naturelle des Poissons Vol. XV*. Paris. Réimpression 1969. A. Asher and Co. Amsterdam, 1–540 pp., Pl. 421–455.

Cuvier, G. and Valenciennes, A. 1846. *Histoire Naturelle des Poissons Vol. XIX*. Paris. Réimpression 1969. A. Asher and Co. Amsterdam, 1–544 pp., Pl. 554–590.

Daeschler, E.B., Shubin, N.H., Thomson, K.S. and Amaral, W.W. 1994. A Devonian tetrapod from North America. Science 265:639–642.

Daget, J. 1950. Revision des Affinités Phylogénétiques des Polyptérídés. Mem. de L'Inst. Franc. d'Afriq. Noire 11:1–178.

Daldorff, L. 1797. Natural History of *Perca scandens*. In: Trans. Linn. Soc., London 3:62–63.

Dall, W. and Milward, E. 1969. Water intake, gut absorption and sodium fluxes in amphibious and aquatic fishes. Comp. Biochem. Physiol. 30:247–260.

Danforth, C.H. 1916. The relation of coronary and hepatic arteries in the common ganoids. Am. J. Anat. 19:391–400.

Daniels, C.B., Orgeig, S. and Smits, A.W. 1995. The evolution of the vertebrate pulmonary surfactant. Physiol. Zool. 68:539–566.

Darwin, C. 1859. *The Origin of Species By Means of Natural Selection.* Murray, London.

Das, B.K. 1927. The bionomics of certain air-breathing fishes of India, together with an account of the development of their air-breathing organs. Phil. Trans. R. Soc., London B216:183–217.

Das, B.K. 1930. Observations on a new mode of aerial respiration in an estuarine gobiid. Proc. Indian Sci. Cong. p. 255.

Das, B.K. 1933. On the bionomics, structure and physiology of respiration in an estuarine air-breathing fish, *Pseudapocrytes lanceolatus.* (Bloch and Schneider). Curr. Sci. 6:389–393.

Das, B.K. 1934. The habits and structure of *Pseudapocryptes lanceolatus*, a fish in the first stages of structural adaptation to aerial respiration. Proc. R. Soc. 115:422–431.

Das, J.R., Saha, N. and Ratha, B.K. 1991. Tissue distribution and subcellular localization of glutamate dehydrogenase in a freshwater air-breathing teleost, *Heteropneustes fossilis.* Biochem. Syst. Ecol. 19:207–212.

Das, S.M. and Saxena, D.B. 1956. Circulation of the blood in the respiratory region of the fishes *Labeo rohita* and *Ophicephalus striatus*. Copeia 1956:100–109.

Davenport, J. and Sayer, M.D.J. 1986. Ammonia and urea excretion in the amphibious teleost *Blennius pholis* (L.) in sea-water and in air. Comp. Biochem. Physiol. 84A:189–194.

Davenport, J. and Woolmington, A.D. 1981. Behavioural responses of some rocky shore fish exposed to adverse environmental conditions. Mar. Behav. Physiol. 8:1–12.

David, L. 1935. Die Entwicklung der Clariiden und ihre Verbreitung. Rev. Zool. Bot. Afr. 28:80–140.

Davies, P.J., Hedrick, M.S. and Jones, D.R. 1993. Neuromuscular control of the glottis in a primitive air-breathing fish, *Amia calva*. Am. J. Physiol. 264:R204–R210.

Daxboeck, C. and Heming, T.A. 1982. Bimodal respiration in the intertidal fish, *Xiphister atropurpureus* (Kittlitz). Mar. Behav. Physiol. 9:23–33.

Daxboeck, C., Barnard, D.K. and Randall, D.J. 1981. Functional morphology of the gills of the bowfin *Amia calva* L., with special reference to their significance during air exposure. Respir. Physiol. 43:349–364.

Day, F. 1868. Observations on some freshwater fishes of India. Proc. Zool. Soc., London, 5/14 1868:274–288.

Day, F. 1877. On amphibious and migratory fishes of Asia. Proc. Zool. Soc., London, 1/18 1877:198–215.

de Almeida-Val, V.M.F. and Hochachka, P.W. 1995. Air-breathing fishes: Metabolic biochemistry of the first diving vertebrates. In: Hochachka, P.W. and Mommsen, T.P. (eds.) *Biochemistry and Molecular Biology of Fishes, Vol. 5.* Elsevier, New York, pp. 45–55.

de Beaufort, L.F. 1909. Die Schwimmblase der Malacopterygii. Morphol. Jahrb. 39:526–642.

Dean, B. 1895. *Fishes, Living and Fossil.* Macmillan, New York, 300 pp.

Dehadrai, P.V. 1960. Respiratory function of the swim-bladder of *Notopterus* (Lacepede). Nature 185:929.

Dehadrai, P.V. 1962a. Respiratory function of the swimbladder of *Notopterus* (Lacepede). Proc. Zool. Soc., London. 139:341–357.

Dehadrai, P.V. 1962b. Observations on certain physiological reactions in *Ophiocephalus striatus* exposed to air. Life Sci. 11:653–657.

Dehadrai, P.V. and Tripathi, S.D. 1976. Environment and ecology of freshwater air-breathing teleosts. In: Hughes, G.M. (ed.) *Respiration of Amphibious Vertebrates.* Academic Press, London, pp. 39–72.

Dejours, P. 1981. *Principles of Comparative Respiratory Physiology*, 2d ed. Elsevier/North-Holland, Amsterdam, 265 pp.

Dejours, P. 1988. *Respiration in Water and Air: Adaptations—Regulation—Evolution.* Elsevier, Amsterdam, 179 pp.

Dejours, P. 1994. Environmental factors as determinants in bimodal breathing: An introductory overview. Am. Zool. 34:178–183.

DeLaney, R.G. and Fishman, A.P. 1977. Analysis of lung ventilation in the aestivating lungfish *Protopterus aethiopicus*. Am. J. Physiol. 233:R181–R187.

DeLaney, R.G., Lahiri, S. and Fishman, A.P. 1974. Aestivation of the African lungfish *Protopterus aethiopicus*: Cardiovascular and respiratory functions. J. Exp. Biol. 61:111–128.

DeLaney, R.G., Laurent, P., Galante, R., Pack, A.I. and Fishman, A.P. 1983. Pulmonary mechanoreceptors in the dipnoi lungfish *Protopterus* and *Lepidosiren*. Am. J. Physiol. 244:R418–R428.

DeLaney, R.G., Lahiri, S., Hamilton, R. and Fishman, A.P. 1977. Acid-base balance and plasma composition in the aestivating lungfish (*Protopterus*). Am. J. Physiol. 232:R10–R17.

DeLaney, R.G., Shub, C. and Fishman, A.P. 1976. Hematologic observations on the aquatic and estivating African lungfish, *Protopterus aethiopicus*. Copeia 1976:423–434.

Dence, W.A. 1933. Notes on a large bowfin (*Amia calva*) living in a mud puddle. Copeia 1933:35.

Denison, R.H. 1941. The soft anatomy of *Bothriolepis*. J. Paleontology 15:553–561.

Denison, R.H. 1956. A review of the habitat of the earliest vertebrates. Fieldiana: Geol. 11:359–457.

Denny, M.W. 1993. *Air and Water*. Princeton Univ. Press, New Jersey, 341 pp.

Dépêche, J., Gilles, R., Daufresne, S. and Chiapello, H. 1979. Urea content and urea production via the ornithine-urea cycle pathway during the ontogenic development of two teleost fishes. Comp. Biochem. Physiol. 63A:51–56.

Deyst, K.A. and Liem, K.F. 1985. The muscular basis of aerial ventilation of the primitive lung of *Amia calva*. Respir. Physiol. 59:213–223.

Dheer, J.M.S., Dheer, T.R. and Mahajan, C.L. 1987. Haematological and haematopoietic response to acid stress in an air-breathing freshwater fish *Channa punctatus*. J. Fish Biol. 30:577–588.

Dickson, K.A. and Graham, J.B. 1986. Adaptations to hypoxic environments in the erythrinid fish *Hoplias microlepis*. Environ. Biol. Fishes 15:301–308.

Dobson, G.E. 1874. Notes on the respiration of some species of Indian freshwater fishes. Proc. Zool. Soc., London 1874:312–321.

Donnelly, B.G. 1973. Aspects of behavior in the catfish, *Clarias gariepinus* (Pisces: Clariidae), during periods of habitat desiccation. Arnoldia 6(9):1–8.

Donnelly, B.G. 1978. Evidence of fish survival during habitat dessication (sic) in Rhodesia. J. Limnol. Soc. S. Africa 4:75–76.

Donnelly, D.G. 1966. Shoaling, communication, and social hunting in the catfish *Clarias ngamensis*. Piscator 20:54–55.

Dorn, E. 1968. Schwimmblasenbau und Lebensweise der Osteoglossiden. Verh. Dtsch. Zool. Ges. 31:370–380.

Dorn, E. 1972. Der Feinbau des Mundatmungsorgans von *Electrophorus* (Zitteraal). Verh. Dtsch. Zool. Ges. 66:101–105.

Dorn, E. 1983. Über die Atmungsorgane einiger luftatmender Amazonasfische. Amazoniana 7:375–395.

Driedzic, W.R., Phleger, C.F., Fields, J.H.A. and French, C. 1978. Alterations in energy metabolism associated with the transition from water to air breathing in fish. Can. J. Zool. 56:730–735.

Dubale, M.S. 1951. A comparative study of the extent of gill-surface in some representative Indian fishes, and its bearing on the origin of the air-breathing habit. J. Univ. Bombay (Nat.Sci.) 19:90–101.

Dube, S.C. and Munshi, J.S.D. 1973. A quantitative study of the erythrocyte and hemoglobin in the blood of an air-breathing fish *Anabas testudineus* (Bloch) in relation to its body size. Folia Haemat. 100:436–446.

Dube, S.C. and Munshi, J.S.D. 1974. Studies on the blood-water diffusion barrier of secondary gill lamellae of an air-breathing fish, *Anabas testudineus* (Bloch). Zool. Anz. 193:35–41.

Dunn, J. 1983. Air-breathing fishes. Waters, J. Vancouver Aquarium 6:20–24.

Dunn, J.F., Davison, W., Maloiy, G.M.O., Hochachka, P.W. and Guppy, M. 1981. An ultrastructural and histochemical study of the axial musculature in the African lungfish. Cell Tissue Res. 220:599–609.

Dunn, J.F., Hochachka, P.W., Davison, W. and Guppy, M. 1983. Metabolic adjustments to diving and recovery in the African lungfish. Am. J. Physiol. 245:R651–R657.

Dutta, H.M. and Munshi, J.S.D. 1985. Functional morphology of air-breathing fishes: A review. Proc. Indian Acad. Sci. (Anim. Sci.) 94:359–375.

Eaton, T.H., Jr. 1960. The aquatic origin of tetrapods. Trans. Kansas Acad. Sci. 63(3):115–120.

Ebeling, A.W. and Alpert, J.S. 1966. Retarded growth of the paradisefish, *Macropodus opercularis* (L.), in low environmental oxygen. Copeia 1966:606–610.

Ebeling, A.W., Bernal, P. and Zuleta, A. 1970. Emersion of the amphibious Chilean clingfish *Sicyases sanguineus*. Biol. Bull. 139:115–137.

Eclancher, B. 1975. Contrôle de la respiration chez les poissons téléostéens: Réactions respiratoires à des changements rectangulaires de l'oxygénation du milieu. C. R. Acad. Sci. Paris 280(D):307–310.

Eclancher, B. and Dejours, P. 1975. Contrôle de la respiration chez les poissons téléostéens: Existence de chémorécepteurs physiologiquement analogues aux chémorécepteurs des vertébrés supérieurs. C. R. Acad. Sci., Paris 280(D):451–453.

Eddy, F.B., Bamford, O.S. and Maloiy, G.M.O. 1980. Sodium and chloride balance in the African catfish *Clarias mossambicus*. Comp. Biochem. Physiol. 66A:637–641.

Edwards, D.G. and Cech, Jr., J.J. 1990. Aquatic and aerial metabolism of the juvenile monkeyface prickleback, *Cebidichthys violaceus*, an intertidal fish of California. Comp. Biochem. Physiol. 96A:61–65.

Eger, W.H. 1971. Ecological and physiological adaptations of intertidal clingfishes (*Teleostei, Gobiesocidae*) in the northern Gulf of California. Ph.D. Dissertation, Dept. of Biol. Sci., Univ.of Arizona.

Eggert, B. 1929a. Bestimmungstabelle und Beschreibung des Arten der family *Periophthalmus*. Z. Wiss. Zool. 133:298–410.

Eggert, B. 1929b. Die Gobiidenflosse und ihre Anpassung an das Landleben. Z. Wiss. Zool. 133:411–440.

Eggert, B. 1935. Beitrag zur Systematik, Biologie und geographischen Verbreitung der Periophthalminae. Zool. Jahrb., Abteilung F. Systematik, Okol. und Geogr. 67:29–114.

Eiseley, L. 1957. *The Immense Journey*. Random House, New York, 210 pp.

Eldon, G.A. 1979a. Habitat and interspecific relationships of the Canterbury mudfish, *Neochanna burrowsius* (Salmoniformes: Galaxiidae). New Zealand J. Mar. Freshwater Res. 13:111–119.

Eldon, G.A. 1979b. Breeding, growth, and aestivation of the Canterbury mudfish, *Neochanna burrowsius* (Salmoniformes: Galaxiidae). New Zealand J. Mar. Freshwater Res. 13:331–346.

Eldon, G.A. 1979c. Food of the Canterbury mudfish, *Neochanna burrowsius* (Salmoniformes: Galaxiidae). New Zealand J. Mar. Freshwater Res. 13:255–261.

Eldon, G.A., Howden, P.J. and Howden, D.B. 1978. Reduction of a population of Canterbury mudfish, *Neochanna burrowsius* (Galaxiidae), by drought. New Zealand J. Mar. Freshwater Res. 12:313–321.

Elkan, E. 1968. Mucopolysaccharides in the anuran defence against desiccation. J. Zool., London 155:19–53.

Elsen, M. 1976. La vessie gazeuse et l'organe labyrinthique des *Anabantidae*. Acad. Royale Des Sciences, Belgique, Cl. Des Sci. 62:49–79.

Erman, M. 1808. Untersuchungen uber das Gas in der Schwimmblase der fische, und über die Mitwirkung des Darmkanals zum Respirationsgeschafte bei der Fischart *Cobitis fossilis*, (Schlammpitzger). Annalen der Physik und Chemie 30:113–160.

Evans, M. 1929. Some notes on the anatomy of the electric eel, *Gymnotus electrophorus*, with special reference to a mouth-breathing organ and the swim-bladder. Proc. Zool. Soc. II:17–23.

Fange, R. 1976. Gas exchange in the swimbladder. In: Hughes, G.M. (ed.) *Respiration of Amphibious Vertebrates*. Academic Press, London, pp. 189–211.

Fange, R. 1983. Gas exchange in fish swim bladder. Rev. Physiol. Biochem. Pharmacol. 97:112–158.

Farber, J. and Rahn, H. 1970. Gas exchange between air and water and the ventilation pattern in the electric eel. Respir. Physiol. 9:151–161.

Farmer, M. 1979. The transition from water to air breathing: Effects of CO_2 on hemoglobin function. Comp. Biochem. Physiol. 62A:109–114.

Farmer, M., Fyhn, H.J., Fyhn, U.E.H. and Noble, R.W. 1979. Occurrence of root effect hemoglobins in Amazonian fishes. Comp. Biochem. Physiol. 62A:115–124.

Farrell, A.P. 1978. Cardiovascular events associated with air-breathing in two teleosts, *Hoplerythrinus unitaeniatus* and *Arapaima gigas*. Can. J. Zool. 56:953–958.

Farrell, A.P. and Randall, D.J. 1978. Air-breathing mechanics in two Amazonian teleosts, *Arapaima gigas* and *Hoplerythrinus unitaeniatus*. Can. J. Zool. 56:939–945.

Favaretto, A.L.V., Petenusci, S.O., Lopes, R.A. and Sawaya, P. 1981. Effect of exposure to air on haematological parameters in *Hypostomus regani* (Pisces: Loricariidae), teleost with aquatic and aerial respiration. I. Red cells. Copeia 1981:918–920.

Feder, M.E. and Burggren, W.W. 1985. Cutaneous gas exchange in vertebrates: Design, patterns, control and implications. Bio. Rev. 60:1–45.

Feldman, J.L., Smith, J.C., Ellenberger, H.H., Connelly, C.A., Liu, G., Greer, J.J., Lindsay, A.D. and Otto, M.R. 1990. Neurogenesis of respiratory rhythm and pattern: Emerging concepts. Am. J. Physiol. 259:R879–R886.

Fenwick, J.C. and Lam, T.J. 1988. Calcium fluxes in the teleost fish tilapia (*Oreochromis*) in water and in both water and air in the marble goby (*Oxyeleotris*) and the mudskipper (*Periophthalmodon*). Physiol. Zool. 61:119–125.

Fernandes, M.N. and Perna, S.A., 1995. Internal morphology of the gill of a loricariid fish, *Hypostomus plecostomus*: arterio-arterial vasculature and muscle organization. Can. J. Zool. 73:2259–2265.

Fernandes, M.N., Rantin, F.T., Kalinin, A.L. and Moron, S.E. 1994. Comparative study of gill dimensions in three erythrinid species in relation to their respiratory function. Can. J. Zool. 72:160–165.

Fernandes, M.N., Sanches, J.R. and Rantin, F.T. 1995. Effect of long and short-term changes in water temperature on the cardiac and respiratory responses to hypoxia of the air-breathing fish, *Hypostomus regani*. Physiol. Zool. 68:67.

Fields, J.H.A., Driedzic, W.R., French, C.J. and Hochachka, P.W. 1978a. Some kinetic properties of skeletal muscle pyruvate kinase from air-breathing and water-breathing fish of the Amazon. Can. J. Zool. 56:751–758.

Fields, J.H.A., Driedzic, W.R., French, C.J. and Hochachka, P.W. 1978b. Kinetic properties of glutamate dehydrogenase from the gills of *Arapaima gigas* and *Osteoglossum bicirrhosum*. Can. J. Zool. 56:809–813.

Fish, G.R. 1956. Some aspects of the respiration of six species of fish from Uganda. J. Exp. Biol. 33:186–195.

Fishman, A.P., DeLaney, R.G. and Laurent, P. 1985. Circulatory adaptation to bimodal respiration in the dipnoan lungfish. J. Appl. Physiol. 59:285–294.

Fishman, A.P., Galante, R.J. and Pack, A.I. 1989. Diving physiology: Lungfish. In: Wood, S.C. (ed.) *Comparative Pulmonary Physiology, Current Concepts*. Dekker, New York, pp. 645–676.

Fishman, A.P., Pack, A.I., DeLaney, R.G. and Galante, R.J. 1987. Estivation in *Protopterus*. In: Bemis, W.E., Burggren, W.W. and Kemp, N.E. (eds.) *The Biology and Evolution of Lungfishes*. Alan R. Liss, Inc., New York, pp. 237–248.

Förg. 1853. Remarques sur l'appareil pulmonaire du *Gymnarchus niloticus*. (Extracted from a letter to G.L. Duvernoy communicated to the French Academy by G. Duvernoy.) Annales Sci. Naturelles, Zool. 20:151–162.

Forselius, S. 1957. Studies of anabantid fishes. I. A qualitative description of the reproductive behavior in territorial species investigated under laboratory conditions with special regard to the genus *Colisa*. An introduction. Zool. Bridgrag Uppsala 32:93–301.

Forster, R.P. and Goldstein, L. 1966. Urea synthesis in the lungfish: Relative importance of purine and ornithine cycle pathways. Science 153:1650–1652.

Foxon, G.E.H. 1933a. Pelvic fins of the *Lepidosiren*. Nature 131:732–733.

Foxon, G.E.H. 1933b. Pelvic filaments of *Lepidosiren*. Nature 131:913–914.

Foxon, G.E.H. 1950. A description of the coronary arteries in dipnoan fishes and some remarks on their importance from the evolutionary standpoint. J. Anat. 84:121–131.

Foxon, G.E.H. 1955. Problems of the double circulation in vertebrates. Biol. Rev. 30:196–228.

French, C.J. and Hochachka, P.W. 1978. Lactate dehydrogenase isozymes from heart and white muscle of water-breathing and air-breathing fish from the Amazon. Can. J. Zool. 56:769–773.

Frey, B.J. and van Aardt, W.J. 1995. Haemoglobin multiplicity and temperature sensitivity of the oxygenation reaction of stripped haemolysates and haemoglobins of the mudfish *Labeo capensis* and the catfish *Clarias gariepinus*. Physiol. Zool. 68:98.

Fullarton, M.H. 1931. Notes on the respiration of *Lepidosiren*. Proc. Zool. Soc., London 1931:1301–1306.

Fyhn, U.E.H., Fyhn, H.J., Davis, B.J., Powers, D.A., Fink, W.L. and Garlick, R.L. 1979. Hemoglobin heterogeneity in Amazonian fishes. Comp. Biochem. Physiol. 62A:39–66.

Galdames-Portus, M.I., Noble, R.W., Farmer, M., Powers, D.A., Riggs, A., Brunori, M., Fyhn, H.J. and Fyhn, U.E.H. 1979. Studies of the functional properties of the hemoglobins of *Osteoglossum bicirrhosum* and *Arapaima gigas*. Comp. Biochem. Physiol. 62A:145–154.

Gans, C. 1970. Strategy and sequence in the evolution of the external gas exchangers of ectothermal vertebrates. Forma et Functio 3:61–104.

Garden, M.D. 1775. An account of the *Gymnotus electricus* or electric eel. Phil. Trans. R. Soc., London 95:102–110.

Gardiner, B. 1980. Tetrapod ancestry: A reappraisal. In: Panchen, A.L. (ed.) *The Terrestrial Environment*. Academic Press, London, pp. 177–185.

Garey, W.F. 1962. Cardiac responses of fishes in asphyxic environments. Biol. Bull. 122:362–368.

Garey, W.F. 1972. Determination of normal blood pH of fishes. Respir. Physiol. 14:180–181.

Garey, W.F. and Rahn, H. 1970. Normal arterial gas tensions and pH and the breathing frequency of the electric eel. Respir. Physiol. 9:141–150.

Garlick, R.L., Bunn, H.F., Fyhn, H.J., Fyhn, U.E.H., Martin, J.P., Noble, R.W. and Powers, D.A. 1979. Functional studies on the separated hemoglobin components of an air-breathing catfish, *Hoplosternum littorale* (Hancock). Comp. Biochem. Physiol. 62A:219–226.

Garzon, B. and del Castillo, B. 1986. Observaciones sobre la biologia del capitan *Eremophilus mutisii* (Humboldt 1805), en condiciones de laboratorio con fines pisciculas. Bol. Fac. Biol. Mar. 6:5–8.

Gee, J.H. 1976. Buoyancy and aerial respiration: Factors influencing the evolution of reduced swim-bladder volume in some Central American catfishes (Trichomycteridae, Callichthyidae, Loricariidae, Astroblepidae). Can. J. Zool. 54:1030–1037.

Gee, J.H. 1980. Respiratory patterns and antipredator responses in the central mudminnow *Umbra limi*, a continuous, facultative, air-breathing fish. Can. J. Zool. 58:819–827.

Gee, J.H. 1981. Coordination of respiratory and hydrostatic functions of the swimbladder in the central mudminnow, *Umbra limi*. J. Exp. Biol. 92:37–52.

Gee, J.H. 1986. Buoyancy control of four species of eleotrid fishes during aquatic surface respiration. Environ. Biol. Fishes 16:269–278.

Gee, J.H. and Gee, P.A. 1991. Reactions of gobioid fishes to hypoxia: Buoyancy control and aquatic surface respiration. Copeia 1991:17–28.

Gee, J.H. and Gee, P.A. 1995. Aquatic surface respiration, buoyancy control and the evolution of air-breathing in gobies (Gobiidae: Pisces). J. Exp. Biol. 198:79–89.

Gee, J.H. and Graham, J.B. 1978. Respiratory and hydrostatic functions of the intestine of the catfishes *Hoplosternum thoracatum* and *Brochis splendens* (Callichthyidae). J. Exp. Biol. 74:1–16.

Geoffroy St. Hilaire, E. 1802a. Sur les branchies du *Silurus anguillaris*. Paris Soc. Philom., Bull. III:105–112.

Geoffroy St. Hilaire, E. 1802b. Histoire naturelle et description anatomique d'un nouveau genre de poisson du Nil, nommé Polyptére. Paris Mus. Nat. Hist. Ann. I:57–68.

George, J.C. 1953. Observations on the air-breathing habit in *Chiloscyllium*. J. Univ. Bombay (Biology) 21:5.

Gérard, P. 1931. Les sacs aériens des Crossoptérygiens et les poumons des Dipneuestes. Études anatomique et histologique. Arch. Biol. 42:251–278.

Geyer, F. and Mann, H. 1939a. Beiträge zur Atmung der Fische. I. Die Atmung des Ungarischen Hundsfisches (*Umbra lacustris* Grossinger). Zool. Anz. 127:234–245.

Geyer, F. and Mann, H. 1939b. Beiträge zur Atmung der Fische. II. Weiteres zur Atmung des Ungarischen Hundsfisches (*Umbra lacustris* Grossinger). Zool. Anz. 127:305–313.

Ghosh, E. 1934. An experimental study of the asphyxiation of some air-breathing fishes of Bengal. J. R. Asiatic Soc. Bengal. Sci. 29:327–332.

Ghosh, T.K. 1984. Bimodal gas exchange of some air-breathing teleostean fishes. Ph.D. Thesis, Bhagalpur Univ., Bhagalpur, India.

Ghosh, T.K., Moitra, A., Kunwar, G.K. and Munshi, J.S.D. 1986. Bimodal oxygen uptake in a freshwater air-breathing fish, *Notopterus chitala*. Japan. J. Ichthyol. 33:280–285.

Gibson, R.N. 1969. The biology and behaviour of littoral fish. Oceanogr. Mar. Biol. Ann. Rev. 7:367–410.

Gibson, R.N. 1982. Recent studies on the biology of intertidal fishes. Oceanogr. Mar. Biol. Ann. Rev. 20:363–414.

Glass, M.L., Ishimatsu, A. and Johansen, K. 1986. Responses of aerial ventilation to hypoxia and hypercapnia in *Channa argus*, an air-breathing fish. J. Comp. Physiol. 156:425–430.

Goldstein, L., Janssens, P.A. and Forster, R.P. 1967. Lungfish *Neoceratodus forsteri*: Activities of ornithine-urea cycle and enzymes. Science 157:316–317.

Goodrich, E.S. 1930. *Studies on the Structure and Development of Vertebrates*. Dover, New York, 837 pp. (1958 Reprint).

Goodyear, C.P. 1970. Terrestrial and aquatic orientation in the star-head topminnow, *Fundulus notti*. Science 168:603–605.

Gordon, M.S. 1970. Patterns of nitrogen excretion in amphibious fishes. J. Exp. Biol. 53:559–572.

Gordon, M.S. 1978. Anerobic energy metabolism in the Australian mudskipper fish, *Periophthalmus vulgaris*. Am. Zool. 18:606.

Gordon, M.S. and Olson, E.C. 1994. *The Invasions of the Land*. Columbia University Press, New York, 312 pp.

Gordon, M.S., Bartholomew, G.A., Grinnell, A.D., Jorgenson, C.B. and White, F.N. 1982. *Animal Physiology: Principles and Adaptations*. Macmillan, New York, 4th ed., 635 pp.

Gordon, M.S., Boetius, I., Evans, D.H., McCarthy, R. and Oglesby, L.C. 1969. Aspects of the physiology of terrestrial life in amphibious fishes. I. The mudskipper, *Periophthalmus sobrinus*. J. Exp. Biol. 50:141–149.

Gordon, M.S., Boetius, J., Boetius, I., Evans, D.H., McCarthy, R. and Oglesby, L.C. 1965. Salinity adaptation in the mudskipper fish *Periophthalmus sobrinus*. Hvalradets Skrifter 48:85–93.

Gordon, M.S., Boetius, J., Evans, D.H. and Oglesby, L.C. 1968. Additional observations on the natural history of the mudskipper fish. Copeia 1968:853–857.

Gordon, M.S., Fischer, S. and Tarifeno, E. 1970. Aspects of the physiology of terrestrial life in amphibious fishes. II. The Chilean clingfish, *Sicyases sanguineus*. J. Exp. Biol. 53:559–572.

Gordon, M.S., Gabaldon, D.J. and Yip, A.Y. 1985. Exploratory observations on microhabitat selection within the intertidal zone by the Chinese mudskipper fish *Periophthalmus cantonensis*. Mar. Biol. 85:209–215.

Gordon, M.S., Ng, W.W. and Yip, A.Y. 1978. Aspects of the physiology of the terrestrial life in amphibious fishes. III. The Chinese mudskipper *Periophthalmus cantonensis*. J. Exp. Biol. 72:57–75.

Gordon, M.S., Schmidt-Nielsen, K. and Kelly, H.M. 1961. Osmotic regulation in the crab-eating frog (*Rana cancrivora*). J. Exp. Biol. 38:659–678.

Gradwell, N. 1971. A photographic analysis of the air breathing behavior of the catfish, *Plecostomus punctatus*. Can. J. Zool. 49:1089–1094.

Graham, J.B. 1970. Preliminary studies on the biology of the amphibious clinid *Mnierpes macrocephalus*. Mar. Biol. 5:136–140.

Graham, J.B. 1971. Aerial vision in amphibious fishes. Fauna 3:14–23.

Graham, J.B. 1973. Terrestrial life of the amphibious fish *Mnierpes macrocephalus*. Mar. Biol. 23:83–91.

Graham, J.B. 1976a. Respiratory adaptations of marine air-breathing fishes. In: Hughes, G.M. (ed.) *Respiration of Amphibious Vertebrates*. Academic Press, London, pp. 165–187.

Graham, J.B. 1976b. Hemoglobin concentrations of air breathing fishes. Am. Zool. 16:192.

Graham, J.B. 1983. The transition to air breathing in fishes: II. Effects of hypoxia acclimation on the bimodal gas exchange of *Ancistrus chagresi* (Loricariidae). J. Exp. Biol. 102:157–173.

Graham, J.B. 1985. Seasonal and environmental effects on the blood hemoglobin concentrations of some Panamanian air-breathing fishes. Environ. Biol. Fishes 12:291–301.

Graham, J.B. 1988. Ecological and evolutionary aspects of integumentary respiration: Body size, diffusion, and the invertebrata. Am. Zool. 28:1031–1045.

Graham, J.B. 1990. Ecological, evolutionary, and physical factors influencing aquatic animal respiration. Am. Zool. 30:137–146.

Graham, J.B. 1994. An evolutionary perspective for bimodal respiration: A biological synthesis of fish air breathing. Am. Zool. 34:229–237.

Graham, J.B. and Baird, T.A. 1982. The transition to air breathing in fishes: I. Environmental effects on the facultative air breathing of *Ancistrus chagresi* and *Hypostomus plecostomus* (Loricariidae). J. Exp. Biol. 96:53–67.

Graham, J.B. and Baird, T.A. 1984. The transition to air breathing in fishes: III. Effects of body size and aquatic hypoxia on the aerial gas exchange of the swamp eel *Synbranchus marmoratus*. J. Exp. Biol. 108:357–375.

Graham, J.B. and Rosenblatt, R.H. 1970. Aerial vision: Unique adaptation in an intertidal fish. Science 169:586–588.

Graham, J.B., Baird, T.A. and Stöckmann, W. 1987. The transition to air breathing in fishes: IV. Impact of branchial specializations for air breathing on the aquatic respiratory mechanisms and ventilatory costs of the swamp eel, *Synbranchus marmoratus*. J. Exp. Biol. 129:83–106.

Graham, J.B., Dudley, R., Aguilar, N.M. and Gans, C. 1995b. Implications of the late Palaeozoic oxygen pulse for physiology and evolution. Nature 375:117–120.

Graham, J.B., Jones, C.B. and Rubinoff, I. 1985. Behavioural, physiological, and ecological aspects of the amphibious life of the pearl blenny *Entomacrodus nigricans* (Gill). J. Exp. Mar. Biol. Ecol. 89:255–268.

Graham, J.B., Kramer, D.L. and Pineda, E. 1977. Respiration of the air breathing fish *Piabucina festae*. J. Comp. Physiol. 122:295–310.

Graham, J.B., Kramer, D.L. and Pineda, E. 1978b. Comparative respiration of an air-breathing and a non-air-breathing characoid fish and the evolution of aerial respiration in characins. Physiol. Zool. 51:279–288.

Graham, J.B., Lai, N.C, Chiller, D. and Roberts, J.L. 1995a. The transition to air breathing in fishes: V. Comparative aspects of cardiorespiratory regulation in *Synbranchus marmoratus* and *Monopterus albus* (Synbranchidae). J. Exp. Biol. 198:1455–1467.

Graham, J.B., Rosenblatt, R.H. and Gans, C. 1978a. Vertebrate air breathing arose in fresh waters and not in the ocean. Evolution 32:459–463.

Gray, I.E. 1954. Comparative study of the gill area of marine fishes. Biol. Bull. 107:219–225.

Greenwood, P.H. 1961. A revision of the genus *Dinotopterus* BLGR. (Pisces, Clariidae) with notes on the comparative anatomy of the suprabranchial organs in the Clariidae. Bull. Br. Mus. 7:215–241.

Greenwood, P.H. 1963. The swimbladder in African Notopteridae (Pisces) and its bearing on the taxonomy of the family. Bull. Br. Mus. 11:379–412.

Greenwood, P.H. 1987. The natural history of African lungfishes. In: Bemis, W.E., Burggren, W.W. and Kemp, N.E. (eds.) *The Biology and Evolution of Lungfishes*. Alan R. Liss, Inc., New York, pp. 163–179.

Greenwood, P.H. and Norman, J.R. 1963. *A History of Fishes*. Ernest Benn Limited, London, 398 pp.

Greenwood, P.H. and Liem, K.F. 1984. Aspiratory respiration in *Arapaima gigas* (Teleostei, Osteoglossomorpha): A reappraisal. J. Zool., London 203:411–425.

Gregory, R.B. 1977. Synthesis and total excretion of waste nitrogen by fish of the *Periophthalmus* (mudskipper) and *Scartelaos* families. Comp. Biochem. Physiol. 57A:33–36.

Gregory, W.K. 1963. *Our Face From Fish to Man*. Hafner, New York, 295 pp.

Griffith, R.W. 1987. Freshwater or marine origin of the vertebrates? Comp. Biochem. Physiol. 87A:523–531.

Griffith, R.W. 1991. Guppies, toadfish, lungfish, coelacanths and frogs: A scenario for the evolution of urea retention in fishes. Environ. Biol. Fishes 32:199–218.

Grigg, G.C. 1965. Studies on the Queensland lungfish, *Neoceratodus forsteri* (Kreft). III. Aerial respiration in relation to habits. Aust. J. Zool. 13:413–421.

Grizzle, J.M. and Thiyagarajah, A. 1987. Skin histology of *Rivulus ocellatus marmoratus*: Apparent adaptation for aerial respiration. Copeia 1987:237–240.

Günther, C.A.L.G. 1871. Description of *Ceratodus*, a genus of ganoid fishes, recently discovered in rivers of Queensland, Australia. Phil. Trans. R. Soc., London 6:511–571.

Günther, C.A.L.G. 1880. *An Introduction to the Study of Fishes*. Adam and Charles Black, Edinburgh, 720 pp.

Haddon, A.C. 1889. Zoological notes from the Torres Straits: Caudal respiration in *Periophthalmus*. Nature 39:285.

Hagedorn, M. 1988. Ecological behavior of a pulse-type electric fish, *Hypopomus occidentalis* (Gymnotiformes, Hypopomidae), in a freshwater stream in Panama. Copeia 1988:324–335.

Hakim, A., Munshi, J.S.D. and Hughes, G.M. 1978. Morphometrics of the respiratory organs of the Indian green snake-headed fish, *Channa punctata*. J. Zool., London 184:519–543.

Halliday, T.R. and Sweatman, H.P.A. 1976. To breathe or not to breathe; the newt's problem. Anim. Behav. 24:551–561.

Halstead, L.B. 1968. *The Pattern of Vertebrate Evolution*. W. H. Freeman, San Francisco, 209 pp.

Halstead, L.B. 1985. The vertebrate invasion of fresh water. Phil. Trans. R. Soc., London 309:243–258.

Hamilton Buchanan, F. 1822. *An Account of the Fishes Found in the River Ganges and Its Branches*. Constable, Edinburgh, 405 pp.

Harms, J.W. 1929. Die Realisation von Genen und die consecutive Adaption. 1: Phasen in der Differenzierung der Anlagen-komplexe und die frage der Landtierwerdung. Z. Wiss. Zool. 133:211–397.

Harms, J.W. 1935. Die Realisation von Genen und die consecutive Adaption. 4. Experimentell hervorgerufener Medienwechsel: Wasser zu Feuchtluft bzw. zu Trockenluft bei Gobiiformes (*Gobius*, *Boleophthalmus* und *Periophthalmus*). Z. Wiss. Zool. 146:417-462.

Harris, V.A. 1960. On the locomotion of the mud-skipper, *Periophthalmus koelreuteri* (Pallas): (Gobiidae). Proc. of the Zool. Soc., London 134:107–135.

Hattingh, J. 1972. Observation on the blood physiology of five South African freshwater fish. J. Fish Biol. 4:555–563.

Hedrick, M.S. and Jones, D.R. 1993. The effects of altered aquatic and aerial respiratory gas concentrations on air-breathing patterns in a primitive fish (*Amia calva*). J. Exp. Biol. 181:81–94.

Hedrick, M.S. and Katz, S.L. 1991. A model of aerial ventilation in an air-breathing fish (*Amia calva*). Am. Zool. 31:68A.

Hedrick, M.S., Burleson, M.L., Jones, D.R. and Milsom, W.K. 1991. An examination of central chemosensitivity in an air-breathing fish (*Amia calva*). J. Exp. Biol. 115:167–174.

Hedrick, M.S., Katz, S.L. and Jones, D.R. 1994. Periodic air-breathing behaviour in a primitive fish revealed by spectral analysis. J. Exp. Biol. 197:429–436.

Heisler, N. 1977. Acid-base equilibrium in a facultative air breathing fish (*Synbranchus marmoratus*) during water breathing and during air breathing. Pflugers Arch. Suppl. 368:R19.

Heisler, N. 1982. Intracellular and extracellular acid-base regulation in the tropical fresh-water teleost fish, *Synbranchus marmoratus* in response to the transition from water breathing to air breathing. J. Exp. Biol. 99:9–28.

Heisler, N. 1986. Mechanisms and limitations of fish acid-base regulation. In: Nilson, S. and Holmgren, S. (eds.) *Fish Physiology: Recent Advances*. Croom-Helm, London, pp. 24–49.

Heisler, N. 1989. Interactions between gas exchange, metabolism, and ion transport in animals: An overview. Can. J. Zool. 67:2923–2935.

Henderson, V. 1975. Scanning electron microscopy of the airbladder in the spotted gar, *Lepisosteus oculatus*. J. Morphol. 147:293–298.

Henninger, G. 1907. Die Labyrinthorgane bei Labyrinthfischen. Zool. Jahrb. 25:252–304.

Herald, E.S. 1961. *Living Fishes of the World*. Doubleday and Company, New York, 304 pp.

Herald, E.S. and Herald, O.W. 1973. The Galapagos four-eyed blenny. Pacific Disc. 26:28–30.

Heymer, A. 1982. Le comportement pseudo-amphibie de *Coryphoblennius galerita* et *Blennius trigloides*. Rev. Francaise Aquariol. 9:91–96.

Hickson, S.J. 1889. *A Naturalist in North Celebes*. Murray, London, 392 pp.

Hill, L.G. 1972. Social aspects of aerial respiration of young gars (*Lepisosteus*). Southwest Nat. 16:239–247.

Hill, L.G., Renfro, J.L. and Reynolds, R. 1972. Effects of dissolved oxygen tensions upon the rate of aerial respiration of young spotted gar *Lepisosteus oculatus* (Lepisosteidae). Southwest Nat. 17:273–278.

Hill, L.G., Schnell, G.D. and Echelle, A.A. 1973. Effects of dissolved oxygen concentrations on locomotory reactions of the spotted gar, *Lepisosteus oculatus* (Pisces: Lepisosteidae). Copeia 1973:119–124.

Hillman, S.S. and Withers, P.C. 1987. Oxygen consumption during aerial activity in aquatic and amphibious fish. Copeia 1987:232–234.

Hochachka, P.W. 1980. *Living Without Oxygen*. Harvard Univ. Press, Cambridge, 181 pp.

Hochachka, P.W. and Hulbert, W.C. 1978. Glycogen "seas," glycogen bodies, and glycogen granules in heart and skeletal muscle of two air-breathing, burrowing fishes. Can. J. Zool. 56:774–786.

Hochachka, P.W. and Somero, G.N. 1984. *Biochemical Adaptation*. Princeton Univ. Press, Princeton, 537 pp.

Hochachka, P.W., Guppy, M., Guderley, H.E., Storey, K.B. and Hulbert, W.C. 1978a. Metabolic biochemistry of water- vs. air-breathing fishes: Muscle enzymes and ultrastructure. Can. J. Zool. 56:736–750.

Hochachka, P.W., Guppy, M., Guderley, H., Storey, K.B. and Hulbert, W.C. 1978b. Metabolic biochemistry of water- vs. air-breathing osteoglossids: Heart enzymes and ultrastructure. Can. J. Zool. 56:759–768.

Hochachka, P.W., Moon, T.W., Bailey, J. and Hulbert, W.C. 1978c. The osteoglossid kidney: Correlations of structure, function, and metabolism with transition to air breathing. Can. J. Zool. 56:820–832.

Hoda, S.M.S. and Akhtar, Y. 1985. Maturation and fecundity of mudskipper *Boleophthalmus dentatus* in the northern Arabian Sea. Indian J. of Fish. 32:64–74.

Holeton, G.F. 1980. Oxygen as an environmental factor of fishes. In: Ali, M.A. (ed.) *Environmental Biology of Fishes*. Plenum, New York, pp. 7–32.

Hora, S.L. 1932. A marine air-breathing fish, *Andamia heteroptera* (Bleeker). Curr. Sci. I:51.

Hora, S.L. 1933. Respiration in fishes. J. Bombay Nat. Hist. Soc. 36:538–560.

Hora, S.L. 1935. Physiology, bionomics, and evolution of air breathing fishes of India. Trans. Natl. Inst. Sci. India I:1–16.

Hora, S.L. and Law, N.C. 1942. Respiratory adaptations of the south Indian homalopterid fishes. J. R. Asiatic Soc. Bengal. Sci. 8:39–48.

Horn, M.H. and Gibson, R.N. 1988. Intertidal fishes. Sci. Am. 256:64–70.

Horn, M.H. and Riegle, K.C. 1981. Evaporative water loss and intertidal vertical distribution in relation to body size and morphology of stichaeoid fishes from California. J. Exp. Mar. Biol. Ecol. 50:273–288.

Horn, M.H. and Riggs, C.D. 1973. Effects of temperature and light on the rate of air-breathing of the bowfin, *Amia calva*. Copeia 1973:653–657.

Howell, B.J. 1970. Acid-base balance in transition from water breathing to air breathing. Am. Physiol. Soc. Sym., Fed. Proc. 29:1130–1134.

Huebner, E. and Chee, G. 1978. Histological and ultrastructural specialization of the digestive tract of the intestinal air-breather *Hoplosternum thoracatum* (Teleost). J. Morphol. 157:301–328.

Huehner, M.K., Schramm, M.E. and Hens, M.D. 1985. On the behavior and ecology of the killifish *Rivulus marmoratus* Poey 1880 (Cyprinodontidae). Florida Sci. 48:1–7.

Huggins, A.K., Skutsch, G. and Baldwin, E. 1969. Ornithine-urea cycle enzymes in teleostean fish. Comp. Biochem. Physiol. 28:587–602.

Hughes, G.M. 1963. *Comparative Physiology of Vertebrate Respiration*. Harvard University Press, Cambridge, 145 pp.

Hughes, G.M. 1972. Morphometrics of fish gills. Respir. Physiol. 14:1–25.

Hughes, G.M. 1973a. Ultrastructure of the lung of *Neoceratodus* and *Lepidosiren* in relation to the lung of other vertebrates. Folia Morphol. 21:155–161.

Hughes, G.M. 1973b. Respiratory responses to hypoxia in fish. Am. Zool. 13:475–489.

Hughes, G.M. 1976. *Respiration of Amphibious Vertebrates*. Academic Press Inc., London, 402 pp.

Hughes, G.M. 1978. A morphological and ultrastructural comparison of some vertebrate lungs. In: Klika, E. (ed.) *XIX. Congressus Morphologicus Symposia*. Charles Univ. Press, Prague, pp. 393–405.

Hughes, G.M. 1984. Measurement of gill area in fishes: Practices and problems. J. Mar. Biol. Assn. U.K. 64:637–655.

Hughes, G.M. and Al-Kadhomiy, N.K. 1986. Gill morphometry of the mudskipper *Boleophthalmus boddarti*. J. Mar. Biol. Assn. U.K. 66:671–682.

Hughes, G.M. and Morgan, M. 1973. The structure of fish gills in relation to their respiratory function. Biol. Rev. 48:419–475.

Hughes, G.M. and Munshi, J.S.D. 1968. Fine structure of the respiratory surfaces of an air-breathing fish, the climbing perch *Anabas testudineus* (Bloch). Nature 219:1382–1384.

Hughes, G.M. and Munshi, J.S.D. 1973a. Nature of the air-breathing organs of the Indian fishes *Channa, Amphipnous, Clarias* and *Saccobranchus* as shown by electron microscopy. J. Zool., London 170:245–270.

Hughes, G.M. and Munshi, J.S.D. 1973b. Fine structure of the respiratory organs of the climbing perch, *Anabas testudineus* (Pisces, Anabantidae). J. Zool., London 170:201–225.

Hughes, G.M. and Munshi, J.S.D. 1978. Scanning electron microscopy of the respiratory surfaces of *Saccobranchus (= Heteropneustes) fossilis* (Bloch). Cell Tissue Res. 195:99–109.

Hughes, G.M. and Munshi, J.S.D. 1979. Fine structure of the gills of some Indian air-breathing fishes. J. Morphol. 160:169–194.

Hughes, G.M. and Munshi, J.S.D. 1986. Scanning electron microscopy of the accessory respiratory organs of the snake-headed fish, *Channa striata* (Bloch) (Channidae, Channiformes). J. Zool., London 209:305–317.

Hughes, G.M. and Singh, B.N. 1970a. Respiration in an air-breathing fish, the climbing perch *Anabas testudineus* Bloch. I. Oxygen uptake and carbon dioxide release into air and water. J. Exp. Biol. 53:265–280.

Hughes, G.M. and Singh, B.N. 1970b. Respiration in an air-breathing fish, the climbing perch *Anabas testudineus*. II. Respiratory patterns and the control of breathing. J. Exp. Biol. 53:281–298.

Hughes, G.M. and Singh, B.N. 1971. Gas exchange with air and water in an air-breathing catfish, *Saccobranchus (= Heteropneustes) fossilis*. J. Exp. Biol. 55:667–682.

Hughes, G.M. and Weibel, E.R. 1976. Morphometry of fish lungs. In: Hughes, G.M. (ed.) *Respiration of Amphibious Vertebrates*. Academic Press, London, pp. 213–232.

Hughes, G.M. and Weibel, E.R. 1978. Visualization of layers lining the lung of the South American lungfish (*Lepidosiren paradoxa*) and a comparison with the frog and rat. Tissue and Cell 10:343–353.

Hughes, G.M., Dube, S.C. and Munshi, J.S.D. 1973b. Surface area of the respiratory organs of the climbing perch, *Anabas testudineus* (Pisces: Anabantidae). J. Zool., London 170:227–243.

Hughes, G.M., Ryan, J.W. and Ryan, N. 1973a. Freeze-fractured lamellate bodies of *Protopterus* lung: A comparative study. J. Physiol. 236:15–16.

Hughes, G.M., Singh, B.R., Guha, G., Dube, S.C. and Munshi, J.S.D. 1974b. Respiratory surface areas of an air-breathing siluroid fish *Saccobranchus (= Heteropneustes) fossilis* in relation to body size. J. Zool., London 172:215–232.

Hughes, G.M., Singh, B.R., Thakur, R.N. and Munshi, J.S.D. 1974a. Areas of the air-breathing surfaces of *Amphipnous cuchia* (Ham.). Proc. Indian Nat. Sci. Acad. 40:379–392.

Hulbert, W.C., Moon, T.W. and Hochachka, P.W. 1978a. The osteoglossid gill: Correlations of structure, function, and metabolism with transition to air breathing. Can. J. Zool. 56:801–808.

Hulbert, W.C., Moon, T.W. and Hochachka, P.W. 1978b. The erythrinid gill: Correlations of structure, function, and metabolism. Can. J. Zool. 56:814–819.

Hulbert, W.C., Moon, T.W., Bailey, J. and Hochachka, P.W. 1978c. The occurrence and possible significance of chloride like cells in the nephron of Amazon fishes. Can. J. Zool. 56:833–844.

Hunter, J. 1861. *Essays and Observations on Natural History, Anatomy, Physiology, Phycology and Geology*, R. Owen (ed.). Hunterian Lectures, British Museum of Natural History, London.

Hyrtl, J. 1845. *Lepidosiren paradoxa*. Bohm. Gesell. Abh. 3:605–668.

Hyrtl, J. 1852a. Über die Schwimmblase von *Lepidosteus osseus*. Wein. Sitz. Ber. Akad. 8:71–72.

Hyrtl, J. 1852b. Ueber das Arterien-System des *Lepidosteus*. Wein. Sitz. Ber. Akad. 8:234–241.

Hyrtl, J. 1853. Über das Labyrinth und die Aortenbogen der Gattung *Ophiocephalus*. Wein. Sitz. Ber. Akad. 10:148–153.

Hyrtl, J. 1854a. Beitrag zur Anatomie von *Heterotis ehrenbergii* C. V. Wien. Denkschr. 8:73–88.

Hyrtl, J. 1854b. Zur Anatomie von *Saccobranchus singio* C. V. Wein. Sitz. Ber. Akad. 11:302–307.

Hyrtl, J. 1856. Anatomische Mittheilungen über *Mormyrus* und *Gymnarchus*. Wien. Denkschr. 12:1–22.

Hyrtl, J. 1858. Über den Amphibienkreislauf von *Amphipnous* und *Monopterus*. Wien Denkschr. 14:39–49.

Hyrtl, J. 1863. Über eine neue Rippenart und uber das Labyrinth von *Polyacanthus hasselti*. Wien. Denkschr. 21:11–16.

Hyrtl, J. 1870. Über die Blutgefass der aussern Kiemendeckelkieme von *Polypterus lapradei* Steindachner. Wien. Akad. Sitzungsb. 60:109–113.

Inger, R.F. 1952. Walking fishes of southeastern Asia travel on land. Chicago Nat. Hist. Mus. Bull. 23:4–5,7.

Inger, R.F. 1957. Ecological aspects of the origin of the tetrapods. Evolution 11:373–376.

Innes, A.J. and Wells, R.M.G. 1985. Respiration and oxygen transport functions of the blood from an intertidal fish, *Helcogramma medium* (Tripterygiidae). Environ. Biol. Fishes 14:213–226.

Ip, Y.K., Chew, S.F. and Lim, R.W.L. 1990. Ammoniagenesis in the mudskipper, *Periophthalmus chrysospilos*. Zool. Sci. 7:187–194.

Ip, Y.K., Chew, S.F. and Low, W.P. 1991. Effects of hypoxia on the mudskipper, *Periophthalmus chrysospilos* (Bleeker, 1853). J. Fish Biol. 38:621–623.

Isaacks, R.E., Kim, H.D. and Harkness, D.R. 1978a. Relationship between phosphorylated metabolic intermediates and whole blood oxygen affinity in some air-breathing and water-breathing teleosts. Can. J. Zool. 56:887–891.

Isaacks, R.E., Kim, H.D. and Harkness, D.R. 1978b. Inositol diphosphate in erythrocytes of the lungfish, *Lepidosiren paradoxa*, and 2,3-diphosphoglycerate in erythrocytes of the armored catfish, *Pterygoplichthys* sp. Can. J. Zool. 56:1014–1016.

Isbrucker, I.J.H. 1980. Classification and catalogue of the mailed *Loricariidae*. Verslagen en technische Cegevens. lst Voor Tax. Zool., Univ. van Amsterdam, No. 22.

Ishimatsu, A. and Itazawa, Y. 1981. Ventilation of the air-breathing organ in the snakehead *Channa argus*. Japan. J. Ichthyol. 28:276–282.

Ishimatsu, A. and Itazawa, Y. 1983a. Difference in blood oxygen levels in the outflow vessels of the heart of an air-breathing fish, *Channa argus*: Do separate blood streams exist in a teleostean heart? J. Comp. Physiol. 149:435–440.

Ishimatsu, A. and Itazawa, Y. 1983b. Blood oxygen levels and acid-base status following air exposure in an air-breathing fish *Channa argus*. Comp. Biochem. Physiol. 74A:787–793.

Ishimatsu, A. and Itazawa, Y. 1993. Anatomy and physiology of the cardiorespiratory system in air-breathing fish, *Channa argus*. In: Singh, B.R. (ed.) *Advances in Fish Research Vol. 1*. Narendra, Delhi, pp. 55–70.

Ishimatsu, A., Itazawa, Y. and Takeda, T. 1979. On the circulatory systems of the snakeheads *Channa maculata* and *C. argus* with reference to bimodal breathing. Japan. J. Ichthyol. 26:167–180.

Itazawa, Y. and Ishimatsu, A. 1981. Gas exchange in an air-breathing fish, the snakehead *Channa argus* in normoxic and hypoxic water and in air. Bull. Japan. Soc. Sci. Fish. 47:829–834.

Iwata, K. 1988. Nitrogen metabolism in the mudskipper, *Periophthalmus cantonensis*: Changes in free amino acids and related compounds in various tissues under conditions of ammonia loading, with special reference to its high ammonia tolerance. Comp. Biochem. Physiol. 91A:499–508.

Iwata, K. and Kakuta, I. 1983. A comparison of catalytic properties of glutamate dehydrogenase from liver and muscle between amphibious *Periophthalmus cantonensis* and water-breathing gobid fishes *Tridentiger obscurus obscurus*. Bull. Japan. Soc. Sci. Fish. 49:1903–1908.

Iwata, K., Kakuta, I., Ikeda, M., Kimoto, S. and Wada, N. 1981. Nitrogen metabolism in the mudskipper, *Periophthalmus cantonensis*, a role of free amino acids in detoxification of ammonia produced during its terrestrial life. Comp. Biochem. Physiol. 68A:589–596.

Jacob, S.S., Nair, N.B. and Balasubramanian, N.K. 1982. Toxicity of certain pesticides found in the habitat to the larvivorous (sic) fishes *Aplocheilus lineatus* (Cuv. & Val.) and *Macropodus cupanus* (Cuv. & Val.). Proc. Indian Acad. Sci. 91:323–328.

Jakubowski, M. 1958. The structure and vascularization of the skin of the pond-loach (*Misgurnus fossilis* L.). Acta Biol. Cracoviensia, Zool. 1:113–127.

Jakubowski, M. 1960. The structure and vascularization of the skin of the eel (*Anguilla anguilla* L.) and the viviparous blenny (*Zoarces viviparus* L.). Acta Biol. Cracoviensia, Zool. 3:1–22.

Janssens, P.A. 1964. The metabolism of the aestivating African lungfish. Comp. Biochem. Physiol. 11:105–117.

Janssens, P.A. 1972. The influence of ammonia on the transition to ureotelism in *Xenopus laevis*. J. Exp. Zool. 182:357–366.

Janssens, P.A. and Cohen, P.P. 1966. Ornithine-urea cycle enzymes in the African lungfish *Protopterus aethiopicus*. Science 152:358–359.

Janssens, P.A. and Cohen, P.P. 1968a. Nitrogen metabolism in the African lungfish. Comp. Biochem. Physiol. 24:879–886.

Janssens, P.A. and Cohen, P.P. 1968b. Biosynthesis of urea in the estivating African lungfish and in *Xenopus laevis* under conditions of water-shortage. Comp. Biochem. Physiol. 24:887–898.

Jasinski, A. 1965. The vascularization of the air bladder in fishes. II. Sevruga (*Acipsenser stellatus* Pallas), grayling (*Thymallus thymallus* L.), pike (*Esox lucius*) and umbra (*Umbra krameri* Walbaum). Acta Biol. Cracoviensia, Zool. 8:199–210.

Jensen, F.B. 1991. Multiple strategies in oxygen and carbon dioxide transport by haemoglobin. In: Woakes, A.J., Grieshaber, M.K. and Bridges, C.R. (eds.) *Physiological Strategies for Gas Exchange and Metabolism*. SEB Ser. 41. Cambridge University Press, New York, pp. 55–78.

Jesse, M.J., Shub, C. and Fishman, A.P. 1967. Lung and gill ventilation of the African lungfish. Respir. Physiol. 3:267–287.

Jeuken, M. 1957. *A Study of the Respiration of Misgurnus fossilis (L.), the Pond Loach*. Uitgeverij Excelsior, Gravenhage, 114 pp.

Jionghua, P., Chenghang, L., Wenbiao, Z., Wensheng, L. and Lixu, F. 1988. The morphological structure of the swim bladder of *Pangasius sutchi* and its air-breathing. Zool. Res. 9:87–93.

Job, T.J. 1941. Life-history and bionomics of the spiny eel, *Mastacembelus pancalas* (Ham.) with notes on the systematics of the Mastacembelidae. Rec. Indian Mus. 43:121–135.

Jobert, C. 1877. Recherches pour servir à l'histoire de la respiration chez les poissons. Ann. Sci. Nat. (Zool.) 5(8):1–4.

Jobert, C. 1878. Recherches anatomiques et physiologiques pour servir à l'histoire de la respiration chez les poissons. Ann. Sci. Nat. (Zool.) 7(5):1–7.

Johansen, K. 1966. Air breathing in the teleost *Symbranchus marmoratus*. Comp. Biochem. Physiol. 18:383–395.

Johansen, K. 1968. Air-breathing fishes. Sci. Am. 219:102–111.

Johansen, K. 1970. Air breathing in fishes. In: Hoar, W.S. and Randall, D.J. (eds.) *Fish Physiology Vol. IV*. Academic Press, New York, pp. 361–411.

Johansen, K. 1972. Heart and circulation in gill, skin and lung breathing. Respir. Physiol. 14:193–210.

Johansen, K. and Lenfant, C. 1967. Respiratory function in the South American lungfish, *Lepidosiren paradoxa* (Fitz). J. Exp. Biol. 46:205–218.

Johansen, K. and Lenfant, C. 1968. Respiration in the African lungfish. *Protopterus aethiopicus*. II. Control of breathing. J. Exp. Biol. 49:453–468.

Johansen, K. and Lenfant, C. 1972. A comparative approach to the adaptability of O_2–Hb affinity. Proc. A. Benzon Symp. IV., Copenhagen, pp. 750–780.

Johansen, K. and Reite, O.B. 1967. Effects of acetylcholine and biogenic amines on pulmonary smooth muscle in the African lungfish, *Protopterus aethiopicus*. Acta Physiol. Scand. 71:248–252.

Johansen, K., Hanson, D. and Lenfant, C. 1970a. Respiration in a primitive air breather, *Amia calva*. Respir. Physiol. 9:162–174.

Johansen, K., Lenfant, C. and Grigg, G.C. 1967. Respiratory control in the lungfish, *Neoceratodus forsteri* (Krefft). Comp. Biochem. Physiol. 20:835–854.

Johansen, K., Lenfant, C. and Hanson, D. 1968a. Cardiovascular dynamics in lungfishes. Z. Vergl. Physiologie 59:157–186.

Johansen, K., Lenfant, C. and Hanson, D. 1970b. Phylogenetic development of pulmonary circulation. Proc. Fed. Am. Soc. Exp. Biol. 29:1135–1140.

Johansen, K., Lenfant, C., Schmidt-Nielsen, K. and Petersen, J.A. 1968b. Gas exchange and control of breathing in the electric eel, *Electrophorus electricus*. Z. Vergl. Physiol. 61:137–163.

Johansen, K., Lykkeboe, G., Weber, R.E. and Maloiy, G.M.O. 1976. Respiratory properties of blood in awake and estivating lungfish, *Protopterus amphibius*. Respir. Physiol. 27:335–345.

Johansen, K., Mangum, C.P. and Lykkeboe, G. 1978b. Respiratory properties of the blood of Amazon fishes. Can. J. Zool. 56:898–906.

Johansen, K., Mangum, C.P. and Weber, R.E. 1978a. Reduced blood O_2 affinity associated with air breathing in osteoglossid fishes. Can. J. Zool. 56:891–897.

Johnston, I.A., Bernard, L.M. and Maloiy, G.M. 1983. Aquatic and aerial respiration rates, muscle capillary supply and mitochondrial volume density in the air-breathing catfish (*Clarias mossambicus*) acclimated to either aerated or hypoxic water. J. Exp. Biol. 105:317–338.

Jolyet, F. and Regnard, P. 1877a. Recherches physiologiques sur la respiration des animaux aquatiques, I. Arch. de Physiol. 2(4):44–62.

Jolyet, F. and Regnard, P. 1877b. Recherches physiologiques sur la respiration des animaux aquatiques, II. Arch. de Physiol. 2(4):584–633.

Jones, D.R. and Milsom, W.K. 1982. Peripheral receptors affecting breathing and cardiovascular function in non-mammalian vertebrates. J. Exp. Biol. 100:59–91.

Jordan, J. 1976. The influence of body weight on gas exchange in the air-breathing fish, *Clarias batrachus*. Comp. Biochem. Physiol. 53A:305–310.

Joshi, B.D. and Tandon, R.S. 1977. Seasonal variations in haematologic values of fresh water fishes. I. *Heteropneustes fossilis* and *Mystus vittatus*. Comp. Physiol. Ecol. 2:22–26.

Karsten, H. 1923. Das Auge von *Periophthalmus koelreuteri*. Z. Naturwiss. 59:115–154.

Katz, S.L. 1996. Ventilatory control in a primitive fish: Signal conditioning via non-linear O_2 affinity. Respir. Physiol. 103:165–175.

Kawamoto, N., Truong, N.V. and Tuy-Hoa, T.T. 1972. Illustrations of some freshwater fishes on the Mekong Delta, Vietnam. Cont. Fac. Agricult., Univ. Cantho. 1:1–53.

Kemp, A. 1987. The biology of the Australian lungfish, *Neoceratodus forsteri*. In: Bemis, W.E., Burggren, W.W. and Kemp, N.E. (eds.) *The Biology and Evolution of Lungfishes*. Alan R. Liss, Inc., New York, pp. 181–198.

Kemp, A. 1996. Role of epidermal cilia in development of the Australian lungfish, *Neoceratodus forsteri* (Osteichthyes: Dipnoi) J. Morph. 228: 203–221.

Kerr, J.G. 1900. The external features in the development of *Lepidosiren paradoxa*, Fitz. Phil. Trans. R. Soc., London 192B:299–330.

Kerr, J.G. 1907. The development of *Polypterus senegalus* Cuv. In: Kerr, J.G. *The Work of John Samuel Budgett*, Budgett Memorial Vol., Edinburgh, pp. 195–290.

Kerr, J.G. 1908. Note on swim-bladder and lungs. Proc. R. Phys. Soc., Edinburgh 17:170–174.

Kerr, J.G. 1910. Note on the posterior vena cava in *Polypterus*. Proc. R. Phys. Soc., Edinburgh 18:102–104.

Kiley, J.P., Eldridge, F.L. and Millhorn, D.E. 1985. The roles of medullary extracellular and cerebrospinal fluid pH in control of respiration. Respir. Physiol. 59:117–130.

Kirsch, R. and Nonnotte, G. 1977. Cutaneous respiration in three freshwater teleosts. Respir. Physiol. 29:339–354.

Kitzan, S.M. and Sweeny, P.R. 1968. A light and electron microscope study of the structure of *Protopterus annectens* epidermis. I. Mucus production. Can. J. Zool. 46:767–772.

Klausewitz, W. 1964. Fische aus dem Roten Meer. VI. Taxionomische und Ökologische Untersuchungen an einigen Fischarten der Küstenzone. Senck. Biol. 45:123–144.

Klausewitz, W. 1967. Über einige Bewegungsweisen der Schlammspringer (*Periophthalmus*). Natur Mus. 97(6):211–222.

Klausewitz, W. 1968a. Some observations on East African mudskippers. Trop. Fish Hobbyist 16:52–65.

Klausewitz, W. 1968b. Wasserscheue fische. Aquarien Mag. 1:6–11.

Klika, E. and Lelek, A. 1967. A contribution to the study of the lungs of the *Protopterus annectens* and *Polypterus senegalensis*. Folia Morphol. 15:168–175.

Klinger, S., Magnuson, J.J. and Gallepp, G.W. 1982. Survival mechanisms of the central mudminnow (*Umbra limi*), fathead minnow (*Pimephales promelas*) and brook stickleback (*Culaea inconstans*) for low oxygen in winter. Environ. Biol. Fishes 7:113–120.

Knoppel, H. 1970. Food of the central Amazonian fishes. Amazoniana 2:257–352.

Kopp, C. 1978. Anatomische und histologische Untersuchungen des Darmtractus der luftatmenden Panzerwelse *Dianema longibarbis* und *Brochis coeruleus* (Teleostei: Siluriformes). Staatsexamenarbeit, Mainz.

Kormanik, G.A. and Evans, D.H. 1988. Nitrogenous waste excretion in the intertidal rock gunnel (*Pholis gunnellus* L.): The effects of emersion. Bull. Mt. Desert Isl. Biol. Lab. 27:33–35.

Kramer, D.L. 1978a. Ventilation of the respiratory gas bladder in *Hoplerythrinus unitaeniatus* (Pisces, Characoidei, Erythrinidae). Can. J. Zool. 56:931–938.

Kramer, D.L. 1978b. Reproductive seasonality in the fishes of a tropical stream. Ecology 59:976–985.

Kramer, D.L. 1978c. Terrestrial group spawning of *Brycon petrosus* (Pisces: Characidae) in Panama. Copeia 1978:536–537.

Kramer, D.L. 1983a. Aquatic surface respiration in the fishes of Panama: Distribution in relation to risk of hypoxia. Environ. Biol. Fishes 8:49–54.

Kramer, D.L. 1983b. The evolutionary ecology of respiratory mode in fishes: An analysis based on the costs of breathing. Environ. Biol. Fishes 9:145–158.

Kramer, D.L. 1987. Dissolved oxygen and fish behavior. Environ. Biol. Fishes 18:81–92.

Kramer, D.L. 1988. The behavioral ecology of air breathing by aquatic animals. Can. J. Zool. 66:89–94.

Kramer, D.L. and Braun, E.A. 1983. Short-term effects of food availability on air-breathing frequency in the fish *Corydoras aeneus* (Callichthyidae). Can. J. Zool. 61:1964–1967.

Kramer, D.L. and Graham, J.B. 1976. Synchronous air breathing, a social component of respiration in fishes. Copeia 1976:689–697.

Kramer, D.L. and McClure, M. 1980. Aerial respiration in the catfish, *Corydoras aeneus* (Callichthyidae). Can. J. Zool. 58:1984–1991.

Kramer, D.L. and McClure, M. 1981. The transit cost of aerial respiration in the catfish *Corydoras aeneus* (Callichthyidae). Physiol. Zool. 54:189–194.

Kramer, D.L. and McClure, M. 1982. Aquatic surface respiration, a widespread adaptation to hypoxia in tropical freshwater fishes. Environ. Biol. Fishes 7:47–55.

Kramer, D.L. and Mehegan, J.P. 1981. Aquatic surface respiration, an adaptive response to hypoxia in the guppy, *Poecilia reticulata* (Pisces, Poeciliidae). Environ. Biol. Fishes 6:299–313.

Kramer, D.L., Lindsey, C.C., Moodie, G.E.E. and Stevens, E.D. 1978. The fishes and the aquatic environment of the central Amazon basin, with particular reference to respiratory patterns. Can. J. Zool. 56:717–729.

Kramer, D.L., Manley, D. and Bourgeois, R. 1983. The effect of respiratory mode and oxygen concentration on the risk of aerial predation in fishes. Can. J. Zool. 61:653–665.

Krogh, A. 1904a. Some experiments on the cutaneous respiration of vertebrate animals. Skand. Arch. Physiol. 16:348–357.

Krogh, A. 1904b. On the cutaneous and pulmonary respiration of the frog. Skand. Arch. Physiol. 15:328–419.

Krogh, A. 1941. *The Comparative Physiology of Respiratory Mechanisms*. Univ. of Pennsylvania Press, Philadelphia, 172 pp.

Krogh, A. and Leitch, I. 1919. The respiratory function of the blood in fishes. J. Physiol. 52:288–300.

Kulakkattolickal, A.T. and Kramer, D.L. 1988. The role of air breathing in the resistance of bimodally respiring fish to waterborne toxins. J. Fish Biol. 32:119–127.

Kushlan, J.A. 1973. Differential responses to drought in two species of *Fundulus*. Copeia 1973:808–809.

Kuttil, A.J. 1963. Studies on the respiration of air breathing fishes. Doctoral Thesis, Univ. Madras, 117 pp.

Lachner, E.A., Robins, C.R. and Courtenay, Jr., W.R. 1970. Exotic fishes and other aquatic organisms introduced into North America. Smithsonian Contrib. Zool. 59:1–29.

Lahiri, S., Szidon, J.P. and Fishman, A.P. 1970. Potential respiratory and circulatory adjustments to hypoxia in the African lungfish. Fed. Proc. 29:1141–1148.

Laming, P.R., Funston, C.W., Roberts, D. and Armstrong, M.J. 1982. Behavioural, physiological and morphological adaptations of the shanny (*Blennius pholis*) to the intertidal habitat. J. Mar. Biol. Assn. U.K. 62:329–338.

Land, M.F. 1987. Vision in air and water. In: Dejours, P., Bolis, L., Taylor, C.R. and Weibel, E.R. (eds.) *Comparative Physiology: Life in Water and on Land*. Springer-Verlag, New York, pp. 289–302.

Landolt, J.C. and Hill, L.G. 1975. Observations of the gross structure and dimensions of the gills of three species of gars (Lepisosteidae). Copeia 1975:470–475.

Larson, H.K. 1983. Notes on the biology of the goby *Kelloggella cardinalis* (Jordan and Seale). Micronesica 19:157–164.

Lauder, G.V. and Liem, K.F. 1981. Prey capture by *Luciocephalus pulcher*: Implications for models of jaw protrusion in teleost fishes. Environ. Biol. Fishes 6:257–268.

Lauder, G.V. and Liem, K.F. 1983a. The evolution and interrelationships of the actinopterygian fishes. Bull. Mus. Comp. Zool. 150:95–197.

Lauder, G.V. and Liem, K.F. 1983b. Patterns of diversity and evolution in ray-finned fishes. In: Davis, R. and Northcutt, R.G. (eds.) *Fish Neurobiology and Behavior*. Univ. of Michigan Press, Ann Arbor, pp. 2–24.

Laurent, P., DeLaney, R.G. and Fishman, A.P. 1978. The vasculature of the gills in the aquatic and aestivating lungfish (*Protopterus aethiopicus*). J. Morphol. 156:173–208.

Laurent, P., Holmgren, S. and Nilsson, S. 1983. Nervous and humoral control of the fish heart: Structure and function. Comp. Biochem. Physiol. 76A:525–542.

Le Moigne, J., Soulier, P., Peyraud-Waitzenegger, M. and Peyraud, C. 1986. Cutaneous and gill O_2 uptake in the European eel (*Anguilla anguilla* L.) in relation to ambient PO_2 10–400 torr. Respir. Physiol. 66:341–354.

Lee, C.G.L., Low, W.P. and Ip, Y.K. 1987. Na+, K+ and volume regulation in the mudskipper, *Periophthalmus chrysospilos*. Comp. Biochem. Physiol. 87A:439-448.

Lefebvre, L. and Spahn, D. 1987. Gray kingbird predation on small fish (*Poecilia* sp.) crossing a sandbar. Wilson Bull. 99:291–292.

Leggett, W.C. and Frank, K.T. 1990. The spawning of the capelin. Sci. Am. 5:102–107.

Leiner, M. 1938. *Die Physiologie der Fischatmung*. Akad. Verlagsges, Leipzig, 134 pp.

Lele, S.H. 1932. The circulation of blood in the air-breathing chambers of *Ophiocephalus punctatus* Bloch. J. Linn. Soc., London 38:49–54.

Lele, S.H. and Kulkarni, R.D. 1939. The skeleton of *Periophthalmus barbarus* (Linn.) II. Branchial arches, vertebral column and appendicular skeleton. Biol. Sci. 7:123–134.

Lenfant, C. and Johansen, K. 1967. Gas exchange and response to environment in the lungfish *Protopterus aethiopicus* (Heckel). Fed. Proc. 26:441.

Lenfant, C. and Johansen, K. 1968. Respiration in the African lungfish *Protopterus aethiopicus*. I. Respiratory properties of blood and normal patterns of breathing and gas exchange. J. Exp. Biol. 49:437–452.

Lenfant, C. and Johansen, K. 1972. Gas exchange in gill, skin, and lung breathing. Respir. Physiol. 14:211–218.

Lenfant, C., Johansen K. and Hanson, D. 1970. Bimodal gas exchange and ventilation-perfusion relationship in lower vertebrates. Proc. Fed. Am. Soc. Exp. Biol. 29:1124–1129.

Lenfant, C., Johansen, K. and Grigg, G.C. 1966. Respiratory properties of blood and pattern of gas exchange in the lungfish, *Neoceratodus forsteri* (Krefft). Respir. Physiol. 2:1–21.

Leydig, F. 1853. Einige histologische Beobachtungen über den Schlammpeitzger (*Cobitis fossilis*). Arch. F. Anat., Phys. U. Wiss. Medicin von J. Müller, pp. 3–8.

Leydig, F. 1854. Histologische Bermerkungen über den *Polypterus bichir*. Siebold und Kolliker, Z. 5:40–74.

Liem, K.F. 1961. Tetrapod parallelisms and other features in the functional morphology of the blood vascular system of *Fluta alba* Zuiew (Pisces: Teleostei). J. Morphol. 108:131–143.

Liem, K.F. 1963a. Comparative osteology and phylogeny of the *Anabantoidei* (Teleostei, Pisces). Illinois Biol. Monogr. 30:1–149.

Liem, K.F. 1963b. Sex reversal as a natural process in the synbranchiform fish *Monopterus albus*. Copeia 1963:303–312.

Liem, K.F. 1967a. Functional morphology of the head of the anabantoid teleost fish *Helostoma temmincki*. J. Morphol. 121:135–158.

Liem, K.F. 1967b. A morphological study of *Luciocephalus pulcher* with notes on gular elements in other recent teleosts. J. Morphol. 121:103–134.

Liem, K.F. 1967c. Functional morphology of the integumentary, respiratory, and digestive systems of the synbranchoid fish *Monopterus albus*. Copeia 1967:375–388.

Liem, K.F. 1980. Air ventilation in advanced teleosts: Biomechanical and evolutionary aspects. In: Ali, M.A. (ed.) *Environmental Physiology of Fishes*. Plenum, New York, pp. 57–91.

Liem, K.F. 1981. Larvae of air-breathing fishes as countercurrent flow devices in hypoxic environments. Science 211:1177–1179.

Liem, K.F. 1984. The muscular basis of aquatic and aerial ventilation in the air-breathing teleost fish *Channa*. J. Exp. Biol. 113:1–18.

Liem, K.F. 1987. Functional design of the air ventilation apparatus and overland excursions by teleosts. Fieldiana: Zool. 37:1–29.

Liem, K.F. 1988. Form and function of lungs: The evolution of air breathing mechanisms. Am. Zool. 28:739–759.

Liem, K.F. 1989a. Respiratory gas bladders in teleosts: Functional conservatism and morphological diversity. Am. Zool. 29:333–352.

Liem, K.F. 1989b. Functional design and diversity in the feeding morphology and ecology of air breathing teleosts. Prog. in Zool. 35:487–500.

Liem, K.F. 1990. Aquatic *versus* terrestrial feeding modes: Possible impacts on the trophic ecology of vertebrates. Am. Zool. 30:209–221.

Liem, K.F., Eclancher, B. and Fink, W.L. 1984. Aerial respiration in the banded knife fish *Gymnotus carapo* (Teleostei: Gymnotoidei). Physiol. Zool. 57:185–195.

Ling, N. and Dean, T.L. 1995. Flexible metabolic depression accompanies seasonal emmersion in the New Zealand black mudfish, *Neochanna diversus*—is it induced by metabolic hypoxia? Physiol. Zool. 68:124.

Little, C. 1983. *The Colonisation of Land: Origins and Adaptations of Terrestrial Animals*. Cambridge University Press, Cambridge, 290 pp.

Little, C. 1990. *The Terrestrial Invasion*. Cambridge University Press, Cambridge, 304 pp.

Liu, W.S. 1993. Development of the respiratory swimbladder of *Pangasius sutchi*. J. Fish Biol. 42:159–167.

Loftus, W. 1979. Synchronous aerial respiration by the walking catfish in Florida. Copeia 1979:156–158.

Lomholt, J.P. 1993. Breathing in the aestivating African lungfish, *Protopterus amphibius*. In: Singh, B.R. (ed.) *Advances in Fish Research Vol. 1*. Narendra, New Delhi, pp. 17–34.

Lomholt, J.P. and Johansen, K. 1974. Control of breathing in *Amphipnous cuchia*, an amphibious fish. Respir. Physiol. 21:325–340.

Lomholt, J.P. and Johansen, K. 1976. Gas exchange in the amphibious fish, *Amphipnous cuchia*. J. Comp. Physiol. 107:141–157.

Lomholt, J.P., Johansen, K. and Maloiy, G.M.O. 1975. Is the aestivating lungfish the first vertebrate with suctional breathing? Nature 257:787–788.

Long, J.A. 1995. *The Rise of Fishes: 500 Million Years of Evolution*. Johns Hopkins Univ. Press, Baltimore, 223 pp.

Louisy, P. 1987. Observations sur l'émersion nocturne de deux blennies Méditerranéennes: *Coryphoblennius galerita* et *Blennius trigloides* (Pisces, Perciformes). Cybium 11:55–73.

Low, W.P., Ip, Y.K. and Lane, D.J.W. 1990. A comparative study of the gill morphometry in the mudskippers—*Periophthalmus chrysospilos, Boleophthalmus boddaerti* and *Periophthalmodon schlosseri*. Zool. Sci. 7:29–38.

Low, W.P., Lane, D.J.W. and Ip, Y.K. 1988. A comparative study of terrestrial adaptations of the gills in three mudskippers—*Periophthalmus chrysospilos, Boleophthalmus boddaerti,* and *Periophthalmodon schlosseri*. Biol. Bull. 175:434–438.

Lowe-McConnell, R.H. 1969. The cichlid fishes of Guyana, S. America, with notes on their ecology and breeding behaviour. Zool. J. Linn. Soc. 48:255–302.

Lowe-McConnell, R.H. 1975. *Fish Communities in Tropical Freshwaters*. Longman, London, 337 pp.

Lowe-McConnell, R.H. 1987. *Ecological Studies In Tropical Fish Communities*. Cambridge University Press, Cambridge, 382 pp.

Lüling, K.H. 1958. Über die Atmung, amphibishche Lebensweise und Futteraufnahme von *Synbranchus marmoratus* (Pisces, Synbranchidae). Bonn. Zool. Beitr. 9:68–94.

Lüling, K.H. 1964a. Zur Biologie und Ökologie von *Arapaima gigas* (Pisces, Osteoglossidae). Z. Morphol. Okol. Tiere 54:436–530.

Lüling, K.H. 1964b. Über die Atmung des *Hoplerythrinus unitaeniatus* (Pisces, Erythrinidae). Bonn. Zool. Beitr. 15:90–102.

Lüling, K.H. 1969. Auf Fischfang in den Urwäldern am Rio Chapare und Rio Chipiriri in Ostbolivien. Aqua Terra 6:56–60.

Lüling, K.H. 1971. Ökologische Beobachtungen und Untersuchungen am Biotop des *Rivulus beniensis* (Cyprinodontidae). Neotropischen Fauna 6:163–193.

Lüling, K.H. 1973. *Sudamerikanische Fische und ihr Lebensraum*. Englebert Pfriem Verlag, Wuppertal.

Lüling, K.H. 1980. Biotop, Begleitfauna und amphibische Lebenweise von *Synbranchus marmoratus* (Pisces, Synbranchidae) in Seitengewässern des mittleren Paraná (Argentinien). Bonn. Zool. Beitr. 31:111–143.

Lupu, H. 1908. Régénération de l'épithélium intestinal du *Cobitis fossilis*. Ann. Scient. de l'Univ. de Jassy, 5:182–(191)247.

Lupu, H. 1911. Nouvelles contributions à l'étude de la respiration intestinale du *Cobitis fossilis*. Ann. Scient. de l'Univ. de Jassy 6:302–309.

Lupu, H. 1914. Recherches histo-physiologiques sur l'intestin du *Cobitis fossilis*. Ann. Scient. de l'Univ. de Jassy 8:52–116.

Lupu, H. 1925. Nouvelles contributions à l'étude du sang de *Cobitis fossilis*. Ann. Scient. de l'Univ. de Jassy 14:60–110.

Macintosh, D.J. 1979. Predation of fiddler crabs (*Uca* spp.) in estuarine mangroves. In: Srivastava, P.B.L. (ed.) *Biotrop Special Publication No. 10, Mangrove and Estuarine Vegetation in South East Asia*. Serdang, Malaysia, pp. 101–110.

Macnae, W. 1968a. Mudskippers. African Wild Life 22:241–248.

Macnae, W. 1968b. A general account of the fauna and flora of mangrove swamps and forests in the Indo-west Pacific region. Adv. Mar. Biol. 6:73–270.

Macnae, W. and Kalk, M. 1962. The ecology of the mangrove swamps at Inhaca Island, Mocambique. J. of Animal Ecol. 50:19–34.

Magid, A.M.A. 1966. Breathing and function of the spiracles of *Polypterus senegalus*. Anim. Behav. 14:530–533.

Magid, A.M.A. 1967. Observations on the venous system of three species of *Polypterus* (Pisces). J. Zool., London 152:19–30.

Magid, A.M.A. 1971. The ability of *Clarias lazera* (Pisces) to survive without air-breathing. J. Zool., London 163:63–72.

Magid, A.M.A. and Babiker, M.M. 1975. Oxygen consumption and respiratory behaviour of three Nile fishes. Hydrobiologia 46:359–367.

Magid, A.M.A., Vokac, Z. and Ahmed, N.E.D. 1970. Respiratory function of the swim-bladders of the primitive fish *Polypterus senegalus*. J. Exp. Biol. 52:27–37.

Magnus, D.B.E. 1963. *Alticus saliens*, ein amphibisch lebender Fisch. Natur. u. Mus. 93:128–132.

Magnus, D.B.E. 1966. Bewegungsweisen des amphibischen Schleimfisches *Lophalticus kirkii magnusi* Klausewitz (Pisces, Salariidae) im Biotop. Verh. Deutsch. Zool. Ges. 1965:542–555.

Magnuson, J.J., Beckel, A.L., Mills, K. and Brandt, S.B. 1985. Surviving winter hypoxia: Behavioral adaptations of fishes in a northern Wisconsin winterkill lake. Environ. Biol. Fishes 14:241–250.

Magnuson, J.J., Keller, J.W., Beckel, A.L. and Gallepp, G.W. 1983. Breathing gas mixtures different from air: An adaptation for survival under the ice of a facultative air-breathing fish. Science 220:312–314.

Mahajan, C.L. and Dheer, J.M. 1979. Autoradiography and differential hemoglobin staining as aids to the study of fish hematology. Experientia 35:834–835.

Maina, J.N. and Maloiy, G.M.O. 1985. The morphometry of the lung of the African lungfish (*Protopterus aethiopicus*): Its structural-functional correlations. Proc. R. Soc., London 224B:399–420.

Maina, J.N. and Maloiy, G.M.O. 1986. The morphology of the respiratory organs of the African air-breathing catfish (*Clarias mossambicus*): A light, electron and scanning microscopic study, with morphometric observations. J. Zool., London 209A:421–445.

Malvin, G.M. 1988. Microvascular regulation of cutaneous gas exchange in amphibians. Am. Zool. 28:999–1007.

Mangum, C.P., Haswell, M.S., Johansen, K. and Towle, D.W. 1978. Inorganic ions and pH in the body fluids of Amazon animals. Can. J. Zool. 56:907–916.

Manickam, P. and Natarajan, G.M. 1985. Observations on the blood parameters of two air breathing mudskippers of south India. J. Curr. Biosci. 2:19–22.

Mark, E.L. 1890. Studies on *Lepidosteus*. Part I. Bull. Mus. Comp. Zool. 19:1–127.

Marlier, G. 1938. Considérations sur les organes accessoires servant à la respiration aérienne chez les téléostéens. Ann. Soc. R. Zool. de Belgique 69:163–185.

Marshall, C. and Schultze, H. 1992. Relative importance of molecular, neontological, and paleontological data in understanding the biology of the vertebrate invasion of land. J. Mol. Evol. 35:93–101.

Martin, K.L.M. 1991. Facultative aerial respiration in an intertidal sculpin, *Clinocottus analis* (Scorpaeniformes: Cottidae). Physiol. Zool. 64:1341–1355.

Martin, K.L.M. 1993. Aerial release of CO_2 and respiratory exchange ratio in intertidal fishes out of water. Environ. Biol. Fishes 37:189–196.

Martin, K.L.M. 1995. Time and tide wait for no fish: Intertidal fishes out of water. Environ. Biol. Fishes 44:165–181.

Martin, K.L.M. 1996. An ecological gradient in air-breathing ability among marine cottid fishes. Physiol. Zool. 69:1096–1113.

Martin, K.L.M. and Lighton, J.R.B. 1989. Aerial CO_2 and O_2 exchange during terrestrial activity in an amphibious fish, *Alticus kirki* (Blenniidae). Copeia 1989:723–727.

Martin, K.L.M., Berra, T.M. and Allen, G.R. 1993. Cutaneous aerial respiration during forced emergence in the Australian salamanderfish, *Lepidogalaxias salamandriodes*. Copeia 1993:875–879.

Marusic, E.T., Balbontin, F., Galli-Gallardo, S.M., Garreton, M., Pang, P.K.T. and Griffith, R.W. 1981. Osmotic adaptations of the Chilean clingfish, *Sicyases sanguineus*, during emersion. Comp. Biochem. Physiol. 68A:123–126.

Mast, S.O. 1915. The behavior of *Fundulus*, with especial reference to overland escape from tide-pools and locomotion on land. J. of Animal Behav. 5:341–350.

Mattias, A.T., Moron, S.E., and Fernandes, M.N. 1996. Aquatic respiration during hypoxia of the facultative air-breathing *Hoplerythrinus unitaeniatus*. A comparison with the water-breathing *Hoplias malabaricus*. In: Val, A.L., Almeida-Val, V.M.F., and Randall, D.J. (eds.) *Physiology and Biochemistry of Fishes of the Amazon*. Instituto Nacional de Pesquisas de Amazonia, Manaus Brazil, pp.203–211.

McBean, R.L. and Goldstein, L. 1967. Ornithine-urea cycle activity in *Xenopus laevis*, adaptation in saline. Science 157:931–932.

McCutcheon, F.M. and Hall, F.G. 1937. Hemoglobin in the amphibia. J. Cell Comp. Physiol. 9:191–197.

McDowall, R.M. 1978. *New Zealand Freshwater Fishes: A Guide and Natural History*. Heinemann Educational, Auckland, 230 pp.

McKenzie, D.J. and Randall, D.J. 1990. Does *Amia calva* estivate? Fish Physiol. Biochem. 8:147–158.

McKenzie, D.J., Aota, S. and Randall, D.J. 1991b. Ventilatory and cardiovascular responses to blood pH, plasma PCO_2, blood O_2 content, and catecholamines in an air-breathing fish, the bowfin (*Amia calva*). Physiol. Zool. 64:432–450.

McKenzie, D.J., Burleson, M.L. and Randall, D.J. 1991a. The effects of branchial denervation and pseudobranch ablation on cardioventilatory control in an air-breathing fish. J. Exp. Biol. 161:347–365.

McMahon, B.R. 1969. A functional analysis of the aquatic and aerial respiratory movements of an African lungfish, *Protopterus aethiopicus*, with reference to evolution of the lung-ventilation mechanism in vertebrates. J. Exp. Biol. 51:407–430.

McMahon, B.R. 1970. The relative efficiency of gaseous exchange across the lungs and gills of an African lungfish *Protopterus aethiopicus*. J. Exp. Biol. 52:1–15.

McMahon, B.R. and Burggren, W.W. 1987. Respiratory physiology of intestinal air breathing in the teleost fish *Misgurnus anguillicaudatus*. J. Exp. Biol. 133:371–393.

Meredith, A.S., Davie, P.S. and Forster, M.E. 1982. Oxygen uptake by the skin of the Canterbury mudfish, *Neochanna burrowsius*. New Zealand J. Zool. 9:387–390.

Merrick, J.R and Schmida, G.E. 1984. *Australian Freshwater Fishes: Biology and Management*. Griffin Press, Netley, South Australia, 409 pp.

Milsom, W.K. 1989a. Comparative aspects of vertebrate pulmonary mechanics. In: Wood, S.C. (ed.) *Comparative Pulmonary Physiology, Current Concepts*. Dekker, New York, pp. 587–620.

Milsom, W.K. 1989b. Mechanisms of ventilation in lower vertebrates: Adaptations to respiratory and nonrespiratory constraints. Can. J. Zool. 67:2943–2955.

Milsom, W.K. 1990. Mechanoreceptor modulation of endogenous respiratory rhythms in vertebrates. Am. J. Physiol. 259:R898–R910.

Milsom, W.K. 1991. Intermittent breathing in vertebrates. Ann. Rev. Physiol. 53:87–105.

Milsom, W.K. and Jones, D.R. 1985. Characteristics of mechanoreceptors in the air-breathing organ of the holostean fish, *Amia calva*. J. Exp. Biol. 117:389–399.

Milward, N.E. 1974. Studies on the taxonomy, ecology and physiology of Queensland mudskippers. Ph.D. Thesis, University of Queensland, 276 pp.

Mishra, A.P. and Singh, B.R. 1979. Oxygen uptake through water during early life of *Anabas testudineus* (Bloch). Hydrobiologia 66:129–133.

Mishra, N., Pandey, P.K. and Munshi, J.S.D. 1977a. Some aspects of haematology of an air breathing Indian mud eel, *Amphipnous cuchia*. Japan. J. Ichthyol. 24:176–181.

Mishra, N., Pandey, P.K., Munshi, J.S.D. and Singh, B.R. 1977b. Haematological parameters of an air-breathing mud eel, *Amphipnous cuchia* (Ham.) (Amphipnoidae; Pisces). J. Fish Biol. 10:567–573.

Mitchell, J. 1864. On the climbing habits of *Anabas scandens*. Ann. Mag. Nat. Hist. 13:117–119.

Mittal, A.K. and Banerjee, T.K. 1974. Structure and keratinization of the skin of a fresh-water teleost *Notopterus notopterus* (Notopteridae, Pisces). J. Zool., London 174:341–355.

Mittal, A.K. and Munshi, J.S.D. 1971. A comparative study of the structure of the skin of certain air-breathing fresh-water teleosts. J. Zool., London 163:515–532.

Mohamed, H.A. 1980. Ecological and physiological studies on an air-breathing fish, *Alticus kirkii magnusi* inhabiting rocky shores of the Red Sea, Egypt. M. Sci. Thesis, Univ. Cairo, Egypt.

Moitra, A., Singh, O.N. and Munshi, J.S.D. 1989. Microanatomy and cytochemistry of the gastro-respiratory tract of an air-breathing cobitidid fish, *Lepidocephalichthys guntea*. Japan. J. Ichthyol. 36:227–231.

Mommsen, T.P. and Walsh, P.J. 1989. Evolution of urea synthesis in vertebrates: The piscine connection. Science 243:72–75.

Mommsen, T.P. and Walsh, P.J. 1991. Urea synthesis in fishes: Evolutionary and biochemical perspectives. In: Hochachka, P.W. and Mommsen T.P. (eds.) *Biochemistry and Molecular Biology of Fishes Vol. I*. Elsevier, New York, pp. 137–163.

Mommsen, T.P. and Walsh, P.J. 1992. Biochemical and environmental perspectives on nitrogen metabolism in fishes. Experientia 48:583–593.

Morii, H. 1979. Changes with time of ammonia and urea concentrations in the blood and tissue of mudskipper fish, *Periophthalmus cantonensis* and *Boleophthalmus pectinirostris* kept in water and on land. Comp. Biochem. Physiol. 64A:235–243.

Morii, H., Nishikata, K. and Tamura, O. 1978. Nitrogen excretion of mudskipper fish *Periophthalmus cantonensis* and *Boleophthalmus pectinirostris* in water and on land. Comp. Biochem. and Physiol. 60A:189–193.

Morii, H., Nishikata, K. and Tamura, O. 1979. Ammonia and urea excretion from mudskipper fish, *Periophthalmus cantonensis* and *Boleophthalmus pectinirostris* transferred from land to water. Comp. Biochem. Physiol. 63A:23–28.

Morris, C. 1885. On the air-bladder of fishes. Proc. Acad. Nat. Sci., Philadelphia 1885:124–135.

Morris, C. 1892. The origin of lungs, a chapter in evolution. Am. Nat. 26:975–986.

Morris, S. and Bridges, C.R. 1994. Properties of respiratory pigments in bimodal breathing animals: Air and water breathing by fish and crustaceans. Am. Zool. 34:216–228.

Mott, J.C. 1951. The gross anatomy of the blood vascular system of the eel *Anguilla anguilla*. Proc. Zool. Soc., London 120:503–518.

Moussa, T.A. 1956. Morphology of the accessory air-breathing organs of the teleost, *Clarias lazera* (C. and V.). J. Morphol. 98:125–160.

Moussa, T.A. 1957. Physiology of the accessory respiratory organs of the teleost, *Clarias lazera* (C. and V.). J. Exp. Zool. 136:419–454.

Müller, J. 1842. Beobachtungen über die Schwimmblase der Fische, mit Bezug auf einige neue Fischgattungen. Archiv für Anatomie, Physiol. Wiss. Medicin, pp. 307–329.

Müller, J. 1844. Ueber den Bau und die Grenzen der Ganoiden und über das natürliche System der Fishe. Berlin, Abhandl. pp. 117–216.

Munshi, J.S.D. 1961. The accessory respiratory organs of *Clarias batrachus* (Linn.). J. Morphol. 109:115–139.

Munshi, J.S.D. 1962a. On the accessory respiratory organs of *Heteropneustes fossilis* (Bloch). Proc. R. Soc., Edinburgh 68:128–146.

Munshi, J.S.D. 1962b. On the accessory respiratory organs of *Ophiocephalus punctatus* (Bloch) and *O. striatus* (Bloch). J. Linn. Soc. London, Zool. 44:616–626.

Munshi, J.S.D. 1968. The accessory respiratory organs of *Anabas testudineus* (Bloch) (Anabantidae, Pisces). Proc. Linn. Soc., London 179:107–126.

Munshi, J.S.D. 1976. Gross and fine structure of the respiratory organs of air-breathing fishes. In: Hughes, G.M. (ed.) *Respiration of Amphibious Vertebrates*. Academic Press, London, pp.73–104.

Munshi, J.S.D. 1985. The structure, function and evolution of the accessory respiratory organs of air-breathing fishes of India. Fortsch. Zool. 30:353–366.

Munshi, J.S.D. 1993. Structure and function of the air-breathing organs of *Heteropneustes fossilis*. In: Singh, B.R. (ed.) *Advances in Fish Research Vol. 1*. Narendra, New Delhi, pp. 99–138.

Munshi, J.S.D. and Dube, S.C. 1973. Oxygen uptake capacity of gills in relation to body size of the air-breathing fish, *Anabas testudineus* (Bloch). Acta Physiol. Acad. Sci. Hungaricae 44:113–123.

Munshi, J.S.D. and Hughes, G.M. 1992. *Air-Breathing Fishes of India*. Oxford and IBH, New Dehli, 338 pp.

Munshi, J.S.D. and Mishra, N. 1974. Structure of the heart of *Amphipnous cuchia* (Ham.) (Amphipnoidae; Pisces). Zool. Anz. 193:228–239.

Munshi, J.S.D. and Singh, B.N. 1968. A study of the gill-epithelium of certain freshwater teleost fishes with special reference to the air-breathing fishes. Indian J. Zool. 9:91–107.

Munshi, J.S.D., Hughes, G.M., Gehr, P. and Weibel, E.R. 1989. Structure of the air-breathing organs of a swamp mud eel, *Monopterus cuchia*. Japan. J. Ichthyol. 35:453–465.

Munshi, J.S.D., Olson, K.R., Ghosh, T.K. and Ojha, J. 1990. Vasculature of the head and respiratory organs in an obligate air-breathing fish, the swamp eel *Monopterus* (= *Amphipnous*) *cuchia*. J. Morphol. 203:181–201.

Munshi, J.S.D., Olson, K.R., Ojha, J. and Ghosh, T.K. 1986. Morphology and vascular anatomy of the accessory respiratory organs of the air-breathing climbing perch, *Anabas testudineus* (Bloch). Am. J. Anat. 176:321–331.

Munshi, J.S.D., Pandey, B.N., Pandey, P.K. and Ojha, J. 1978. Oxygen uptake through gills and skin in relation to body weight of an air-breathing siluroid fish, *Saccobranchus* (= *Heteropneustes*) *fossilis*. J. Zool., London 184:171–180.

Munshi, J.S.D., Roy, P.K., Ghosh, T.K. and Olson, K.R. 1994. Cephalic circulation in the air-breathing snakehead fish, *Channa punctata*, *C. gachua*, and *C. marulius* (Ophiocephalidae, Ophiocephaliformes). Anat. Rec. 238:77–91.

Munshi, J.S.D., Sinha, A.L. and Ojha, J. 1976. Oxygen uptake capacity of gills and skin in relation to body weight of the air-breathing siluroid fish, *Clarias batrachus* (Linn.). Acta Physiol. Acad. Sci. Hungaricae 48:23–33.

Murdy, E.O. 1989. A taxonomic revision and cladistic analysis of the oxudercine gobies (Gobiidae: Oxudercinae). Rec. Australian Mus., Supp. 11:1–93.

Mustafa, S. and Mubarak, K.V.A. 1980. Air breathing of *Colisa fasciata* (Schneider). Proc. Indian Acad. Sci. (Ani. Sci.) 89:21–24.

Myers, G.S. 1942. The "lungs" of *Bothriolepis*. Stanford Ichthyol. Bull. 2:134–136.

Nakamura, K. 1994. Air breathing abilities of the common carp. Fish. Sci. 60:271–274.

Natarajan, G.M. 1978. Observations on the oxygen consumption in Indian air-breathing fishes: I. Oxygen consumption in the climbing perch, *Anabas scandens* (Cuvier). Comp. Physiol. Ecol. 3:246–248.

Natarajan, G.M. 1979. Oxygen consumption in *Channa gachua* (Ham.). Geobios 6:30–31.

Natarajan, G.M. 1980. Bimodal gas exchange and some blood parameters in the Indian air-breathing fish, *Osphromenus* (sic) *olfax* (Day). Curr. Sci. 49:371–372.

Natarajan, G.M. 1981. Effect of lethal (Lc 50/48 hrs) concentration of metasystox on selected oxidative enzymes, tissue respiration and histology of gills of the fresh water air-breathing fish, *Channa striatus* (Bleeker). Curr. Sci. 50:985–989.

Natarajan, G.M. 1983. Effect of sublethal concentration of matasystox on selected oxidative enzymes, tissue respiration, and hematology of the freshwater air-breathing fish, *Channa striatus* (Bleeker). Pestic. Biochem. Physiol. 21:194–198.

Natarajan, G.M. 1984a. Effect of lethal (LC50/48h) concentration of metasystox on some selected enzyme systems in the air-breathing fish, *Channa striatus*. Comp. Physiol. Ecol. 9:29–32.

Natarajan, G.M. 1984b. Diurnal rhythm of bimodal oxygen uptake in an airbreathing loach, *Lepidocephalus thermalis* (Val.). Curr. Sci. 53:394–396.

Natarajan, G.M. and Rajulu, G.S. 1982. Bimodal oxygen uptake and some blood parameters in the bubble nest builder, *Trichogaster pectoralis* (Regan). Curr. Sci. 51:948–950.

Natarajan, G.M. and Rajulu, G.S. 1983. Bimodal oxygen uptake in the amphibious mud-skipper *Periophthalmus koelreuteri* (Pallas) (Gobiidae: Teleostei). J. Anim. Morphol. Physiol. 30:177–180.

Neill, W.T. 1950. An aestivating bowfin. Copeia 1950:240.

Nelson, J.S. 1994. *Fishes of the World*, 3rd ed. Wiley & Sons, New York, 600 pp.

Nikinmaa, M. 1990. *Vertebrate Red Blood Cells: Adaptations of Function to Respiratory Requirements*. Springer-Verlag, Berlin, 262 pp.

Nikinmaa, M., Kunnamo-Ojala, T. and Railo, E. 1986. Mechanisms of pH regulation in lamprey (*Lampetra fluviatilis*) red blood cells. J. Exp. Biol. 122:355–367.

Nonnotte, G. 1984. Cutaneous respiration in the catfish *Ictalurus melas* R. Comp. Biochem. Physiol. 78A:515–517.

Nonnotte, G. and Kirsch R. 1978. Cutaneous respiration in seven sea-water teleosts. Respir. Physiol. 35:111–118.

Nursall, J.R. 1981. Behavior and habitat affecting the distribution of five species of sympatric mudskippers in Queensland. Bull. Mar. Sci. 31:730–735.

Nysten, M. 1962. Étude anatomique des rapports de la vessie aérienne avec l'axe vertébral chez *Pantodon buchholzi* Peters. Ann. Mus. R. l'Afrique Cent. Sci. Zool. 8:187–220.

Oduleye, S.O. 1977. Unidirectional water and sodium fluxes and respiratory metabolism in the African lungfish, *Protopterus annectens*. J. Comp. Physiol. 119:127–139.

Oglialoro, C.M. 1947. Vascolarizzazione della mucosa bucco-faringea e respirazione accessoria attraverso itale regione in alcuni teleostei marini. Boll. Soc. Italiana de Biol. Sper. 23:990–992.

Ojha, J. and Munshi, J.S.D. 1974. Morphometric studies on the gill and skin dimensions in relation to body weight in a fresh-water mud eel, *Macrognathus aculeatum* (Bloch). Zool. Anz. 193:364–381.

Ojha, J., Dandotia, O.P. and Munshi, J.S.D. 1977. Oxygen consumption of an amphibious fish *Colisa fasciatus* in relation to body weight. Polskie Arch. Hydrobiol. 24:547–553.

Ojha, J., Dandotia, O.P., Patra, A.K. and Munshi, J.S.D. 1978. Bimodal oxygen uptake in relation to body weight in a freshwater murrel, *Channa* (= *Ophiocephalus*) *gachua*. Z. Tierphysiol., Tierernahrg. Futtermittelkde. 40:57–66.

Ojha, J., Mishra, N., Saha, M.P. and Munshi, J.S.D. 1979. Bimodal oxygen uptake in juveniles and adults amphibious (sic) fish, *Channa* (= *Ophiocephalus*) *marulius*. Hydrobiologia 63:153–159.

Olson, K.R. 1981. Morphology and vascular anatomy of the gills of a primitive air-breathing fish, the bowfin (*Amia calva*). Cell Tissue Res. 218:499–517.

Olson, K.R. 1991. Vasculature of the fish gill: Anatomical correlates of physiological functions. J. Electron Microscopy Technique 19:389–405.

Olson, K.R. 1994. Circulatory anatomy in bimodally breathing fish. Am. Zool. 34:280–288.

Olson, K.R., Lipke, D., Munshi, J.S.D., Moitra, A., Ghosh, T.K., Kunwar, G., Ahmad, M., Roy, P.K., Singh, O.N., Nasar, S.S.T., Pandey, A., Oduleye, S.O. and Kullman, D. 1987. Angiotensin-converting enzyme in organs of air-breathing fish. Gen. Comp. Endocrinol. 68:486–491.

Olson, K.R., Munshi, J.S.D., Ghosh, T.K. and Ojha, J. 1986. Gill microcirculation of the air-breathing climbing perch, *Anabas testudineus* (Bloch): Relationships with the accessory respiratory organs and systemic circulation. Am. J. Anat. 176:305–320.

Olson, K.R., Munshi, J.S.D., Ghosh, T.K. and Ojha, J. 1990. Vascular organization of the head and respiratory organs of the air-breathing catfish, *Heteropneustes fossilis*. J. Morphol. 203:165–179.

Olson, K.R., Roy, P.K., Ghosh, T.K. and Munshi, J.S.D. 1994. Microcirculation of gills and accessory respiratory organs from the air-breathing snakehead fish, *Channa punctata*, *C. gachua*, and *C. marulius*. Anat. Rec. 238:92–107.

Orgeig, S. and Daniels, C.B. 1995. The evolutionary significance of pulmonary surfactant in lungfish (Dipnoi). Am. J. Respir. Cell Mol. Biol. 13:161–166.

Owen, R. 1841. Description of the *Lepidosiren annectens*. Trans. Linn. Soc., London 18:327–361.

Owen, R. 1846. Lectures on the comparative anatomy and physiology of the vertebrate animals. Longman, Brown, Green, and Longmans, London, 308 pp.

Pack, A.I., Galante, R.J. and Fishman, A.P. 1990. Control of interbreath interval in the African lungfish. Am. J. Physiol. 259:R139–R146.

Pack, A.I., Galante, R.J. and Fishman, A.P. 1992. Role of lung inflation in control of air breath duration in African lungfish (*Protopterus annectens*). Am. J. Physiol. 262:R879–R884.

Packard, G.C. 1974. The evolution of air-breathing in Paleozoic gnathostome fishes. Evolution 28:320–325.

Packard, G.C. 1976. Devonian amphibians: Did they excrete carbon dioxide via skin, gills, or lungs? Evolution 30:270–280.

Paine, R.T. and Palmer, A.R. 1978. *Sicyases sanguineus*, a unique trophic generalist from the Chilean intertidal zone. Copeia 1978:75–81.

Panchen, A.L. (ed.). 1980. *The Terrestrial Environment and the Origin of Land Vertebrates*. Academic Press, London, 633 pp.

Pandey, B.N., Pandey, P.K., Munshi, J.S.D. and Choudhary, B.P. 1971. Studies on blood and haemoglobin in relation to metabolic rate in an air-breathing siluroid fish, *Heteropneustes fossilis* (Bloch). Comp. Physiol. Ecol. 2:226–228.

Pandey, K. and Shukla, J.P. 1982. Behavioural studies of freshwater fishes during a solar eclipse. Environ. Biol. Fishes 7:63–64.

Pandian, T.J. and Vivekanandan, E. 1976. Effects of feeding and starvation on growth and swimming activity in an obligatory air-breathing fish. Hydrobiologia 49:33–39.

Parker, W.N. 1892. On the anatomy and physiology of *Protopterus annectens*. Trans. R. Irish Acad. 30:109–230.

Parsons, C.W. 1930. The conus arteriosus in fishes. Q. J. Micro. Sci. 73:145–176.

Patra, A.K., Biswas, N., Ojha, J. and Munshi, J.S.D. 1978. Circadian rhythm in bimodal oxygen uptake in an obligatory air-breathing swamp eel. Indian J. Exp. Biol. 16:808–809.

Pattle, R.E. 1976. The lung surfactant in the evolutionary tree. In: Hughes, G.M. (ed.) *Respiration of Amphibious Vertebrates*. Academic Press, London, pp. 233–255.

Pearse, A.S. 1932. Observations on the ecology of certain fishes and crustaceans along the bank of the Malta River at Port Canning. Rec. Indian Mus., Calcutta 34:289–298.

Pelster, B., Bridges, C.R. and Grieshaber, M.K. 1988. Physiological adaptations of the intertidal rockpool teleost *Blennius pholis* L., to aerial exposure. Respir. Physiol. 71:355–374.

Perez, J.E. 1979. Respiracion aerea y acuatica en peces de la especie *Hoplosternum littorale*. I. Parametros sanguineos. Acta Cient. Venezolana 30:314–317.

Perez, J.E. 1980. Respiracion aerea y acuatica en peces de la especie *Hoplosternum littorale*. II. Afinidad de sus hemoglobinas por el oxigeno. Acta Cient. Venezolana 31:449–455.

Perna, S.A. and Fernandes, M.N., 1996. Gill morphology of the facultative air-breathing loricariid fish, *Hypostomus plecostomus* (Walbaum) with special emphasis on aquatic respiration. Fish. Physiol. Biochem. 15:213–220.

Perry, S.F. and Wood, C.M. 1989. Control and coordination of gas transfer in fishes. Can. J. Zool. 67:2961–2970.

Peters, H.M. 1978. On the mechanism of air ventilation in anabantoids (Pisces: Teleostei). Zoomorphologie 89:93–123.

Peters, W. 1846. Über eine neue Gattung von Labyrinthfischen aus Quellimane. Arch. Anat. Physiol. Wiss. Med. 1846:480–482.

Peters, W. 1853. Über das Kiemengerüst der Labyrinthfische. Arch. Anat. Physiol. Wiss. Med. 1853:427–430.

Pettit, M.J. and Beitinger, T.L. 1980. Thermal responses of the South American lungfish, *Lepidosiren paradoxa*. Copeia 1980:130–136.

Pettit, M.J. and Beitinger, T.L. 1981. Aerial respiration of the brachiopterygian fish, *Calamoichthys calabaricus*. Comp. Biochem. Physiol. 68A:507–509.

Pettit, M.J. and Beitinger, T.L. 1985. Oxygen acquisition of the reedfish, *Erpetoichthys calabaricus*. J. Exp. Biol. 114:289–306.

Phleger, C.F. and Saunders, B.S. 1978. Swim-bladder surfactants of Amazon air-breathing fishes. Can. J. Zool. 56:946–952.

Piiper, J. 1982. Respiratory gas exchange at lungs, gills and tissues: Mechanisms and adjustments. J. Exp. Biol. 100:5–22.

Piiper, J. 1989. Factors affecting gas transfer in respiratory organs of vertebrates. Can. J. Zool. 67:2956–2960.

Piiper, J. and Scheid, P. 1992. Gas exchange in vertebrates through lungs, gills, and skin. News Physiol. Sci. 7:199–203.

Piiper, J., Gatz, R.N. and Crawford, Jr., E.C. 1976. Gas transport characteristics in an exclusively skin-breathing salamander, *Desmognathus fuscus* (Plethodontidae). In: Hughes, G.M. (ed.) *Respiration of Amphibious Vertebrates*. Academic Press, London, pp. 339–356.

Pinter, H. 1986. *Labyrinth Fish*. Barron's, Woodbury, N.Y., 144 pp.

Poey, F. 1856. A los lepidopteras y cocodrilas. Mem. Hist. Nat. de Isla Cuba 11:136.

Poey, F. 1858. Observations on different points of natural history of the island of Cuba, with reference to the ichthyology of the United States. Ann. Lyceum Nat. Hist. New York 6:133.

Poll, M. 1932. Au sujet du *Phractolaemus spinosus* Pellegrin

(= *Phractolaemus ansorgii* Boulenger). Rev. Zool. Bot. Afr. 21:287–290.

Poll, M. 1962. Étude sur la structure adulte et la formation des sacs pulmonaires des *Protoptères*. Ann. Mus. R. l'Afrique Cent. Sci. Zool. 108:130–172.

Poll, M. and Deswattines, C. 1967. Étude systématique des appareils respiratoire et circulatoire des Polypteridae. Ann. Mus. R. l'Afrique Cent. Sci. Zool. 158:1–63.

Poll, M. and Nysten, M. 1962. Vessie natatoire pulmonoïde et pneumatisation des vertèbres chez *Pantodon buchholzi* Peters. Bull. des Séances 8:434–454.

Ponniah, A.G. and Pandian, T.J. 1977. Surfacing activity and food utilization in the air-breathing fish *Polyacanthus cupanus* exposed to constant PO_2. Hydrobiologia 53:221–227.

Potter, G.E. 1927. Respiratory function of the swim bladder in *Lepidosteus*. J. Exp. Zool. 49:45–67.

Poulin, R., Wolf, N.G. and Kramer, D.L. 1987. The effect of hypoxia on the vulnerability of guppies (*Poecilia reticulata*, Poeciliidae) to an aquatic predator (*Astronotus ocellatus*, Cichlidae). Environ. Biol. Fishes 20:285–292.

Power, M.E. 1984. Depth distributions of armored catfish: Predator-induced resource avoidance? Ecology 65:523–528.

Powers, D.A. 1980. Molecular ecology of teleost fish hemoglobins: Strategies for adapting to changing environments. Am. Zool. 20:139–162.

Powers, D.A., Fyhn, H.J., Fyhn, U.E.H., Martin, J.P., Garlick, R.L. and Wood, S.C. 1979a. A comparative study of the oxygen equilibria of blood from 40 genera of Amazonian fishes. Comp. Biochem. Physiol. 62A:67–85.

Powers, D.A., Martin, J.P., Garlick, R.L., Fyhn, H.J. and Fyhn, U.E.H. 1979b. The effect of temperature on the oxygen equilibria of fish hemoglobins in relation to environmental thermal variability. Comp. Biochem. Physiol. 62A:87–94.

Pradhan, V. 1961. A study of blood of a few Indian fishes. Proc. Indian Acad. Sci. 54:251–256.

Prasad, M.S. 1988. Morphometrics of gills during growth and development of the air-breathing habit in *Colisa fasciatus* (Bloch and Schneider). J. Fish Biol. 32:367–381.

Prasad, M.S. and Prasad, P. 1985. Larval behavior during growth and development of air-breathing habit in *Colisa fasciatus* (Bloch and Schneider). Indian J. Fish. 32:185–191.

Prasad, M.S. and Singh, B.R. 1984. Oxygen uptake through water during early life in *Colisa* (= *Trichogaster*) *fasciatus* (Bloch and Schneider). Polskie Arch. Hydrobiol. 31:153–159.

Purser, G.L. 1926. *Calamoichthys calabaricus* (J.A. Smith). Part I. The alimentary and respiratory systems. Trans. R. Soc., Edinburgh 54:767–784.

Pusey, B.J. 1986. The effect of starvation on oxygen consumption and nitrogen excretion in *Lepidogalaxias salamandroides* (Mees). J. Comp. Physiol. B. 156:701–705.

Pusey, B.J. 1989a. Aestivation in the teleost fish *Lepidogalaxias salamandroides* (Mees). Comp. Biochem. Physiol. 92A:137–138.

Pusey, B.J. 1989b. Gas exchange during aestivation in *Lepidogalaxias salamandroides*. Fishes of the Sahul 5:219–223.

Pusey, B.J. 1990. Seasonality, aestivation and the life history of the salamander fish *Lepidogalaxias salamandroides* (Pisces: Lepidogalaxiidae). Environ. Biol. Fishes 29:15–26.

Qasim, S.Z. 1957. The biology of *Blennius pholis* L. (Teleostei). Proc. Zool. Soc., London 128:161–208.

Quekett, J. 1844. On a peculiar arrangement of blood-vessels in the air-bladder of fishes, with some remarks on the evidence which they afford of the true function of that organ. Trans. R. Micro. Soc., London 1:99–108.

Rahn, H. 1966a. Aquatic gas exchange: Theory. Respir. Physiol. 1:1–12.

Rahn, H. 1966b. Development of gas exchange: Evolution of the gas transport system in vertebrates. Proc. R. Soc. Med. 59:493–494.

Rahn, H. and Garey, W.F. 1973. Arterial CO_2, O_2, pH, and HCO_3^- values of ectotherms living in the Amazon. Am. J. Physiol. 225:735–738.

Rahn, H. and Howell, B.J. 1976. Bimodal gas exchange. In: Hughes, G.M. (ed.) *Respiration of Amphibious Vertebrates*. Academic Press, London, pp. 271–285.

Rahn, H., Rahn, K.B., Howell, B.J., Gans, C. and Tenney, S.M. 1971. Air breathing of the garfish (*Lepisosteus osseus*). Respir. Physiol. 11:285–307.

Rama Samy, M. and Reddy, T.G. 1978. Bimodal oxygen uptake of an air-breathing fish *Channa gachua* (Hamilton). Indian J. Exp. Biol. 16:693–695.

Ramaswamy, M. 1983. Changes in glycogen levels in an air-breathing fish, *Channa gachua* (Hamilton) following aerial exposure. Curr. Sci. 52:35–37.

Ramaswamy, M. and Reddy, T.G. 1977. Air-breathing in the catfish, *Mystus vittatus* (Bloch). Curr. Sci. 46:563–564.

Ramaswamy, M. and Reddy, T.G. 1983. Ammonia and urea excretion in three species of air-breathing fish subjected to aerial exposure. Proc. Indian Acad. Sci. 92:293–297.

Randall, D. 1982. The control of respiration and circulation in fish during exercise and hypoxia. J. Exp. Biol. 100:275–288.

Randall, D.J. 1994. Cardio-respiratory modeling in fishes and the consequences of the evolution of airbreathing. Cardioscience 5:167–171.

Randall, D.J. and Daxboeck, C. 1984. Oxygen and carbon dioxide transfer across fish gills. In: Hoar, W.S. and Randall, D.J. (eds.) *Fish Physiology Vol. X, Part A*. Academic Press, London, pp. 263–314.

Randall, D.J., Burggren, W.W., Farrell, A.P. and Haswell, M.S. 1981a. *The Evolution of Air Breathing in Vertebrates*. Cambridge University Press, Cambridge, 133 pp.

Randall, D.J., Cameron, J.N., Daxboeck, C. and Smatresk, N. 1981b. Aspects of bimodal gas exchange in the bowfin, *Amia calva* L. (Actinopterygii: Amiiformes). Respir. Physiol. 43:339–348.

Randall, D.J., Farrell, A.P. and Haswell, M.S. 1978a. Carbon dioxide excretion in the pirarucu (*Arapaima gigas*), an obligate air-breathing fish. Can. J. Zool. 56:977–982.

Randall, D.J., Farrell, A.P. and Haswell, M.S. 1978b. Carbon dioxide excretion in the jeju, *Hoplerythrinus unitaeniatus*, a facultative air-breathing teleost. Can. J. Zool. 56:970–973.

Randall, D.J., Wood, C.M., Perry, S.F., Bergman, H., Maloiy, G.M.O., Mommsen, T.P. and Wright, P.A. 1989. Urea excretion as a strategy for survival in a fish living in a very alkaline environment. Nature 337:165–166.

Rao, H.S. and Hora, S.L. 1938. On the ecology, bionomics, and systematics of the blennid fishes of the genus *Andamia* Blyth. Rec. Indian Mus. Calcutta 40:377–401.

Rauther, M. 1910. Die akzessorischen Atmungsorgane der Knochenfische. Ergeb. Fortschr. Zool. 2:517–585.

Rauther, M. 1914. Über die respiratorische Schwimmblase von *Umbra*. Zool. Jahrb. Allg. Zool. Physiol. 34:339–364.

Rauther, M. 1922. Zur Kenntnis der Polypteridenlunge. Mitt. Nat. Stuttgart 91:290–297.

Rauther, M. 1937. Die pneumatischen Darmanhänge. In: *Bronns Klassen und Ordnungen des Tierreiches* 6, Abt. I. Buch 2. Akad. Verl. M.B.H., Leipzig, pp. 759–826.

Ravindranath, K., Rao, K.G. and Kamal, M.Y. 1985. A case of lerneosis and its control in an air-breathing fish culture system. Curr. Sci. 54:885–886.

Richards, F.A. 1957. Oxygen in the Ocean. In: Hedgpeth, J.W. (ed.)

Treatise on Marine Ecology and Paleoecology, Vol. 1 Ecology. Nat'l. Research Council, Nat'l. Acad. Sci., Washington D.C., pp. 185–238.

Richter, H. 1935. Die Luftatmung und die akzessorischen Atmungsorgane von *Gymnotus electricus* L. Gegen. Morphol. Jahrb. 75:469–476.

Riggs, A.F. 1951. Properties of tadpole and frog hemoglobins. J. Gen. Physiol. 35:23–40.

Roberts, J.L. and Graham, J.B. 1985. Adjustments of cardiac rate to changes in respiratory gases by a bimodal breather, the Panamanian swamp eel, *Synbranchus marmoratus*. Am. Zool. 25:51A.

Roberts, T.R. and Vidthayanon, C. 1991. Systematic revision of the Asian catfish family Pangasiidae, with biological observations and descriptions of three new species. Proc. Acad. Nat. Sci., Philadelphia 143:97–144.

Robertson, J.D. 1957. The habitat of the early vertebrates. Biol. Rev., Cambridge Phil. Soc. 32:156–187.

Robertson, J.I. 1913. The development of the heart and vascular system of *Lepidosiren paradoxa*. Q. J. Micro. Sci. 59:53–132.

Romer, A.S. 1958a. Tetrapod limbs and early tetrapod life. Evolution 12:365–369.

Romer, A.S. 1958b. *The Vertebrate Story*. Univ. of Chicago Press, Chicago, 437 pp.

Romer, A.S. 1964. *The Vertebrate Body*. W.B. Saunders Company, Philadelphia, 627 pp.

Romer, A.S. 1966. *Vertebrate Paleontology*. Univ. Chicago Press, Chicago, 468 pp.

Romer, A.S. 1967. Major steps in vertebrate evolution. Science 158:1629–1637.

Romer, A.S. 1972. Skin breathing—primary or secondary? Respir. Physiol. 14:183–192.

Rosen, D.E. and Greenwood, P.H. 1976. A fourth neotropical species of synbranchid eel and the phylogeny and systematics of synbranchiform fishes. Bull. Am. Mus. Nat. Hist. 157:1–69.

Rosen, D.E., Forey, P.L., Gardiner, B.G. and Patterson, C. 1981. Lungfishes, tetrapods, paleontology, and plesiomorphy. Bull. Am. Mus. Nat. Hist. 167:159–276.

Rowntree, W.S. 1903. On some points in the visceral anatomy of the Characinidae, with an enquiry into the relations of the ductus pneumaticus in Physostomi generally. Trans. Linn. Soc., London, Zool., Second Ser. 9:8–83.

Roy, P.K. and Munshi, J.S.D. 1986. Morphometrics of the respiratory organs of a freshwater major carp, *Cirrhinus mrigala* in relation to body weight. Japan. J. Ichthyol. 33:269–279.

Roy, R. and Das, A.B. 1989. Seasonal reorganization of nitrogen metabolism in an Indian air-breathing teleosts (sic), *Channa punctatus* (Bloch). J. Biosci. 14:183–187.

Rozemeijer, M.J.C. and Plaut, I. 1993. Regulation of nitrogen excretion of the amphibious blenniidae *Alticus kirki* (Günther, 1868) during emersion and immersion. Comp. Biochem. Physiol. 104A:57–62.

Sacca, R. and Burggren, W. 1982. Oxygen uptake in air and water in the air-breathing reedfish *Calamoichthys calabaricus*: Role of skin, gills and lungs. J. Exp. Biol. 97:179–186.

Sagemehl, M. 1885. Beitrage zur vergleichenden Anatomie der Fische III. Das Cranium der Characiniden nebst allgemeinen Bermerkungen über die mit einem Weber'schen Apparat versehenen Physostomen familien. Morphol. Jahrb. 10:1–119.

Saha, N. and Ratha, B.K. 1987. Active ureogenesis in a freshwater, air-breathing teleost, *Heteropneustes fossilis*. J. Exp. Zool. 241:137–141.

Saha, N. and Ratha, B.K. 1989. Comparative study of ureogenesis in freshwater, air-breathing teleosts. J. Exp. Zool. 252:1–8.

Saha, N. and Ratha, B.K. 1990. Alterations in excretion pattern of ammonia and urea in a freshwater air-breathing teleost, *Heteropneustes fossilis* (Bloch) during hyper-ammonia stress. Indian J. Exp. Biol. 28:597–599.

Saha, T.K., Arya, M.B. and Das, A.B. 1994. Alterations in the patterns of nitrogenous excretion in a freshwater air breathing teleost, *Channa punctatus* (Bloch), during ammonia-stress. Indian J. Exp. Biol. 32:196–199.

Saint-Paul, U. and Soares, G.M. 1988. Ecomorphological adaptation to oxygen deficiency in Amazon floodplains by serrasalmid fish of the genus *Mylossoma*. J. Fish Biol. 32:231–236.

Salih, M.S. and Al-Jaffery, A. 1980. Morphological study of *Boleophthalmus dentatus* Cuvier and Valenciennes, 1837 with reference to its feeding habits. I. Habitat, external features and integument. Bull. Biol. Res. Cen., Baghdad 12:11–45.

Santhanam, R., Sukumaran, N. and Natarajan, P. 1987. *A Manual of Fresh-Water Aquaculture*. Oxford and IBH, New Delhi, 193 pp.

Santos, C.T.C., Fernandes, M.N. and Severi, W. 1994. Respiratory gill surface area of a facultative air-breathing loricariid fish, *Rhinelepis strigosa*. Can. J. Zool. 72:2009–2015.

Satchell, G.H. 1971. *Circulation in Fishes*. Cambridge University Press, Cambridge, 131 pp.

Satchell, G.H. 1976. The circulatory system of air-breathing fish. In: Hughes, G.M. (ed.) *Respiration of Amphibious Vertebrates*. Academic Press, London, pp. 105–123.

Satchell, G.H. 1991. *Physiology and Form of Fish Circulation*. Cambridge University Press, Cambridge, 235 pp.

Saul, W.G. 1975. An ecological study of fishes at a site in upper Amazonian Ecuador. Proc. Acad. Nat. Sci., Philadelphia 127:93–134.

Sawaya, P. 1946. Sôbre a biologia de alguns peixes de respiracao aérea (*Lepidosiren paradoxa* Fitzinger e *Arapaima gigas* Cuvier). Bol. Fac. Univ. San Paulo (Zool.) 11:255–285.

Saxena, D.B. 1963. A review on ecological studies and their importance in the physiology of air-breathing fishes. Ichthyologica 2:116–128.

Sayer, M.D.J. 1988. An investigation of the pattern of nitrogen excretion in *Blennius pholis*. Comp. Biochem. Physiol. 89A:359–363.

Sayer, M.D.J. and Blackstock, N. 1988. The relationship between nitrogen output and changes in mucous cell function in the amphibious teleost *Blennius pholis* L., during aerial exposure. J. Fish Biol. 32:765–776.

Sayer, M.D.J. and Davenport, J. 1991. Amphibious fish: Why do they leave water? Rev. Fish Biol. Fisheries 1:159–181.

Schaeffer, B. 1965. The rhipidistian-amphibian transition. Am. Zool. 5:267–276.

Schaller, F. and Dorn, E. 1973. Atemmechanismus und Kreislauf des Amazonasfisches Pirarucu. Naturwissenschaften 60:303.

Schmalhausen, I.I. 1968. *The Origin of Terrestrial Vertebrates*. Academic Press, New York, 314 pp.

Schmidt, P.J. 1942. A fish with feather-like gill-leaflets. Copeia 1942:98–100.

Scholander, P.F., Bradstreet, E. and Garey, W.F. 1962. Lactic acid response in the grunion. Comp. Biochem. Physiol. 6:201–203.

Schöttle, E. 1931. Morphologie und Physiologie der Atmung bei wasser-, schlamm-, und landlebenden Gobiiformes. Z. Wiss. Zool. 140:1–114.

Schultze, H.P. and Arsenault, M. 1985. The panderichthyid fish *Elpistostege*, a close relative of tetrapods? Palaeontology 28:293–309.

Schuster, S. 1989. The volume of air within the swimbladder and breathing cavities of the anabantoid fish *Colisa lalia* (Perciformes, Belontiidae). J. Exp. Biol. 144:185–198.

Schwartz, E. 1969. Luftatmung bei *Pantodon buchholzi* und ihre Beziehung zur Kiemenatmung. Z. Vergl. Physiol. 65:324–330.

Seghers, B.H. 1978. Feeding behavior and terrestrial locomotion in the cyprinodontid fish, *Rivulus hartii* (Boulenger). Verh. Int. Verein. Limnol. 20:2055–2059.

Seghers, B.H. and Nielsen, M.A. 1982. Adaptations of the cyprinodontid fish, *Rivulus hartii*, to ephemeral streams. Newsletter Int. Assoc. Fish. Ethol. 1:15.

Senna, A. 1924. Sull'organo respiratorio soprabranchiale degli Ofiocefalidi e sua semplificazione in *Parophiocephalus* subgen. n. Monitore Zool. Italiana 35:149–160.

Sheel, M. and Singh, B.R. 1981. O_2 uptake through water during early life of *Heteropneustes fossilis* (Bloch). Hydrobiologia 78:81–86.

Shelton, G. and Boutilier, R.G. 1982. Apnoea in amphibians and reptiles. J. Exp. Biol. 100:245–273.

Shelton, G. and Croghan, P.C. 1988. Gas exchange and its control in non-steady-state systems: The consequences of evolution from water to air breathing in the vertebrates. Can. J. Zool. 66:109–123.

Shelton, G., Jones, D.R. and Milsom, W.K. 1986. Control of breathing in ectothermic vertebrates. In: Cherniack, N.S. and Widdicombe, J.G. (eds.) *Handbook of Physiology* Sec. 3. The Respiratory System. Vol. 2. Control of Breathing. Physiological Society, Bethesda, MD. pp 857–909.

Shen, S., Yang, T.H. and Lin, J.J. 1986. *A Review of the Blenniid Fishes in the Waters Around Taiwan and Its Adjacent Islands*. Taiwan Mus. Spec. Pub., Ser. 5, 74 pp.

Shih, H.J. 1940. On the foods of *Monopterus*. Sinensia 11:573–576.

Shlaifer, A. 1941. Additional social and physiological aspects of respiratory behavior in small tarpon. Zoologica 26:55–60.

Shlaifer, A. and Breder, C.M. 1940. Social and respiratory behavior of small tarpon. Zoologica 25:493–514.

Siau, H. and Ip, Y.K. 1987. Activities of enzymes associated with phosphoenolpyruvate metabolism in the mudskippers, *Boleophthalmus boddaerti* and *Periophthalmodon schlosseri*. Comp. Biochem. Physiol. 88B:119–125.

Simroth, H. 1891. *Die Entstehung der Landtiere*. Engelmann, Leipzig, 492 pp.

Singer, T.D. and Ballantyne, J.S. 1991. Metabolic organization of a primitive fish, the bowfin (*Amia calva*). Can. J. Fish. Aquat. Sci. 48:611–618.

Singh, B.N. 1976. Balance between aquatic and aerial respiration. In: Hughes, G.M. (ed.) *Respiration of Amphibious Vertebrates*. Academic Press, London, pp. 125–164.

Singh, B.N. and Hughes, G.M. 1971. Respiration of an air-breathing catfish *Clarias batrachus* (Linn.). J. Exp. Biol. 55:421–434.

Singh, B.N. and Hughes, G.M. 1973. Cardiac and respiratory responses in the climbing perch *Anabas testudineus*. J. Comp. Physiol. 84:205–226.

Singh, B.N. and Munshi, J.S.D. 1968. On the respiration organs of an air breathing fish *Periophthalmus vulgaris* (Eggert). Indian Sci. Congress Assoc. Proc. 55:511.

Singh, B.N. and Munshi, J.S.D. 1969. On the respiratory organs and mechanics of breathing in *Periophthalmus vulgaris* (Eggert). Zool. Anz. 183:92–110.

Singh, B.R. 1988. Development of the respiratory epithelium of the neo-morphic air breathing organ in some teleosts and its evolutionary significance—a SEM study. Zool. Anz. 221:399–410.

Singh, B.R. 1993. Development, origin and evolution of the neomorphic air breathing organs in teleosts. In: Singh, B.R. (ed.) *Advances in Fish Research Vol. 1*. Narendra, New Delhi, pp. 1–16.

Singh, B.R. and Mishra, A.P. 1980. Development of the air breathing organ in *Anabas testudineus* (Bloch). Zool. Anz. 205:359–370.

Singh, B.R. and Thakur, R.N. 1979. Oxygen uptake capacity of the amphibious fish—*Amphipnous cuchia* (Ham.) (Symbranchiformes, Amphipnoidae). Acta Physiol. Acad. Sci., Hungaricae 54:13–21.

Singh, B.R., Guha, G. and Munshi, J.S.D. 1974. Mucous cells of the respiratory sac in an air-breathing fish *Saccobranchus fossils* (Bloch). La Cellule 70:7–15.

Singh, B.R., Mishra, A.P. and Singh, I. 1984. Development of the air-breathing organ in the mud-eel, *Amphipnous cuchia* (Ham.). Zool. Anz. 213:395–407.

Singh, B.R., Mishra, A.P., Prasad, M.S. and Singh, I. 1989b. Development of the neo-morphic air-breathing organs in three genera of estuarine gobies. Gegen. Morphol. Jahrb. 135:529–540.

Singh, B.R., Mishra, A.P., Sheel, M. and Singh, I. 1982a. Development of the air breathing organ in the cat fish, *Clarias batrachus* (Linn). Zool. Anz. 208:100–111.

Singh, B.R., Prasad, M.S. and Mishra, A.P. 1986. Oxygen uptake through water during early life in *Channa striatus* (Bloch). Polskie Arch. Hydrobiol. 33:97–104.

Singh, B.R., Prasad, M.S. and Mishra, A.P. 1989a. Development of the neo-morphic air breathing organ in the hill stream genus, *Amblyceps* (Blyth). Zool. Anz. 223:331–340.

Singh, B.R., Prasad, M.S., Yadav, A.N. and Kiran, U. 1991. The morphology of the respiratory epithelium of the air breathing organ of the swamp-eel, *Monopterus albus* (Zuiew). A SEM study. Zool. Anz. 226:27–32.

Singh, B.R., Sheel, M. and Mishra, A.P. 1981. Development of the air breathing organ in the cat fish, *Heteropneustes fossilis* (Bloch). Zool. Anz. 206:215–226.

Singh, B.R., Singh, I., Mishra, A.P. and Prasad, M.S. 1988. Development of the air breathing organ in the snake-headed fish, *Channa striatus* (Bloch). Z. Mikrosk.-Anat. Forsch. 102:332–344.

Singh, B.R., Singh, R.P. and Mishra, A.P. 1982b. Development of the air breathing organ in the snake-headed fish, *Channa punctatus* (Bloch). Zool. Anz. 208:428–439.

Singh, B.R., Thakur, R.N. and Yadav, A.N. 1976. Changes in the blood parameters of an air breathing fish during different respiratory conditions. Folia Haemat. 2:216–225.

Singh, B.R., Yadav, A.N., Ojha, J. and Munshi, J.S.D. 1981. Gross structure and dimensions of the gills of an intestinal air-breathing fish (*Lepidocephalichthys guntea*). Copeia 1981:224–229.

Singh, B.R., Yadav, A.N., Prasad, M.S., Mishra, A.P. and Singh, I. 1990. Neo-morphic organ for CO_2 elimination in air breathing teleosts. Eur. Arch. Biol. 101:257–267.

Singh, R.A. and Singh, S.N. 1986. Liver arginase in air-breathing and non-air-breathing freshwater teleost fish. Biochem. Syst. Ecol. 14:239–241.

Singh, R.P., Prasad, M.S., Mishra, A.P. and Singh, B.R. 1982. Oxygen uptake through water during early life in *Channa punctatus* (Bloch) (Pisces; Ophiocephaliformes). Hydrobiologia 87:211–215.

Sinha, A.L. 1977. On the structure and function of the respiratory apparatus of certain teleostean fishes. Ph.D. Thesis, Bhagalpur University, Bhagalpur, India, 118 pp.

Siret, J.R., Carmena, A.O. and Callejas, J. 1976. Erythrokinetic study in the fish manjuari (*Atracosteus tristoechus*). Comp. Biochem. Physiol. 55A:127–128.

Skobe, Z., Garant, P.R. and Albright, J.T. 1970. Ultrastructure of a new cell in the gills of the air-breathing fish *Helostoma temmincki*. J. Ultrastruct. Res. 31:312–322.

Sleggs, G.F. 1990. Observations upon the economic biology of the capelin (*Mallotus villosus* O.F. Müller). Rep. Newfoundland Fish. Res. Comm. 1:38–42.

Smatresk, N.J. 1988. Control of the respiratory mode in air-breathing fishes. Can. J. Zool. 66:144–151.

Smatresk, N.J. 1990. Chemoreceptor modulation of endogenous respiratory rhythms in vertebrates. Am. J. Physiol. 259:R887–R897.

Smatresk, N.J. 1994. Respiratory control in the transition from water to air breathing in vertebrates. Am. Zool. 34:264–279.

Smatresk, N.J. and Azizi, S.Q. 1987. Characteristics of lung mechanoreceptors in spotted gar, *Lepisosteus oculatus*. Am. J. Physiol. 252:R1066–R1072.

Smatresk, N.J. and Cameron, J.N. 1982a. Respiration and acid-base physiology of the spotted gar, a bimodal breather. I. Normal values, and the response to severe hypoxia. J. Exp. Biol. 96:263–280.

Smatresk, N.J. and Cameron, J.N. 1982b. Respiration and acid-base physiology of the spotted gar, a bimodal breather. II. Responses to temperature change and hypercapnia. J. Exp. Biol. 96:281–293.

Smatresk, N.J. and Cameron, J.N. 1982c. Respiration and acid-base physiology of the spotted gar, a bimodal breather. III. Response to transfer from fresh water to 50% sea water, and control of ventilation. J. Exp. Biol. 96:295–306.

Smatresk, N.J. and Smits, A.W. 1991. Effects of central and peripheral chemoreceptor stimulation on ventilation in the marine toad, *Bufo marinus*. Respir. Physiol. 83:223–238.

Smatresk, N.J., Burleson, M.L. and Azizi, S.Q. 1986. Chemoreflexive responses to hypoxia and NaCN in longnose gar: Evidence for two chemoreceptor loci. Am. J. Physiol. 251:R116–R125.

Smith, D.G. and Gannon, B.J. 1978. Selective control of branchial arch perfusion in an air-breathing Amazonian fish *Hoplerythrinus unitaeniatus*. Can. J. Zool. 56:959–964.

Smith, D.G., Duiker, W. and Cooke, I.R.C. 1983. Sustained branchial apnea in the Australian short-finned eel, *Anguilla australis*. J. Exp. Zool. 226:37–43.

Smith, H.M. 1945. The fresh-water fishes of Siam, or Thailand. Bull. U.S. Natl. Mus. 188:1–622.

Smith, H.W. 1930. Metabolism of the lungfish *Protopterus aethiopicus*. J. Biol. Chem. 88:97–130.

Smith, H.W. 1931. Observations on the African lung-fish, *Protopterus aethiopicus*, and on evolution from water to land environments. Ecology 12:164–181.

Smith, H.W. 1935a. The metabolism of the lung-fish. II. Effect of feeding meat on metabolic rate. J. Cell. Comp. Physiol. 6:335–349.

Smith, H.W. 1935b. Lung-fish. Aquarium 1:241–243.

Smith, H.W. 1959. *From Fish to Philosopher*. Little, Brown, and Co., Boston, 264 pp.

Smith, R.S. and Kramer, D.L. 1986. The effect of apparent predation risk on the respiratory behavior of the Florida gar (*Lepisosteus platyrhincus*). Can. J. Zool. 64:2133–2136.

Smits, A.W., Orgeig, S. and Daniels, C.B. 1994. Surfactant composition and function in lungs of air-breathing fishes. Am. J. Physiol. 266:R1309–R1313.

Soljan, T. 1932. *Blennius galerita* L., poisson amphibien des zones supralittorale et littorale exposées de l'Adriatique. Acta Adriatica 1:1–14.

Soni, V.C. and George, B. 1986. Age determination and length-weight relationship in the mudskipper *Boleophthalmus dentatus*. Indian J. Fish. 33:231–234.

Sowerby, A.de.C. 1923. The mudskipper. China J. Sci. Arts, Shanghai 1:343–345.

Spencer, W.B. 1892. Note on the habits of *Ceratodus forsteri*. Proc. R. Soc., Victoria 4:81–84.

Spiropulos Piccolo, R. and Sawaya, P. 1981. Blood vascular and respiratory systems of *Synbranchus marmoratus* Bloch, 1795 (Air-breathing teleost fish), some morphological and physiological aspects. Bol. Fisiol. Anim., Univ. Sao Paulo 5:59–73.

Sponder, D.L. and Lauder, G.V. 1981. Terrestrial feeding in the mudskipper *Periophthalmus* (Pisces: Teleostei): A cineradiographic analysis. J. Zool., London 193:517–530.

Spurway, H. and Haldane, J.B.S. 1953. The comparative ethology of vertebrate breathing. Behavior 6:6–34.

Srivastava, A.K. 1968. Studies on the hematology of certain freshwater teleosts. IV. Hemoglobin. Folia Haemat. 90:411–418.

Stebbins, R.C. and Kalk, M. 1961. Observations on the natural history of the mudskipper *Periophthalmus sobrinus*. Copeia 1961:18–27.

Steeger, H.-U. and Bridges, C.R. 1995. A method for long-term measurement of respiration in intertidal fishes during simulated intertidal conditions. J. Fish Biol. 47:308–320.

Steen, J.B. 1971. *Comparative Physiology of Respiratory Mechanisms*. Academic Press, London, 182 pp.

Steffensen, J.F., Lomholt, J.P. and Vogel, W.O.P. 1986. *In vivo* observations on a specialized microvasculature, the primary and secondary vessels in fishes. Acta Zool. 67:193–200.

Stensio, E.A.S. 1948. On the Placodermi of the Upper Devonian of East Greenland. II. Antiarchi: Subfamily Bothriolepidae. Paleozool. Groenlandica 2:1–622.

Sterba, G. 1963. *Freshwater Fishes of the World*. Viking Press, New York, 878 pp.

Stevens, D.E. and Holeton, G.F. 1978a. The partitioning of oxygen uptake from air and from water by the large obligate air-breathing teleost pirarucu (*Arapaima gigas*). Can. J. Zool. 56:974–976.

Stevens, E.D. and Holeton, G.F. 1978b. The partitioning of oxygen uptake from air and from water by erythrinids. Can. J. Zool. 56:965–969.

Stevens, J.K. and Parsons, K.E. 1980. A fish with double vision. Nat. Hist. 89:62–67.

Stiffler, D.F. 1989. Interactions between cutaneous ion-exchange mechanisms and acid-base balance in amphibians. Can. J. Zool. 67:3070–3077.

Stiffler, D.F., Graham, J.B., Dickson, K.A. and Stöckmann, W. 1986. Cutaneous ion transport in the freshwater teleost *Synbranchus marmoratus*. Physiol. Zool. 59:406–418.

Storey, K.B. and Storey, J.M. 1990. Metabolic rate depression and biochemical adaptation in anaerobiosis, hibernation and estivation. Q. Rev. Biol. 65:145–174.

Storey, K.B., Guderley, H.E., Guppy, M. and Hochachka, P.W. 1978. Control of ammoniagenesis in the kidney of water- and air-breathing osteoglossids: Characterization of glutamate dehydrogenase. Can. J. Zool. 56:845–851.

Subramanian, A. 1984. Burrowing behavior and ecology of the crab-eating Indian snake eel *Pisodonophis boro*. Environ. Biol. Fishes 10:195–202.

Sufi, S.M.K. 1956. Revision of the oriental fishes of the family Mastacembelidae. Bull. Raffles Mus. 27:92–146.

Svensson, G.S.O. 1933. Freshwater fishes from the Gambia River. Kungl. Svenska Vetenskapasakad, Stockholm, Ser. 3, 12:1–102.

Swan, H. and Hall, F.G. 1966. Oxygen-hemoglobin dissociation in *Protopterus aethiopicus*. Am. J. Physiol. 210:487–489.

Swan, H., Jenkins, D. and Knox, K. 1968. Anti-metabolic extract from the brain of *Protopterus aethiopicus*. Nature 217:671.

Szidon, J., Lahiri, S., Lev, M. and Fishman, A.P. 1969. Heart and circulation of the African lungfish. Circ. Res. 25:23–38.

Takasusuki, J, Fernandes, M.N. and Severi, W. Manuscript. Aquatic respiration during hypoxia in the loricariid fish, *Rhinelepis strigosa*. A correlation with the air breathing oxygen threshold. Manuscript in review.

Tamura, O. and Moriyama, T. 1976. On the morphological feature of the gill of amphibious and air-breathing fishes. Bull. Fac. Fish., Nagaski Univ. 41:1–8.

Tamura, S.O., Morii, H. and Yuzuriha, M. 1976. Respiration of the amphibious fishes *Periophthalmus cantonensis* and *Boleophthalmus chinensis* in water and on land. J. Exp. Biol. 65:97–107.

Tarnchalanukit, W. 1986. Experimental hybridization between catfishes of the families Clariidae and Pangasiidae in Thailand. Environ. Biol. Fishes 16:317–320.

Taylor, E.W. 1992. Nervous control of the heart and cardiorespiratory interactions. In: Hoar, W.S., Randall, D.J. and Farrell, A.P. (eds.) *Fish Physiology, Vol. XII, Part B*. Academic Press, Inc., San Diego, pp. 343–387.

Taylor, J. 1831. On the respiratory organs and air-bladder of certain fishes of the Ganges. Edinburgh J. Sci. 5:33–51.

Taylor, M. 1914. The development of *Symbranchus marmoratus*. Quart. J. Micro. Sci. 59:1–51.

Teal, J.M. and Carey, F.G. 1967. Skin respiration and oxygen debt in the mudskipper *Periophthalmus sobrinus*. Copeia 1967:677–679.

Thakur, N.K., Dutta, B.R. and Satpathy, B.B. 1987. *Pangasius pangasius* (Ham.)—an addition to the list of air-breathing teleosts. Curr. Sci. 56:315–317.

Thomas, E.I. 1967. Studies on the heart of *Heteropneustes fossilis* (Bloch). Proc. Indian Acad. Sci. 66:53–62.

Thomas, E.I. 1976. The minute anatomy of the heart of *Anabas testudineus* (Cuvier). Zool. Anz. 196:397–404.

Thomson, K.S. 1969. The biology of the lobe-finned fishes. Biol. Rev. 44:91–154.

Thomson, K.S. 1980. The ecology of the Devonian lobe-finned fish. In: Panchen, A.L. (ed.) *The Terrestrial Environment and the Origin of Land Vertebrates*. Academic Press, London, pp. 187–222.

Thomson, K.S. 1991. Where did tetrapods come from? Am. Sci. 79:488–490.

Thomson, K.S. 1993. The origin of the tetrapods. Am. J. Sci. 293-A:33–62.

Thys van den Audenaerde, D.F.E. 1959. Existence d'une vessie natatoire pulmonoïde chez *Phractolaemus ansorgei* BLGR. Rev. Zool. Bot. Afr. 59:364–366.

Thys van den Audenaerde, D.F.E. 1961. L'anatomie de *Phractolaemus ansorgei* BLGR. et la position systématique des Phractolaemidae. Ann. Mus. R. l'Afrique Cent. Sci. Zool. 103:100–167.

Todd, E.S. 1968. Terrestrial sojourns of the long-jaw mudsucker, *Gillichthys mirabilis*. Copeia 1968:192–194.

Todd, E.S. 1971. Respiratory control in the longjaw mudsucker, *Gillichthys mirabilis*. Comp. Biochem. Physiol. 39A:147–163.

Todd, E.S. 1972. Hemoglobin concentration in a new air-breathing fish. Comp. Biochem. Physiol. 42A:569–573.

Todd, E.S. 1973. Positive buoyancy and air-breathing: A new piscine gas bladder function. Copeia 1973:461–464.

Todd, E.S. 1976. Terrestrial grazing by the eastern tropical Pacific goby *Gobionellus sagittula*. Copeia 1976:374–377.

Todd, E.S. and Ebeling, A.W. 1966. Aerial respiration in the longjaw mudsucker *Gillichthys mirabilis* (Teleostei: Gobiidae). Biol. Bull. 130:265–288.

Toews, D.P. and Stiffler, D.F. 1989. The role of the amphibian kidney and bladder in the regulation of acid-base relevant ions. Can. J. Zool. 67:3064–3069.

Trewavas, E. 1962. Fishes of the crater lakes of the northwestern Cameroons. Sond. Aus. Bonner Zool. Beitrage Heft 1/3 (1962):146–192.

Truchot, J.P. and Duhamel-Jouve, A. 1980. Oxygen and carbon dioxide in the marine intertidal environment: Diurnal and tidal changes in rockpools. Respir. Physiol. 39:241–254.

Tytler, P. and Vaughan, T. 1983. Thermal ecology of the mudskipper, *Periophthalmus koelreuteri* (Pallas) and *Boleophthalmus boddarti* (Pallas) of Kuwait Bay. J. Fish Biol. 23:327–337.

Uchida, K. and Fujimoto, M. 1933. Life history and method of culture of the Corean snake-head fish, *Ophicephalus argus*. Bull. Fish. Exp. Stn., Chosen 3:1–91.

Ultsch, G.R. 1987. The potential role of hypercarbia in the transition from water-breathing to air-breathing in vertebrates. Evolution 41:442–445.

Ultsch, G.R. 1989. Ecology and physiology of hibernation and overwintering among freshwater fishes, turtles, and snakes. Biol. Rev. 64:435–516.

Ultsch, G.R. 1996. Gas exchange, hypercarbia and acid-base balance, paleoecology, and the evolutionary transition from water-breathing to air-breathing among vertebrates. Palaeogeog., Palaeoclim., Palaeoecol. 123: 1–27.

Ultsch, G.R., Jackson, D.C. and Moalli, R. 1981. Metabolic oxygen conformity among lower vertebrates: The toadfish revisited. J. Comp. Physiol. 142:439–443.

Vaillant, L. 1895. Sur les habitudes terricoles d'un Siluroïde Africain *Clarias lazera*, Cuvier et Valenciennes. Bull. Mus. d'Hist. Naturelle 1:271–272.

Val, A.L. and de Almeida-Val, V.M.F. 1995. *Fishes of the Amazon and Their Environment*. Springer-Verlag, Berlin, 224 pp.

Val, A.L., de Almeida-Val, V.M.F. and Affonso, E.G. 1990. Adaptive features of Amazon fishes: Hemoglobins, hematology, intraerythrocytic phosphates and whole blood Bohr effect of *Pterygoplichthys multiradiatus* (Siluriformes). Comp. Biochem. Physiol. 97B:435–440.

Valentin, G.G. 1842. Beiträge zur Anatomie des Zitteraales (*Gymnotus electricus*). Allgem. Schweiz. Gesel. 6:1–74.

Van Der Hoeven, J. 1841. Ueber die zellige Schwimmblase des *Lepidosteus*. Arch. Anat. 8:221–223.

Van Dijk, D.E. 1959. A fish that walks, jumps and burrows. Nucleon 1:36–38.

Vandewalle, P. and Chardon, M. 1991. A new hypothesis on the air flow in air breathing in *Clarias gariepinus* (Teleostei, Siluriformes). Belgian J. Zool. 121:73–80.

Vargas, F. and Concha, J. 1957. Metabolismo respiratorio del teleosteo *Sicyases sanquineus*. Investnes. Zool. Chil. 3:146–154.

Vari, R.P. 1995. The Neotropical fish family Ctenoluciidae (Teleostei: Ostariophysi: Chariciformes): Supra and intrafamilial phylogenetic relationships, with a revisionary study. Smithsonian Contrib. Zool. 564:1–97.

Venkateswarlu, T. 1966. A note on the oxygen carrying capacity of the blood in three gobiid fishes. Naturwissenschaften 12:3.

Vetter, R.D., Hwang, H. and Hodson, R.E. 1986. Comparison of glycogen and adenine nucleotides as indicators of metabolic stress in mummichogs. Trans. Am. Fish. Soc. 115:47–51.

Vivekanandan, E. 1976. Effects of feeding on the swimming activity and growth of *Ophiocephalus striatus*. J. Fish Biol. 8:321–330.

Vivekanandan, E. 1977. Surfacing activity and food utilisation in the obligatory air-breathing fish *Ophiocephalus striatus* as a function of body weight. Hydrobiologia 55:99–112.

Vivekanandan, E. and Pandian, T.J. 1979. Erythrocyte count and haemoglobin concentration of some tropical fishes. J. Madurai Kamaraj Univ. (Sci.) 8:71–75.

Vokac, Z., Ahmed, N.E.D. and Magid, A.M.A. 1972. The *in vitro* respiratory properties of blood of the primitive fish *Polypterus senegalus*. J. Exp. Biol. 57:461–469.

Volz, W. 1906. Der Circulations- und Respirationsapparat von *Monopterus javanensis* Lac. Zool. Jahrb. 23:163–186.

Wade, R. A. 1962. The biology of the tarpon *Megalops atlanticus*, with emphasis on larval development. Bull. Mar. Sci. 12:545–622.

Walsh, P.J. and Henry, R.P. 1991. Carbon dioxide and ammonia metabolism and exchange. In: Hochachka, P.W. and Mommsen, T.P. (eds.) *Biochemistry and Molecular Biology of Fishes Vol. 1*. Elsevier, New York, pp. 181–207.

Weber, R.E., Johansen, K., Lykkeboe, G.M. and Maloiy, G.M. 1976b. Oxygen-binding properties of hemoglobins from estivating and active African lungfish. J. Exp. Zool. 199:85–96.

Weber, R.E., Sullivan, B., Bonaventura, J. and Bonaventura, C. 1976a. The hemoglobin system of the primitive fish, *Amia calva*, isolation and functional characterization of the individual hemoglobin components. Biochem. Biophys. Acta 434:18–31.

Weber, R.E., Wood, S.C. and Davis, B.J. 1979. Acclimation to hypoxic water in facultative air-breathing fish: Blood oxygen affinity and allosteric effectors. Comp. Biochem. Physiol. 62A:125–129.

Weibel, E.R. 1970. Morphometric estimation of pulmonary diffusion capacity. I. Model and method. Respir. Physiol. 11:54–75.

Weibezahn, F.H. 1967. Estudios sobre la respiracion aerea en *Hoplerythrinus unitaeniatus* (Spix) (Cypriniformes, Characidae). Bull. Soc. Venezolana Ciencias Naturales 27:178–188.

Weitzman, S.H. 1964. Osteology and relationships of South American characid fishes of subfamilies Lebiasininae and Erythrininae with special reference to subtribe Nannostomina. Proc. U.S. Natl. Mus. 116:127–170.

Wells, N.A. and Dorr, Jr., J.A. 1985. Form and function of the fish *Bothriolepis* (Devonian; Placodermi, Antiarchi): The first terrestrial vertebrate? Michigan Acad. 17:157–173.

Wells, R.M.G., Forster, M.E. and Meredith, A.S. 1984. Blood oxygen affinity in the amphibious fish *Neochanna burrowsius* (Galaxiidae: Salmoniformes). Physiol. Zool. 57:261–265.

West, N.H. and Burggren, W.W. 1984. Factors influencing pulmonary and cutaneous arterial blood flow in the toad, *Bufo marinus*. Am. J. Physiol. 247:R884–R894.

Whitear, M. and Mittal, A.K. 1984. Surface secretions of the skin of *Blennius* (*Lipophrys*) *pholis* L. J. Fish Biol. 25:317–331.

Whiting, H.P. and Bone, Q. 1980. Ciliary cells in the epidermis of the larval Australian Dipnoan, *Neoceratodus*. Zool. J. Linn. Soc. 68:125–137.

Whitmore, T.C. 1981. *Wallace's Line and Plate Tectonics*. Clarendon Press, Oxford, 90 pp.

Wilder, B.G. 1875. Notes on the North American ganoids. I. On the respiratory actions of *Amia* and *Lepidosteus*. Proc. Am. Assoc. Adv. Sci. 24:151–153.

Wilder, B.G. 1877. On the respiration of *Amia*. Proc. Am. Assoc. Adv. Sci. 26:306–313.

Wiley, E.O. 1976. The phylogeny and biogeography of fossil and recent gars (Actinopterygi: Lepisosteidae). U. Kansas Mus. Nat. Hist. Misc. Publ. 64:1–111.

Willem, V. and Boelaert, R. 1937. Les manoeuvres respiratoires de *Periophthalmus*. Bull. L'Acad. R. Sci. Letteres Beaux-Arts de Belgique 23:942–959.

Willmer, E.N. 1934. Some observations on the respiration of certain tropical fresh-water fishes. J. Exp. Biol. 11:283–306.

Wilson, E.O. 1994. *Naturalist*. Island Press, Washington, D.C., 380 pp.

Windley, B.F. 1984. *The Evolving Continents*. Wiley, New York, 399 pp.

Wintrobe, M.M. 1934. Variations in the size and hemoglobin content of erythrocytes in the blood of various vertebrates. Folia Haemat. 51:32–49.

Wolf, N.G. and Kramer, D.L. 1987. Use of cover and the need to breathe: The effects of hypoxia on vulnerability of dwarf gouramis to predatory snakeheads. Oecologia 73:127–132.

Wood, C.M. 1993. Ammonia and urea metabolism and excretion. In: Evans, D.H. (ed.) *The Physiology of Fishes*. CRC Press, Boca Raton, pp. 379–425.

Wood, S.C. and Johansen, K. 1972. Adaptations to hypoxia by increased HbO_2 affinity and decreased red cell ATP concentration. Nature 237:278–279.

Wood, S.C. and Lenfant, C.J.M. 1976. Physiology of fish lungs. In: Hughes, G.M. (ed.) *Respiration of Amphibious Vertebrates*. Academic Press, London, pp. 257–270.

Wood, S.C., Weber, R.E. and Davis, B.J. 1979. Effects of air-breathing on acid-base balance in the catfish, *Hypostomus* sp. Comp. Biochem. Physiol. 62A:185–187.

Wright, W.G. and Raymond, J.A. 1978. Air-breathing in a California sculpin. J. Exp. Zool. 203:171–176.

Wu, C.H. 1984. Electric fish and the discovery of animal electricity. Am. Sci. 72:598–607.

Wu, H.W. and Chang, H. 1945. On the structures of the intestine of the Chinese pond loach with special reference to its adaptation for aerial respiration. Sinensia 16:1–8.

Wu, H.W. and Kung, C.C. 1940. On the accessory respiratory organ of *Monopterus*. Sinensia 11:59–67.

Wu, H.W. and Liu, C.K. 1940. The bucco-pharyngeal epithelium as the principal respiratory organ in *Monopterus javanensis*. Sinensia 11:221–239.

Wu, H.W. and Liu, C.K. 1943. On the blood vascular system of *Monopterus javanensis*, an air-breathing fish. Sinensia 14:62–97.

Yadav, A.N. and Singh, B.R. 1980. The gut of an intestinal air-breathing fish, *Lepidocephalus guntea* (Ham.). Arch. Biol. (Bruxelles) 91:413–422.

Yadav, A.N. and Singh, B.R. 1989. Gross structure and dimensions of the gill in an air-breathing estuarine goby, *Pseudapocryptes lanceolatus*. Japan. J. Ichthyol. 36:252–259.

Yadav, A.N., Prasad, M.S. and Singh, B.R. 1990. Gross structure of the respiratory organs and dimensions of the gill in the mudskipper, *Periophthalmodon schlosseri* (Bleeker). J. Fish Biol. 37:383–392.

Yadav, A.N., Sharma, S.N., Guha, G. and Singh, B.R. 1978. Blood parameters of two hill stream cobitids in relation to their habitat. Hydrobiologia 60:75–79.

Yamaguchi, K., Glahn, J., Scheid, P. and Piiper, J. 1987. Oxygen transfer conductance of human red blood cells at varied pH and temperature. Respir. Physiol. 67:209–223.

Yoshiyama, R.M. and Cech, J.J. 1994. Aerial respiration by rocky intertidal fishes of California and Oregon. Copeia 1994:153–158.

Yoshiyama, R.M., Valpey, C.J., Schalk, L.L., Oswald, N.M., Vaness, K.K., Lauritzen, D. and Limm, M. 1995. Differential propensities for aerial emergence in intertidal sculpins (Teleostei; Cottidae). J. Exp. Mar. Biol. and Ecol. 191:195–207.

Youson, J.H. 1976. Fine structure of granulated cells in the posterior cardinal and renal veins of *Amia calva* L. Can. J. Zool. 54:843–851.

Yu, K.L. and Woo, N.Y.S. 1987a. Metabolic adjustments of an air-breathing teleost, *Channa maculata*, to acute and prolonged exposure to hypoxic water. J. Fish Biol. 31:165–175.

Yu, K.L. and Woo, N.Y.S. 1987b. Changes in blood respiratory properties and cardiovascular function during acute exposure to hypoxic water in an air-breathing teleost *Channa maculata*. J. Fish. Biol. 30:749–760.

Yu, K.L. and Woo, Y.S. 1985. Effects of ambient oxygen tension and temperature on the bimodal respiration of an air-breathing teleost. Physiol. Zool. 58:181–189.

Yu, X-b. 1990. Cladistic analysis of sarcopterygian relationships, with a description of three new genera of Porolepiformes from the Lower Devonian of Yunnan, China. Ph.D. Dissertation, Yale Univ.

Zander, C.D. 1967. Beiträge zur Ökologie und Biologie litoralbewohnender Salariidae und Gobiidae (Pisces) aus dem Roten Meer. "Meteor" Forsch-Ergebn. D2:69–84.

Zander, C.D. 1972a. Beziehungen zwischen Körperbau und Lebensweise bei Blenniidae (Pisces) aus dem Roten Meer. I. Aussere Morphologie. Mar. Biol. 13:238–246.

Zander, C.D. 1972b. Beziehungen zwischen Körperbau und

Lebensweise bei Blenniidae (Pisces) des Roten Meer. II. Bau der Flossen und ihrer Muskulatur. Z. Morphol. Oekol. Tiere 71:299–327.

Zander, C.D. 1974. Beziehungen zwischen Körperbau und Lebensweise bei Blenniidae (Pisces) aus dem Roten Meer. III. Morphologie des Auges. Mar. Biol. 28:61–71.

Zander, C.D. 1983. Terrestrial sojourns of two Mediterranean blennioid fish. Senckenbergiana Marit. 15:19–26.

Zander, C.D. and Bartsch, I. 1972. In situ Beziehungen zwischen Nahrungsangebot und aufgenommener Nahrung bei 5 *Blennius*-Arten (Pisces) des Mittelmeeres. Mar. Biol. 17:77–81.

Zanjani, E.D., Yu, M.L., Perlmutter, A. and Gordon, A.S. 1969. Humoral factors influencing erythropoiesis in the fish (blue gourami) *Trichogaster trichopterus*. Blood 33:573–581.

Zaret, T.M. and Rand, A.S. 1971. Competition in tropical stream fishes: Support for the competitive exclusion principle. Ecology 52:336–342.

Zheng, W.B. and Liu, W.S. 1988. Morphology and histology of the swimbladder and ultrastructure of respiratory epithelium in the air-breathing catfish, *Pangasius sutchi* (Pangasiidae). J. Fish Biol. 33:147–154.

Zimmer, C. 1995. Coming on to the land. Discover 16:118–127.

Zograff, N. 1886. Ueber den sogenannten Labyrinthapparat der Labyrinthfische (Labyrinthici). Biol. Centrablatt. 5(22):679–686.

Zograff, N. 1888. On the construction and purpose of the so-called labyrinthine apparatus of the labyrinthic fishes. Quart. J. Micro. Sci. 28:501–512.

Index

Entries followed by f or t denote figures and tables, respectively.

B

I

J

K

L

M